AF245587

BIOLOGY OF
the Salivary Glands

Edited by
Kathleen Dobrosielski-Vergona

Associate Professor and Chair
Department of Anatomy/Histology
School of Dental Medicine
University of Pittsburgh
Pittsburgh, Pennsylvania

CRC Press
Boca Raton   Ann Arbor   London   Tokyo

Library of Congress Cataloging-in-Publication Data

Biology of the salivary glands / [edited by] Kathleen Dobrosielski-Vergona.
    p. cm.
    Includes bibliographical references and index.
    ISBN 0-8493-8847-3
    1. Salivary glands—Physiology. I. Dobrosielski-Vergona, Kathleen.
    [DNLM: 1. Salivary glands—physiology. WI 230 B615]
QP191.B56 1993
612.3′13—dc20
DNLM/DLC
for Library of Congress

92-22809
CIP

Developed by Telford Press

International Standard Book Number 0-8493-8847-3

Library of Congress Card Number 92-22809
Printed in the United States 1 2 3 4 5 6 7 8 9 0
Printed on acid-free paper

To my parents, Ray and Sophie,
for motivation

To my husband, Ron,
for dedication

To my son, Raymond,
for inspiration

EDITOR

Kathleen Dobrosielski-Vergona, Ph.D., is Associate Professor and Chairman, Department of Anatomy/Histology, School of Dental Medicine, University of Pittsburgh.

Dr. Dobrosielski-Vergona received her B.S. in Biology (1970) and her Ph.D. in Cell Biology (1976) from the University of Pittsburgh. After doing postdoctoral work at the Cancer Research Center of Allegheny General Hospital and the Department of Anatomy and Cell Biology, School of Medicine, University of Pittsburgh, Dr. Dobrosielski-Vergona was appointed an Assistant Professor of Anatomy/Histology in the School of Dental Medicine, University of Pittsburgh. She was promoted to Associate Professor in 1983, to Acting Chairman of the Department of Anatomy/Histology in 1990, and Chairman in 1991.

Dr. Dobrosielski-Vergona is a member of the Pittsburgh Cancer Institute, the Pitt Geriatric Education Center, International Association for Dental Research, Salivary Research and Geriatric Oral Research Groups of the American Association for Dental Research, Western Pennsylvania and Pennsylvania Dental Associations, American Society of Cell Biology, American Association of Anatomists, the Tissue Culture Association, and a founding member of the American Association of Oral Biologists.

She has been the recipient of several research and teaching grants from a variety of agencies, including the National Institutes of Health and the American Cancer Society. Among other awards, she has received a Public Health Service Research Fellowship Award from the National Institute on Aging, the Achievement of Excellence Award from the Geriatric Education Center of Pennsylvania, the Outstanding Instructor Award from the American Student Dental Association, and Honorary Membership to Omicron Kappa Upsilon, the National Dental Honorary Fraternity. She has been a selected biographee in *Who's Who in the East, American Men and Women in Science, Who's Who of Emerging Leaders in America,* and *Two Thousand Notable American Women.*

Dr. Dobrosielski-Vergona has presented invited lectures and published articles on age-related alterations in structure and function. Currently, Dr. Dobrosielski-Vergona is investigating the role of glucose-6-phosphatase and gamma glutamyl transpeptidase in salivary glands.

PREFACE

Salivary glands provide a model system to study many biological phenomena. The variety of topics included in this book reveal the multiple disciplines that utilize salivary glands to expand our knowledge and understanding of the development, structure, and regulation of living systems.

The first three chapters consider structural details of the salivary glands in human and animal systems. Such information, provided by Drs. Saracco and Crabill, is the basis for surgical and pathological human studies. The complex heterogeneity of these glands becomes more readily recognized by the classifications and histochemical properties detailed by Dr. Pinkstaff, allowing the identification of the unique tissues and cells that comprise the salivary system. Finally, Drs. Phillips, Tandler, and Nagato consider the role of salivary glands in revealing aspects of evolutionary diversity, even among closely related species.

The secretion of mucins and antimicrobial proteins, as well as the transport of water and ions, is a major function of salivary glands. Chapters 4, 5, and 6 explore the current research underway in these areas, as well as the role of the salivary gland secretions in nutrition. Dr. Castle reviews the contributions of individual cell processes, e.g., synthesis, storage, and exocytosis to the secretion of salivary proteins. The transport of ions and fluid within the acinar and ductal components of the salivary glands are explored by Dr. Turner. Finally, the extensive literature on the nutritional role of salivary glands is reviewed by Dr. Etzel.

The stimuli for salivary gland secretion are considered in Chapters 7, 8, and 9. The various mechanisms that connect specific stimuli with the expected responses are described by Drs. Baum, Ambudkar, and Horn. Further, Dr. Quissell elucidates the research into the mechanisms of the stimulus-response coupling phenomenon. Abnormal function of the salivary glands, as occurs during various diseases, is reviewed by Dr. Rossie.

Molecular genetics has proved an exciting and efficient approach to understanding biological systems. Dr. Kousvelari demonstrates the progress made in salivary glands biology by studying oncogene expression. Pathologies of salivary glands frequently involve alterations in normal cell proliferation and regeneration. In Chapter 11, Dr. Humphreys-Beher considers the regulatory role of plasma membrane proteins in cell proliferation. Dr. P. C. and P. A. Denny provide a detailed review of the kinetics of the cell populations in salivary glands, which may lead to corrective therapies for abnormal division by these cells.

Drs. Redman and Quissell in Chapter 13, refer to the "layers of complexity" that often hinder *in vivo* studies. The attractiveness of better controlled *in vitro* experiments often provides a more direct experimental approach. The extensive efforts to establish salivary gland cells in culture with *in vivo* phenotypes are reviewed by Drs. Redman, Oliver, Patton, and Wellner. The reader is reminded of the shortcomings of such model systems, as well.

The last four chapters focus on the development and differentiation of salivary glands. Dr. Cutler emphasizes the significance of extracellular components, including collagen and proteoglycans, in the morphogenesis of salivary glands. The "signature" of cells, enabling their recognition by characteristic phenotypes, is described by Dr. Ball. Such markers allow an estimate of a cell's stage of differentiation. Dr. Scott, in Chapter 18, discusses the structural changes that occur in both minor and major salivary glands during aging. The glands are not stable in their cellular makeup but continue to undergo alterations in cellular proportions throughout the life span. The final chapter considers functional changes in the flow, composition and immune properties of human and rodent saliva with age. Recent studies, from unmedicated individuals, generally reveal no loss in the ability of major salivary glands to produce saliva. Rodent studies, on the other hand, indicate an age-related decline in salivary flow. Finally, compositional and immunologic changes with age in both humans and rodents are presented.

In summary, the efforts of the contributors to this volume provide a broad range of topics on several biological aspects of the salivary glands in humans as well as other mammals. From the molecular events at the level of oncogenes, to the gross anatomic details of the human parotid gland, the *Biology of Salivary Glands* is a smorgasbord of information that can be sampled to satisfy inquiries from a variety of disciplines. *Bon appetit!*

ACKNOWLEDGMENTS

I am most grateful to Ms. Kimberly Barna and Mrs. Mary Ann Miaczynski for their editing and typing, and Mr. Peter Draus, for his expertise in computer operations. The continuous help and patience of Dr. P. Kelly, of Telford Press, sustained my fortitude through the compilation of this manuscript.

CONTRIBUTORS

Indu S. Ambudkar, Ph.D.
Senior Investigator
Clinical Investigations and Patient
  Care Branch
National Institute of Dental Research
National Institutes of Health
Bethesda, MD

William D. Ball, Ph.D.
Associate Professor
College of Medicine, Department of
  Anatomy and the Cancer Center
Howard University
Washington, D.C.

Bruce J. Baum, D.M.D., Ph.D.
Clinical Director
National Institute of Dental Research
National Institutes of Health
Bethesda, MD

David Castle, Ph.D.
Associate Professor
Department of Anatomy/Cell
  Biology
University of Virginia Medical
  Center
Charlottesville, VA

Edward V. Crabill, Ph.D.
Professor
Department of Anatomy/Histology
School of Dental Medicine
University of Pittsburgh
Pittsburgh, PA

Leslie S. Cutler, D.D.S., Ph.D.
Interim Vice President & Provost,
Health Affairs and Executive
  Director
University of Connecticut Health
  Center
Farmington, CT

Paul Denny, Ph.D.
Professor
School of Dentistry
University of Southern California
Los Angeles, CA

Patricia Denny, B.A., M.A.
Research Laboratory Technician
School of Dentistry
University of Southern California
Los Angeles, CA

Kenneth R. Etzel, Ph.D.
Assistant Professor
Department of Microbiology/
  Chemistry
School of Dental Medicine
University of Pittsburgh
Pittsburgh, PA

Anne Field, B.D.S., M.D.S.
Senior Lecturer
School of Dental Surgery
University of Liverpool
Liverpool, England

Valerie J. Horn, D.D.S., Ph.D.
Special Dental Postdoctoral Fellow
Oral Pathology Research Laboratory
Department of Veterans Affairs
  Medical Center
VA Medical Center
Washington, D.C.

Michael Humphreys-Beher, Ph.D.
Associate Professor
Department of Oral Biology
University of Florida
Gainesville, FL

Eleni Kousvelari, D.D.S., M.Sc., D.Sc.
Senior Staff Fellow
National Institute of Dental Research
National Institutes of Health
Bethesda, MD

Toshikazu Nagato, D.D.S., Ph.D.
Associate Professor
Department of Oral and Maxillofacial Surgery
Ehime University School of Medicine
Shizukawa, Shigenobu, Onsengun
Ehime, Japan

Constance Oliver, M.S., Ph.D.
Scientific Officer
Office of Naval Research
Arlington, VA

Lauren L. Patton, D.D.S.
Assistant Professor
Department of Dental Ecology
School of Dentistry
University of North Carolina
Chapel Hill, NC

Carleton J. Phillips, B.S., M.A., Ph.D.
Professor and Chairman
Department of Biological Sciences
Illinois State University
Normal, IL

Carlin A. Pinkstaff, Ph.D.
Professor
Department of Anatomy
School of Dentistry
University of West Virginia
Morgantown, WV

David Quissell, Ph.D.
Professor and Chair
Basic Sciences and Oral Research Department
U.C.H.S.C.
University of Colorado
Denver, CO

Robert S. Redman, D.D.S., M.S.D., Ph.D.
Chief
Oral Pathology Research Laboratory
Department of Veterans Affairs Medical Center
Washington, D.C.

Karen Rossie, D.D.S., M.S.
Assistant Professor
Department of Diagnostic Services
School of Dental Medicine
University of Pittsburgh
Pittsburgh, PA

Charles G. Saracco, D.D.S.
Clinical Associate Professor
Department of Anatomy/Histology
School of Dental Medicine
University of Pittsburgh
Pittsburgh, PA

John Scott, B.D.S., Ph.D., F.D.S., FRCPath
Professor
Dean of Dental Studies
School of Dental Surgery
University of Liverpool
Liverpool, England

Bernard Tandler, B.S., A.M., Ph.D.
formerly, Professor
School of Dentistry
Case Western Reserve University
Cleveland, OH

R. James Turner, Ph.D.
Chief
Membrane Biology Section, Clinical
  Investigation and Patient Care
  Branch
National Institute of Dental Research
National Institutes of Health
Bethesda, MD

K. Dobrosielski-Vergona, Ph.D.
Associate Professor and Chair
Department of Anatomy/Histology
School of Dental Medicine
University of Pittsburgh
Pittsburgh, PA

Robert B. Wellner, Ph.D.
Research Physiologist
U.S. Army Medical Research
  Instituteof Infectious Diseases
Fort Detrick
Frederick, MD

TABLE OF CONTENTS

Anatomy of the Human Salivary Glands

Charles G. Saracco and Edward V. Crabill

INTRODUCTION

Human saliva is produced by three pairs of major salivary glands and several minor mucous and serous salivary glands located in specific parts of the oral cavity. The large major salivary glands are paired and include the parotid glands, the submandibular glands, and the sublingual glands. The minor salivary glands include the glands of the tongue (anterior and posterior lingual glands), the labial glands, the buccal glands, the molar glands, the incisive glands, and the palatine glands.

Saliva has several functions. In addition to moistening the oral tissues as an aid for speech, and functioning as an aqueous solvent necessary for taste, it acts as a masticatory wetting agent which assists swallowing. Certain glands initiate the digestive process by secreting an enzyme. Also, saliva has an antibacterial action which inhibits or prevents the onset of dental caries.

It is generally accepted that of the total volume of saliva formed in 1 day, 60 to 70% is secreted by the submandibular glands, 25 to 35% by the parotid glands, and 5% or less by the sublingual glands. The minor salivary glands contribute approximately 5 to 8% of the total daily production (Hand, 1986).

THE PAROTID GLAND

The parotid gland is the largest of the salivary glands. The gland is irregular in shape, but roughly resembles an inverted pyramid. Its weight in the adult varies from 15 to 30 g, and it is approximately 6 cm in length (superior to

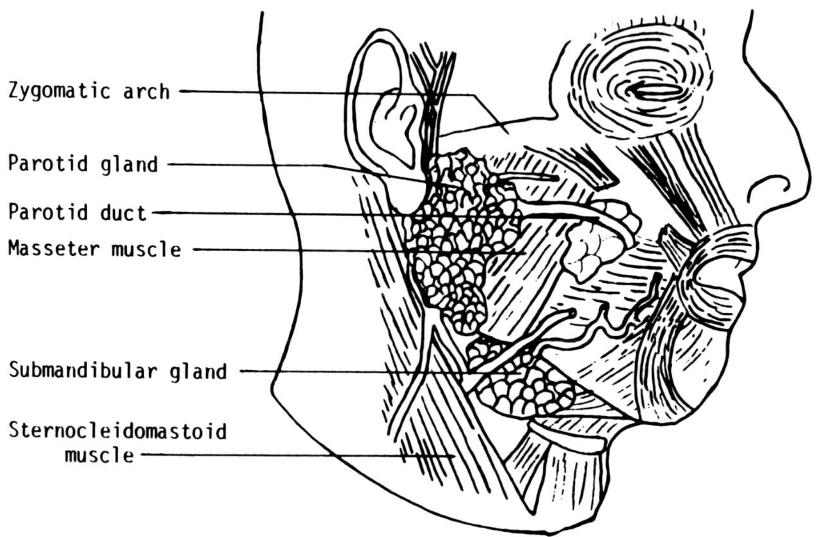

Figure 1. Lateral view of the parotid and submandibular glands.

inferior), and 3 to 4 cm in width. It is located on the side of the face in the retromandibular fossa (Figure 1). More specifically, the "bed" or recess which the gland occupies is located anterior and inferior to the external acoustic meatus, inferior to the zygomatic arch, posterior to the ramus of the mandible and the masseter muscle, anterior to the mastoid process and the superior part of the sternocleidomastoid muscle, and lateral to the styloid process. The gland presents three surfaces (lateral, anteromedial, and posteromedial), four borders (superior, anterior, posterior, and medial), and an apex.

The Lateral Surface

The lateral surface of the parotid gland is located immediately deep to the skin and superficial fascia of the head. The gland is enclosed by a bilaminar extension of the outer (investing) layer of the deep cervical fascia, the parotid portion of the parotideomasseteric fascia. The outer lamina is more dense than the inner lamina, but both are attached to the capsule of the gland by fibrous connective tissue, which contributes to making surgical removal of the gland difficult. The lateral surface of the gland is lateral to that portion of the posterior part of the masseter muscle which covers the temporomandibular joint. It is also lateral to the anterior part of the mastoid process and to the sternocleidomastoid muscle. A detached portion of the parotid gland, the accessory parotid gland, which is present in about 20% of all individuals, is located more anteriorly on the lateral surface of the masseter muscle.

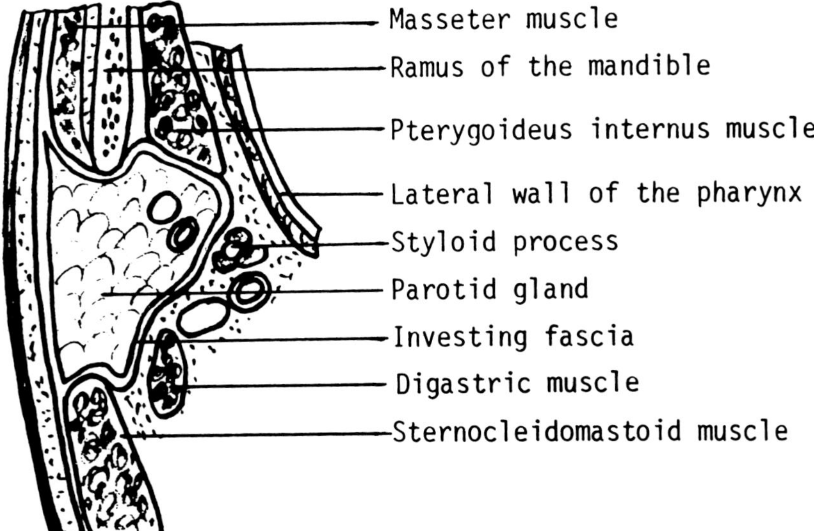

Figure 2. Horizontal section of the parotid gland.

The Anteromedial Surface

The anteromedial surface of the gland (Figure 2) is grooved by the posterior border of the ramus of the mandible and the masseter muscle. The medial portion of this surface projects anteriorly to contact the posterior border of the internal pterygoid muscle inferiorly, but higher up it extends between the medial surface of the ramus and the lateral surface of the internal pterygoid muscle to form a deep pole of the gland. The part of the gland which connects the deep pole and the larger superficial portion is the isthmus. It is from the deep pole that the maxillary artery leaves the gland to enter the pterygo-mandibular space of the infratemporal fossa.

The Posteromedial Surface

The posteromedial surface of the gland (Figure 2) is grooved by the mastoid process and the sternocleidomastoid muscle. This part of the gland contacts the posterior belly of the digastric muscle, the styloid process, and the styloid muscles. The external carotid artery grooves this surface before it enters the gland, and the facial nerve enters the gland on the superior part of this surface. The posteromedial surface meets the anteromedial surface of the gland to form a vertical medial margin. Above, this medial margin contacts the styloid process, while inferiorly it can extend as far medially as the wall of the pharynx.

The Apex

The apex of the parotid gland (Figure 1) is located between the angle of the mandible and the sternocleidomastoid muscle, and usually extends below the plane of the lower border of the mandible. The retromandibular vein leaves the medial surface of the apex to enter the upper posterior part of the carotid triangle of the neck. Also leaving the apex of the gland is the cervical branch of the facial nerve.

The Superior Border

The superior border is always inferior to the zygomatic arch. From this border the temporal branch of the facial nerve leaves the gland. Along the posterior part of this border both the auriculotemporal nerve and the superficial temporal artery leave the gland, while the superficial temporal vein enters the gland at this site.

The Anterior Border

The anterior border of the parotid gland is located immediately lateral to the posterior part of the masseter muscle. The parotid duct leaves this border to pass anteriorly, deep to the accessory parotid gland when present. Also, from this border the zygomatic, upper buccal, lower buccal, and marginal mandibular branches of the facial nerve leave the gland, along with the transverse facial artery.

The Posterior Border

The posterior border lies anterior and slightly lateral to the sternocleido-mastoid muscle. No neural or vascular structures enter or leave the gland by way of this border.

The Medial Border

The medial border of the gland (Figure 2) is lateral to the styloid process superiorly, while inferior to the styloid process, it is lateral to the wall of the pharynx. The muscles which originate from the styloid process separate the parotid gland from the carotid sheath.

The Parotid Duct

The parotid duct leaves the anterior border of the gland to pass anteriorly, deep to the accessory parotid gland (when present), across the lateral surface of the masseter muscle at the level of the junction of the muscle's upper and middle thirds. At the anterior border of the masseter muscle the duct turns medially and, penetrating the buccal fat (suctorial) pad and buccinator muscle, empties into the buccal vestibule opposite the upper second molar tooth.

Structures Within the Parotid Gland

The external carotid artery enters the posteromedial surface of the gland to divide into its two terminal branches, the maxillary and superficial temporal arteries, at the level of the neck of the mandible. The maxillary artery gives rise to two branches within the gland, the deep auricular and the anterior tympanic arteries. The deep auricular artery passes through the gland and penetrates the cartilaginous external acoustic meatus to supply the lateral surface of the tympanic membrane; in its course it sends a branch to the temporomandibular joint. After passing through the gland, the anterior tympanic artery enters the petrotympanic fissure to supply the medial surface of the tympanic membrane. Within the gland the superficial temporal artery gives rise to the transverse facial artery and to small branches which supply the temporomandibular joint and the sternocleidomastoid muscle. Also within the gland the superficial temporal vein joins the maxillary vein to form the retromandibular vein.

The facial nerve enters the superior portion of the posteromedial surface and within the gland it gives rise to its two divisions. The upper, or temporofacial division branches into the temporal, zygomatic, and upper buccal branches. The lower, or cervicofacial division, divides into the lower buccal, marginal mandibular, and cervical branches. The major sensory nerve within the gland is the auriculotemporal nerve. After entering the anteromedial surface of the gland this nerve curves upward and laterally to leave the superior border posterior to the superficial temporal artery.

The concept proposed by McWhorter (1917), that the divisions of the facial nerve pass between separate superficial and deep parts of the parotid gland around an isthmus which connects the two parts, has not been confirmed. Although numerous masses of glandular tissue (isthmuses) which connect the superficial and deep portions of the gland and which pass through the plane of the facial nerve and its branches have been described, studies of the duct patterns within these masses by McKenzie (1948) show that these are not sites of contact between superficial and deep parts of the parotid gland, or routes of passage of major tributaries of the parotid duct, but rather that they are integral components of both "deep" and "superficial" portions of the gland. McKenzie's observations indicate that, within the parotid gland, glandular tissue, duct tributaries, and facial nerve branches are all intimately interwoven.

The arterial supply to the parotid gland is provided by glandular branches of the several arteries which pass through its substance, namely the external carotid, posterior auricular, superficial temporal, transverse facial, and maxillary arteries. The veins which drain the gland empty chiefly into the retromandibular vein. The lymphatic vessels of the gland drain primarily to the deep parotid nodes embedded within the gland and to the superficial and deep cervical nodes.

THE SUBMANDIBULAR GLAND

The submandibular gland (Figure 1) is roughly egg shaped and approximately half the size of the parotid gland. It varies considerably in size, being on average 4 to 5 cm in length, and 7 to 10 g in weight. It is located in the submandibular fossa on the medial surface of the body of the mandible below the mylohyoid line. If the gland is large it will overlap the digastric muscle and may extend from the submandibular fossa posteriorly on to the medial surface of the internal pterygoid muscle. The bed, or recess, of the gland also includes the mylohyoid and hyoglossus muscles medially and superiorly, and the skin and superficial fascia inferiorly and laterally.

The gland is comprised of a larger superficial portion located external to the mylohyoid muscle, and a smaller, deep portion located deep to this muscle. The superficial portion of the gland lies within the submaxillary fascial space, while the deep portion lies within the sublingual space. The gland presents three surfaces — lateral, medial, and inferior, and three poles — anterior superficial, anterior deep, and posterior. The posterior pole of the gland is more rounded than both the superficial and deep anterior poles.

The structures related deeply to the superficial portion of the gland, in addition to the muscles mentioned above, include the anterior and posterior bellies of the digastric muscle, the stylohyoid muscle, the mylohyoid nerve, the hypoglossal nerve, the vena comitans nervi hypoglossi, and the submental artery. The platysma muscle, facial vein, submental vein, and the cervical branch of the facial nerve are related to the inferior surface of the superficial portion of the gland. The cervical part of the facial artery lies in a groove on the superior posterior portion of the gland, and it is from this artery that the gland receives its blood supply.

The superficial portion of the submandibular gland, like the parotid gland, is enclosed by a bilaminar reflection of the investing layer of the deep cervical fascia. The fascial space which the gland occupies is bounded by lateral and medial lamellae of the investing fascia as it passes from the hyoid bone to the lower border of the mandible. Unlike the situation with the parotid gland, there is no fibrous connection between the investing fascia and the capsule of the submandibular gland. Therefore, surgical removal of this gland is less difficult than that of the parotid gland.

The deep part of the submandibular gland (Figure 3) projects medially from the middle portion of the medial surface of the superficial part to pass anteriorly between the mylohyoid and hyoglossus muscles. This part of the gland is located in the sublingual fascial space and is presumed to not have a deep cervical fascial investment as does the superficial part. In this regard it is similar to the sublingual

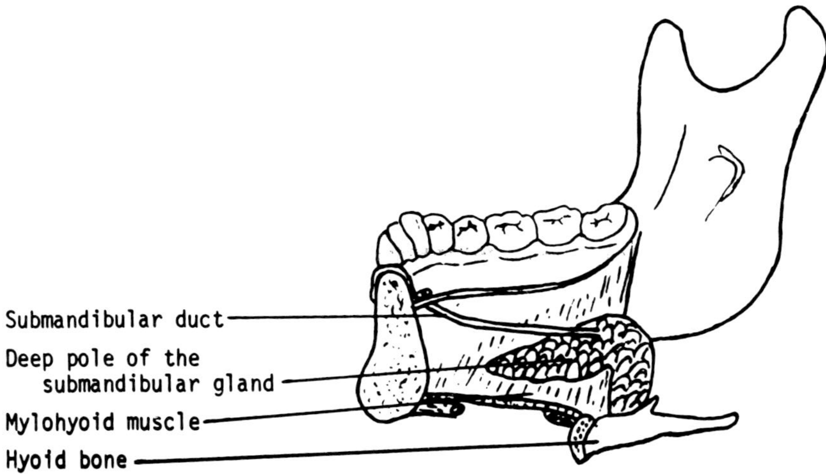

Figure 3. Deep part of the submandibular gland.

gland, which has no fascial covering. The submandibular duct leaves the deep part of the gland and, passing anteriorly, medial to the sublingual gland and lateral to the genioglossus muscle, it empties into the sublingual fossa of the oral cavity at the sublingual papilla. The sublingual papilla is located inferior and immediately lateral to the lingual frenum. The duct leaves the submandibular gland at the level of the anterior border of the hyoglossus muscle or adjacent to the lower second molar tooth.

The arteries which supply the gland are derived from the facial and lingual arteries, while the veins of the gland are tributary to the submental and facial veins. The lymphatic drainage is to the nearby submandibular nodes.

THE SUBLINGUAL GLAND

The sublingual gland (Figure 4) is the smallest of the major salivary glands. It is an elongated, flattened (almond shaped) gland averaging about 3 to 4 cm in length and only 2 to 3 g in weight. Because of its position just beneath the mucous membrane of the floor of the mouth, it is responsible for forming the sublingual eminence in this part of the oral cavity. The gland is thin and convex laterally (Figure 5), with the posterior inferior end being widest. Also, the gland is located between the genioglossus muscle and the body of the mandible (Figure 6), where it occupies the sublingual fossa of the mandible superior to the mylohyoid line. Anteriorly, it contacts the opposite sublingual gland in the area of the mandibular symphysis. Inferiorly, it rests upon the mylohyoid muscle, and posteriorly it is in contact with the deep portion of the submandibular gland. The medial surface of the gland is in contact with the lingual nerve and the submandibular duct.

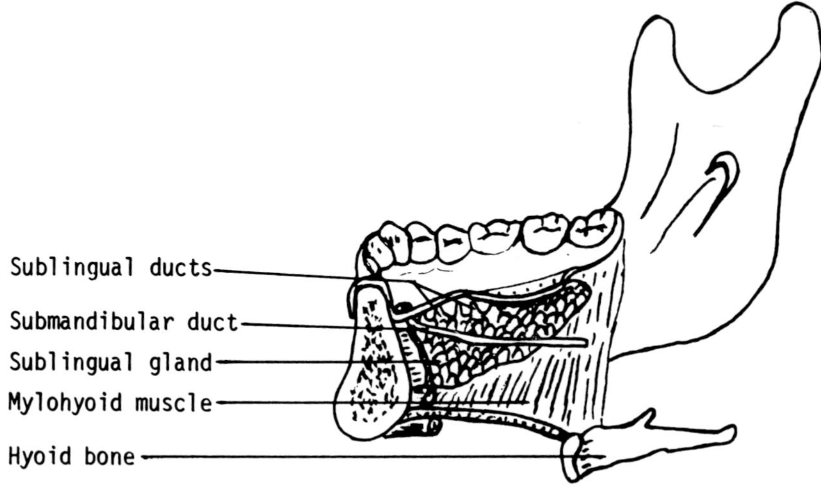

Figure 4. Medial view of the sublingual gland.

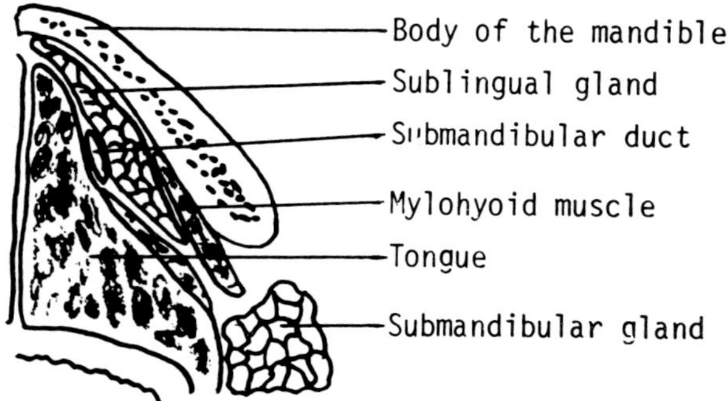

Figure 5. Horizontal section of the sublingual gland.

The sublingual gland, unlike the other two major salivary glands, is not ensheathed by an extension or reflection of the deep cervical fascia; instead, it is surrounded by the loose connective tissue of the sublingual space.

The sublingual gland usually empties its secretions into the sublingual fold by way of 8 to 20 small ducts. The sublingual fold is a raised portion of mucous membrane which extends laterally from the sublingual papilla over the sublingual eminence. These ducts are called the ducts of Ravinus. Sometimes several ducts from the inferior and lateral portions of the gland join together to form the duct of Bartholin, which then joins the submandibular duct near the sublingual papilla. All of the secretions of the sublingual and submandibular glands flow into the sublingual fossa in the floor of the mouth below the tongue.

The arteries to this gland are branches of the sublingual and submental

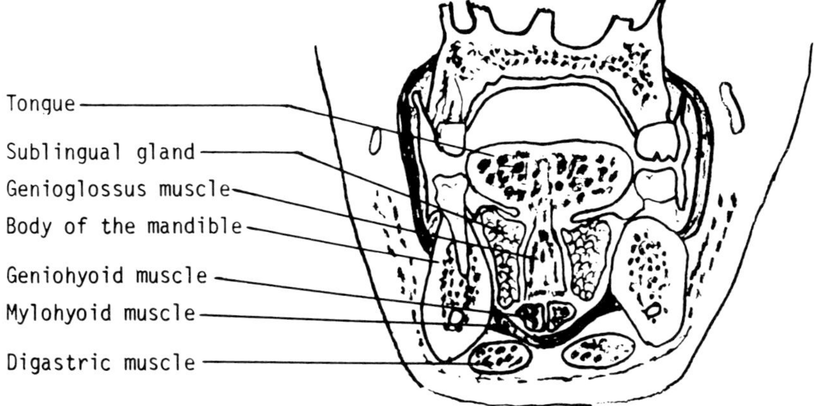

Figure 6. Coronal section of the sublingual gland.

arteries; the veins accompany these arteries to empty into tributaries of the vena comitans nervi hypoglossi and facial vein, respectively. The lymphatic drainage is to the superior deep cervical nodes.

THE ANTERIOR LINGUAL GLANDS

The anterior lingual glands are located between muscle fibers in the anterior inferior portion of the tongue near the apex. These glands, which are approximately 20 mm long, are composed of right and left masses of glandular tissue. Three or four small ducts from each glandular mass open on the underside of the apex of the tongue, thereby depositing the secretions of these glands into the sublingual fossa. The secretions of these glands are both mucous and serous. Additionally, a few small, purely mucous glands are present on the tip and lateral margins of the anterior portion of the tongue.

THE POSTERIOR LINGUAL GLANDS

These lingual glands are of two types, strictly mucous and strictly serous, and their distribution is restricted to specific parts of the tongue. The posterior lingual mucous glands are, for the most part, located on the sides of the tongue, lateral and posterior to the sulcus terminalis, and on the dorsum of the posterior third of the tongue in association with the lingual tonsils. The posterior lingual serous glands are only associated with the circumvallate papillae. These glands, which empty into the circular sulcus surrounding each circumvallate papilla, help to distribute food substances to the taste buds located on the walls of each sulcus.

THE LABIAL GLANDS

The labial glands are found in both the upper and lower lips. They are located in the submucosal tissue between the labial mucous membrane and the orbicularis oris muscle. They are mucous glands, and their ducts empty into the vestibule of the oral cavity.

THE BUCCAL GLANDS

The buccal glands are similar to the labial glands except that they are located between the mucous membrane and the buccinator muscle. Their ducts also empty into the vestibule of the oral cavity.

THE MOLAR GLANDS

The molar glands typically consist of a mass of three or four glands located between the buccinator muscle and the anterior border of the masseter muscle. Their ducts pierce the buccinator muscle to open into the vestibule of the oral cavity opposite the molar teeth.

THE INCISIVE GLANDS

The incisive glands are located in the floor of the mouth in the area of the lingual frenum. Their ducts also empty into the sublingual fossa.

THE PALATINE GLANDS

The palatine glands are located in the submucosal tissue of both the nasal and oral surfaces of the soft palate, and in the oral surface of the hard palate posterior to a line between the first molar teeth. The midline and periphery of the soft palate are void of these glands.

INNERVATION OF THE SALIVARY GLANDS

The superior and inferior salivatory nuclei of the brain stem are the parasympathetic nuclei concerned with salivary gland secretion. Visceral efferent fibers emanating from the superior salivatory nucleus are associated with cranial nerve VII, while fibers from the inferior salivatory nucleus are associated with cranial nerve IX.

The preganglionic parasympathetic fibers leaving the superior salivatory nucleus become components of the nervus intermedius, the general visceral efferent and general visceral afferent portion of the facial nerve. The fibers of this nerve enter the internal acoustic meatus in company with the special visceral efferent fibers of the facial nerve and pass into the facial canal within the temporal bone. At the geniculate ganglion some parasympathetic fibers leave the nervus intermedius by way of the greater petrosal nerve which enters the middle cranial fossa by passing through the hiatus of the facial canal. The greater petrosal nerve courses anteriorly and medially in the middle cranial fossa to enter the foramen lacerum. On the anterior border of this foramen the nerve enters the pterygoid canal to pass anteriorly into the pterygopalatine fossa. Within the pterygopalatine fossa, fibers of this nerve synapse in the pterygopalatine ganglion with postganglionic parasympathetic neurons which are distributed, by way of branches of the maxillary division of the trigeminal nerve, to minor salivary glands of the hard palate, soft palate, and upper lip.

The remaining preganglionic parasympathetic fibers of the nervus intermedius continue from the geniculate ganglion as components of the chorda tympani nerve, which accompanies the facial nerve through the descending portion of the facial canal. The chorda tympani nerve leaves the facial canal just proximal to the stylomastoid foramen, passes anteriorly and superiorly through the middle ear cavity, and then continues inferiorly through the petrotympanic fissure into the infratemporal fossa to join the lingual branch of the mandibular division of the trigeminal nerve. After accompanying the lingual nerve into the floor of the mouth, the preganglionic fibers of the chorda tympani leave the lingual nerve to enter the submandibular ganglion, where they synapse with postganglionic parasympathetic neurons. Some of these postganglionic secretomotor fibers pass directly to the submandibular gland, while others join the lingual nerve in the vicinity of the submandibular ganglion to be distributed, by way of branches of the lingual nerve, to the sublingual gland, the incisive glands in the floor of the mouth, and the anterior lingual glands. Preganglionic parasympathetic fibers from the inferior salivatory nucleus, after joining the glossopharyngeal nerve, accompany this nerve's tympanic branch onto the medial wall of the middle ear, where they contribute to the formation of the tympanic plexus. The fibers leave the tympanic plexus as components of the lesser petrosal nerve, which passes to the otic ganglion for synapse with postganglionic parasympathetic neurons. These postganglionic neurons leave the ganglion to immediately join the undivided portion of the mandibular division of the trigeminal nerve. The postganglionic secretomotor fibers are then distributed to the parotid gland by way of the auriculotemporal nerve, to the minor glands of the lower lip by way of the inferior alveolar nerve and its mental branch, and to the buccal and molar glands by way of the long buccal nerve. The majority of the posterior lingual glands are in all probability supplied by a parasympathetic pathway associated with the inferior salivatory nucleus and the glossopharyngeal nerve, but the course of the preganglionic and postganglionic neurons, as well as the location of the ganglion itself, has yet to be determined.

Whether sympathetic nerves have a salivary secretomotor function in humans is controversial. Kuntz and Richins (1946) reported that sympathetic nerve fibers in the salivary glands mainly innervate blood vessels rather than any of the glandular elements. However, in 1974 two groups of investigators (Harrop and Garrett, and Hodgson and Spiers) reported that, in the rat parotid gland, adrenergic fibers were secretomotor.

The cell bodies of the preganglionic sympathetic neurons associated with the salivary glands are located in the intermediolateral cell column of the upper thoracic portion of the spinal cord. Fibers of these neurons pass to the superior cervical sympathetic ganglion, where synapse with all the postganglionic neurons that travel to the salivary glands, both major and minor, occurs. Postganglionic sympathetic fibers to all of the salivary glands, except the palatine glands, are derived from the external carotid plexus on the external carotid artery. Sympathetic fibers are then carried to the salivary glands by secondary plexuses associated with those branches of the external carotid artery which supply the glands. Postganglionic sympathetic fibers reach the minor salivary glands of the palate by a different route. After leaving the superior cervical sympathetic ganglion and participating in forming the internal carotid plexus, these fibers accompany the deep petrosal nerve and nerve of the pterygoid canal to enter the pterygopalatine ganglion, from which they are distributed to the palate by branches of the maxillary division of the trigeminal nerve.

DEVELOPMENTAL ANATOMY OF THE SALIVARY GLANDS

All salivary glands, major and minor, develop as proliferations of oral epithelium into the underlying mesenchyme. The epithelial outgrowths give rise to the ducts and alveolar components of the glands, while the stroma and blood vessels are derived from mesenchymal tissues. Based upon their sites of origin within the primitive oral cavity, it seems probable that the majority of the salivary glands are formed from epithelial tissue which is endodermal in origin, while the labial glands certainly, and the parotid glands probably, are derived from ectodermal epithelium.

The Parotid Gland

At approximately the middle of the 6th week of development (10 to 12 mm C.R. length), an elongated epithelial furrow appears in the cheek region of the primitive oral cavity; this groove runs posteriorly from the angle of the oral fissure along the internal junction of the maxillary process and mandibular arch.

The furrow becomes converted into a tube by the fusion of its edges and, except at its anterior end, loses its continuity with the oral epithelium. The tube, which grows posteriorly into the tissues of the cheek, persists as the duct of the parotid gland, while from its distal end solid cell cords, which will become the intraglandular ducts, begin to proliferate and branch extensively into the pre-auricular mesenchyme. As the cell cords cease growing and branching, knob-like clusters of cells, which will become the acinar parts of the gland, appear at their terminal ends. The cell cords become hollowed out in the 6th month, and during the subsequent 2 months progressive differentiation of the cellular components of the canalized cords and terminal knobs proceeds sequentially throughout the branching ducts and acinar portions of the gland. Although the gland is capable of secretion during the 9th month, full maturation of its cells most likely is not attained until after birth.

As a result of forward growth and continued fusion of the maxillary and mandibular processes, the opening of the mouth is moved anteriorly and its size reduced, causing the orifice of the main duct in the adult to be located several centimeters posterior to the angle of the mouth.

The Submandibular Gland

The development of the submandibular gland exhibits a departure from the usual in that the formation of the glandular elements precedes that of the excretory duct. Late in the 6th week (13 to 14 mm C.R. length), an epithelial outgrowth can be detected, opposite the anterior part of the tubotympanic recess, in the posterior part of a narrow sulcus located between the base of the developing tongue and the floor of the primitive oral cavity. The gland bud rapidly invades the local mesenchymal tissues, increasing in size by forming branching, nonpatent, intraglandular ducts and alveolar precursors.

As the gland begins to grow and the tongue continues to increase in size, the submandibular duct develops as a modification of the lingual portion of the floor of the primitive oral cavity. The longitudinal groove alongside the base of the tongue becomes larger and deeper, thereby incorporating the site of origin of the gland into its floor. Posteriorly, in the region of the developing gland, the opposed edges of this groove begin to fuse and separate from the surface epithelium; this process effectively separates the primordia of the intraglandular ducts from the oral epithelium so that these ducts become continuous with the excretory duct of the gland. Fusion of the edges of the sulcus proceeds slowly from posterior to anterior; therefore, the termination of the duct does not reach its adult location in the anterior part of the floor of the mouth until the 9th or 10th week. As the fusion progresses anteriorly, the termination of the duct is always open to the oral cavity, and the lumen of the duct appears to be patent from the very beginning. Development of the main duct usually is well along before any of the ducts within the gland develop a lumen. Cell differentiation and maturation of the branching ducts and acini are similar to that seen in the parotid gland.

The Sublingual Gland

This gland arises somewhat later than the submandibular gland, usually during the 7th week (18 to 20 mm C.R. length). It is formed by a series of epithelial outgrowths from the floor of the mouth located in a line lateral to the sulcus which gives rise to the submandibular duct. With continued growth of the gland primordia, the primitive secretory elements come to be closely packed together, enveloped in a common connective sheath, while the ducts retain their original connections to the floor of the mouth. Frequently, some of the epithelial outgrowths which appear in the anterior part of the floor of the mouth are located in the sulcus which will become the submandibular duct. Therefore, some of the sublingual ducts (e.g., the Duct of Bartholin) come to open either into the submandibular duct or very near its termination.

REFERENCES

1. Hand, A. H., Salivary glands, in *Orban's Oral Histology and Embryology*, S. N. Bohaskar, Ed., C. V. Mosby, St. Louis, 1986, 354.
2. Harrop, T. J. and Garrett, J. R., Effects of preganglionic sympathectomy on secretory changes in parotid acinar cells of rats on eating, *Cell Tissue Res.*, 154, 135, 1974.
3. Hodgson, C. and Spiers, R. L., The effect of preganglionic cervical sympathectomy on the amylase content of parotid glands in fasted and fed rats, *J. Physiol. Lond.*, 237, 56, 1974.
4. Kuntz, A. and Richins, C. A., Components and distribution of the nerves of the parotid and submandibular glands, *J. Comp. Neurol.*, 85, 21, 1946.
5. McKenzie, J., The parotid gland in relation to the facial nerve, *J. Anat.*, 82, 183, 1948.
6. McWhorter, G. L., The relations of the superficial and deep lobes of the parotid gland to the ducts and to the facial nerve, *Anat. Rec.*, 12, 149, 1917.

Cytology, Histology, and Histochemistry of Salivary Glands: An Overview

Carlin A. Pinkstaff

By definition, the salivary glands are those glands that empty their secretory products into the oral cavity. Saliva, the final product of salivary glands, is formed by units of secretory cells, often called secretory endpieces, and a system of ducts. Primary saliva is formed by cells of the secretory endpieces and is modified by cells of the duct system as it passes toward the oral cavity. The primary saliva is much more than a simple fluid in that it may contain mucin, enzymes, immunoglobulins, growth factors, etc. The composition of the final saliva, as well as the primary saliva, is quite possibly unique for each salivary gland and is determined by the nature of the secretory endpiece cells and ductal elements of that gland. Salivary glands are increasingly being used to study many biological phenomena, e.g., mucin secretion, protein secretion, epithelial transport of water and ions, exocytosis and endocytosis, growth factors, etc. and it is imperative that investigators be aware of differences between the salivary glands in the species they are studying and those of other species if they intend to compare the results they obtain. There are remarkable differences between the salivary glands of different species, as well as between different salivary glands of the same species. Salivary-gland diversity, with respect to both form and function, is not generally appreciated by individuals who are not involved in the study of salivary glands. The morphological diversity of salivary glands is well illustrated in several monographs (Tandler, 1978; Young and van Lennep, 1978; Wilborn and Shackleford, 1980; Pinkstaff, 1980).

The primary goal of this overview is not to add another review of the morphology of the major salivary glands to the literature but to show how histochemistry has added to our understanding of major salivary gland structure and function. By applying histochemical staining methods to tissue sections, an investigator can study biochemical phenomena at the cellular level. The presence of an enzyme may be determined in an homogenate of a salivary gland; however, only an appropriate histochemical method can localize that enzyme to a particular duct segment or to a certain cell type within that duct segment. By using both histochemical and biochemical methods in a complementary study, an investigator can gain considerable insight into biological processes occurring in salivary glands.

Salivary glands have been subdivided into major (extrinsic) and minor (intrinsic) salivary glands. Major salivary glands empty their secretory products into the oral cavity via one or more extraglandular excretory ducts, while minor salivary glands lie in the lamina propria or submucosa of the oral cavity and empty their products into the oral cavity via numerous short excretory ducts. The following discussion will deal only with the major salivary glands of man, other primates, and mice and rats. There are three pairs of major salivary glands in most mammals; the parotid (PG), submandibular (SMG), and sublingual (SLG) gland/glands. The major salivary glands are usually classified as compound, tubuloacinar glands (Tandler, 1978; Wilborn and Shackleford, 1980). Describing a gland as compound tubuloacinar gland implies that there are two different types of secretory endpieces, tubules and acini; the term compound means that the secretory endpieces empty into a branched duct system.

The cells of the secretory endpieces are pyramidal shaped and empty their secretory products into central lumens of acini or tubules. Controversy exists as to the types of cells forming the secretory endpieces. Examination of sections of salivary gland secretory endpieces stained with hematoxylin and eosin allows one to describe two types of secretory cells, serous and mucous. When histochemical staining methods for glycoconjugates are used it is possible to identify a third cell type, the seromucous cell. Traditionally, the serous cell has been described as a cell that is primarily a protein secreting cell and contains little, if any, mucin. Histochemical staining methods for glycoconjugates have shown that many serous cells do contain appreciable concentrations of neutral and/or acidic glycoconjugates and Munger (1964) has referred to such cells as seromucous cells. On the other hand, other authors say that seromucous cells must contain appreciable concentrations of acidic glycoconjugates; they believe that cells containing appreciable concentrations of neutral glycoconjugates but lacking acidic glycoconjugates should still be called serous cells. This controversy has been reviewed by Pinkstaff (1980, 1981), Sakai (1989), and Hagelqvist et al. (1991). The classification scheme of Munger will be used in this overview even though it has not been accepted by all investigators. The author is of the opinion that there are relatively few, if any, salivary gland secretory endpiece cells that are truly serous.

Mucous cells are relatively easy to recognize since most contain both acidic and neutral glycoconjugates. The mucous tubules in some mammalian SLG and SMG may have groups of secretory cells that appear to cap their blind ends, the demilunes. In most species, the demilunes are usually either serous or seromucous but they have been shown to be mucous cells in carnivore salivary glands. Demilune cells may deliver their secretory products into the lumens of mucous tubules via intercellular canaliculi that occur between mucous tubule cells (Young and van Lennep, 1978; Tandler, 1978) or by thin processes of demilune cells that pass between mucous cells to reach the lumens (Tandler, 1978; Pinkstaff, 1980).

Historically, most authors have referred to the acinar cells of human, mouse, and rat PG as serous cells. With the advent of histochemical staining methods for the demonstration of glycoconjugates, particularly lectin-binding methods, it has been possible to show that these cells do synthesize and secrete glycoconjugate-containing secretory granules. This has resulted in an increasing number of authors who refer to PG acinar cells as seromucous cells. Old terminology, however, dies hard and there are still many authors that refer to such cells as serous cells. Whether one calls these cells serous or seromucous, it is important to realize that many cells said to be serous in nature do contain appreciable concentrations of neutral and/or acidic glycoconjugates. Nearly 40 years ago, Junqueira et al. (1951) contended that the PG acinar cells of many species, including man, mouse, and rat, were not serous cells when they said, "Contrairement à l'opinion commune la parotide ne paraît donc pas être une glande séreuse." Junqueira and co-workers based their conclusions on periodic acid-Schiff (PAS) staining of PG acinar cells. Munger (1964) corroborated the observations of Junqueira and co-workers and called cells containing appreciable neutral or acidic glycoconjugates, seromucous cells.

Lectin binding to sugar residues of glycoconjugates in mouse, rat, and human PG acinar cells has been observed using both light and electron microscopic methods. For example, *N*-acetylglucosamine, as shown by WGA (wheat germ agglutinin, *Triticum vulgare*) labeling, is present in secretory granules of human and mouse PG acinar cells (Laden et al., 1984; Jezernik and Pipan, 1989a) and α-fucose, demonstrated by UEA-I (*Ulex europaeus* agglutinin) labeling, is present in human PG acinar cells (Born et al., 1987). In recent years, numerous reports of lectin-binding studies on mouse and rat PG have appeared in the literature; the recent monograph by Masahiko Mori (1991) should be consulted for a review of many of these studies. The excellent light microscopical studies of Schulte and Spicer (1983a, 1984) on mouse and rat salivary glands should be consulted; these reports clearly show lectin labeling of glycoconjugates of mouse and rat PG acinar cells as do many other studies. In addition, relatively high concentrations of fucose, fructose, and mannose have been demonstrated in mouse PG acinar cells using autoradiographic techniques (Watanabe and Kihara, 1986). A review of studies of a variety of nonhuman primate PG glands stained by nonlectin staining methods suggests that acinar cells of PG of many primates are seromucous cells (Wilcox and Pinkstaff, 1982). Based upon the results obtained with histochemical staining methods, it is appropriate to refer to human, mouse, and rat PG acinar cells as seromucous cells.

The human SMG has been described as a mixed gland that contains serous acini and mucous tubules; mucous tubules are often capped by serous demilunes. The mucous tubules in human SMG contain neutral, acidic sulfated, and acidic nonsulfated glycoconjugates as revealed by traditional histochemical staining methods (Leppi and Spicer, 1966; Eversole, 1972). Lectin-labeling studies clearly show the presence of a variety of sugar residues in SMG mucous tubules (Laden et al., 1984; Nakai et al., 1985). Glycoconjugates in the human SMG serous acini and demilunes have been shown by PAS and/or Alcian blue (AB) staining methods (Munger, 1964; Leppi and Spicer, 1966; Harrison et al., 1987) and lectin-binding methods (Laden et al., 1984; Nakai et al., 1985). These cells can therefore be referred to as seromucous cells. Seromucous acini and demilunes are also present in the SMG of several nonhuman primates (Leppi and Spicer, 1966; Wilcox and Pinkstaff, 1982). It should be noted that the mucous glycoproteins in human SMG seromucous cells may be different from those in the mucous tubules (Harrison et al., 1987) and the same is true with respect to these compounds in similar cells of some nonhuman primate SMG (Leppi and Spicer, 1966; Wilcox and Pinkstaff, 1982).

The SMG of mice and rats are quite different from those of man and most nonhuman primates in that mucous tubules or seromucous demilunes are usually not present. The acini of mouse and rat SMG are usually referred to as serous acini but are PAS-positive (Spicer and Duvenci, 1964) and are labeled by various lectins (Schulte and Spicer, 1983a, 1984; Takai et al., 1986; Mori, 1991) and may be called seromucous acini. It should be noted that while both mouse and rat SMG acinar cells are labeled by various lectins, there may be differences between labeling in the glands of the two species and between sexes of the same species (Schulte and Spicer, 1984; Schulte, 1987). For example, serous acini in rat SMG contain glycoconjugates with terminal β-galactose residues which are lacking in mouse SMG submandibular glands, on the other hand, penultimate β-galactose residues are present in mouse submandibular glands but absent in rat SMG acini (Schulte and Spicer, 1984). Benau et al. (1990) have recently demonstrated the activity of galactosyltransferase in acinar cells of rat SMG using a tetrazolium salt staining method; this approach may be useful in determining enzymes responsible for the addition of sugar residues to salivary gland glycoproteins, particularly those at sites other than at the terminal ends of glycoprotein molecules. Spicer et al. (1987) have also shown that acinar cells of SMG of different strains of mice may be labeled differently by some lectins.

The human SLG is a complex organ that consists of a large, sometimes absent, major (greater) SLG and a group of smaller independent minor (lesser) SLG that are always present. The minor SLG will not be discussed in this overview. The major human SLG is a mixed gland containing mucous tubules, occasional seromucous acini, and seromucous demilunes (Riva et al., 1988). Results obtained using traditional staining methods (PAS, Alcian blue, High iron diamine, etc.) have clearly shown that SLG mucous tubules contain neutral, acidic sulfated and acidic nonsulfated glycoconjugates (Leppi and Spicer, 1966).

Labeling of mucous tubules by lectins has shown the presence of several different sugar residues in SLG mucous tubular glycoconjugates (Laden et al., 1984). The seromucous acini and demilunes of human SLG appear to contain appreciable concentrations of sialic acid-containing glycoconjugates as shown by traditional staining methods (Leppi and Spicer, 1966; Eversole, 1972) and lectin-labeling methods (Laden et al., 1984). The human SLG have not been examined as thoroughly as have been the PG and SMG and the author is unaware of any studies on human SLG using electron microscopic cytochemical methods for demonstrating glycoconjugates. Reports on glycoconjugate staining studies of SLG of nonhuman primates can be found in Leppi and Spicer (1966) and Wilcox and Pinkstaff (1982).

The SLG of mice and rats are very similar to those of man in that they contain mucous tubules, seromucous demilunes and occasional seromucous acini. The mucous tubules contain neutral and acidic nonsulfated glycoconjugates but lack sulfated glycoconjugates. Sialic acid-containing glycoconjugates have been demonstrated in both rat and mouse SLG mucous tubules using traditional staining methods (Spicer and Duvenci, 1964). Seromucous cells of mouse and rat SLG contain glycoconjugates as shown by PAS staining and lectin-labeling of sugar residues. It should be emphasized that these comments refer to mice and rats in general but there may be differences in staining between strains of mice and rats, for example, Spicer et al. (1987) reported DBA (*Dolichos biflorus* agglutinin)-labeling of α-D-*N*-acetylgalactosamine in SLG seromucous demilunes of C57BL/6J mice but no such staining occurred in *Mus castaneus* or *M. hortulanus* SLG seromucous demilunes. Some studies have reported very good correlation between biochemical and lectin-labeling studies in identifying some types of carbohydrates present in salivary glands (Schulte, 1987; Herp et al., 1988).

It is beyond the scope of this overview to describe all of the ultrastructural features of mucous- and seromucous-secreting secretory endpiece cells. It is possible to illustrate one way in which the ultrastructure of these cells vary by briefly mentioning the wide diversity of secretory granule ultrastructure. Secretory granules of mucous cells tend to be rather plain and are often composed of little more than an electron-lucent flocculent material. Such is the case with respect to secretory granules of SLG mucous tubules of mouse (Poldermans et al., 1986), rat (Pinkstaff, 1980), and human (Riva et al., 1988) and human SMG mucous tubules (Tandler, 1987). On the other hand, secretory granules of seromucous cells vary greatly in their ultrastructural appearance. The secretory granules of the mouse SMG seromucous acinar cells are usually reported to closely resemble the granules of mucous cells (Caramia, 1966). When specimens are fixed and stained with standard electron microscopic methods, the secretory granules of rat PG acinar cells are usually described as rather plain, moderately electron-dense granules; however, it is possible to demonstrate more elaborate substructure when sections are stained with cytochemical methods for glycoconjugates (Simson, 1977). Secretory granules of mouse PG seromucous cells are often described as being similar to those of rat PG acinar cells. Some authors, however, have shown mouse PG secretory granules that have a slightly more electron-dense rim with occasional dense central cores or spherules (Thomopoulos et al., 1987; Meisel et al., 1988).

Secretory granules of rat SMG acinar cells show varied substructure. Many authors have described the secretory granules in rat SMG acinar cells as being electron lucent with relatively little substructure (Lundquist and Norberg, 1988); however, granules with a crystal-like structure have also been described (Luzzatto et al., 1968). Simson et al. (1978) have shown that fixatives also influence rat SMG secretory granule substructure. Rat SMG secretory granule substructure can also be influenced by administration of chemical agents to the rat, e.g., lamellar inclusions have been seen in secretory granules of rat SMG acinar cells after isoproterenol administration (Simson et al., 1978). The fine structure of secretory granules of seromucous cells of human PG acini has been described by several authors (Chaudhry et al., 1987; Tandler, 1987); the granules vary from light to moderately electron dense with a very electron dense spherule placed either centrally or eccentrically in the granule. Mature secretory granules in seromucous cells of the human SMG are generally described as having a tripartite substructure consisting of an eccentrically located electron-dense spherule that was often found near the midpoint of a less electron dense cresenteric material. The remainder of the granule matrix was even less electron dense than the cresenteric-shaped component (Tandler, 1987). It is interesting that secretory granules seen in serous (seromucous?) cells of SMG of four species of macaques are similar to those seen in seromucous cells of human SMG glands (Kagayama, 1971; Nagato and Tandler, 1986b). The most common secretory granule in seromucous cells of the human SLG contains an electron dense spherule surrounded by fairly electron lucent material (Riva et al., 1988). Reports by Young and van Lennep (1978), Pinkstaff (1980), and Phillips and Tandler (1987) may be consulted by those interested in secretory granule substructure.

Several studies have reported the localization of specific chemical constituents to the regions of secretory granules. For example, Machino et al. (1986) have localized amylase throughout the matrix of secretory granules from human submandibular acini while lysozyme was localized only over the central electron-dense core of the granule. On the other hand, Takano et al. (1991) reported the localization of amylase over electron-dense cores and areas of medium density of human submandibular acinar secretory granules but they failed to observe labeling over electron-lucent areas of granules. Similar results have been observed in secretory granules of human parotid gland by Zimmer et al. (1984); they localized amylase to the dense central cores of human parotid secretory granules.

Future studies of secretory granule ultrastructure may enable us to localize other specific chemical constituents to certain areas of the granule but if we are to make the most of comparative studies we must be aware of factors that may influence granule substructure. Some of the factors which may influence secretory granule substructure include immersion vs. perfusion fixation, fixative solution used, fasted vs. nonfasted animals, administration of sialagogues prior to killing, cytochemical staining methods used in the study, young vs. old animals, males vs. females, etc. Some of the factors that may influence the substructure of salivary gland secretory granules have been discussed by Nagato and Tandler (1986b) and Phillips and Tandler (1987). It is apparent that granule substructure should only be compared when identical experimental conditions have been used by investigators.

In most mammalian salivary glands there are intercalated or intercalary ducts (ID) that receive the products of the secretory endpiece cells. The ID empty into striated ducts (SD) and the SD empty into excretory ducts (ED). The ED coalesce to form a main excretory duct (MED), or in some glands into multiple MED. The major salivary glands are lobulated organs, and ID and SD normally occur within the lobules (intralobular ducts) while the ED usually lie between lobules (interlobular ducts) but it is not unusual to find ED within lobules. The ID-SD-ED system is seen in human and nonhuman primate major salivary glands. In some rodents, including mice and some rats, there is an additional duct segment called the granular duct (GD) interposed between the ID and SD of the SMG.

The ID vary in length from only a few cells to several cells and are formed by nearly squamous to cuboidal cells. For many years, ID were simply considered to be conduits that served to connect secretory endpieces (acini or tubules) to SD, or to GD when present, but ID are being studied more closely because of their possible roles in modification of primary saliva as it passes through this duct segment. Secretory granules in ID of mouse, rat, and human PG and SMG are usually seen in cells near the secretory endpieces (Caramia, 1966; Qwarnström and Hand, 1983; Lantini et al., 1988); in most strains of mice, granular ID in SMG are found only in females (Caramia, 1966). Granules are usually not present in ID of mouse or rat SLG and few, if any, granules are said to be present in ID of human SLG (Riva et al., 1988). Periodic acid-Schiff-positive granules have been seen in ID of human PG and SMG (Harrison et al., 1987) and rat SMG (Qwarnström and Hand, 1983). The secretory granules in human PG and SMG are also stained by electron microscopic staining methods for glycoconjugates (Harrison et al., 1987) and Born et al. (1987) have shown that secretory granules in human PG gland acini contain α-L-fucose. The secretory granules in ID of mouse PG contain glycoconjugates with varied sugar residues (Jezernik and Pipan, 1989a); however, not all ID cells of mouse PG contain secretory granules and those that lack granules also lack osmium tetroxide stained Golgi complexes (Jezernik and Pipan, 1989b). Secretory granules in ID of human PG and SMG are seen in association with thiamine pyrophosphatase-labeled Golgi complexes (Harrison et al., 1988). The presence of glycoconjugates in ID cells of some mouse and rat major salivary glands has been shown by lectin labeling at the light microscopic level (Schulte and Spicer, 1983a,b, 1984; Spicer et al., 1987; Mori,

1991). Other materials may also be secreted by ID cells, e.g., epidermal growth factor has been localized in granules of ID cells in SMG of female A/J mice (Tanaka et al., 1984), and Smith and Toms (1986) have shown the presence of immunoreactive insulin-like material in ID cells of rat PG and SMG.

Cells of the ID of rat PG are capable of endocytosis of foreign proteins which have been retrogradely perfused through the ferritin and bovine serum albumin and glycosylated bovine serum albumin (Hand et al., 1987; Lotti and Hand, 1989). In addition, Hand et al. (1987) observed uptake of two acinar proteins, amylase, and protein B_1 by ID cells; they suggested an endocytotic mechanism for the removal of normal, modified, or abnormal proteins from the primary saliva. Secretory granules in ID, and endocytotic capacity of these ducts, strongly suggests that these ducts serve as more than simple conduits connecting secretory endpieces and larger ducts. It should also be noted that ID are believed to be capable of supplying progenitor cells that differentiate into new acinar or duct cells (Zajicek et al., 1985, 1989; Schwartz-Arad et al., 1988).

Granular ducts are present in mouse and rat submandibular glands but are not present in human submandibular glands. Unfortunately, the GD has also been referred to by many other names, e.g., granular convoluted tubule, convoluted granular tubule, granular tubule, secretory tubule, serous tubule, etc. (Pinkstaff, 1980). Granular duct is certainly the most appropriate term for this ductal segment. Names for this duct segment that contain "tubule" should not be used since such names imply that the GD is a modified tubule rather than a modified duct segment.

The GD in the submandibular glands of most strains of mice are responsible for a marked sexual dimorphism of this gland. Cells of the GD are columnar cells; they contain few basally localized mitochondria, large concentrations of granular endoplasmic reticulum, and numerous apical secretory granules. There are usually more GD in male mouse SMG and they are usually larger and contain more granules than those in female mouse SMG (Caramia, 1966). Granular ducts in rat SMG resemble the GD in male mouse SMG but there does not appear to be a morphological sexual dimorphism of these ducts in rat SMG (Tamarin and Sreebny, 1965).

The secretory granules of GD in mouse and rat SMG stain with the PAS method, are Alcian blue (pH 2.5) negative, and exhibit no autographic labeling by S^{35} (Spicer and Duvenci, 1964); thus, the granules of GD in mouse and rat SMG contain neutral glycoconjugates but lack acidic glycoconjugates. Neutral glycoconjugates have been localized in granules of GD cells of rat SMG using electron microscopic staining methods (Neiss, 1986). Some general observations can be made with respect to some sugar residues that appear to be common to granules of GD of both rats and mice, irrespective of sex or strain. Glycoconjugates in GD of both mice and rats are labeled by PNA (*Arachis hypogaea*, peanut agglutinin), which has binding specificity to β-D-galactose-(1 → 3)-D-*N*-acetylgalactosamine, and by RCA-I (*Ricinus communis* agglutinin), which binds to β-D-galactose > α-D-galactose. However, it should be noted that most authors have observed no staining or ± staining of these granules when DBA (*Dolichos biflorus* agglutinin), which specifically binds to α-D-*N*-acetylgalactosamine, is used. These results, as well as results with other lectins, have been described in the following reports (Naito et al., 1983; Schulte and Spicer, 1983a, 1984; Mori 1991). Lectin-labeling studies have sometimes provided some rather unexpected results; for example, Schulte and Spicer (1983a) noted that approximately 10% of the cells of mouse GD were labeled by LTA (*Lotus tetragonolobus* agglutinin), which specifically binds to α-L-fucose, but no LTA-labeled cells were found in GD ducts of rat SMG (Schulte and Spicer, 1984). Conversely, a population of cells was noted in rat GD that was similar to the population of LTA-labeled cells in mouse SMG but these cells were labeled by GSA I-B$_4$ [*Griffonia (Bandeiraea) simplicifolia* agglutinin] which specifically binds to α-D-galactose. The incorporation of fucose into secretory glycoproteins of GD of mouse SMG has also been shown by autoradiographic methods (Lima and Haddad, 1981). Even though the GD of mouse and rat SMG contain populations of similarly appearing cells, the sugar residues in the cells are quite different. It is apparent that the GD of mice and rats do contain sugar residues that can be labeled by a variety of lectins, and it is very likely that the macromolecules containing these sugar residues are released into the saliva.

The SMG of mice and rats contain numerous biologically active polypeptides (Barka, 1980) and several have been localized to the GD cells, e.g., erythropoietin in mouse and rat (Fava-de-Moraes et al., 1979), renin in mouse (Bing et al., 1980), tonin in rat (Ledoux et al., 1982) and various members of the kallikrein family in rat (Ørstavik et al., 1980; Berg et al., 1992). Two growth factors have been localized immunocytochemically in GD cells of rodent SMG. Nerve growth factor (NGF) was the first growth factor to be detected and was characterized and initially purified from SMG of male mice (Thoenen et al., 1985). Localization of NGF in GD cells of mouse SMG is well documented (Olson et al., 1987), and *in situ* hybridization has been used to localize NGF mRNA in GD cells of mouse SMG (Ayer-Le Lievre et al., 1989). NGF has been localized in GD of SMG glands of some other mammals (Nishiyama and Saito, 1987) but while the rat SMG have GD, it is generally said that they have very low levels of NGF, if any. Recently, Olson et al. (1987) have reported a few weakly NGF-positive cells in GD of male rat SMG; however, NGF could not be detected in these glands using bioassay methods. Epidermal growth factor (EGF) has been localized immunohistochemically in GD of mouse and rat SMG (Mori et al., 1983; Raaberg et al., 1988) and mRNA for EGF has been localized in mouse GD (Gresik et al., 1985). Barka (1980) should be consulted for a list of other growth factors and biologically active polypeptides that have been reported to be present in mouse and/or rat SMG. Future studies may show other biologically active polypeptides occurring in GD cells of the SMG of these species.

In human PG and SMG, the ID empty into SD. The transition from one duct type to another may be rather abrupt but there are often short transitional duct segments where cells may have features of both duct types. In human PG and

SMG there are transitional cells that exist between the ID and SD; these columnar cells contain numerous mitochondria that are dispersed throughout the cells. In the human SLG, ducts containing numerous mitochondria, but lacking folded plasmalemmas, generally join the ID and few typical SD are present (Riva et al., 1988). The intralobular SD of human PG and SM are morphologically identical and contain occasional short basal cells, with the vast majority of cells being tall columnar cells that have very intricate interdigitating basal plasmalemmal folds (Riva et al., 1976; Tandler and Riva, 1986); intralobular SD in rodent salivary glands are morphologically very similar to those in human major salivary glands (Caramia, 1966; Tamarin and Sreebny, 1965; Pinkstaff, 1980). Elongate mitochondria are present in the cytoplasm between the plasmalemmal folds of the columnar cells. This arrangement of basal plasmalemmal infolding and mitochondria is often cited as being typical of epithelia involved in electrolyte and/or fluid transport (Hand, 1987; Tandler, 1987). A role for SD cells in fluid and electrolyte transport is also supported by the immunohistochemical demonstration of Na^+,K^+-ATPase in basolateral regions of SD cells in mouse, rat, and human PG, SMG, and SLG (Winston et al., 1988), and by histoenzymological demonstration of ouabain-sensitive Na^+,K^+-ATPase in SD cells of rat PG and mouse SMG (Sims-Sampson et al., 1984). The role of Na^+,K^+-ATPase in salivary gland ductal function is discussed by Hand (1987). Carbonic anhydrase (CA) occurs most commonly, and in abundance, in epithelia that is involved in electrolyte and water transport (Spicer et al., 1984). It is not surprising that isoenzymes of CA (Ca-I and CA-II) have been localized in SD cells of human PG and SMG (Noda et al., 1986a) and in SD of mouse and rat SMG (Hennigar et al., 1983; Noda et al., 1986b). Recently, Nishita et al. (1989) have shown CA-III to be present in SD of rat SMG. Occasional cells in the SD of human PG (Riva et al., 1976; Chaudhry et al., 1987), human SMG (Wilborn and Shackleford, 1980), SMG of rat (Tamarin and Sreebny, 1965), and SMG of mouse (Caramia, 1966) occur that lack plasmalemmal folds, have numerous mitochondria, and have an electron-dense cytoplasmic matrix. These are the so-called dark cells and will be discussed in more detail in the section on main excretory ducts.

The apical regions of SD cells in human PG and SMG contain small vesicles and occasional small secretory granules (Chaudhry et al., 1987). Such vesicles and secretory granules have also been observed in SD cells of the SMG of two species of macaques (Nagato and Tandler, 1986b), and these authors have suggested that the vesicles, both in macaques and in man, may contain kallikrein which has been localized in SD cells of human SMG using immunohistochemistry (Ørstavik et al., 1980). The apically located vesicles may contain secretory immunoglobulins; Nakamura et al. (1982) have used immunocytochemical methods to localize secretory component, IgA, IgM, and J chains in apically placed vesicles in SD cells of human PG and SMG. Lectin-labeling studies of human PG and SMG have shown labeling of sugar residues in SD. Laden et al. (1984) have shown labeling of a population of SD cells in the human SMG with DBA, and Vigneswaran et al. (1989) have shown strong labeling of SD in human major salivary glands with UEA-I and other lectins. Lending support to the idea that SD cells are capable of synthesizing secretory glycoconjugates is the presence of a thiamine pyrophosphatase-reactive Golgi complexes in SD cells of human PG and SMG as shown by Harrison et al. (1988).

The luminal and/or subluminal regions of SD cells of human PG and SMG are often stained by methods for the demonstration of glycoconjugates. Some of the enzymes reported to be present in these ducts are glycoenzymes (enzymes that contain covalently bonded carbohydrate), and these enzymes may be responsible for staining reactions that occur near the duct lumens, e.g., kallikrein is a glycoenzyme and is present in SD cells of human PG (Ørstavik et al., 1980). Two other glycoproteins, epithelial membrane antigen and carcinoembryonic antigen, have been localized in SD cells of PG or SMG (Stead et al., 1988) and may contribute to luminal staining with methods that stain glycoconjugates. An extremely interesting question is whether human SMG contain EGF/γ-urogastrone. There is also controversy as to which structures in human SMG gland contain EGF if it is present. Mori et al. (1987) and Mori (1991) have reported EGF in SD cells but not in acinar cells; however, Poulsen et al. (1986) localized EGF in acinar cells but not in SD cells. Kasselberg et al. (1985) have reported EGF in acini and occasional SD cells but Beerstecher et al. (1988) and Mori (1991) have denied the existence of EGF in either acinar cells or SD cells of human SMG. Recently, Tatemoto et al. (1988) have reported human epidermal growth factor (hEGF)/γ-urogastrone in ID cells of human SMG. The discrepancies between the results reported by these groups of investigators may reflect differences in antibody specificity or other aspects of the techniques used by the different groups. Several investigators have attempted to detect a variety of proteins in SD cells, e.g., lysozyme, lactoferrin, glial fibrillary acidic protein, S-100 protein, γ-enolase (neuron specific enolase), α_1-antichymotrypsin and α_1-antitrypsin (Korsrud and Brandtzaeg, 1982; Stead et al., 1988; Mori, 1991).

Lectin-labeling studies of mouse and rat PG and SMG have clearly shown the presence of glycoconjugates in SD cells of both species and in both glands (Schulte and Spicer, 1983a, 1984). These studies have shown that SD in both glands contain some sugar residues in common, e.g., β-galactose and α-fucose; however, α-galactose is present in the apical cytoplasm of SD in rats but not in mice. Flint et al. (1986) found two minor populations of cells in SD of rat and mouse PG and SMG that contained α-galactose. One population consisted of infrequent basal cells that had no access to the ductal lumen, while the other population consisted of columnar or pyramidal-shaped cells which made contact with the basement membrane and reacted the ductal lumen; the majority of cells in these ducts lacked α-galactose. The SD cells of rat and mouse PG and SMG do contain glycoconjugates and it is almost certain that some of these compounds are exported, as has been suggested by autoradiographic studies of L-^{3}H-fucose incorporation into SD cells of rat SMG and PG (Hand, 1979). The incorporation of ^{3}H-fucose into secretory glycoproteins in SD cells of rat PG and SMG was rapid; in the SD cells of PG, silver grains appeared over the Golgi complexes within 3.5 to 20 min after injection of ^{3}H-fucose and, by 2 to 4 h, grains were concentrated over secretory granules in the apical cytoplasm of the duct cells. Hand obtained similar results when rat SMG were examined.

In addition to their role in fluid and electrolyte transport which modifies saliva, the SD cells also secrete materials into the saliva as it flows through that ductal segment, e.g., secretory glycoproteins, enzymes, etc. Endocytosis of proteins by ID cells of rat PG was mentioned earlier in this overview and SD cells of rat PG also have this capacity. Hand and co-workers (Hand et al., 1987; Lotti and Hand, 1989) have shown that these cells are involved in endocytosis of infused foreign proteins and normal acinar proteins that are present in the saliva.

Information on SD in SLG is limited because the SLG has not been studied as extensively as the other major salivary glands. Typical SD cells with basal striations caused by basal plasmalemmal folds and parallel arranged mitochondria are said to be few in number or rare in human SLG (Tandler and Riva, 1986; Riva et al., 1988). Most of the intralobular ducts in the human SLG contain numerous mitochondria dispersed throughout the cytoplasm, but these cells lack basal plasmalemmal folds. The SD cells of the human SLG are said to contain CA; Noda et al. (1986a) report staining for both CA-I and CA-II. Striated duct cells in mouse SLG also contain CA-I and CA-II but only CA-II is present in SD cells of the rat SLG (Hennigar et al., 1983). Immunohistochemical staining of Na^+,K^+-ATPase has shown that this enzyme is present in SD cells of mouse, rat, and human SLG (Winston et al., 1988).

Lectin labeling of apical areas and basal striations of SD cells of human SLG occurs with UEA-I (Laden et al., 1984). The apical cytoplasmic regions of SD cells of mouse and rat SLG contain glycoconjugates that have α-fucose or β-galactose as terminal sugars; in SD cells of rat SLG there are also glycoconjugates that have α-galactose as the terminal sugar (Schulte and Spicer, 1983a, 1984). Flint et al. (1986) have shown two populations of cells in SD of rat SLG that contain β-galactose as the terminal sugar residue as shown by labeling with GSA I-B$_4$. The two cell types are infrequently seen basal cells which do not reach the lumen of the SD and tall columnar or pyramidal cells that reach the lumen and are in contact with the basement membrane; however, the majority of cells in these ducts do not stain with GSA I-B$_4$. Similar populations of cells can be seen in SD of mouse SLG but they are stained for α-fucose by LTA and are not stained by GSA I-B$_4$ (Schulte and Spicer, 1983a).

Granulated cells, resembling those seen in GD of mouse SMG, have been seen in SD of mouse SLG of some species and strains of mice. Immunohistochemical staining methods have shown EGF and renin to be present in granulated cells in SD of mouse SLG, but the authors did not determine whether these two compounds occurred in the same cells or in different cells (Gresik and Barka, 1983). NGF has also been localized in granulated cells of SLG striated ducts (Hazen-Martin et al., 1987; Olson et al., 1987). Gresik and Barka (1983) suggest that finding these granulated cells in both SMG and SLG of mice is related to the embryological development of the two glands. They describe how the two glands arise from the same area of oral epithelium in the floor of the mouth and how they further develop in a common mesenchymal bed until the time that the cells destined to give rise to the acinar cells of the two glands begin to undergo cytodifferentiation.

Striated ducts empty their contents into intralobular ED, and intralobular ED coalesce to form interlobular ED. The ED of the human PG and SMG contain the same cell types seen in SD — light, dark, and basal cells (Wilborn and Shackleford, 1980), as do the ED of rat SMG (Tamarin and Sreebny, 1965). These ducts, like SD, appear to be involved in fluid and electrolyte transport. Immunocytochemical staining has localized Na^+,K^+-ATPase in ED epithelium of mouse, rat, and human major salivary glands (Winston et al., 1988) and Sims-Sampson et al. (1984) showed that the Na^+,K-ATPase in the ED of mouse SMG was ouabain sensitive. Lantini et al. (1990) have localized K^+-dependent *p*-nitro phenylphosphatase activity to lateral plasmalemmal folds of human parotid and submandibular gland ED. Carbonic anhydrases I, II, and III have also been localized in ED of the SLG and SMG of rats and mice (Hennigar et al., 1983; Noda et al., 1986b; Spicer et al., 1990) and in ED of all human major salivary glands (Noda et al., 1986a). Kallikrein activity has been localized in ED of human major salivary glands but the authors noted that activity appeared to be lower than that seen in SD (Ørstavik et al., 1980).

Sugar residues of glycoconjugates in tall columnar or pyramidal cells of the ED of human SMG have been shown to contain α-D-*N*-acetylgalactosamine by DBA labeling (Laden et al., 1984). Excretory ducts of human PG contain α-L-fucose shown by UEA-I labeling (Born et al., 1987). Columnar cells of ED of human SLG have been shown to contain α-L-fucose based upon UEA-I labeling (Laden et al., 1984). Rat and mouse major salivary gland ED glycoconjugates have been extensively studied by Schulte and Spicer (1983a, 1984); a brief summary of this work follows. Cells in ED in the major salivary gland of both mice and rats have been shown to have glycoconjugates containing α-galactose, β-galactose, and α-fucose as terminal sugars. The cells that are labeled predominately consist of two cell types, basal cells and tall columnar or pyramidal cells. It is tempting to suggest that the tall columnar or pyramidal cells might be dark cells. Overall, the staining pattern of the cells in the ED closely resembles the staining pattern of cell populations seen in SD. The ED clearly play a role in the modification of the saliva before it reaches the oral cavity.

There have been relatively few studies of the extraglandular MED which pass from the major salivary glands to the oral cavity. The light microscopic appearance of the human parotid MED has been described by Rother (1963a) and Takeda (1987). The epithelial layer of the mucosa contains tall columnar cells that have varying numbers of cilia, basally located cuboidal cells, and occasional goblet cell-like mucous secreting cells. The columnar cells show cytoplasmic PAS staining of neutral glycoconjugates, and the goblet cells contain AB-positive acidic glycoconjugates (Rother, 1963b).

Rother (1963a) described the light microscopic appearance of the MED of the human SMG as being very similar to the MED of the PG. Electron microscopic examination of the epithelium of the MED of the human SMG has shown that it is pseudostratified and consists of cuboidal basal cells, a few goblet cells, and two types of tall columnar cells; a few of the tall columnar cells are ciliated but the majority are nonciliated (Testa-Riva et al., 1981). The authors referred to the tall columnar, nonciliated cells as principal cells. The principal cells have folded basolateral plasmalemmas that interdigitate with adjacent cells; these cells have well-developed Golgi complexes and there are periodic acid-silver proteinate-stained vesicles in the apical cytoplasm. Fusion of these stained vesicles with the apical plasmalemma has been seen which suggests exocytosis of this material. The authors believe that the MED changes the nature of the SMG saliva by fluid and electrolyte modification and by adding secretory material from both principal cells and goblet cells. The tall columnar, nonciliated cells contain considerable PAS-positive neutral glycoconjugates (Rother, 1963b; Perra et al, 1988); some of the PAS-positive material was glycogen since diastase pretreatment prior to PAS staining resulted in decreased PAS reactivity (Perra et al., 1988). An AB-positive glycocalyx was apparent on the apical plasma-lemmas (Rother, 1963b; Perra et al., 1988). Perra et al. (1988) also showed that the principal cells of the MED of the SMG contain enzymes involved in the metabolism of male and female steroid hormones, e.g., 3β-hydroxysteroid dehydrogenase and 17β-hydroxysteroid dehydrogenase. In addition, they observed prostaglandin synthetase activity in principal cells and they suggested that prostaglandin synthesized by this enzyme is involved in electrolyte transport occurring in this duct. It is evident that modification of the saliva occurs in the MED of the human SMG.

The ultrastructural appearance of the epithelium of the MED of the rat PG has been described by Sato and Miyoshi (1988). These authors described six cell types in the MED of the rat PG — basal cells, cuboidal cells, Types I and II light cells, dark cells, and tuft cells. The arrangement of cells in this duct is unusual in that all cell types, except some cuboidal cells, contact the basement membrane. The two types of light cells, dark cells, and tuft cells extend from the basement membrane to the ductal lumen. The basal cells resemble other such cells that are found in intraglandular excretory ducts and in the MED of the major glands of other species. The majority of cells in the MED of the PG are cuboidal cells; they occur as a stratified cuboidal layer that is two to three cells thick. Therefore, the epithelium of the MED of the PG in some places is stratified cuboidal epithelium; however, in other areas it is pseudostratified. Dark cells and both types of light cells have basal plasmalemmal infoldings; no such folds were seen in tuft cells. Type I light cells and dark cells had mitochondria arranged between the plasmalemmal folds but Type II light cells lacked the plasmalemmal fold-mitochondria relationship. In the Type II light cells the mitochondria were large and were found in the perinuclear cytoplasm. Tuft cells have thick microvilli that are much longer than microvilli seen on the other columnar cells. Electron-lucent vesicles, which may be related to secretory activity, were apically located in all of the columnar cells.

The epithelial lining of the MED of the rat SMG has been examined by several investigators and each investigator has classified the cells making up this duct in a different way. In the simplest classification, the duct epithelium is said to consist of three cell types — basal, light, and dark cells (Tamarin and Sreebny, 1965; Shackleford and Schneyer, 1971). Higashi et al. (1989) have also described three cells types — basal, epithelial (light cells), and brush (dark cells) cells. Sato and Miyoshi (1988) described five cells types in the SD of the rat SMG — Type I light cells, Type II light cells, dark cells, tuft cells and basal cells. They said that their dark cells were similar to those reported by Tamarin and Sreebny (1965), while their tuft cells were comparable to the dark cells of Shackleford and Schneyer (1971) and the brush cells of Higashi et al. (1989). Their Type I light cells resembled those reported by Tamarin and Sreebny (1965) and their Type II light cells resembled those of Shackleford and Schneyer (1971); they suggested that Type I and Type II light cells might be the same cell type at different stages in its life cycle. Knauf et al. (1983) have attempted to classify the cell types of the MED of the rat SMG using light microscopic staining methods, the PAS reaction, and Halmi staining for basophilia. Based upon staining results, they described five cells types — Type I/basal cells, Type IIa/true light cells, Type IIb/ potential dark cells, Type IIIa dark cells, and Type IIIb dark cells in the MED of the rat SMG. Type I, Type IIa and Type IIb cells were PAS and Halmi negative; Type IIIa cells were PAS positive, but Halmi negative; and Type IIIb cells were PAS and Halmi positive. In metabolic alkalosis, Type IIb cells were said to increase in number while Type IIIa and Type IIIb dark cells decreased in number. In metabolic acidosis, Type IIb cells decreased in number while Type IIIa and Type IIIb dark cells increased in number. The authors suggested a constant population of Type IIb, Type IIIa, and Type IIIb cells that exhibited a shift in staining affinity when disturbances occurred in acid-base balance; they believed that these three cell types constituted a dark cell system. Knauf and co-authors did not attempt to compare their reported cell types to those reported by other authors. A compromise needs to be worked out with respect to the classification of the cell types in the MED of the rat SMG, e. g., the term brush cell is probably more appropriate than tuft cells since these cells resemble brush cells described in other organ systems. Brush or tuft cells do not appear to be related to fluid and electrolyte transport, at least not to the extent of that of the light cells, since they lack the plasmalemmal foldings and associated mitochondria characteristic of light cells. The concentration of apically located vesicles suggests a secretory function for brush cells but it is also possible that the vesicles may be related to absorption which is suggested by the large apical microvilli. These cells may be one of the two types of dark cells reported to exhibit PAS staining (Knauf et al., 1983). Higashi et al. (1989) suggested a chemoreceptive function for brush cells in the MED of the rat SMG and that the vesicles might be related to endocytotic processes. The dark cells which do have characteristics similar to those of striated duct cells may be involved in fluid and electrolyte transport. If one considers the diversity of cell types that occurs in MED, it is evident that this segment of the duct system does more than simply connect SD and the oral cavity. Basal, light, and dark cells have also been reported to occur in the MED of the mouse SMG (Hanker et al., 1977) and in MED of other species (Pinkstaff, 1980).

The dark cells of striated, intraglandular ED and MED constitute perhaps the greatest enigma for salivary gland morphologists. Even though many investigators have reported their presence in duct epithelium of salivary glands from a wide range of species (Young and van Lennep, 1978; Pinkstaff, 1980), it has been suggested that they are artifacts occurring in inadequately fixed tissue (Tandler and Riva, 1986; Riva et al., 1990). It may be possible that some duct cells may undergo densification due to inadequate fixation, but it does not explain why all cells in that particular duct do not undergo at least some degree of densification! The more prominent apical microvilli of dark cells in the MED of the rat SMG could hardly result from inadequate fixation. In addition, reported changes in the incidence of dark cells in MED of the rat SMG from animals that were undergoing metabolic acidosis or metabolic alkalosis would not be expected to result from tissue fixation, either adequate or inadequate. At the present time, many questions about dark cells are still unanswered.

The secretory endpieces and ductal segments of some salivary glands have myoepithelial cells (MEC) located between their cells and their respective basal lamina. Myoepithelial cells are apparently associated with secretory endpieces and ID of all human major salivary glands. Myoepithelial cell processes that extend from MEC at the junction between ID and SD are also said to be present on SD of human PG (Cutler et al., 1977), but an association of MEC with ED has not been clearly established. Additional details on MEC of human salivary glands can be found in reports by Nilsen and Donath (1981), Chaudhry et al. (1987), and Riva et al. (1988). The secretory endpieces of rat SLG and SMG are embraced by MEC and ID also have associated MEC (Tamarin, 1966; Drenckhahn et al., 1977; Nagato et al., 1980). In the rat PG, there are MEC on ID but their presence on acini is questionable (Drenckhahn et al., 1977; Nagato et al., 1980). Most authors say MEC do not occur on SD or ED of rat major salivary glands (Tamarin and Sreebny, 1965; Drenckhahn et al., 1977). Myoepithelial cells are associated with the secretory endpieces and ID of the major salivary glands of mice (Caramia, 1966; Poldermans et al., 1986) and with the GD of the mouse SMG (Caramia, 1966).

Myoepithelial cells are generally believed to be nonsecretory cells that possess morphological features characteristically seen in contractile cells such as smooth muscle cells, e.g., myofilaments, caveolae and attachment plaques (Garrett and Emmelin, 1979; Chaudhry et al., 1987). Nexuses, or gap junctions, have also been reported between myoepithelial cells and it has been suggested that the nexuses may be involved in synchronization of MEC contraction (Nagato and Tandler, 1986a). Actual contraction of MEC in adult salivary glands has not been reported but Morgan et al. (1979) have reported contraction of elements in cultured rudiments of mouse submandibular glands as early as 18 days postconception and they suggested MEC contraction prior to the time when MEC can be identified morphologically. Salivary gland MEC are said to have numerous functions which are related to their purported contractility; detailed discussion of MEC functions in salivary glands can be found in reviews by Young and van Lennep (1977) and Garrett and Emmelin (1979).

Lending support to a contractile function for MEC cells are reports of the immunocytochemical localization of actin and/or myosin in human major salivary gland MEC (Palmer, 1986) and rodent salivary gland MEC (Drenckhahn et al., 1977). Human salivary gland MEC possess adenosine triphosphatase (ATPase) activity which also may be related to MEC contractility (Garrett and Emmelin, 1979; Palmer, 1986, Nagashima and Ono, 1985) but they lack alkaline phosphatase activity. Mouse and rat salivary gland MEC lack ATPase activity but show alkaline phosphatase activity (Tamarin, 1966). Detailed discussions of the localization of phosphatases in salivary gland MEC of several species can be found in Young and van Lennep (1977, 1978), Garrett and Emmelin (1979), and Pinkstaff (1980).

In recent years, there have been numerous studies pertaining to the immunocytochemical localization of a variety of proteins in MEC of salivary glands. Many studies on cytokeratin expression have been undertaken in an effort to explain the histogenesis of salivary gland tumors; this approach is illustrated well in a study of cytokeratin expression in cystadenolymphomas (Orito et al., 1989). Striated ducts in normal glands showed strong staining of basal cells with a broad range monoclonal antibody (PKK 1) and basal cells in tumor specimens also revealed strong staining with this antibody. The authors believed that these staining patterns showed that ductal basal cells contributed to the histogenesis of this tumor. The localization of some proteins in MEC is controversial with some authors reporting their presence and others their absence, e.g., glial fibrillary acidic protein, vimentin, and S-100 protein. Some proteins that are usually said to be absent in MEC are desmin, astroprotein, lactoferrin, involucrin, epithelial membrane antigen, carcinoembryonic antigen, lysozyme, filamin, amylase, fibronectin, secretory component, IgA, IgM, IgG, carbonic anhydrases I and II, and cytochrome P-450$_{C21}$. Only a few reports of lectin labeling of MEC in human or rodent salivary glands have appeared and in no instance was mention made of labeled secretory granules. Histochemical studies support the contention that MEC are primarily nonsecretory cells.

Hopefully this brief overview has shown how histochemistry has been used, and can be used, to study the salivary glands. Histochemical methods have been used to study other aspects of salivary gland biology which could not be discussed in this work, e.g., development of salivary glands, innervation of salivary glands, effects of aging on salivary glands, salivary gland tumor pathology, etc. Many investigators are using histochemical methods to study salivary gland biology. The applications and the horizons in such studies are virtually unlimited.

REFERENCES

1. Ayer-Le Lievre, C., Ebendal, T., Olson, L., Seiger, A., and Persson, H., Detection of nerve growth factor and its mRNA by separate and combined immunohistochemistry and *in situ* hybridization in mouse salivary glands, *Histochem. J.*, 21, 1, 1989.

2. Barka, T., Biologically active polypeptides in submandibular glands, *J. Histochem. Cytochem.*, 28, 836, 1980.

3. Beerstecher, H. J., Huiskens-van der Meij, C., and Warnaar, S. O., An immunohistochemical study performed with monoclonal and polyclonal antibodies to mouse epidermal growth factor, *J. Histochem. Cytochem.*, 36, 1153, 1988.

4. Benau, D. A., Schumacher, W. G., McGuire, E. J., Fitzpatrick-McElligott, S., Storey, B. T., and Roth, S., Light micoscopic localization of glycocyltransferase activities in cells and tissues, *J. Histochem. Cytochem.*, 38, 23, 1990.

5. Berg, T., Wassdal, I., and Sletten, K., Immunohistochemical localization of rat submandibular gland esterse B (homologous to the RSKG-7 kallikrein gene) in relation to other serine proteases of the kallikrein family, *J. Histochem. Cytochem.*, 40, 83, 1992.

6. Bing J., Poulsen, K., Hackenthal, E., Rix, E., and Taugner, R., Renin in the submaxillary gland: a review, *J. Histochem. Cytochem.*, 28, 874, 1980.

7. Born, I. A., Zimmer, K.-P., Schwechheimer, K., Maier, H., and Möller, P., Binding sites of *Ulex europaeus*-Lectin I in human parotid gland. A light-microscopic and ultrastructural study using the immunoperoxidase technique and immunocryoultramicrotomy, *Cell Tissue Res.*, 248, 455, 1987.

8. Caramia, F., Ultrastructure of mouse submaxillary gland. I. Sexual differences, *J. Ultrastruct. Res.*, 16, 505, 1966.

9. Chaudhry, A. P., Cutler, L. S., Yamane, G. M., Labay, G. R., Sunderraj, M., and Manak, J. R., Jr., Ultrastructure of normal human parotid gland with special emphasis on myoepithelial distribution, *J. Anat.*, 152, 1, 1987.

10. Cutler, L. S., Chaudhry, A., and Innes, D. J., Jr., Ultrastructure of the parotid duct. Cytochemical studies of the striated duct and papillary cysttadenomalymphomatosum of the human parotid gland, *Arch. Pathol. Lab. Med.*, 101, 420, 1977.

11. Drenckhahn, D., Gröschel-Stewart, U., and Unsicker, K., Immunofluorescence-microscopic demonstration of myosin and actin in salivary glands and exocrine pancreas of the rat, *Cell Tissue Res.*, 183, 273, 1977.

12. Eversole, L. R., The mucoprotein histochemistry of human mucous acinar cell containing salivary glands: submandibular and sublingual glands, *Arch. Oral Biol.*, 17, 43, 1972.

13. Fava-de-Moraes, F., Zangheri, E. O., and Doine, A. I., Immunohistochemical localization of erythropoietin in the rat and mouse submandibular gland, *Histochem. J.*, 11, 97, 1979.

14. Flint, F. F., Schulte, B. A., and Spicer, S. S., Glycoconjugate with terminal a galactose. A property common to basal cells and a subpopulation of columnar cells of numerous epithelia in mouse and rat. *Histochemistry*, 84, 387, 1986.

15. Garrett, J. R. and Emmelin, N., Activities of salivary myoepithelial cells: a review, *Med. Biol.*, 57, 1, 1979.

16. Gresik, E. W. and Barka, T., Epidermal growth factor, renin, and protease in hormonally responsive duct cells of the mouse sublingual gland, *Anat. Rec.*, 205, 169, 1983.

17. Gresik, E. W., Gubits, R. M., and Barka, T., In situ localization of mRNA for epidermal growth factor in the submandibular gland of the mouse, *Histochem. J. Cytochem.*, 33, 1235, 1985.

18. Hagelqvist, E., Ahlner, B. H., and Lind, M. G., Morphology and histochemistry of rabbit submandibular glands, *Acta Oto-Laryngol. Suppl.*, 480, 1, 1991.

19. Hand, A. R., Synthesis of secretory and plasma membrane glycoproteins by striated duct cells of rat salivary glands as visualized by radioautography after ^{3}H-fucose injection, *Anat. Rec.*, 195, 317, 1979.

20. Hand, A. R., Functional ultrastructure of the salivary glands, in *The Salivary System*, L. M. Sreebny, Ed., CRC Press, Boca Raton, FL, 1987, 43.

21. Hand, A. R., Coleman, R., Mazariegos, M. R., Lustmann, J., and Lotti, L. V., Endocytosis of proteins by salivary gland duct cells, *J. Dent. Res.*, 66, 412, 1987.

22. Hanker, J. S., Preece, J. W., Burkes, E. J., Jr., and Romanovicz, D. K., Catalase in salivary gland striated and excretory duct cells. I. The distribution of cytoplasmic and particulate catalase and the presence of catalase-positive rods, *Histochem. J.*, 9, 711, 1977.

23. Harrison, J. D., Auger, D. W., and Badir, M. S., Ultrastructural phosphatase histochemistry of submandibular and parotid glands of man, *Histochem. J.*, 20, 117, 1988.

24. Harrison, J. D., Auger, D. W., Paterson, K. L., and Rowley, P. S. A., Mucin histochemistry of submandibular and parotid salivary glands of man: light and electron microscopy, *Histochem. J.*, 19, 555, 1987.

25. Hazen-Martin, D. J., Landreth, G., and Simson, J. A. V., Immunocytochemical localization of nerve growth factor in mouse salivary glands, *Histochem. J.*, 19, 210, 1987.

26. Hennigar, R. A., Schulte, B. A., and Spicer, S. S., Immunolocalization of carbonic anhydrase isozymes in rat and mouse salivary and exorbital lacrimal glands, *Anat. Rec.*, 207, 605, 1983.

27. Herp, A., Borelli, C., and Wu, A. M., Biochemistry and lectin binding properties of mammalian salivary mucous glycoproteins, *Adv. Exp. Med. Biol.*, 228, 395, 1988.

28. Higashi, K., Gomi, T., Soeda, M., Sasa, S., Kimura, A., and Kikuchi, Y., New morphological aspects of the brush cells in the main excretory ducts of the rat submandibular glands, *Zool. Sci.*, 6, 675, 1989.

29. Jezernik, K. and Pipan, N., Cytochemical localization of carbohydrates in intercalated duct and acinar cells of mouse parotid gland, *Histochem. J.*, 21, 131, 1989a.

30. Jezernik, K. and Pipan, N., Reduction capacity of untreated and of repeatedly isoproterenol treated parotid gland of the mouse, *Histochemistry*, 92, 531, 1989b.

31. Junqueira, L. C., Sessot, A., and Nahas, L., Sur la nature des cellules sécrétantes de la glande parotide, *Bull. Microsc. Appl.*, 1, 133, 1951.

32. Kagayama, M., The fine structure of the monkey submandibular gland with a special reference to intra-acinar nerve endings, *Am. J. Anat.*, 131, 185, 1971.

33. Kasselberg, A. G., Orth, D. N., Gray, M. E., and Stahlman, M. T., Immunocyto-chemical localization of human epidermal growth factor/urogastrone in several human tissues, *J. Histochem. Cytochem.*, 33, 315, 1985.

34. Knauf, H., Lübcke, R., Röttger, P., Baumann, K., and Richet, G., Relation of dark cells to the transport of H^+/HCO_3^- and K^+ ions: a microperfusion study in the rat submaxillary duct, *Kid. Internatl.*, 23, 350, 1983.

35. Korsrud, F. R. and Brandtzaeg, P., Characterization of epithelial elements in human major salivary glands by functional markers: localization of amylase, lactoferrin, lysozyme, secretory component, and secretory immunoglobulins by paired immunofluorescence staining, *J. Histochem. Cytochem.*, 30, 657, 1982.

36. Laden, S. A., Schulte, B. A., and Spicer, S. S., Histochemical evaluation of secretory glycoproteins in human salivary glands with lectin-horseradish peroxidase conjugates, *J. Histochem. Cytochem.*, 32, 965, 1984.

37. Lantini, M. S., Valentino, L., and Riva, A., A granular cell in the proximal intercalated duct of human parotid and submandibular glands, *J. Submicrosc. Cytol. Pathol.*, 20, 147, 1988.

37a. Lantini, M. S., Proto, E., Puxeddu, P., Riva, A., and Testa Riva, F., Fine structure of excretory ducts of human salivary glands, *J. Submicrosc. Cytol. Pathol.*, 22, 465, 1990.

38. Ledoux, S., Gutkowska, J., Garcia, R., Thibault, G., Cantin, M., and Genest, J., Immunohistochemical localization of tonin in rat salivary glands and kidney, *Histochemistry*, 76, 329, 1982.

39. Leppi, T. J. and Spicer, S. S., The histochemistry of mucins in certain primate salivary glands, *Am. J. Anat.*, 118, 833, 1966.

40. Lima, T. G. and Haddad, A., Light- and electron-microscopic radioautographic study of glycoprotein secretion in the granular duct of the submandibular gland of the male mouse, *Cell Tissue Res.*, 220, 405, 1981.

41. Lotti, L. V. and Hand, A. R., Endocytosis of native and glycosylated bovine serum albumin by duct cells of the rat parotid gland, *Cell Tissue Res.*, 255, 333, 1989.

42. Lundquist, P.-G. and Norberg, L. E., Salivary gland morphology after α-adrenergic and cholinergic stimulation. An ultrastructural study. *Acta Otolaryngol. Suppl.*, 447, 14, 1988.

43. Luzzatto, A. C., Procicchiani, G., and Rosati, G., Rat submaxillary gland: an electron microscope study of the secretory granules of the acinus, *J. Ultrastruct. Res.*, 22, 185, 1968.

44. Machino, M., Morioka, H., and Tachibana, M., Amylase and lysozyme differentiate their localization within the serous secretory granule of the human salivary gland, *Acta Histochem. Cytochem.*, 19, 329, 1986.

45. Meisel, D. L., Skobe, Z., Prostak, K. S., and Shklar, G., A light and electron microscope study of aging parotid and submandibular salivary glands of Swiss-Webster mice, *Exp. Gerontol.*, 23, 197. 1988.

46. Morgan, W. D., Williams, J. E., Lee, C. W., and Dawe, C. J., Microcinematographic demonstration of synchronous and asynchronous myoepithelial contractions in mouse submandibular gland rudiments in organotypic culture, *In Vitro*, 15, 1013, 1979.

47. Mori, M., *Histochemistry of the Salivary Glands*, CRC Press, Boca Raton, FL, 1991.

48. Mori, M., Hamada, K., Naito, R., Tsukitani, K., and Asano, K., Immunohistochemical localization of epidermal growth factor in rodent submandibular glands, *Acta Histochem. Cytochem.*, 16, 536, 1983.

49. Mori, M., Naito, R., Tsukitani, K., Okada, Y., Hayashi, T., and Kato, K., Immunohistochemical distribution of human epidermal growth factor in salivary gland tumours, *Virchows Arch. A*, 411, 499, 1987.

50. Munger, B. L., Histochemical studies on seromucous- and mucous-secreting cells of human salivary glands, *Am. J. Anat.*, 115, 411, 1964.

51. Nagashima, Y. and Ono, K., Myoepithelial cell substructure in the submandibular gland of man, *Anat. Embryol.*, 171, 259, 1985.

52. Nagato, T. and Tandler, B., Gap junctions in rat sublingual gland, *Anat. Rec.*, 214, 71, 1986a.

53. Nagato, T. and Tandler, B., Ultrastructure of the submandibular gland in 2 species of macaques, *Acta Anat.*, 126, 255, 1986b.

54. Nagato, T., Yoshida, H., Yoshida, A., and Uehara, Y., A scanning electron microscope study of myoepithelial cells in exocrine glands, *Cell Tissue Res.*, 209, 1, 1980.

55. Naito, R., Takai, Y., Tsukitani, K., Asano, K., and Mori, M., Use of lectins for differential localization of secretory materials of granular convoluted tubules and ducts in the submandibular gland, *Acta Histochem. Cytochem*, 16, 483, 1983.

56. Nakai, M., Tsukitani, K., Tatemoto, Y., Hikosaka, N., and Mori, M., Histochemical studies of obstructive adenitis in human submandibular salivary glands. II. Lectin binding and keratin distribution in the lesions, *J. Oral Pathol.*, 14, 671, 1985.

57. Nakamura, T., Nagura, H., Watanabe, K., Komatsu, N., Uchikoshi, S., Kobayashi, K., and Nabeshima, J., Immunocytochemical localization of secretory imunoglobulins in human parotid and submandibular glands, *J. Electron Microsc.*, 31, 151, 1982.

58. Neiss, W. F., Ultracytochemistry of intracellular membrane glycoconjugates, *Adv. Anat. Embryol. Cell Biol.*, 99, 1, 1986.

58a. Nilsen, R. and Donath, K., Actin containing cells in normal human salivary glands. An immunohistochemical study, *Virchows Arch. A*, 391, 315, 1981.

59. Nishita, T., Oshige, H., Matsushita, H., Kano, Y., and Asari, M., The immunohistolocalization of carbonic anhydrase III in the submandibular gland of rats and hamsters, *Histochem. J.*, 21, 8, 1989.

60. Nishiyama, N. and Saito, H., Distribution of nerve growth factor activity in submandibular and prostate glands of various mammals, *Biomed. Res.*, 8, 61, 1987.

61. Noda, Y., Sumitomo, S., Hikosaka, N., and Mori, M., Immunohistochemical observations on carbonic anhydrase I and II in human salivary glands and submandibular obstructive adenitis, *J. Oral Pathol.*, 15, 187, 1986a.

62. Noda, Y., Sumitomo, S., Orito, T., and Mori, M., Immunohistochemical localization of carbonic anhydrase I and II in submandibular salivary glands of the mouse, rat, hamster and guinea pig, *Arch. Oral Biol.*, 31, 795, 1986b.

63. Olson, L., Ayer-Le Lievre, C., Ebendal, T., and Seiger, A., Nerve growth factor-like immunoreactivities in rodent salivary glands and testis, *Cell Tissue Res.*, 248, 275, 1987.

64. Orito, T., Shinohara, H., Okada, Y., and Mori, M., Heterogeneity of keratin expression in epithelial tumor cells of adenolymphoma in paraffin sections, *Path. Res. Pract.*, 184, 600, 1989.

65. Ørstavik, T. B., Nustand, K., and Brandtzaeg, P., Localization of glandular kallikreins in rat and man, in *Enzymatic Release of Vasoactive Peptides*, F. Gross and G. Vogel, Eds., Raven Press, New York, 1980, 137.

66. Palmer, R. M., The identification of myoepithelial cells in human salivary glands. A review and comparison of light microscopical methods, *J. Oral Pathol.*, 15, 221, 1986.

67. Perra, M. T., Puxeddu, P., and Sirigu, P., A histochemical study of the main excretory duct of the human submandibular gland, *Arch. Oral Biol.*, 33, 525, 1988.

68. Phillips, C. J. and Tandler, B., Mammalian evolution at the cellular level, in *Current Mammalogy, Vol. 1*, H. H. Genoways, Ed., Plenum Publishing, New York, 1987, 1.

69. Pinkstaff, C. A., The cytology of salivary glands, *Internatl. Rev. Cytol.*, 63, 141, 1980.

70. Pinkstaff, C. A., Histochemical characterization of salivary gland secretion, in *Saliva and Salivation, Advances in Physiological Sciences, Vol. 28*, T. Zelles, Ed., Akademiai Kiado, Budapest, 1981, 135.

71. Poldermans, J. E., Bos-Vreugdenhil, A. P., and De Lange, G. L., Release mechanism in mucous cells of the mouse sublingual salivary gland, *J. Biol. Buccale*, 14, 207, 1986.

72. Poulsen, S. S., Nexø, E., Olsen, P. S., Hess, J., and Kirkegaard, P., Immunohistochemical localization of epidermal growth factor in rat and man, *Histochemistry*, 85, 389, 1986.

73. Qwarnström, E. E. and Hand, A. R., A granular cell at the acinar-intercalated duct junction of the rat submandibular gland, *Anat. Rec.*, 206, 181, 1983.

74. Raaberg, L., Nexø, E., Mikkelsen, J. D., and Poulsen, S. S., Immunohistochemical localisation and developmental aspects of epidermal growth factor in the rat, *Histochemistry*, 89, 351, 1988.

75. Riva, A., Testa-Riva, F., Del Fiacco, M., and Lantini, M. S., Fine structure and cytochemistry of the intralobular ducts of the human parotid gland, *J. Anat.*, 122, 627, 1976.

76. Riva, A., Tandler, B., and Testa Riva, F., Ultrastructural observations on human sublingual gland, *Am. J. Anat.*, 181, 385, 1988.

76a. Riva, A., Lantini, M. S., and Testa Riva, F., Normal human salivary glands, in *Ultrastructure of the Extraparietal Glands of the Digestive System*, A. Riva and P. M. Motta, Eds., Kluwer Academic Publishers, Boston, 1990, 53.

77. Rother, P., Die unterschiede im bau des ductus parotideus und ductus submandibularis, *Anat. Anz.*, 112, 172, 1963a.

78. Rother, P., Die sekretorische funktion des ductus parotideus und ductus submandibularis, *Acta Histochem.*, 15, 325, 1963b.

79. Sakai, T., Major ocular glands (harderian gland and lacrimal gland) of the musk shrew (*Suncus Murinus*) with a review on the comparative anatomy and histology of the mammalian lacrimal glands, *J. Morphol.*, 201, 39, 1989.

80. Sato, A. and Miyoshi, S., Ultrastructure of the main excretory duct epithelia of the rat parotid and submandibular glands with a review of the literature, *Anat. Rec.*, 220, 239, 1988.

81. Schulte, B. A., Genetic and sex-related differences in the structure of submandibular glycoconjugates, *J. Dent. Res.*, 66, 442, 1987.

82. Schulte, B. A. and Spicer, S. S., Light microscopic detection of sugar residues in glycoconjugates of salivary glands and the pancreas with lectin-horseradish peroxidase conjugates. I. Mouse, *Histochem. J.*, 15, 1217, 1983a.

83. Schulte, B. A. and Spicer, S. S., Light microscopic histochemical detection of terminal galactose and *N*-acetylgalactosamine residues in rodent complex carbohydrates using a galactose oxidase-Schiff sequence and peanut lectin-horseradish peroxidase conjugate, *J. Histochem. Cytochem.*, 31, 19, 1983b.

84. Schulte, B. A. and Spicer, S. S., Light microscopic detection of sugar residues in glycoconjugates of salivary glands and the pancreas with lectin-horseradish peroxidase conjugates. II. Rat, *Histochem. J.*, 16, 3, 1984.

85. Schwartz-Arad, D., Arber, L., Arber, N., Zajicek, G., and Michaeli, Y., The rat parotid gland — a renewing cell population, *J. Anat.*, 161, 143, 1988.
86. Shackleford, J. M. and Schneyer, L. H., Ultrastructural aspects of the main excretory duct of rat submandibular gland, *Anat. Rec.*, 169, 679, 1971.
87. Simson, J. A. V., The influence of fixation on the carbohydrate cytochemistry of rat salivary gland secretory granules, *Histochem. J.*, 9, 645, 1977.
88. Simson, J. A. V., Dom, R. M., Sannes, P. L., and Spicer, S. S., Morphology and cytochemistry of acinar secretory granules in normal and isoproterenol-treated rat submandibular glands, *J. Microsc.*, 113, 185, 1978.
89. Sims-Sampson, G., Gresik, E. W., and Barka, T., Histochemical localization of ouabain-sensitive, K^+-dependent p-nitrophenylphosphatase (Na^+-K^+-ATPase) activity in the submandibular gland of the mouse: effect of androgen, thyroid hormone or postnatal age, *Anat. Rec.*, 210, 53, 1984.
90. Smith, P. H. and Toms, B. B., Immunocytochemical localization of insulin- and glucagonlike peptides in rat salivary glands, *J. Histochem. Cytochem.*, 34, 627, 1986.
91. Spicer, S. S. and Duvenci, J., Histochemical characteristics of mucopolysaccharides in salivary and exorbital lacrimal glands, *Anat. Rec.*, 149, 333, 1964.
92. Spicer, S. S., Sens, M. A., Hennigar, R. A., and Stoward, P. J., Implications of the immunohistochemical localization of the carbonic anhydrase isozymes for their function in normal and pathologic cells, *Ann. N.Y. Acad. Sci.*, 429, 382, 1984.
93. Spicer, S. S., Erlandsen, S. L., Wilson, A. C., Hammer, M. F., Hennigar, R. A., and Schulte, B. A., Genetic differences in the histochemically defined structure of oligosaccharides in mice, *J. Histochem. Cytochem.*, 35, 1231, 1987.
93a. Spicer, S. S., Ge, Z.-H., Tashian, R. E., Hazen-Martin, D. J., and Schulte, B. A., Comparative distribution of carbonic anhydrase isozymes III and II in rodent tissues, *Am. J. Anat.*, 187, 55, 1990.
94. Stead, R. H., Qizilbash, A. H., Kontozoglou, T., Daya, A. D., and Riddell, R. H., An immunohistochemical study of pleomorphic adenomas of the salivary gland: glial fibrillary acidic protein-like immunoreactivity identifies a major myoepithelial component, *Human Pathol.*, 19, 32, 1988.
95. Takai, Y., Murase, N., Hosaka, M., Sumitomo, S., Noda, Y., and Mori, M., Comparison of lectin binding patterns in salivary glands of mice and rats with special reference to different fixatives used, *Acta Histochem.*, 78, 31, 1986.
96. Takano, K., Bogert, M., Malamud, D., Lally, E., and Hand, A. R., Differential distribution of salivary agglutinin and amylase in the Golgi apparatus and secretory granules of human salivary gland acinar cells, *Anat. Rec.*, 230, 307, 1991.
97. Takeda, Y., Histoarchitecture of the human parotid duct. Light-microscopic study, *Acta Anat.*, 128, 291, 1987.
98. Tamarin, A., Myoepithelium of the rat submaxillary gland, *J. Ultrastruct. Res.*, 16, 320, 1966.
99. Tamarin, A. and Sreebny, L. M., The rat submaxillary salivary gland. A correlative study by light and electron microscopy, *J. Morphol.*, 117, 295, 1965.
99a. Tanaka, T., Sakano, A., and Takahashi, A., Immunohistochemical localization of epidermal growth factor in the intercalated duct secretory granules of the submandibular glands of adult female A/J mice, *Acta Histochem. Cytochem.*, 17, 565, 1984.
100. Tandler, B., Salivary glands and the secretory process, in *Textbook of Oral Biology*, J. H. Shaw, E. A. Sweeney, C. C. Cappuccino, and S. M. Meller, Eds., W. B. Saunders Co., Philadelphia, 1978, 547.

101. Tandler, B., Structure of the human parotid and submandibular glands, in *The Salivary System*, L. M. Sreebny, Ed., CRC Press, Boca Raton, FL, 1987, 21.

102. Tandler, B. and Riva, A., Salivary Glands, in *Human Oral Embryology and Histology*, I. A. Mjor and O. Fejerskov, Eds., Munksgaard, Copenhagen, 1986, 243.

103. Tatemoto, Y., Tsukitani, K., Oosumi, H., Mori, M., Kobayashi, K., Kurobe, M., and Hayashi, K., Immunohistochemical localization of human epidermal growth factor/g-urogastrone in submandibular glands and in their obstructive lesions, *Acta Histochem. Cytochem.*, 21, 291, 1988.

104. Testa-Riva, F., Puxeddu, P., Riva, A., and Diaz, G., The epithelium of the excretory duct of the human submandibular gland: a transmission and scanning electron microscopic study, *Am. J. Anat.*, 160, 381, 1981.

105. Thoenen, H., Korsching, S., Heumann, R., and Acheson, A., Nerve growth factor, in *Growth Factors in Biology and Medicine*, D. Evered, J. Nugent, and J. Whelan, Eds., Pitman, London, 1985, 113.

106. Thomopoulos, G. N., Schulte, B. A., and Spicer, S. S., Postembedment staining of complex carbohydrates: influence of fixation and embedding procedures, *J. Electron Microsc. Tech.*, 5, 17, 1987.

107. Vigneswaran, N., Haneke, E., and Hornstein, O. P., A comparative lectin histo-chemical study of major and minor salivary glands with special reference to the labial glands, *Arch. Oral Biol.*, 34, 739, 1989.

108. Watanabe, M. and Kihara, T., Review: carbohydrate distribution in mammal; whole-body autoradiographic approach, *Acta Histochem Cytochem*, 19, 161, 1986.

109. Wilborn, W. H. and Shackleford, J. M., Microanatomy of human salivary glands, in *The Biologic Basis of Dental Caries. An Oral Biology Textbook*, L. Menaker, Ed., Harper and Row Publishers, Hagerstown, PA, 1980, 3.

110. Wilcox, W. B. and Pinkstaff, C. A., Histology and mucosubstance histochemistry of the major salivary glands in Galago senegalensis, in *The Lesser Bushbaby (Galago) as an Animal Model: Selected Topics*, D. E. Haines, Ed., CRC Press, Boca Raton, FL, 1982, 199.

111. Winston, D. C., Hennigar, R. A., Spicer, S. S., Garrett, J. R., and Schulte, B. A., Immunohistochemical localization of Na^+,K^+-ATPase in rodent and human sali-vary and lacrimal glands, *J. Histochem. Cytochem.*, 36, 1139, 1988.

112. Young, J. A. and van Lennep, E. W., Morphology and physiology of salivary myoepithelial cells, in *Gastrointestinal Physiology II*, R. K. Crane, Ed., University Park Press, Baltimore, 1977, 105.

113. Young, J. A. and van Lennep, E. W., *The Morphology of Salivary Glands*, Academic Press, London, 1978.

114. Zajicek, G., Yagil, C., and Michaeli, Y., The streaming submandibular gland, *Anat. Rec.*, 213, 150, 1985.

114a. Zajicek, G., Schwartz-Arad, D., Arber, N., and Michaeli, Y., The streaming of the submandibular gland II: parenchyma and stroma advance at the same velocity, *Cell Tissue Kinet.*, 22, 343, 1989.

115. Zimmer, K. P., Caselitz, J., Seifert, G., and Grenner, G., Immunoelectron micros-copy of amylase in the human parotid gland. Ultrastructural localization by use of both the protein A-gold and the biotin-avidin-gold technique, *Virchows Arch.*, A 404, 187, 1984.

Evolutionay Divergence of Salivary Gland Acinar Cells: A Format for Understanding Molecular Evolution

Carleton J. Phillips, Bernard Tandler, and Toshikazu Nagato

INTRODUCTION

Traditionally, salivary gland secretory endpieces are said to consist of some combination of serous, mucous, or seromucous secretory cells. These terms, derived from histological and histochemical staining, serve to superficially categorize the endpiece cells in terms of their secretory products.

Acinar cells are all designed around a common theme. Secretory cells are more or less similar in that they all are polarized and regulated. They all utilize a phylogenetically ancient intracellular process that involves the synthesis, modification, and packaging of proteins in membrane-bound secretory granules. Once they have been prepared within the intracellular milieu, the secretory granules are stored temporarily in the apical cytoplasm. Then, upon stimulation, the contents of the secretory granules are released into a lumen. At this level of consideration, all acinar cells are essentially the same, even though the products might vary somewhat. However, from our perspective, having examined the ultrastructure of acinar cells in the parotid, submandibular, and sublingual salivary glands of hundreds of species of mammals, acinar cells appear to be extremely diversified. There is such great diversity in the structure of acinar cells and their secretory products that the traditional histological and histochemical groupings are both overly simplistic and misleading. In fact, it may be true that salivary gland epithelium consists of the most diversified group of cells in mammals.

In what ways are acinar cells diversified and what is the biological significance of the extraordinary diversification exhibited by these cells? What factors have influenced the process of diversification? How did acinar cells come to be so diversified? These are the types of questions that come to mind when one first realizes the extent to which salivary gland acinar cells can vary. At the same time, a consideration of these questions quickly leads one to three conclusions: (1) salivary glands are highly unusual, perhaps unique, organs; (2) salivary glands can evolve very rapidly; and (3) evolutionary diversification at the cellular level is a complicated phenomenon.

Salivary glands are unusual because they play multiple roles in the lives of mammals. Saliva is an extremely complex fluid containing a huge array of substances that have the potential of affecting many aspects of the life of an individual mammal. It is apparent from the other chapters in this book, and from the literature on salivary glands, that saliva can affect digestion, oral microflora, tissue maintenance, and even behavior (Young and van Lennep, 1978; Fadem, 1986; Aloe et al., 1986; Phillips and Tandler, 1987). Salivary glands appear to evolve extremely rapidly. The idea that an organ has evolved rapidly makes sense only if one can make reasonable comparisons to other organs. Such interspecific comparisons make it possible to establish a relative rate of evolutionary diversification. When salivary glands are compared interspecifically, they obviously are more diversified than are other organ systems for which interspecific data are available. It thus is apparent that the salivary glands are not conserved. In fact, there even are examples of unique versions of salivary glands that have evolved within families of mammals. Insofar as we know, this amount of divergence is unprecedented in any other organ system.

Evolutionary diversification at the cellular level is a complex process. In its simplest form, such diversification involves a gradual modification of secretory molecules. Realistically, however, evolutionary diversification involves a variety of mechanisms that affect posttranslational modification, extracellular signaling, gene and cell recruitment, and causes alterations in the secretory process and duplication of entire glands.

This chapter is organized around our approach to the study of secretory cell evolution. It offers an overview of the thought processes that underlie this type of research. After explaining our chiropteran model, we review some of our evidence of morphological and chemical diversity and also discuss mechanisms of cellular diversification. Finally, our title calls attention to the idea that studies of salivary gland acinar cells might serve as a basis for improving our understanding of molecular evolution. The field of molecular evolution is relatively new and is widely regarded as distinct from the more traditional ways of investigating the evolutionary process. Molecular data, primarily in the form of DNA sequences or amino acid composition of polypeptides, sometimes are difficult to reconcile with data obtained by interspecific morphological comparisons. More important, perhaps, molecular data sometimes are difficult to interpret in terms of Darwinian evolution. We think that analysis of acinar cell diversity might help solve these problems because salivary gland acinar cells appear to be diversified in ways that reflect phylogeny. Moreover, these phenotypically diversified cells provide the critical environment in which genes are expressed.

THE MODEL

In order to logically approach some of the issues outlined in the foregoing paragraphs, it was necessary for us to find a means of constructing and then testing hypotheses about cellular evolution. Our criteria were straightforward. We needed a group of related species whose salivary glands differed in some recognizable way at the species level. It also was important to select a group of mammals that had been studied intensively by systematic biologists. Ideally, an independently established and widely accepted phylogenetic framework would be available for most of the species. The topology of such a framework, or tree, then could be used to detect trends or to test evolutionary predictions about salivary gland diversification (Figure 1).

Among the orders of mammals, we found that bats (Order Chiroptera) meet all of our criteria. To nonspecialists, "the bat" is often regarded as an arcane animal. In reality, however, bats represent insightful examples of the Class Mammalia; the 800 or so species of bats compose nearly 23% of existing mammalian species. This alone makes them "typical" among mammals. Moreover, their ability to obtain and assimilate a great variety of nutrients is reflected in the evolutionary history of bats. Although many species of bats are insectivorous, there also are species that are carnivorous, frugivorous, nectarivorous, sangivorous, or omnivorous. Within these dietary categories many species are highly selective. For instance, there are bats whose diets consist mostly of particular species of frogs, or fish, or other bats, or scorpions, or certain species of beetles, or the pollen and nectar in certain flowers, avian or mammalian bloods, or certain species of figs or other tropical fruits (Hill and Smith, 1984; Fleming, 1988).

Not surprisingly, virtually every aspect of chiropteran anatomy and physiology in some way correlates with feeding habits. This correlation of morphology and ecology is referred to as ecomorphology. Examples of ecomorphological relationships that have been studied in bats include a variety of traits that are morphologically correlated to diet. The list includes brain size, dental morphology, aspects of the somatosensory system, wing shape, eyes, digestive tract and

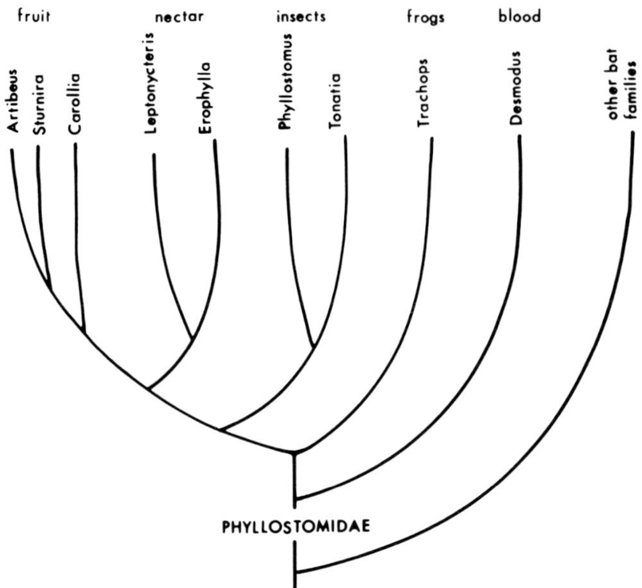

Figure 1. A phylogenetic tree showing the evolutionary relationships of nine genera of phyllostomid bats. This tree is based on chromosomal, morphological, and immunological (albumin proteins) data (Honeycutt and Sarich, 1987; Baker et al., 1989). Generalized information about diets is given above the genera.

gross and microanatomy (Eisenberg and Wilson, 1978; Findley and Wilson, 1982; Phillips et al., 1977, 1984; Forman et al., 1979; Mennone et al., 1986; Studholme et al., 1987).

The variety of ecomorphological features exhibited by bats has made chiropteran evolution and systematics a favorite topic for biologists. The origin of bats, and their general history over time, is still widely debated (Hill and Smith, 1984; Pettigrew et al., 1989; Novacek, 1992). Even so, the evolutionary history of most families and genera of bats has been established fairly well through a combination of techniques, which include karyology, protein analyses, histology, and morphometrics (Hill and Smith, 1984; Honeycutt and Sarich, 1987; Baker et al., 1988, 1989). Because the phylogenetic relationships of many bats are known, this information serves as a reference point for interpretation of interspecifically variable characteristics of salivary glands. An appreciation of the evolutionary history of these animals is the essential ingredient in "making sense" of secretory cell comparisons.

The evolutionary relationships of nine of the genera that we discuss herein are summarized in a simplified phylogenetic tree shown in Figure 1. The topology of the tree is based on a consensus, or congruence, of morphological, chromosomal, and immunological data sets (Honeycutt and Sarich, 1987; Baker et al., 1989). As can be seen from the tree, vampire bats (represented by the genus *Desmodus*) are an early side-branch of the family Phyllostomidae. The three representative fruit bats (*Artibeus, Sturnira,* and *Carollia*) share a common ancestry with bats that are specialized for including pollen and nectar, along with insects, in their diets (represented by *Leptonycteris* and *Erophylla*). The shared ancestor of these two divergent groups probably was an insectivorous or omnivorous species. The fringe-lipped bat, *Trachops,* is a specialized offshoot from an early phyllostomid (Figure 1). This unusual bat is related to two other genera of carnivorous bats (*Vampyrum* and *Chrotopterus*), but is unique among members of the Phyllostomidae in that it feeds on tropical frogs. The "other" families of microchiropteran bats are highly diversified, but most of the species are insectivorous. These other families can serve as a reference point, or "out group," for interpretation of salivary gland data from the various phyllostomids.

Bats have the same three major salivary glands — parotid, submandibular, and sublingual — that are found in other mammals (Phillips et al., 1977). Bats differ, however, in frequently having two versions of the submandibular gland. In many species of microchiropteran bats there is both a principal and an accessory submandibular gland (Robin, 1981; Pinkstaff et al., 1982); the term principal has been reserved for the gland that most resembles the mixed type of gland seen in other species of mammals. In some bats the principal submandibular salivary gland is the larger of the two submandibular glands, whereas in others it is the smaller. In some instances, the histology of the accessory submandibular gland is highly modified (Phillips and Tandler, 1987; Phillips et al., 1987; Tandler et al., 1990a). In addition to the major salivary glands, bats also have numerous small salivary glands located in various places in the oral epithelium. A few species of bats are known to have salivary glands at the angle of the mouth. In these species, the glands can be detected by the presence of a large pad-like structure that becomes apparent when the mouth is agape (Bhatnagar, 1988). In *Diaemus youngi,* one of the species of vampire bats, these glands emit a liquid that is said to have a mustelid (skunk-like) odor (Greenhall, 1988).

In order for bat salivary glands to serve as a useful model it also was essential for the glands to vary interspecifically in ways that could be detected by means of histology, histochemistry, and transmission electron microscopy. As we have summarized previously, this clearly is the case (Phillips et al., 1977; Phillips and Tandler, 1987; Tandler et al., 1989, 1990a).

Our selection of bat salivary glands as a model system for secretory cell evolution required us to obtain specimens of bats by collecting them in their native habitats in Africa, Asia, and North and South America. More than 170 species, representing both suborders and most families of bats, have been obtained and examined since 1980. What is the advantage of examining so many species from within a single order? The answer is twofold. First, from a practical perspective it must be remembered that bats are so diversified it is necessary to study a fairly large number of species just to obtain some sense of the overall diversity. Second, a large sample of species is necessary if we are to have an opportunity to gain some insight into the incidence of unusual features or novelties.

In order to obtain adequate comparative ultrastructural data, we developed a special triple aldehyde fixative that would produce acceptable results under virtually any situation in the field (Forman and Phillips, 1988). This fixative was developed (Phillips, 1985) from a similar fixative originally described by Kalt and Tandler (1971); the field fixative consists of 1% formaldehyde (titrated from paraformaldehyde), 3% glutaraldehyde, 0.5% acrolein, 2.5% dimethylsulfoxide (DMSO), and 1 mM CaCl$_2$ in 0.05 M cacodylate buffer with sucrose (pH 7.2). After overnight fixation, diced tissues are transferred to fresh 0.05 M cacodylate buffer and stored therein until refrigeration is available (Phillips, 1985). The tissues then are transferred to 3% glutaraldehyde in 0.05 M cacodylate and stored at 4°C. For standard transmission electron microscopy (TEM), tissues are rinsed in buffered sucrose, post-fixed in 2% OsO$_4$ for 2 h, washed in distilled water, and dehydrated. For cytochemical investigations, we reduce the concentration of glutaraldehyde to 1% and fix for 1 to 3 h rather than overnight. For light microscopic histochemical comparisons we field fix tissues in 4% formaldehyde buffered with 0.05 M cacodylate, with 0.1 M sucrose (pH 7.2) and 1% cetylpyridinium chloride added (Forman and Phillips, 1988). Salivary glands selected for immunohistochemical comparison are fixed overnight in Bouin's and then transferred to 70% ethanol for storage.

EVIDENCE OF DIVERGENCE

Secretory Granule Ultrastructure

Polarized, regulated secretory cells are ideally suited for comparative ultrastructural analysis because they usually are organized in a uniform way and because the secretory product is temporarily stored in the apical cytoplasm (Kelly, 1985). This is in contrast to constitutive secretory cells in which the newly synthesized products are quickly transported to the plasma membrane and released by exocytosis (Tartakoff and Vassalli, 1978). Because in regulated secretory cells the product is briefly stored, it is possible to examine it with the TEM after fixation. It also is possible to distinguish among "new" and "old" stored products on the basis of cytoplasmic location. Secretory products are exported in a chronological fashion; therefore, we know that products that are positioned closer to the Golgi complex are the most recently packaged. In making interspecific comparison of the contents of secretory granules, it is important to differentiate among new and old granules.

Ultrastructurally, stored secretory products are extremely variable in appearance. At one end of the spectrum there are secretory granules that are uniformly electron dense; at the other extreme one finds granules that are pale and essentially featureless. Classically, the former type of granule is referred to as a serous granule, whereas the latter is regarded as a mucous granule. Neither of the terms tells us anything significant about the chemical details of the product. For example, a serous granule can be rich in mucosubstances as well as in enzymatic and nonenzymatic proteins (Pinkstaff et al., 1982).

In terms of electron microscopy, the most interesting secretory granules are those that exhibit some form of internal substructure. It is known that substructure is a consequence of the chemical nature of the granule contents. Although it probably is true that the fixation process influences the microscopic appearance of substructure, the differences in appearance due to fixation have been shown to be minor (Tandler and MacCallum, 1972).

A number of investigations have revealed that the various components of secretory granules tend to be segregated rather than mixed (e.g., Ravazzola and Orci, 1980; Kousvelari et al., 1982; Machino et al., 1986). In other words, after secretory granules are packaged at the Golgi complex the various proteinaceous and carbohydrate ingredients tend to remain physically separated into zones or layers within the secretory granule. In some instances, as in the uniformly electron-dense serous granules, segregation is not visually detectable by standard TEM, but can be revealed by other forms of microscopy (Rothman et al., 1989). In other instances, granules have some visible form of substructure that reflects the segregation phenomenon (Kousvelari et al., 1982).

Variations in the appearance of secretory granule substructure probably are the result of a variety of factors. A list of likely factors would include differences in protein content, i.e., a particular substance present in one granule but not in another; differences in protein secondary or tertiary structure or charge, i.e., the same proteins but different conformation; differences in relative proportions of granule contents; and stearic relationships of the constituent molecules.

What is the morphological evidence of divergence in secretory products? Even a casual glance at the literature on salivary gland ultrastructure reveals that secretory granules are highly variable (Young and van Lennep, 1978). However, the only meaningful interspecific comparisons are those that have been made between homologous cells. Cells are regarded to be homologous if it reasonably can be argued that they were derived from a common evolutionary ancestry. As an example, acinar cells in the parotid salivary gland in different species are probably homologous, whereas parotid acinar cells are not homologous to acinar cells in the submandibular salivary gland. Thus, interspecific comparisons among parotid acinar cell secretory products can reveal patterns of evolutionary divergence in parotid acinar cells. Comparisons among nonhomologous cells may be valuable, but this type of comparison cannot give insight into the mechanisms and patterns of evolutionary divergence.

A superficial comparison of acinar cell secretory granules in parotid salivary glands of various species of microchiropteran bats reveals great interspecific variation (Figures 2 to 7). Likewise, the secretory granules in demilunar cells or acinar cells of submandibular glands also vary widely when homologous cells

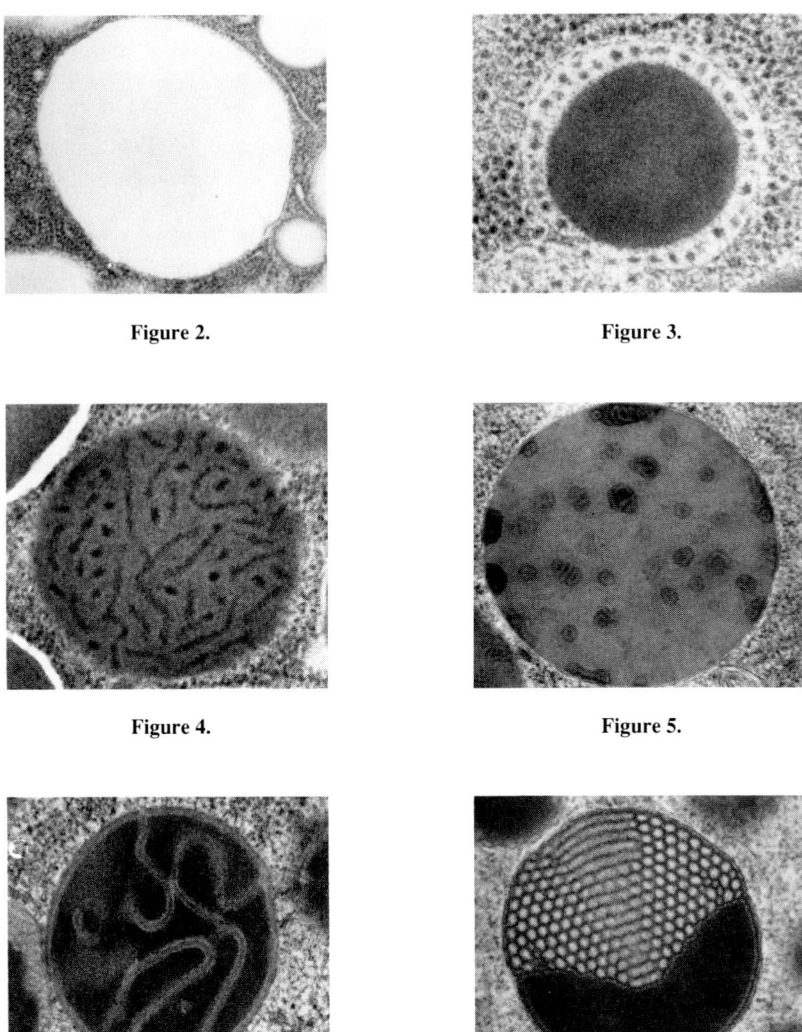

Figure 2.   Figure 3.

Figure 4.   Figure 5.

Figure 6.   Figure 7.

Figures 2 to 7. Transmission electron micrographs (TEM) of mature parotid salivary gland acinar cell secretory granules in a variety of bat species. (2) *Artibeus jamaicensis* (frugivore), × 22,000. (3) *Sturnira ilium* (frugivore), × 47,500. (4) *Tadarida brasiliensis* (insectivore), × 36,800. (5), *Pteronotus quadridens* (insectivore), × 27,000. (6) *Eptesicus lynni* (insectivore), × 32,800. (7) *Phyllostomus hastatus* (carnivore/insectivore), × 28,500.

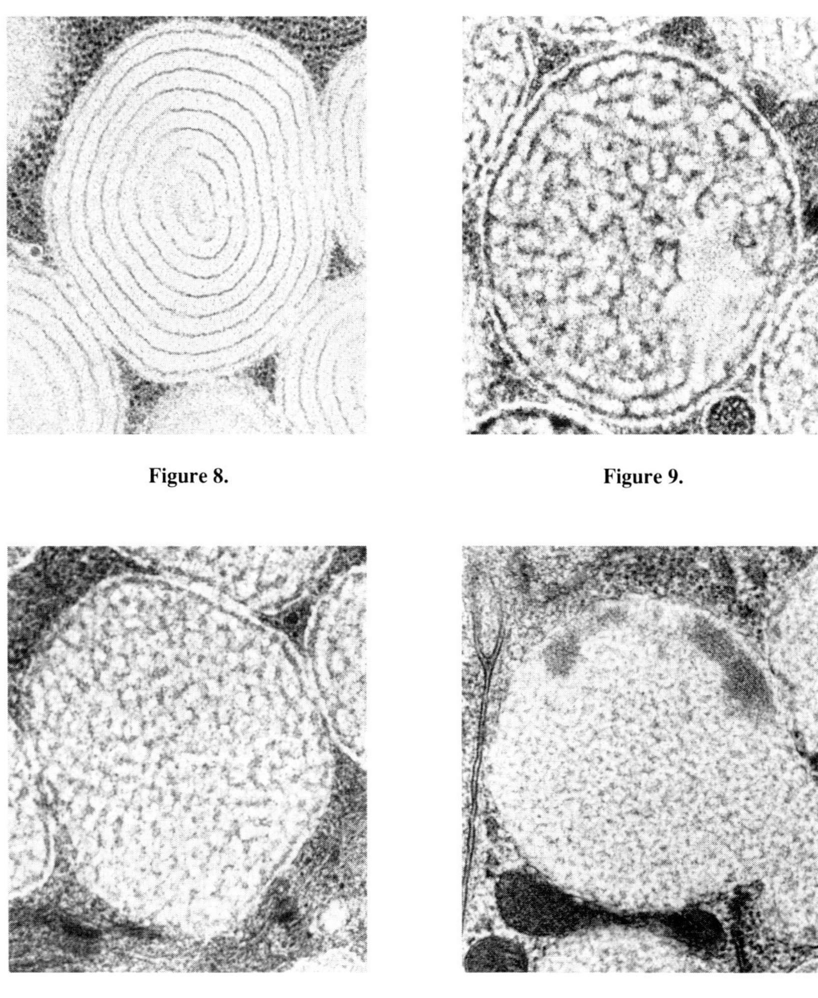

Figure 8.

Figure 9.

Figure 10.

Figure 11.

Figures 8 to 11. TEM comparisons of mature secretory granules in the seromucous demilunar cells in the submandibular salivary glands in four species of *Artibeus* (a frugivorous phyllostomid bat, see Figure 1). (8) *Artibeus cinereus*, × 29,300. (9) *A. lituratus*, × 32,500. (10) *A. jamaicensis*, × 38,800. (11) *A. concolor*, × 31,000.

are compared (Figures 8 to 11). But is this dramatic variation in ultrastructural appearance biologically meaningful in some aspect of the lives of bats?

Originally, the significance of ultrastructural variation in secretory granules was addressed by testing several hypotheses. First, we hypothesized that granule ultrastructure is correlated with diet. The evolutionary history of bats and their variety of dietary habits made it relatively simple to test this hypothesis. The parotid acinar secretory granules were compared across 100 species of microchiropteran bats representing the full range of dietary habits — insectivorous, carnivorous, sanguivorous, frugivorous, and nectarivorous. In terms of secretory granule ultrastructure, frugivorous and nectarivorous species differ from the others. In the species of microchiropteran bats that eat insects, flesh, or blood, and thus have high protein diets, the parotid secretory granules may be classified as serous (Phillips et al., 1987). Although far from identical, these secretory granules nevertheless all share an electron-dense, concentrated matrix (Figures 4 to 7). In species that feed on a lower protein diet of plant materials, the parotid acinar secretory granules have a paler matrix (Figures 2 and 3); in the stenodermatine fruit bats, the granules are electron-lucent (Figure 2).

Another hypothesis was that secretory granule morphology would have a genetic basis. We tested this hypothesis at several levels. For example, we were able to compare secretory granules in homologous cells in two to twenty individuals from more than 100 species. When we did this, no individual variation was detected. We also tested for geographic variation by examining secretory granule ultrastructure in individuals collected from widely separated locations. However, we did not detect any geographic variation. Additionally, we found that secretory granule ultrastructure in microchiropteran bats is essentially a marker, or "fingerprint," for species (Phillips and Tandler, 1987; Tandler et al., 1990a).

We further hypothesized that the extent of interspecific similarity in secretory granule ultrastructure would be proportional to the genetic distance between species. In other words, secretory granules would be more similar in closely related species than in distantly related species. In one sense, this hypothesis would be difficult to test because of the subjective nature of the comparison. On the other hand, the morphology of some secretory granules is so regular that departures from the basic theme would be obvious. With nearly 170 species available to us, it was not difficult to select microchiropteran bats suitable for testing this hypothesis.

Phyllostomid fruit bats in the genus *Artibeus* (Figure 1) are geographically widespread in the Neotropics. Systematists have used morphometric data, electrophoretic comparison of proteins, and restriction site mapping of mitochondrial DNA (mtDNA) to estimate the genetic relationships among and within the species of *Artibeus* (Straney et al., 1979; Koop and Baker, 1983; Pumo et al., 1988; Owen, 1988; Phipps et al., 1991). We compared the ultrastructure of the secretory granules produced by the endpiece secretory cells in the principal submandibular gland in five species of *Artibeus* (Tandler et al., 1986). The serous cells in all five species produce ultrastructurally identical electron-dense granules. By way of contrast, the secretory granules in the seromucous demilunar cells vary interspecifically. On the basis of demilunar cell secretory granule substructure (Figures 8 to 11), we placed the secretory products in three groups. These groupings, in turn, matched the prevailing opinion about the genetic distances among the five species. The two most closely related species, *Artibeus jamaicensis* and *A. lituratus* were virtually identical. On the basis of mtDNA analysis and the molecular clock hypothesis, it can be estimated that these species might have diverged nearly two million years ago (Phillips et al., 1991). *A. concolor* clearly diverged earlier than *A. lituratus* (Straney et al., 1979; Koop and Baker, 1983), and the demilunar secretory granules differ appreciably from those produced in the former two species (Figure 11). Finally, two of the species, *A. cinereus* and *A. phaeotis*, are closely related to each other but are distantly related to the other three (Koop and Baker, 1983). Indeed, there are some systematists who regard *A. cinereus* and *A. phaeotis* as members of a separate genus, *Dermanura* (Owen, 1988). Interestingly, the substructure of the demilunar cell secretory granules also sets *A. cinereus* and *A. phaeotis* apart from the others (Figure 8 and Tandler et al., 1986); this finding might support Owen's (1988) hypothesis. In summary, the extent of interspecific ultrastructural difference in secretory granule substructure in at least some instances correlates with genetic distance between species.

Chemical Evidence

Because saliva is a chemically complex product, it has not been characterized for any species other than human beings and certain laboratory animals. Nevertheless, there are data that indicate that saliva produced by homologous glands can be chemically divergent. For example, the glycoconjugates produced by the submandibular salivary gland have been shown to exhibit structural variation in glycoprotein saccharide moieties (Schulte, 1987). In fact, examples of histochemical variation have been reported in both interspecific and interstrain comparisons (Schulte, 1987). Moreover, sexual polymorphism and intraindividual heterogeneity also have been reported (Schulte, 1987).

We hypothesized that ultrastructural variation in the appearance of secretory granules would, in some way, correlate with chemical variation. One test of this hypothesis could be made by comparing the parotid acinar cell secretory product among various species of phyllostomid bats (Phillips et al., 1987; Tandler et al., 1988). This was possible because at least some data currently are available on the enzyme content of these granules (Junqueira et al., 1973). On this basis, it appears that the pale, electron-lucent secretory granules in phyllostomid fruit bats (Figures 2 and 3) are correlated with low enzyme content (Phillips et al., 1987). Related species that feed on insects of other vertebrates have electron-dense secretory granules with high enzyme content. The functional significance of a lowered enzyme content is not immediately apparent, but it may be noteworthy that these fruit bats have a relatively low protein diet in comparison to the diets of insectivorous/carnivorous species.

Ultrastructural and histochemical comparisons of the glycoprotein components in submandibular gland secretory products in three species of the East African house bat, *Scotophilus*, were made in order to further test the hypothesis that ultrastructural and chemical variation are correlated (Phillips et al., submitted). The salivary glycoproteins appear to have several roles in oral ecology; one

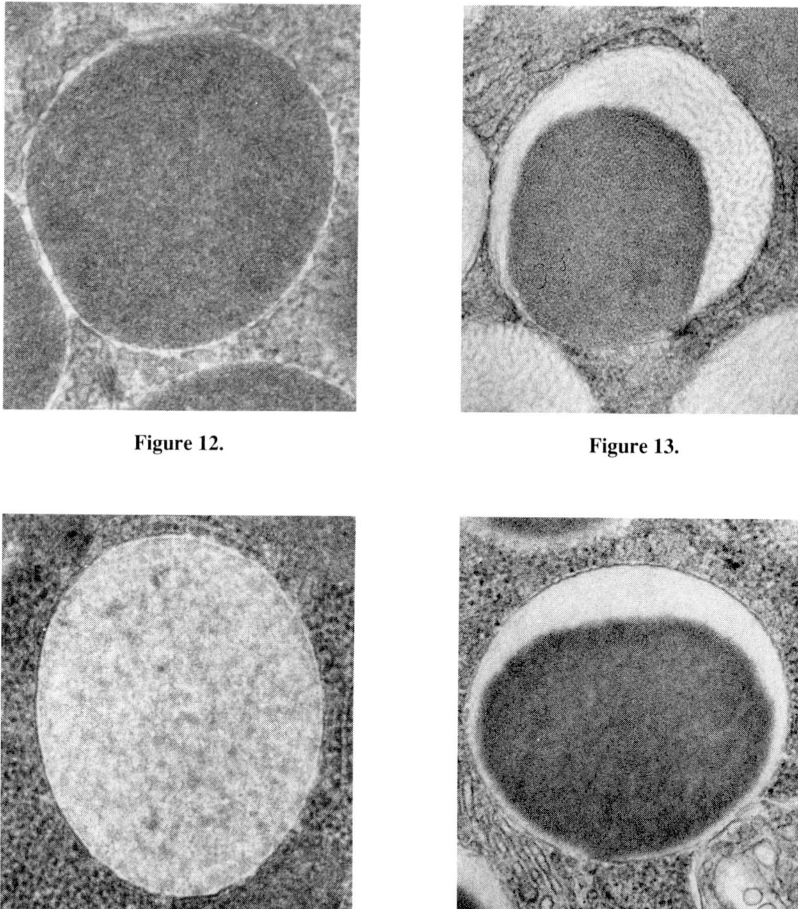

Figure 12.   **Figure 13.**

Figure 14.   **Figure 15.**

Figures 12 to 17. TEM comparisons of secretory granules in demilunar and mucuous tubule cells in three species of house bat, *Scotophilus*. (12) *Scotophilus dingani*, demilunar cell, × 42,000. (13) *S. dingani*, mucous tubule cell, × 30,000. (14) *S. nux*, demilunar cell, × 44,000. (15) *S. nux*, mucous tubule cell, × 32,000. (16) *S. borbonicus*, demilunar cells, × 33,600. (17) *S. borbonicus*, mucous tubule cell, × 39,000.

of the most important of these involves interaction with the oral microflora. Some glycoproteins might serve as antibacterial agents (Eggert et al., 1987), whereas others might compete with bacteria by binding to cell surface receptors. In the latter instance, the glycoproteins would block potentially pathogenic bacteria that might attach to the apical surfaces of mucosal epithelial cells. Bacterial colonization apparently is preceded by such attachment (Schulte, 1987).

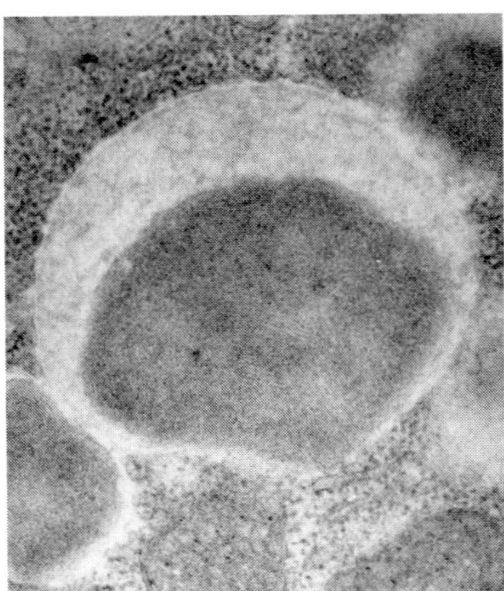

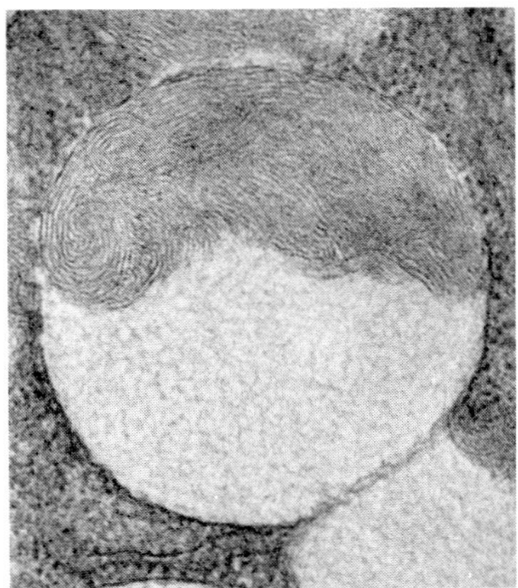

<table>
<tr><td align="center">**Figure 16.**</td><td align="center">**Figure 17.**</td></tr>
</table>

House bats of the genus *Scotophilus* are insectivorous feeders; they are members of the family Vespertilionidae, which has a world-wide distribution. In the three species that we compared, *Scotophilus dingani, S. borbonicus,* and *S. nux,* the principal submandibular salivary gland consists of mucous tubules with mucous demilunes. In *S. dingani* the mucous tubule secretory cells have granules with homogeneously speckled, moderately dense matrix and a dense inclusion (Figure 13). In *S. borbonicus* the homologous cells have secretory granules with a large, fibrillar inclusion. The lucent matrix of the granule contains a meshwork of fine thread-like material (Figure 17). The secretory granules in both of these species contain sulfated glycoconjugates, PAPD$^+$ neutral glycoconjugates, and sialic acid-containing glycoconjugates. In contrast, in *S. nux* the secretory granules have a lucent, featureless matrix and a large, dense inclusion (Figure 15). In terms of glycoconjugates, these secretory granules differ from those in the other two species in containing considerable sialic acid-containing glycoconjugates, no sulfated glycoconjugates, and no PAPD$^+$ neutral glycoconjugates. The secretory granules in the demilunar cells also are interspecifically variable in these three bats. The ultrastructural variation that is readily apparent probably is due to differences in the structural characteristics or types of the glycoconjugates that are present within the granules (Figures 12, 14, 16). Thus, ultrastructural variation in secretory granules is indicative of chemical variation.

Divergence in Cell and Membrane Structure

Until fairly recently, cells were viewed as somewhat formless membrane-bound containers that housed various metabolic activities. Today, it is widely understood that cell architecture is hardly accidental (Gershon et al., 1985; Phillips and Tandler, 1987). Indeed, cells typically are highly structured; the morphological differences among cell types largely are the result of differences in their cytoskeleton (Gershon et al., 1985; Phillips and Satir, 1988). Likewise, cell membranes no longer are viewed as continuous structures. It now is recognized that there are specialized regions, or microdomains, within plasma membranes (Palade, 1985). Finally, it is known that the internal localization of organelles also is determined in some specific way. This is especially true of polarized, regulated secretory cells in which the positions of the granular endoplasmic recticulum (GER), smooth endoplasmic reticulum (SER), nucleus, Golgi complex, and mitochondria are predictable for any particular cell type (Phillips and Tandler, 1987). Most important, it is clear from studies of morphological differentiation of salivary gland acinar cells that correct polarity, appropriate forms of contact between adjacent cells, and establishment of the correct membrane microdomains are prerequisites to secretory function (Tamaki et al., 1989).

What factors determine the cell architecture and membrane microdomains? That is, how are these features encoded in the genome and how is a cell assembled? This consideration is far from trivial. The current thinking is that some type of "directed" assembly is probably involved (Grimes and Aufderheide, 1991). Furthermore, if homologous cells differ interspecifically in terms of architecture, then what effect would these differences have on both the secretory process and the secretory products? What about differences within membrane microdomains? What would be the effect of differences in membrane surface features, surface area, or membrane receptors? Earlier in this chapter we mentioned that salivary gland acinar cells appear to "house" the secretory process. In reality, however, the highly ordered character of polarized acinar cells clearly suggests that these cells do more than merely provide a container in which the secretory process takes place. Indeed, these cells provide an interactive "environment" in which the process is carried out.

We think that consideration of comparative cell architecture and membrane microdomains, i.e., cell phenotypic characteristics, is extremely important to the development of an understanding of the evolutionary process at the molecular level (Phillips and Tandler, 1987). After all, gene expression occurs within particular cellular environments. Moreover, as we discuss later in this chapter (see Extracellular Signaling), consideration of the cellular environment provides us with examples of how phenotypes can directly influence both gene expression and the structural and functional nature of gene products. Consequently, the point can be made that cell phenotypic characteristics are not equivalent in importance to other phenotypic features. Instead, the phenotypic characteristics of cells must be regarded as special because of their direct interactive relationship to gene expression and the nature and use of gene products.

Our concept of cell phenotypes is important because it implies that what natural selection "sees" often is a result of a combination of genetically encoded polypeptides and cell phenotypic characteristics rather than just the direct product of a particular gene. In fact, natural selection might operate through modification of cell phenotypes (and consequently the internal cellular milieu) more often than it operates by direct selection on individual protein coding genes. The significance of this idea is discussed further in the section on Relaxed Constraints.

Architecture of salivary gland acinar cells can be used to distinguish three types of cells: serous, mucous, or seromucous. It is generally known that each of these three types of cells has its own ultrastructural characters. For example, in mucous cells the nuclei typically are heterochromatic, irregular in outline, and basally displaced. In serous cells the nuclei typically are round in outline, more nearly euchromatic, cradled in GER, and almost always centrally positioned. From this level of comparison alone it is obvious that the major categories of secretory product are produced in architecturally distinct cells.

By comparing homologous cells in a wide variety of interrelated species, we have been able to address two previously unanswered questions. What types of variation in cell architecture and membrane ultrastructure are found among homologous cells? Once again, are there consistent patterns to this variation?

Regardless of its fluidity, the cell membrane is not a randomized structure. Indeed, the cell membrane is highly regionalized (Palade, 1985). Regionalization allows for chemical diversification of the outer cell surface and facilitates diversification among interior membranes. In eukaryotic cells there are at least ten well-defined, chemically distinct, membrane regions, or microdomains (Palade, 1985).

The evolution of diversity in salivary gland acinar cells has involved many, or perhaps all, of the membrane microdomains. Sometimes, aspects of evolution of microdomains can be visualized with TEM. For example, if the basal cell membrane of salivary gland acinar cells were to be compared ultrastructurally, it is apparent that this region of the cell surface can be highly modified (Tandler et al., 1990a). Although systematic comparisons have not been undertaken, data from studies of human and dog salivary glands show that the acinar cell basal membrane can harbor a Na^+-K^+-ATPase (Nakagaki et al., 1978; Tandler and Riva, 1986). Diversity in the ultrastructure of the basal membrane is significant because transport processes across this membrane are likely to affect the nature of the product elaborated by the cell.

The parotid acinar cells can be used as an excellent model of basal membrane diversity. In most species of insectivorous or carnivorous microchiropteran bats, the basal membrane is highly infolded (Figure 18) and thus has a relatively great surface area. In many instances, the pattern of infolding and the details of its ultrastructural appearance are distinctive at the family level. Thus, one can easily distinguish among parotid acinar cells of species of mormoopid, phyllostomid, and vespertilionid bats just by comparing the ultra-structure of the basal membrane.

Evidence of dramatic evolutionary change in the basal membrane can be visualized by ultrastructural comparison among species of phyllostomid bats (Tandler et al., 1990a). As previously mentioned, dietary diversity is the key

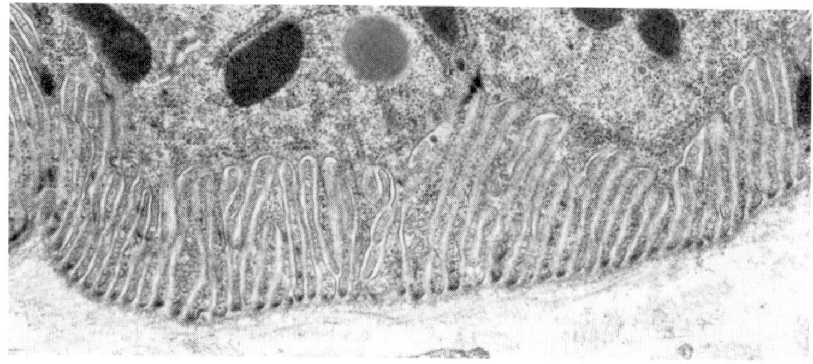

Figure 18.

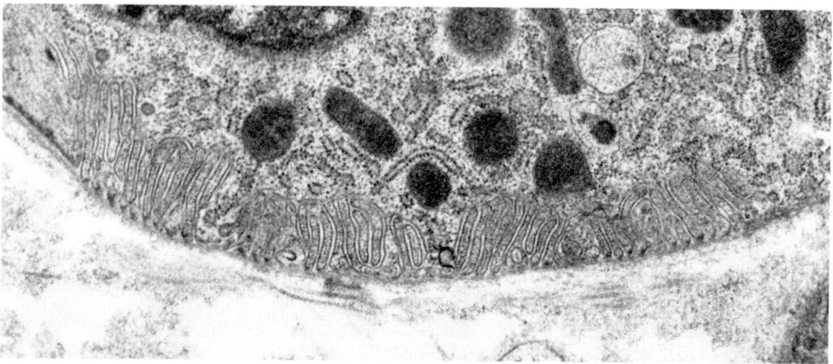

Figure 19.

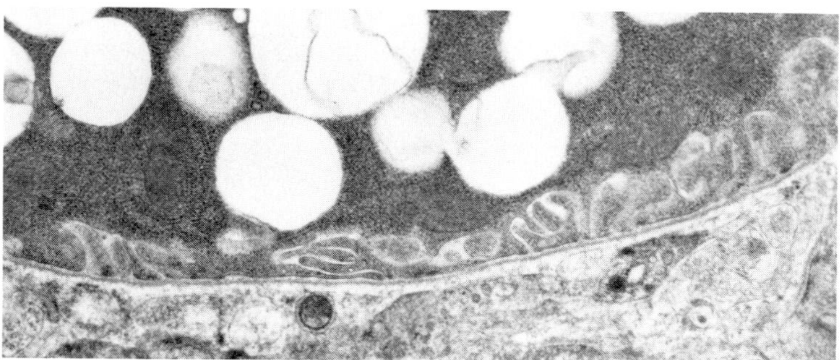

Figure 20.

Figures 18 to 20. Interspecific TEM comparisons of basal cell membrane in acinar cells in the parotid salivary gland. (18) *Eptesicus brasiliensis*, an insectivorous vespertilionid species, × 13,300. (19) *Tonatia bidens*, a primarily insectivorous phyllostomid species, × 13,000. (20) *Artibeus jamaicensis*, a frugivorous phyllostomid species, × 13,300.

characteristic of this large family of neotropical bats (about 120 species). From systematic research it is apparent that the ancestor of the existing species that compose the family was most likely insectivorous. Similarly, the least derived living members of the family are either insectivorous or carnivorous. In these less derived species, such as *Tonatia bidens* (Figure 1), the basal membrane of the parotid acinar cells typically is moderately infolded (Figure 19). During the evolution of the Phyllostomidae, there were groups of species that diverged from this basic plan. For instance, the vampire bats represent extreme specialization in diet (Greenhall and Schmidt, 1988). In the common vampire bat, *Desmodus rotundus*, the basal membrane has been highly modified into a slightly infolded arrangement (Tandler et al., 1990a,b). Likewise, in the parotid acinar cells of the various frugivorous phyllostomids (Figure 1), the basal membrane has lost its folds almost entirely (Figure 20). What do changes in the ultrastructural appearance of a membrane microdomain signify? In the example cited above, the ultrastructural differences in basal membrane would only document with certainty a dramatic change in cell surface area. However, because a notable degree of interspecific diversification has occurred in this region of acinar cell membrane, it would be interesting to also know more about surface receptors and enzymes situated within the membrane in these animals.

The GER of the acinar cells offers another example of diversity in a membrane microdomain. The club-footed bat, *Tylonycteris*, is an insectivorous vespertilionid genus. In this bat, the GER in the parotid acinar cells differs strikingly from the GER in homologous cells in other vespertilionids. In *Tylonycteris*, the GER appears to be structurally subdivided so that some regions are dramatically different from other regions (Tandler et al., 1990a). The unusual, subdivided appearance of the GER results from the fact that some areas have augmented numbers of ribosomes on the cytoplasmic aspect and electron-dense material within the subjacent cisternae. The functional significance of this modified microdomain is unknown but it would be reasonable to hypothesize that it involves regional differences in protein synthesis (Tandler et al., 1990a). This in turn raises the possibility that proteins synthesized for export can be altered as a result of modification of membrane microdomains.

Mitochondria offer another interesting example of interspecific variation within acinar cells. In inter- and intraspecific comparisons, it is readily apparent that mitochondrial morphology is consistent within cell types among individual bats within a species, but that it can vary dramatically between species or between cell types within species (Tandler et al., 1990a). In some instances mitochondria form cup shapes and are nested one inside the other. In other instances the mitochondria have numerous, parallel, prismatic cristae, each periodically attached to its neighbors by intercisternal bridges (Tandler et al., 1988). Sometimes mitochondria have paracrystalline plates within their cristae (Tandler et al., 1990a). In some species, a subpopulation of unusually large mitochondria is found in the acinar cells. One example of this is seen in *Artibeus jamaicensis*, where the large mitochondria exhibit a few stacks of cristae (Figure 21). In seromucous acinar cells in the accessory submandibular salivary gland of the long-haired fruit bat, *Stenonycteris lanosus*, which is a member of the suborder Megachiroptera, there are unique giant mitochondria (Figure 22) that often contain threadlike structures and 5-nm helical filaments (Tandler et al., in press).

The significance of the great diversity in mitochondrial morphology is still unknown, but the available data suggest that all of these forms are fully functional. It is possible, of course, that mitochondrial morphology relates in some way to the physiological needs of a particular cell type. It probably is significant that although they are morphologically diversified, the various mitochondrial morphotypes nevertheless are consistent within a cell type within a species. In other words, the morphotypes do not appear to be transient or spurious. Recent investigations of the expression of mitochondrial DNA (mtDNA) suggest the possibility of tissue specific or time-dependent expression of some of the genes that encode proteins that form part of the hydrophobic shell of the NADH dehydrogenase complex (Attardi et al., 1989). Additionally, it has been shown that mitochondrial gene expression differs between oxidative and glycolytic muscle fibers (Williams, 1986). The idea of tissue-specificity is further suggested from investigations of the effects of certain deleterious mutations in mtDNA. Generally speaking, these mutations appear to have tissue-specific phenotypic effects (Singh et al., 1989). Ultimately, we may find that mitochondria have evolved and diversified along with other components of the acinar cells.

MECHANISMS OF DIVERGENCE

To what can we attribute the diversity seen in chiropteran salivary gland acinar cells? Can the obvious patterns of variation be explained in some easy way? The traditional view of evolution involves a process of accumulation of "point mutations." This seemingly slow means of modifying gene products leads to the concept of gradualism. However, there is no indication that salivary gland evolution has been gradual. Indeed, the diversity in mammalian salivary glands (Young and van Lennep, 1978), especially the diversity seen within the order Chiroptera (Phillips and Tandler, 1987; Tandler et al., 1990a), could only have resulted from a rapid process. Moreover, there is no reason to think that salivary gland evolution can be explained by merely evoking a commonly used neo-Darwinian scenario in which base-pair mutations in nuclear protein coding genes result in nonneutral amino acid substitutions in the exported protein. Furthermore, few of the salivary proteins are bona fide secretory products in the sense that they are strictly the synthesized product of a gene. Indeed, it is well established that protein diversity can be produced by alternative splicing of

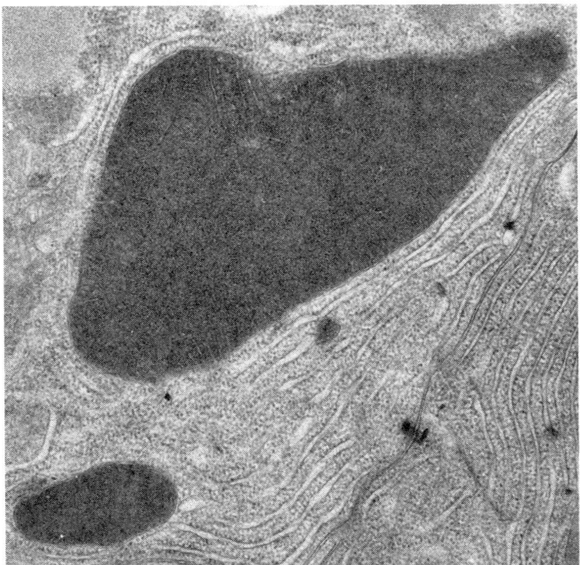

Figure 21.

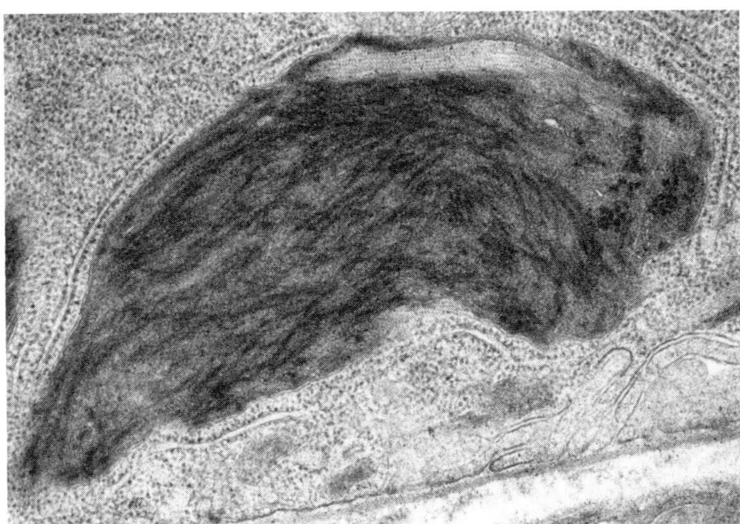

Figure 22.

Figures 21 and 22. TEM of two examples of highly unusual mitochondrial morphotypes. (21) Greatly enlarged mitochondrion next to "normal" profile in a submandibular salivary gland acinar cell in *Artibeus jamaicensis*, × 25,300. (22) Giant mitochondrion with dense matrical ribbons in a seromucous cell in the accessory submandibular gland in *Stenonycteris lanosus*, × 28,000.

mRNAs. An interesting example of this can be seen in the evolution of a human β-crystallin gene (Hogg et al., 1986). Proteins can be further diversified through posttranslational modification, as discussed below.

If traditional Darwinian gradualism is difficult to reconcile with the dramatic diversity in salivary gland acinar cells, then how do we explain the data? Moreover, what does salivary gland diversity tell us about the mechanisms of evolution? We think that in the case of salivary glands, natural selection primarily (but not totally) has been focused on mechanisms that promote diversification. In other words, selection has favored successful mechanisms regardless of their ultimate underlying genetic or epigenetic bases. Our theory thus is an expansion of the more traditional, — but still modern, — way of looking at the evolutionary process, in which specific proteins are selected on a step-by-step basis. On the other hand, this theory does have several advantages. For instance, it could provide us with a variety of testable hypotheses. More important, it also offers us a foundation for explaining the rapid diversification of a fully integrated organ system. Finally, although based on salivary glands, this approach to evolutionary biology might have broader utility in that it could help explain how saltatorial evolution could occur at the cellular level. Indeed, there already are indications that α/β barrel enzyme families have diversified by several means, including various forms of gene permutation (Luger et al., 1989; Farber and Petsko, 1990).

Relaxed Constraints

The proline-rich proteins (PRPs), as well as some lesser known salivary proteins, have been shown to vary widely when compared in terms of amino acid primary structure (Bennick, 1987; Mamula et al., 1988). The large amount of variation between and within species suggest that many amino acid substitutions are functionally neutral. In other words, a variety of substitutions probably can occur without affecting the secondary or tertiary structure in a way that will affect the functional integrity of the protein. This phenomenon is true for many proteins; frequently, the amino acid sequence of substantial regions of a protein can be highly variable (Perutz, 1983; Nei, 1987; Beato, 1989). At certain positions in the polypeptide chain, amino acids can be freely substituted (any amino acid will do), whereas in other positions, only substitutions within categories, i.e., one hydrophobic amino acid for another, will be tolerated without affecting the functional characteristics of a protein. Such substitutions are said to be neutral insofar as selection is concerned. This phenomenon appears to be widespread among proteins and in part is the basis for Kimura's (1968) Neutral Theory of evolution, which argues that a large proportion of molecular variation within populations is functionally neutral (see Nei, 1987 for an expanded discussion of this theory). Kimura's ideas create a conceptual problem for anyone inclined to explain salivary gland acinar cell diversity in terms of

Darwinian selection on a protein by protein basis. One would expect the primary structure (amino acid sequence) of secretory proteins to be interspecifically variable if large regions of the proteins are "allowed" to differ. Selection would not "see" these differences among proteins and, thus, the amino acid composition would be relatively unconstrained and could vary widely. The portions of the gene that encode the amino acids for the variable regions of such proteins are said to be under relaxed constraints. However, there are two reasons why relaxation of constraints, i.e., relaxed selection, is not the primary mechanism in salivary gland secretory cell diversity. Relaxation of constraints — a type of laissez faire evolution — would not promote phenotypic or genotypic diversity. It would, however, allow for diversity. Base-pair substitutions, additions, or deletions in protein-coding nuclear DNA appear to occur at some average rate, which differs among orders of mammals (Nei, 1987; Beintema et al., 1989). This mutation rate would be unaffected by relaxed constraints on the encoded gene product. Moreover, it is difficult to imagine how selection affects many of the components of saliva. Exceptions to this would include enzymes essential for neonate survival or possible pheromonal molecules in saliva that would affect breeding behavior and, ultimately, reproductive success. However, it is less clear how many of the acidic, basic, and phosphorylated PRPs and other proteins that are part of saliva might directly affect individual survival or reproductive success. Most likely, their evolutionary history is best relegated to the less definite category of group selection.

In summary, salivary proteins or large portions of their respective component domains might be under relaxed constraints but this does not account for the diversity in acinar cells or their products. Moreover, the idea that most amino acid substitutions are functionally neutral (Kimura's Neutral Theory) compounds the difficulty in understanding rapid diversity through selection on a protein by protein basis. This leads us to seek their mechanisms to explain patterns of diversity. In particular we now can stress the respective roles of variation in cell phenotypic characteristics (including the intracellular milieu) and forms of mutation other than nucleotide base-pair substitutions and small additions and deletions in the protein-coding DNA.

Promotion of Protein Diversity at the Chromosomal Level

There are some data that strongly suggest that diversity in salivary proteins sometimes has been promoted through unequal crossing over (Dickinson et al., 1989). In this example, Dickinson et al. (1989) used mRNA isolation and cDNA cloning to investigate a class of salivary proteins that they termed "spot proteins." These proteins, which are produced in submandibular salivary glands, are highly variable in murid rodents. The pattern of amino acid substitutions in the spot proteins indicated that these proteins were under relaxed constraints (Dickinson et al., 1989). However, in addition to relaxed constraints on the proteins, cDNA sequences also suggested that rapid divergence had been promoted by a gene conversion event and by unequal crossing over involving spot-protein DNA sequences.

Posttranslational Modifications

Much of the diversity in salivary proteins probably results directly from some type of posttranslational modification of gene products (Bennick, 1987; Mamula et al., 1988). As an example, in human beings the diversified PRPs are derived from a small number of genes clustered together in chromosome band 12p13.2 (Mamula et al., 1985). Many of the highly variable PRPs that eventually are found in human parotid gland saliva are apparently the result of posttranslational cleavage of a relatively small number of large precursor molecules (Mamula et al., 1988). The acidic PRPs in human saliva probably result from cleavages of a few precursors by kallikrein or a kallikrein-like enzyme. Likewise, the diversified phosphorylated PRPs might result from a protein kinase that phosphorylates a limited number of precursors (Bennick, 1987).

The extremely diverse salivary glycoproteins contain a medium length (~200 amino acid) polypeptide chain that serves as a backbone for a molecule consisting of 40 to 70% carbohydrate. To varying degrees these glycoproteins also can be sulfated and can contain fatty acids (Levine et al., 1987). Evolutionary diversification of glycoconjugates has been documented in many groups of mammals (Pinkstaff, 1980) and clearly has occurred among species of bats. As described in a previous section, histochemical techniques have revealed variation in glycoconjugate products of both demilunar and acinar cells in the submandibular salivary gland of the house bat, *Scotophilus*. Diversification of secretory proteins through variations in glycosylation is of special interest because many of these products appear to interact in some way with oral microflora (Eggert et al., 1987). Indeed, at least some of the glycoconjugates may have an antibacterial function (Eggert et al., 1987). With this in mind, it is easy to understand why it would be important for the glycoproteins to be so greatly diversified. The interspecific histochemical differences that we have documented in bats could likewise be regarded as the result of an interplay between glycoproteins and bacteria during the history of a species. Furthermore, given the significance of the oral microflora, it is tempting to speculate that there has been a concerted selection for mechanisms that promote glycoprotein diversification.

In summary, posttranslational modification of proteins (through cleavage, phosphorylation, and glycosylation) alleviates total reliance upon step-by-step, gradual accumulation of beneficial base-pair mutations in protein-coding DNA sequences. To judge from available data, posttranslational modification may be a major source of diversification in exported salivary gland proteins. Moreover, evolutionary studies of mammalian ribonuclease have revealed that they can be diversified both through alterations in the enzymatic control of glycosylation and as a result of amino acid substitutions that eliminate glycosylation sites (Beintema et al., 1989). Such a model might apply to salivary glycoproteins as well.

Extracellular Signaling

Salivary gland acinar cells are responsive to a variety of extracellular signals (e.g., Humphreys-Behr, 1985; Kousvelari et al., 1987; Johnson et al., 1987). This responsiveness is dependent upon the presence, density, and intramembrane distribution of cell surface receptors as well as patterns of innervation. Extracellular signals have the potential of creating diversity by affecting gene expression, rates of protein synthesis, and posttranslational modification of proteins.

In studies on isolated parotid acinar cell secretory granules, the β-adrenergic agonist isoproterenol has been shown to have a quantitative effect on secretory proteins (Arvan and Castle, 1986). In particular, α-amylase was reduced and replaced as the principal component by other polypeptides that were enriched in proline, glycine, and glutamine (Arvan and Castle, 1986). Another good example of the role of extracellular signaling involves N-linked protein glycosylation in rat and human parotid acinar cells. Stimulation of β-adrenoreceptors can cause a significant alteration in the oligosaccharide structures in secretory glycoproteins (Humphreys-Beher, 1985; Kousvelari et al., 1987). It thus is clear that responsiveness of the glycosylation process to extracellular signals can result in the diversification of set of basic glycoproteins produced within an acinar cell. This is important because it also has been shown that receptor-mediated modifications in glycosylation can affect the function of a glycoprotein (Olden et al., 1985). Additionally, Kousvelari et al. (1987) have pointed out that the functional characteristics of glycoproteins secreted by parotid acinar cells potentially can be altered through nerve stimulation.

Yet another example of the effects of extracellular signaling, which is highly relevant to our data from bats, concerns the direct effect of diet on PRP biosynthesis by the parotid gland. Mehansho et al. (1983) showed that an increase in dietary tannins affects gene expression in the rat parotid acinar cells. Specifically, they found that the glands hypertrophy and that certain PRPs are secreted copiously. Similar effects of tannin and the β-agonist isoproterenol on salivary glands, suggest that β-adrenoreceptors might be involved in both cases (Mehansho et al., 1983).

The PRPs have been shown to have a very high affinity for tannin. Mehansho et al. (1983) pointed out that salivary PRPs could bind to tannin and, thus, the tannin consumed in the diet would be less likely to bind to dietary proteins. Tropical fruit bats face two problems that are of interest to our discussion. First, unlike their insectivorous relatives, they feed on plant material containing tannins. Second, many of the available plant materials are not necessarily high in protein; therefore, large volumes must be consumed and processed quickly in order for them to assimilate adequate amounts of dietary protein. It thus is not surprising that the ultrastructure of the parotid acinar cells in phyllostomid fruit bats differs strikingly from the ultrastructure in insectivorous/carnivorous phyllostomids (Figures 2 to 7). Indeed, the ultrastructural appearance of the parotid acinar cell secretory granules in phyllostomid fruit bats (Figure 2) mimics that which can be experimentally induced in rats by treatment with thyroxine (Johnson et al., 1987). In the rats, the pale secretory granules induced by thyroxine treatment reflect (1) increased output and (2) significantly increased levels of salivary PRPs (Johnson et al., 1987). Collectively, all of these data sets strongly indicate that the ultrastructural appearance of phyllostomid fruit bat parotid secretory granules (Figures 2, 3, and Phillips et al., 1987) is both related to diet (possibly tannin content) and may be the result of responsiveness of the acinar cells to extracellular signaling. In considering the evolution of the parotid acinar cells in *Artibeus, Sturnira, Carollia*, and other fruit bats (Figure 1) (Phillips et al., 1987; Tandler et al., 1988, 1990a), our attention is drawn to the possible role of receptors positioned in basal or basso-lateral cell membrane. The ultrastructural distinctiveness of this membrane microdomain, which we discussed previously (Figures 18 to 20), takes on enhanced significance in the present context.

The picture that emerges when considering effects of extracellular signaling on secretory product is indeed dramatic. First, extracellular signaling can promote diversification by affecting glycosylation (and probably other post-translational modifications as well). Second, extracellular signaling can have a temporal effect in which glycoproteins are modified transiently. It is immediately apparent that rapid evolutionary change in secreted products could be accomplished through changes in density of receptors. There are data that show that β-receptor density in salivary gland acinar cells is influenced by adrenal and thyroid hormones and that receptor density in turn influences protein synthesis (Pointon and Banerjee, 1979; Johnson et al., 1987).

Finally, it also is clear that positions of nerve terminals (Figure 23) or the presence, absence, and locations of gap junctions (Figure 24) used in communication among acinar cells (Nagato and Tandler, 1988) could provide another means for rapid evolutionary divergence (Phillips and Tandler, 1987). Insofar as innervation is concerned, interspecific comparisons among bats have documented species differences (Tandler et al., 1989, 1990a) as well as differences among acinar cells within a particular gland (Gilley et al., submitted). In the latter instance, it is reasonably clear that in the submandibular gland in phyllostomid fruit bats the serous acinar cells are very highly innervated, whereas the seromucous demilunar cells virtually lack direct contact with nerves (Figure 23). The impact of this difference in innervation is unknown but it is reasonable to think that such a dramatic difference between two groups of secretory cells could affect the secretory process and, perhaps, the products as well.

Regulation of Gene Expression

If we are correct in thinking that evolution has favored mechanisms that promote diversity in salivary glands, then we might expect that (1) unexpected products might appear in the saliva of some species and (2) the same gene(s) might be expressed in nonhomologous cells in different species. Salivary gland acinar cell products can be diversified through variability, or changes, in the regulation of gene expression. One aspect of this has been termed gene recruit-

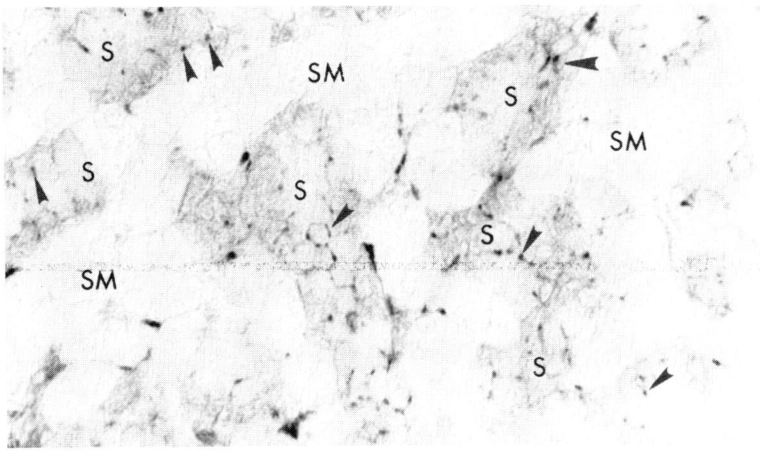

Figure 23. Light micrograph (LM) of fine nerve fibers and nerve terminals in the submandibular salivary gland of *Artibeus jamaicensis*. Note the concentration of nerves and terminals (arrowheads) associated with serous acinar cells (S) and relative paucity of terminals associated with seromucous demilunar cells (SM) PAP preparation with antineuron-specific enolase. × 385.

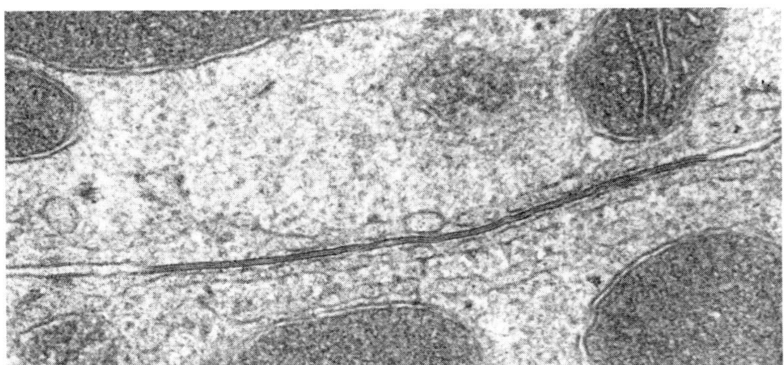

Figure 24. TEM example of a gap junction between two seromucous cells in the submandibular gland of *Tonatia sylvicola*. × 48,800.

ment. As a result of gene recruitment, a product usually produced elsewhere in the body could be added to saliva by the selective expression of a gene, or genes, that ordinarily would not be expressed in salivary gland cells. In some cases, gene recruitment simply involves changes in the regulation of expression of certain genes, whereas in others duplicated copies of genes might be recruited. The latter mechanism can allow for product divergence. While the ancestral version continues to encode some necessary polypeptide, the duplicate can provide new versions of the product. One of the better known examples involves the duplication and recruitment of the lysozyme c gene for expression in the stomach of ruminant mammals (Dobson et al., 1984; Irwin and Wilson, 1989, 1990).

We hypothesized that the lysozyme c gene might have been recruited for expression in the salivary glands of some bats. We selected lysozyme c because of its antibacterial properties and previous investigations had revealed that it is a component of the serous parotid acinar secretory product in human beings (Machino et al., 1986). Immunohistochemical localization of lysozyme c within secretory granules can provide good indirect evidence of gene expression in a particular cell type. However, the presence of immunoreactivity within the granules does not say anything about the details of expression.

In contrast to human beings, the lysozyme c gene does not appear to be expressed in parotid glands in insectivorous molossid bats. Furthermore, secretory granules in the parotid acinar cells in many phyllostomid bats also fail to show any immunoreactivity to antilysozyme c (Figure 25). Immunohistochemistry suggests that lysozyme c is, however, an important component of the secretory product of parotid intercalated duct cells in phyllostomid bats (Figure 25). So, although both human and phyllostomid bat saliva contain lysozyme c, in at least some species it is being produced in completely different cell types. The gene is expressed in the parotid acinar cell in human beings and in the parotid intercalated duct cells in many phyllostomids.

The production of lysozyme c in the parotid intercalated duct cells in phyllostomid bats also is interesting for another reason. In an earlier histological overview of the parotid glands in this family of bats it was pointed out that the highly derived pollen and nectar feeding species (the subfamily Glossophaginae) were characterized by exceedingly long parotid intercalated ducts (Phillips et al., 1977). This was particularly true for species of the genus *Leptonycteris*, which feeds on nectar of the agave plant (Figure 1). Nectar is a watery mixture rich in sugars, an ideal substrate for bacterial growth in the oral cavity. One possible indirect effect of this unusual diet can be seen in the teeth of glossophagines, in which enamel and dentin frequently are eroded (Phillips, 1971). It thus appears reasonable to think that the lysozyme c gene has been recruited for the purpose of adding a potent antibacterial component of the saliva. Moreover, from an evolutionary perspective, it is noteworthy that the potential amount of lysozyme c in the saliva could have been increased by greatly increasing the number of intercalated duct cells in comparison to the numbers seen in the parotid glands in insectivorous and carnivorous species.

Gene recruitment clearly is one of the means by which salivary gland secretory products have been diversified. However, it also should be noted that in the case of phyllostomid bats, this process has involved the recruitment of a nonacinar cell into the secretory process. Parotid salivary gland intercalated duct cells are not necessarily secretory, but in some phyllostomid bats the intercalated ducts effectively constitute an elongation of the secretory endpiece.

The expression of the plasminogen activator gene in both the principal and accessory submandibular salivary glands of the common vampire bat, *Desmodus rotundus*, is another interesting example of gene recruitment. The evolutionary origin of the vampire bats has been the subject of much debate among biologists;

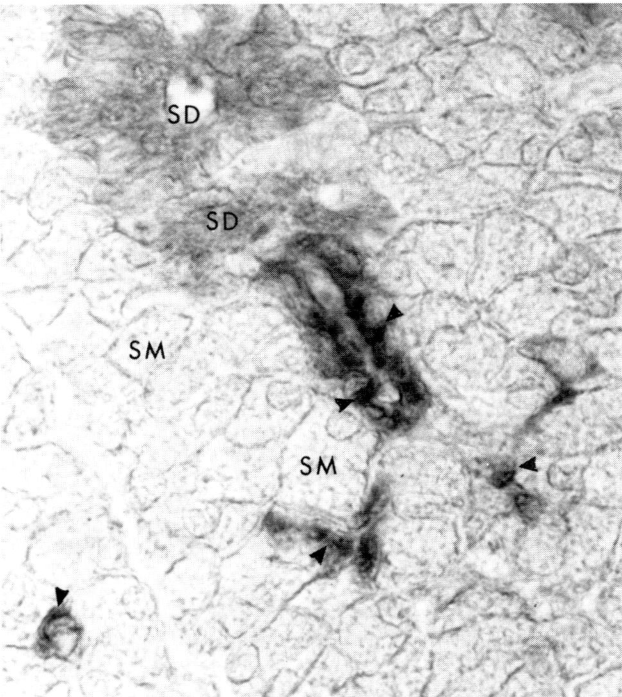

Figure 25. LM immunohistochemistry showing lysozyme c-like immunoreactivity (arrowheads) in intercalated ducts in the parotid salivary gland in *Artibeus jamaicensis*. Note the absence of immunoreactivity in the seromucous parotid acinar cells (SM) and striated ducts (SD). PAP preparation with antiserum against human lysozyme c. × 825.

at present it is accepted that these unusual animals constitute a specialized subfamily of phyllostomids (Forman et al., 1968; Baker et al., 1988). Apparently, the vampire bats diverged early in the history of the family (Figure 1).

Vampire bats are adapted for feeding on blood of other mammals or birds. The upper incisors are greatly enlarged and are kept razor sharp through a form of tooth-to-tooth sharpening known as thegosis (Phillips and Steinberg, 1976). Certain regions of skin that surround the nose leaf have been associated with infrared reception (Kürten and Schmidt, 1982; Schäfer et al., 1988). The stomach has been modified into a tube-like structure but otherwise is somewhat similar to that in other species in the family (Forman et al., 1979; Yamada et al., 1984). The salivary glands of the common vampire bat have been investigated histologically, histochemically, and ultrastructurally (DiSanto, 1960; Tandler et al., 1989, 1990a,b). Not surprisingly, the saliva contains factors that inhibit platelet aggregation and coagulation and activate plasminogen (Hawkey, 1988).

The plasminogen activator (PA) in vampire bat saliva recently was isolated and characterized by Gardell et al. (1989). Three slightly different cDNAs were recovered from a pooled sample of principal and accessory submandibular

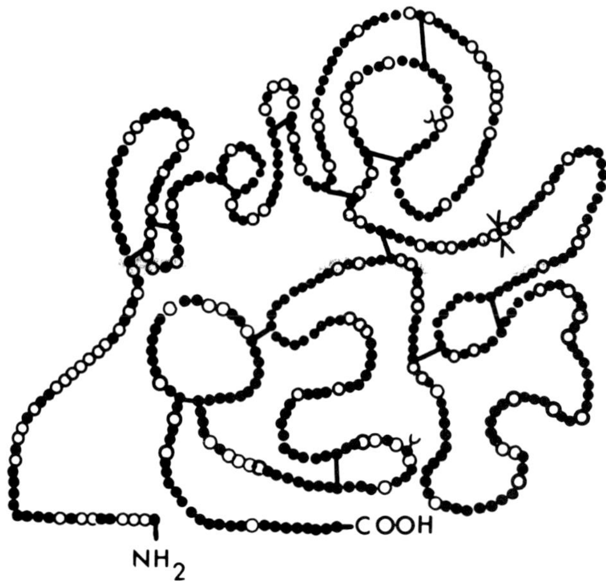

Figure 26. Diagramatic representation of the vampire bat salivary plasminogen activator (PA) and its signal peptide and prosequence. Solid circles represent amino acids conserved in comparison to human tissue PA; open circles represent amino acid substitutions. Predicted disulfide bridges between cysteines are shown as solid lines; possible N-linked glycosylation sites are shown as brackets on stems; and the plasmin-sensitive processing site is marked with Vs. Adapted from Gardell et al., *J. Biol. Chem.*, 264, 17947, 1989.

salivary glands from *Desmodus rotundus*. One of the bat PA proteins (Figure 26) was found to be extremely active in the presence of fibrin I. In fact, in the presence of fibrin I, vampire bat PA activity is stimulated 45,000-fold. By comparison, human tissue PA is stimulated only 250-fold in the presence of fibrin I (Gardell et al., 1989). Judging from the cDNA sequences, the gene encoding the vampire bat salivary PAs differs from the human tissue PA gene in two important ways; bat PA lacks a kringle 2 domain and a plasmin-sensitive processing site (Gardell et al., 1989). The gene for vampire bat PA thus might have originated by duplication of a tissue PA gene followed by divergence. If so, this would then explain the apparent enhanced or tailored product produced in *Desmodus*. It appears that gene duplication might be a fairly common source of "new" genes.

Diversification of the Secretory Process

In at least several instances, acinar cell secretory products have been diversified through diversification of the secretory process itself. In the first example discussed below, large amounts of phospholipid are added to the secretory granules through a modification of the packaging process at the Golgi complex (Phillips and Tandler, 1987). In the second example, an alternative synthesis process involving SER appears to provide an unconventional steroid product for export from an acinar cell (Nagato et al., 1984).

Bats of the genus *Miniopterus* are members of the Vespertilionidae, which is a geographically widespread family of insectivorous microchiropteran species. We examined the salivary glands in species representing 12 genera within this family and found that some species of *Miniopterus* exhibit a unique feature in their accessory submandibular glands. In *Miniopterus* this gland consists of mucous acini with serous demilunes. In the mucous cells, the apical cytoplasm typically is filled with granules of variable size (up to 2 μm). These granules are highly unusual in that they are bound by multiple layers of membrane (Figure 27). Examination of many such granules in a large sample of the mucous cells revealed that extra layers of membrane are added onto the forming granules in the vicinity of the Golgi complex (Tandler and Phillips, 1985; Phillips and Tandler, 1987). When the mature secretory granules reach the inner face of the apical cell membrane, only the outermost granule membrane fuses with it. Consequently, the secretory product that reaches the lumen includes a very large amount of membrane. This unusual packaging process thus results in a secretory product that is very rich in phospholipids.

At first glance it may seem that the unique modification of the secretory process seen in the accessory submandibular of *Miniopterus* represents some types of spurious abnormality in the secretory process. However, this clearly is not the case because we have found similar versions of this system in two of three species of *Miniopterus* collected from East Africa and Thailand. Thus, this modification in the secretory process most likely traces back to a time before the divergence of two widely distributed species in which it occurs. Data such as these can shed light on evolutionary relationships among congeneric species as well as the evolution of modified secretory processes. Finally, what is the possible significance of this particular type of secretory product diversification? The role of lipids in saliva has not been determined with certainty, but some data suggest that they are involved in the interactions between salivary proteins and bacteria, that they facilitate penetration of the mucosa by lipophilic substances (Murty et al., 1984), or that they provide a protective coating for the mucosa and teeth.

The second example of diversification in the secretory process involves a phyllostomid bat of the genus *Tonatia*. The seromucous acinar cells in the principal submandibular salivary gland in adult males of *Tonatia sylvicola* contain an unusual form of crystalloid SER (Nagato et al., 1984). This SER consists of a hexagonal array of membranous tubules that are about 75 nm in diameter (Figure 28). Linear filamentous structures that are 11 to 12 nm in thickness form a highly regular lattice around the array of tubules (Figure 29) (Phillips and Tandler, 1987). The chemical nature of the product of this SER can be inferred from somewhat similar structures in certain steroidogenic cells of the adrenal cortex and ovary. The fact that this type of SER is not found in females

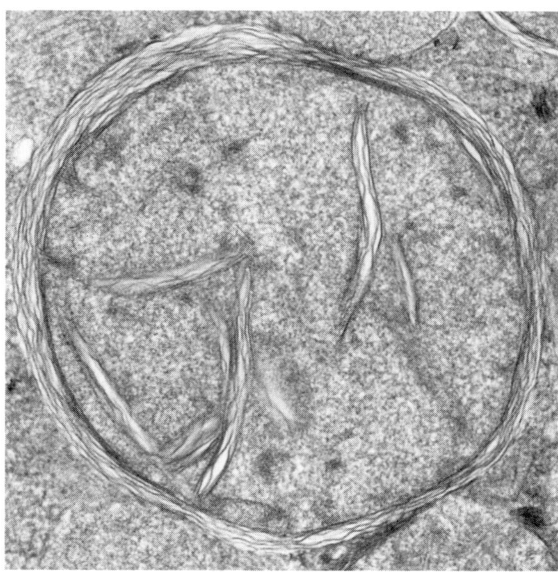

Figure 27. TEM of mucous secretory granule in the accessory submandibular gland in *Miniopterus schreibersi*. Note the multiple layers of membrane that encase the granule. × 48,600.

of *T. sylvicola* or in males of the related species, *T. bidens*, strongly suggests that the steroid product relates in some way to sexual selection, territoriality, or species recognition in *T. sylvicola* (Nagato et al., 1984). The possibility that in some species saliva contains molecules that influence behavior also has been reported in another phyllostomid, the short-tailed fruit bat, *Carollia perspicillata*. In this species, the adult males lick the roof of their cave roost apparently as a means of marking their territory (Fleming, 1988). Similar "scent marking" behavior with saliva has been described in marsupials (Fadem, 1986). The salivary glands positioned at the angle of the mouth in one species of vampire bat, *Diaemus youngi*, also seem to be involved in behavior. These glands apparently are seasonally enlarged and are reported to emit a strong-smelling secretion (Greenhall, 1988). Furthermore, in certain rodents and a strain of domestic pig, it has been shown that saliva can influence sexual behavior (Gray et al., 1984; Friedle and Fischer, 1984; Perry et al., 1980; Booth, 1984).

The foregoing examples illustrate that product diversification can occur by means of (1) alteration in the "standard" secretory process and (2) addition of an alternative secretory system. An alteration of the secretory process, such as that seen in *Miniopterus*, might have originated as an "abnormality" that conferred some direct benefit to the species in which it originated.

Gland Duplication

As we mentioned previously, microchiropteran bats frequently have two pairs of submandibular salivary glands, the principal and the accessory. Because

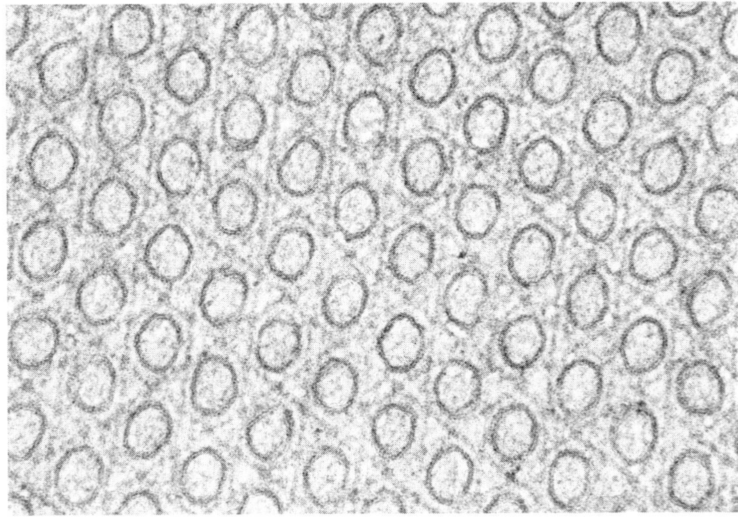

Figure 28. TEM of an unusual array of tubular smooth endoplasmic reticulum (SER) in a seromucous cell in the principal submandibular gland in *Tonatia sylvicola.* × 92,500.

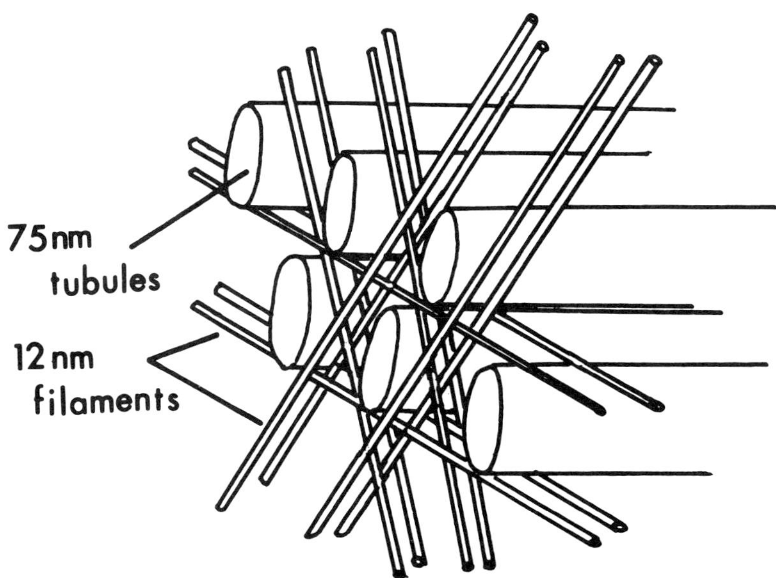

Figure 29. Diagrammatic representation of the unusual SER seen in the principal submandibular gland in *Tonatia sylvicola.*

this arrangement is found in all nine of the families that we have examined, it is logical to think that the submandibular gland must have been duplicated early in the history of microchiropteran bats. Supernumerary salivary glands of various types might be fairly common among mammals. In human beings, a small accessory parotid gland is found in about 21% of Americans and 69% of Japanese (Toh et al., 1984). In some instances this extra gland is histologically identical to the standard parotid, whereas in other cases the acinar cells are mucous rather than serous (Toh et al., 1984). Judging from our survey of microchiropteran bats, the accessory submandibular salivary glands offered, on another level, an evolutionary opportunity remarkably similar to that offered by gene duplication. In many species of bats the accessory submandibular salivary gland differs considerably from the principal gland. One notable example involves several genera of carnivorous, frog-eating bats. Within the Neotropical family Phyllostomidae, the fringe-lipped bat, *Trachops cirrhosus*, is unusual in that it commonly feeds on frogs (Figure 1). Feeding on frogs poses some special problems for this species because the integument in many anurans contains toxic alkaloids. Several years ago it was shown that *Trachops* uses acoustic means to help it discriminate among the mating cells of different species of frogs (Tuttle and Ryan, 1981). Recently, unique facial papillae in *Trachops* have been shown to house a sensory system that includes abundant free nerve endings, that possibly serve as chemo- or thermoreceptors (Weiss et al., submitted). In *Trachops*, the accessory submandibular has large follicles (Figure 30) and thus is thyroid like in appearance (Phillips et al., 1987). These follicles are formed by the "terminal" acinar cells, which differ considerably from any traditional form of salivary gland acinar cell. Unlike typical acinar secretory cells, most of the cells in *Trachops* lack any indication of secretory granules. Instead, the cytoplasm near the apical surface contains a small number of electron lucent, amorphous to ribbon-like bodies that possibly are discharged into the lumen by exocytosis (Tandler et al., 1990a). Does this unique type of salivary gland have some relationship to the diet of *Trachops*? To test this idea, species of the genera *Megaderma* and *Cardioderma* were collected in Thailand and Kenya, respectively. These bats were selected for comparison to *Trachops* because species in both genera are known to include anurans in their diets. Furthermore, *Megaderma* and *Cardioderma* are members of the family Megadermatidae and, thus, their dietary habits have evolved independently from that of the Neotropical *Trachops cirrhosus* (Phillips et al., 1987; Phillips and Tandler, 1987). In both *Megaderma* and *Cardioderma*, the accessory submandibular salivary glands are histologically similar to the thyroid-like gland in *Trachops*. Apparently, the accessory submandibular salivary gland has been adapted to meet the needs of a particular diet in all of these bats. Moreover, the histology of the gland is a remarkable example of evolutionary convergence.

In *Erophylla sezekorni*, a pollen- and nectar-feeding bat that occurs only in the Greater Antilles (Figure 1), the accessory submandibular glands exhibit

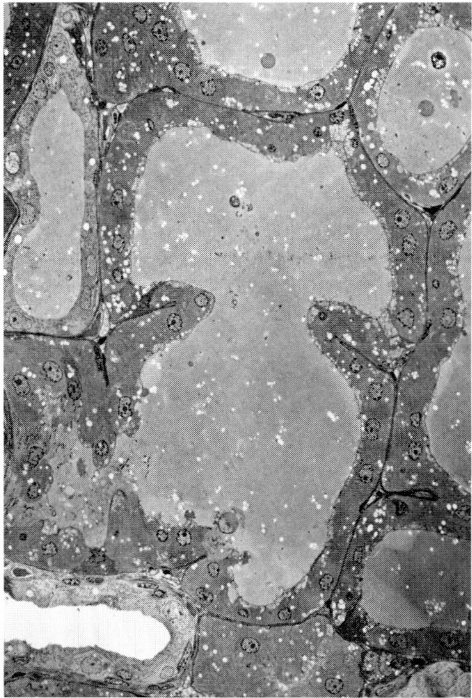

Figure 30. Very low magnification electron micrograph of the follicle-like endpieces in the accessory submandibular gland of *Trachops cirrhosus*. Ducts are present in both the upper and lower left-hand corners. × 745.

extreme diversification that possibly relates in some way to sexual behavior. The large, green-colored accessory submandibular gland in adult male specimens is so extremely modified that it hardly can be described as a "salivary" gland (Figures 31 and 32). Females of this species only have a minuscule version of this gland. The "green" gland of males consists of elongated tubules rather than acini. Ultrastructurally, the cells that compose these tubules (Figure 32) bear little resemblance to conventional acinar cells (Tandler et al., 1990a). Unusual accumulations of peroxisomes in a spiral or circular arrangement characterizes the cytoplasm of these cells. Similar arrangements have been described in the uropygial glands of ducks, where the unusual peroxisome arrangement has been linked to lipid metabolism (Zaar and Gorgas, 1985). The function of the green gland in *Erophylla sezekorni* is unknown, but its obvious sexual dimorphism suggests that it is influenced by androgenic hormones. It seems likely that this gland plays some role in reproductive behavior rather than in diet. As we discussed previously, the involvement of salivary glands in some aspect of behavior does not appear to be rare, either in bats or other mammals.

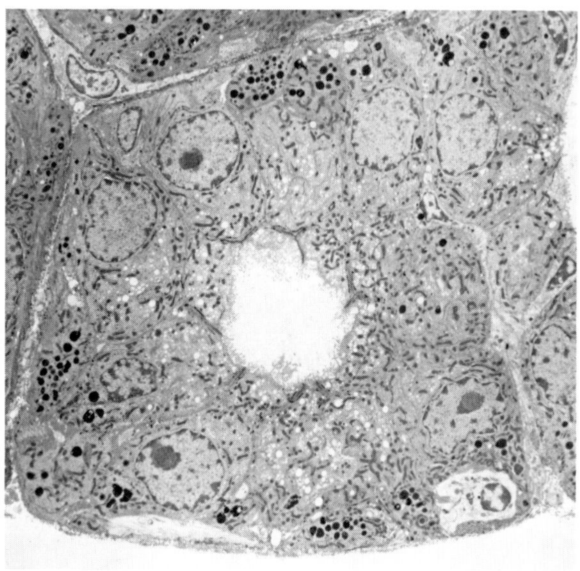

Figure 31. Survey electron micrograph of a secretory endpiece in the unusual "green" accessory submandibular gland in *Erophylla sezekorni.* × 2500.

CONCLUSIONS

Salivary gland acinar cells, and, indeed, salivary glands in general, are highly diversified. This level of diversity and the patterns that can be detected within groups of mammals, make salivary glands an outstanding model for investigating molecular evolution and interpreting it at the cellular level. The extent of the diversification in salivary gland cell structure and secretory products cannot be explained just in terms of relaxed constraints on secretory proteins or by accumulations of "point mutations." Evocation of a standard (and restrictive) set of neo-Darwinian ideas about evolution fall short of providing an adequate explanation. Instead, in mammalian salivary glands, natural selection appears to have favored mechanisms that promote diversity, regardless of their possible underlying genetic, or epigenetic, bases. Mechanisms that promote diversity clearly could have been advantageous in the sense that they laid the foundation for rapid evolutionary change. What would be the value of rapid diversification in salivary glands and their products? One answer is that rapid diversification helped make it possible for mammals to successfully exploit an ever-expanding variety of nutrient resources. Another answer is that diversification of products made it possible for mammals to match their saliva against the oral microflora, which adapt rapidly and are strongly influenced by diet. Finally, there is the intriguing role that saliva may play in social behavior. Scent marking, sexual behavior, and species recognition are characteristics that can be species or individual specific. Therefore, any of these could require rapid evolution of salivary gland products. Because of the multiplicity of mechanisms that promote

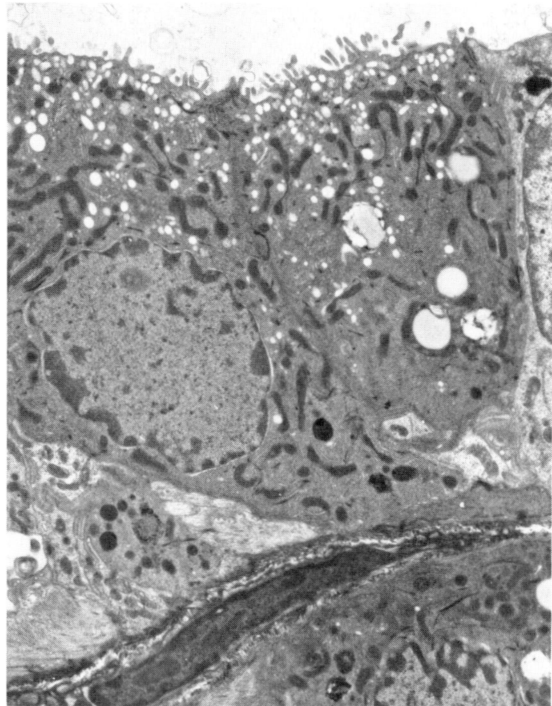

Figure 32. Highly modified acinar secretory cells in the accessory submandibular gland in *Erophylla sezekorni*. × 6800.

diversity, the salivary glands might be viewed as a potential "test bed." Salivary glands constitute an ongoing natural experiment involving variant molecules, variations in gene expression, variations in interactions with other organs and the nervous system, and the testing of modified versions of cell structure and the secretory process. In summary, evolutionary diversity is characteristic of salivary glands. There have been modifications to virtually every aspect of salivary gland structure and function. It should be noted in closing that salivary glands might prove to be among the most valuable assets to the study of evolutionary biology. Potentially, data derived from comparative investigations of salivary glands might shed light on several important questions, including the issue of whether or not there are mechanisms that allow for punctuated evolution. Studies of salivary glands also could contribute new ideas about how integrated organ systems have evolved.

ACKNOWLEDGMENTS

The research program summarized in this chapter has been supported by a great variety of sources and people. In particular we are pleased to mention NIH grant DE 07648 (Tandler and Phillips), Hofstra University HCLAS grants

(Phillips), and the Research Corporation (Phillips). A number of people have been extremely helpful to us: Hugh H. Genoways, University of Nebraska, and Duane Schlitter, Carnegie Museum of Natural History, have played prominent roles in our field work in Suriname, the Antilles, Thailand, India, and Kenya; Carlin A. Pinkstaff, West Virginia University Medical School, Arthur R. Hand, University of Connecticut Dental School, and Kuniaki Toyoshima, Kyushu Dental College, have collaborated in aspects of our research. Finally, we also wish to recognize the valuable editorial assistance provided by Allison Weiss.

REFERENCES

1. Aloe, L., Alleva, E., Böhm, A., and Levi-Montalcini, R., Aggressive behavior induces release of nerve growth factor from mouse salivary gland into the bloodstream, *Proc. Natl. Acad. Sci. U.S.A.*, 83, 6184, 1986.
2. Arvan, P. and Castle, J. D., Isolated secretion granules from parotid glands of chronically stimulated rats possess an alkaline internal pH and inward-directed H^+ pump activity, *J. Cell Biol.*, 103, 1257, 1986.
3. Attardi, G., Chomyn, A., and Loguercio Polosa, P., Evidence for translational control of mitochondrial gene expression in rat muscle and brain synaptosome mitochondria, in *Advances in Myochemistry*, G. Benzi, Ed., John Libbey Eurotext, Ltd., London, 1989, 55.
4. Baker, R. J., Honeycutt, R. L., and Bass, R. A., Genetics, in *Natural History of Vampire Bats*, A. M. Greenhall and U. Schmidt., Eds., CRC Press, Boca Raton, FL, 1988, 31.
5. Baker, R. J., Hood, C. S., and Honeycutt, R. L., Phylogenetic relationships and classification of the higher categories of the new world bat family Phyllostomidae, *Syst. Zool.*, 38, 228, 1989.
6. Beato, M., Gene regulation by steroid hormones, *Cell*, 56, 335, 1989.
7. Beintema, J. J., Schüller, C., Irie, M., and Carsana, A., Molecular evolution of the ribonuclease superfamily, *Prog. Biophys. Mol. Biol.*, 51, 165, 1988.
8. Bennick, A., Structural and genetic aspects of proline-rich proteins, *J. Dent. Res.*, 66, 457, 1987.
9. Bhatnagar, K. P., Anatomy, in *Natural History of Vampire Bats*, A. M. Greenhall and U. Schmidt., Eds., CRC Press, Boca Raton, FL, 1988, 41.
10. Booth, W. D., Sexual dimorphism involving steroidal pheromones and their binding in the submaxillary salivary gland of the Gottingen miniature pig, *J. Endocrinol.*, 100, 195, 1984.
11. Dickinson, D. P., Mirels, L., Tabak, L. A., and Gross, K. W., Rapid evolution of variants in a rodent multigene family encoding salivary proteins, *Mol. Biol. Evol.*, 6, 80, 1989.
12. DiSanto, P. E., Anatomy and histochemistry of the salivary glands of the vampire bat, *Desmodus rotundus*, *J. Morphol.*, 106, 301, 1960.
13. Dobson, D. E., Prager, E. M., and Wilson, A. C., Stomach lysozymes of ruminants. I. Distribution and catalytic properties, *J. Biol. Chem.*, 259, 11607, 1984.

14. Eggert, F. M., Maenz, L., and Tam, Y.-C., Measuring the interaction of human secretory glycoproteins with oral bacteria, *J. Dent. Res.*, 66, 610, 1987.

15. Eisenberg, J. F. and Wilson, D. E., Relative brain size and feeding strategies in the *Chiroptera, Evolution*, 32, 740, 1978.

16. Fadem, B. H., Chemical communication in gray short-tailed opossums (*Monodelphis domestica*) with comparisons to other marsupials and with reference to monotremes, in *Chemical Signals in Vertebrates, Vol. IV, Ecology, Evolution, and Comparative Biology.*, E. Duvall, D. Muller-Schwarze, and R. M. Silverstein, Eds., Plenum Press, New York, 1986, 587.

17. Farber, G. K. and Petsko, G. A., The evolution of α/β barrel, *TIBS*, 15, 228, 1990.

18. Findley, J. S. and Wilson, D. E., Ecological significance of chiropteran morphology, in *Ecology of Bats*, T. H. Kunz., Ed., Plenum Press, New York, 1982, 243.

19. Fleming, T. H., *The Short-Tailed Fruit Bat—A Study in Plant-Animal Interaction*, University of Chicago Press, Chicago, 1988.

20. Forman, G. L., Baker, R. J., and Gerber, J. D., Comments on the systematic status of vampire bats (family Desmodontidae), *Syst. Zool.*, 17, 417, 1968.

21. Forman, G. L. and Phillips, C. J., *Ecological and Behavioral Methods for the Study of Bats*, T. H. Kunz, Ed., Smithsonian Institute Press, Washington, D.C., 1988, 405.

22. Forman, G. L., Phillips, C. J., and Rouk, C. S., Alimentary Tract, in *Biology of Bats of the New World Family Phyllostomatidae, Part III*, R. J. Baker, J. K. Jones, Jr., and D. C. Carter, Eds., Texas Tech University Press, Lubbock, 1979, 205.

23. Friedle, R. E. and Fischer, R. B., Discrimination of salivary olfactants by male *Mesocricetus auratus*, *Psychol. Rep.*, 55, 67, 1984.

24. Gardell, S. J., Duong, L. T., Diehl, R. E., York, J. D., Hare, T. R., Register, R. B., Jacobs, J. W., Dixson, R. F., and Friedman, P. A., Isolation, characterization and cDNA cloning of a vampire bat salivary plasminogen activator, *J. Biol. Chem.*, 264, 17947, 1989.

25. Gershon, N. D., Porter, K. R., and Trus, B. L., The cytoplasmic matrix: its volume and surface area and the diffusion of molecules through it, *Proc. Natl. Acad. Sci. U.S.A.*, 82, 5030, 1985.

26. Gilley, R. S., Phillips, C. J., and Tandler, B., Interspecific differences in the innervation of the submandibular salivary gland in three species of bats, *Cell Tissue Res.*, submitted.

27. Gray, B., Fischer, R. B., and Meunier, G. F., Preferences for salivary odor cues by female hamsters, *Horm. Behav.*, 18, 451, 1984.

28. Greenhall, A. M., Feeding behavior, in *Natural History of Vampire Bats*, A. M. Greenhall and U. Schmidt, Eds., CRC Press, Boca Raton, FL, 1988, 451.

29. Greenhall, A. M. and Schmidt, U., *Natural History of Vampire Bats*, CRC Press, Boca Raton, FL, 1988.

30. Grimes, G. W. and Aufderheide, K. J., *Cellular Aspects of Pattern Formation: The Problem of Assembly. Monographs in Development Biology, Vol. 22*, Karger Press, New York, 1991, 94.

31. Hawkey, C., Salivary antihemostatic factors, in *Natural History of Vampire Bats*, A. M. Greenhall and U. Schmidt, Eds., CRC Press, Boca Raton, FL, 1988, 133.

32. Hill, J. E. and Smith, J. D., *Bats, A Natural History*, University of Texas Press, Austin, 1986.

33. Hogg, D., Tsiu, L.-C., Gorin, M., and Greitman, M. L., Characterization of the human β-crystalline gene HubA3/A1 reveals ancestral relationships among the β-crystallin superfamily., *J. Biol. Chem.*, 261, 12420, 1986.

34. Honeycutt, R. L. and Sarich, V., Albumin evolution and subfamilial relationships among New World leaf-nosed bats (family Phyllostomidae), *J. Mammal.*, 68, 508, 1987.

35. Humphreys-Beher, M. G., Strain-specific differences in the proline-rich proteins and glycoproteins induced in rat salivary glands by chronic isoprenaline treatment, *Biochem. J.*, 230, 469, 1985.

36. Irwin, D. M. and Wilson, A. C., Multiple cDNA sequences and the evolution of bovine stomach lysozyme, *J. Biol. Chem.*, 264, 11387, 1989.

37. Irwin, D. M. and Wilson, A. C., Concerted evolution of ruminant stomach lysozymes, *J. Biol. Chem.*, 265, 4944, 1990.

38. Johnson, D. A., Alvares, D. F., Etzel, K. R., and Kalu, D. N., Regulation of salivary proteins, *J. Dent. Res.*, 66, 576, 1987.

39. Junqueira, L. C. U., Toledo, A. M. S., and Doine, A. I., Digestive enzymes in the parotid and submandibular glands of mammals, *An. Acad. Bras. Ciênc.*, 45, 629, 1973.

40. Kalt, M. R. and Tandler, B., A study of fixation of early amphibian embryos for electron microscopy, *J. Ultrastruct. Res.*, 36, 633, 1971.

41. Kelly, R. B., Pathways of protein secretion in eukaryotes, *Science*, 230, 25, 1985.

42. Kimura, M., Evolutionary rate at the molecular level, *Nature*, 217, 624, 1968.

43. Koop, B. F. and Baker, R. J., Electophoretic studies of relationships of six species of *Artibeus* (Chiroptera:Phyllostomidae), *Occas. Papers Mus.*, 83, 1, 1983.

44. Kousvelari, E. E., Fox, P. C., and Baum, B. J., Regulatory aspects of N-linked glycoprotein, *J. Dent. Res.*, 66, 552, 1987.

45. Kousvelari, E. E., Oppenheim, F. G., and Cutler, L. S., Ultrastructural localization of salivary acidic proline-rich proteins from *Macaca fasicularis*, *J. Histochem. Cytochem.*, 30, 274, 1982.

46. Krüten, L. and Schmidt, U., Thermoperception in the common vampire bat (*Desmodus rotundus*), *J. Comp. Physiol.*, 146, 223, 1982.

47. Levine, M. J., Reddy, M. S., Tabak, L. A., Loomis, R. E., Bergey, E. J., Jones, P. C., Cohen, R. E., Stinson, M. W., and Al-Hashmi, I., Structural aspects of salivary glycoproteins, *J. Dent. Res.*, 66, 436, 1989.

48. Lunger, K., Hommel, U., Herold, M., Hufsteenge J., and Kirschner, K., Correct folding of circularly permuted variants of a βα barrel enzyme, *Science*, 243, 206, 1989.

49. Machino, M., Morioka, H., and Tachibana, M., Amylase and lysozyme of the human salivary gland, *Acta Histochem. Cytochem.*, 19, 329, 1986.

50. Mamula, P. W., Morley, D. J., Larsen, S. H., and Karn, R. C., Expression of human salivary protein genes, *Biochem. Genet.*, 26, 165, 1986.

51. Mamula, P. W., Heerema, N., Palmer, C. G., Lyons, K. M., and Karn, R. C., Localization of the human salivary protein complex (SPC) to chromosome band 12p13.2, *Cytogenet. Cell Genet.*, 39, 279, 1985.

52. Mehansho, H., Hagerman, A., Clements, S., Butler, L., Rogler, J., and Carlson, D. M., Modulation of proline-rich protein biosynthesis in rat parotid glands by sorghums with high tannin levels, *Proc. Natl. Acad. Sci. U.S.A.*, 80, 3948, 1983.

53. Mennone, A., Phillips, C. J., and Pumo, D. E., Evolutionary significance of interspecific differences in gastrin-like immunoreactivity in the pylorus of phyllostomid bats, *J. Mammal.*, 67, 373, 1986.

54. Murty, V. L. N., Slominiany, B. L., Zdebska, E., Slominiany, A., Mandel, I. D., and Levy, B. M., Lipid composition of marmoset saliva, *Comp. Biochem. Physiol.*, 79A, 41, 1984.

55. Nagato, T. and Tandler, B., Gap junctions in rat sublingual gland, *Anat. Rec.*, 214, 71, 1986.

56. Nagato, T., Tandler, B., and Phillips, C. J., Unusual smooth endoplasmic reticulum in submandibular acinar cells of the male round-eared bat, *Tonatia sylvicola*, *J. Ultrastruct. Res.*, 87, 275, 1984.

57. Nakagaki, I., Goto, T., Saski, S., Imai, Y., Histochemical and cytochemical localization of (Na^+-K^+)-activated adenosine triphosphatase in the acini of dog submandibular glands, *J. Histochem. Cytochem.*, 26, 835, 1978.

58. Nei, M., *Molecular Evolutionary Genetics*, Columbia University Press, New York, 1987.

59. Olden, K., Bernard, B. A., Humphries, M. J., Yeo, T.-K., Yeo, K.-T., White, S. L., Newtown, S. A., Baur, H. C., and Parent, J. B., Functions of glycoprotein glycans, *TIBS*, 10, 78, 1985.

60. Oppenheim, F. G., Hay, D. I., Smith, G. D., and Troxler, R. F., Molecular basis of salivary proline-rich protein and peptide synthesis: cell-free translations and processing of human and macaque statherin mRNAs and partial amino acid sequence of their signal peptides, *J. Dent. Res.*, 66, 462, 1987.

61. Owen, R. D., Phenetic analysis of the bat subfamily stenodermatinae (Chiroptera: Phyllostomatidae), *J. Mammal.*, 69, 796, 1988.

62. Palade, G. E., Differentiated microdomains in cellular membranes: current status, in *The Cell in Contact-Adhesions and Junctions as Morphogenetic Determinants*, G. M. Edelman and J.-P. Thiery, Eds., John Wiley & Sons, New York, 1985, 9.

63. Pertuz, M. F., Species adaptation in a protein molecule, *Mol. Biol. Evol.*, 1, 28, 1983.

64. Perry, G. C., Patterson, R. L. S., MacFie, H. J. H., and Stinson, C. G., Pig courtship behavior: pheromonal property of androstene steroids in male submaxillary secretions, *Anim. Prod.*, 31, 191, 1980.

65. Pettigrew, J. D., Jamieson, B. G. M., Robson, S. K., Hall, L.S., McNally, K. I., and Cooper, H. M., Phylogenetic relations between microbats, megabats and primates (Mammalia: Chiroptera and Primates), *Phil. Trans. Royal Soc. London*, 325, 489, 1989.

66. Phillips, C. J., The dentition of glossophagine bats: development, morphological characteristics, variation, pathology, and evolution, *Univ. Kansas, Mus. Nat. Hist.*, 54, 1, 1971.

67. Phillips, C. J., Field fixation and storage of museum tissue collections suitable for electron microscopy, *Acta Zool. Fenn.*, 170, 90, 1985.

68. Phillips, C. J. and Steinberg, B., Histological and scanning electron microscopic studies of tooth structure and thegosis in the common vampire bat, *Desmodus rotundus*, *Occ. Papers Mus.*, 42, 1, 1976.

69. Phillips, C. J. and Tandler, B., Mammalian evolution at the cellular level, in *Current Mammalogy*, Genoways, H. H., Ed., Plenum Press, New York, 1966, 1.

70. Phillips, C. J., Grimes, G. W., and Forman, G. L., Oral Biology, in *Biology of Bats of the New World Family Phyllostomatidae, Part II*, R. J. Baker, J. K. Jones, Jr., and D. C. Carter, Eds., Texas Tech University Press, Lubbock, 1977, 121.

71. Phillips, C. J., Nagato, T., and Tandler, B., Comparative ultrastructure and evolutionary patterns of acinar secretory product of parotid salivary glands in Neotropical bats, in *Studies in Neotropical Mammalogy: Essays in Honor of Philip Hershkovitz*, B. D. Patterson and R. M. Timm., Eds., *Fieldiana: Zoology*, 39, 213, 1987.

72. Phillips, C. J., Studholme, K. M., and Forman, G. L., Comparative ultrastructure of gastric mucosae in four genera of bats (Mammalia: Chiroptera), with comments on gastric evolution, *Ann. Carnegie Mus.*, 53, 71, 1984.

73. Phillips, C. J., Tandler, B., and Pinkstaff, C. A., Unique salivary glands in two genera of tropical microchiropteran bats: an example of evolutionary convergence in histology and histochemistry, *J. Mammal.*, 68, 235, 1987.

74. Phillips, C. J., Pumo, D. E., Genoways, H. H., Ray, P. E., and Briskey, C. A., Mitochondrial DNA evolution and phylogeography in two fruit bats, *Artibeus jamaicensis* and *A. lituratus.*, in *Latin American Mammology: Topics in History, Biodiversity, and Conservation*, 1991, 97.

75. Phillips, C. J., Tandler, B., Pinkstaff, C. A., and Schlitter, D. A., Electron microscopic and histochemical variation in submandibular gland in three species of house bat, Scotophilus. *J. Mammal.*, submitted.

76. Phillips, M. J. and Satir, P., The cytoskeleton of the hepatocyte: organization, relationships, and pathology, in *The Liver: Biology and Pathobiology*, I. M. Arias, W. B. Jakoby, H. Popper, D. Schachter, and D. A. Shafritz, Eds., Raven Press, New York, 1988, 11.

77. Pinkstaff, C. A., The cytology of salivary glands, *Int. Rev. Cytol.*, 63, 141, 1980.

78. Pinkstaff, C. A., Tandler, B., and Cohan, R. P., Histology and histochemistry of the parotid and the principal and accessory submandibular glands of the little brown bat, *J. Morphol.*, 172, 2, 1982.

79. Pointon, S. E. and Banerjee, S. P., β-adrenergic and muscarinic cholinergic receptors in rat submaxillary glands. Effects of thyroidectomy, *Biochem. Acta*, 583, 129, 1979.

80. Pumo, D. E., Goldin, E. Z., Elliot, B., Phillips, C. J., and Genoways, H. H., Mitochondrial DNA polymorphism in three antillean island populations of the fruit bat, *Artibeus jamaicensis*, *Mol. Biol. Evol.*, 5, 79, 1988.

81. Ravazzola, M. and Orci, L., Glucagon and glicentin immunoreactivity are topologically segregated in the α granule of the human pancreatic A-cell, *Nature*, 284, 66, 1980.

82. Robin H. A., Recherches anatomiques les mammiferes de l'Ordre des chiroptres, *Ann. Sci. Nat. Zool.*, 12, 1, 1881.

83. Rothman, S. S., Iskander, N., Attwood, D., Vladimirsky, Y., McQuaid, K., Grendell, J., Kirz, J., Ade, H., McNulty, I., Kern, D., Chang, T. H. P., and Rarback, H., The interior of a whole and unmodified biological object the zymogen granule — viewed with a high resolution X-ray microscope, *Biochim. Biophys. Acta*, 991, 484, 1989.

84. Schäfer, K., Braun, H. A., and Krüten, L., Analysis of cold and warm receptor activity in vampire bats and mice, *Pflügers Arch.*, 412, 188, 1988.

85. Schulte, B. A., Genetic and sex-related differences in the structure of submandibular glycoconjugates, *J. Dent. Res.*, 66, 442, 1987.

86. Singh, G., Lott, M. T., and Wallace, D. C., A mitochondrial DNA mutation as a cause of Leber's hereditary optic neuropathy, *N. Engl. J. Med.*, 320, 1300, 1989.

87. Straney, D. O., Smith, M. H., Greenbaum, I. F., and Baker, R. J., Biochemical Genetics, in *Biology of Bats of the New World Family Phyllostomatidae. Part III*, R. J. Baker, J. K. Jones, Jr., and D. C. Carter, Eds., Texas Tech University Press, Lubbock, 1979, 157.

88. Studholme, K. M., Yazulla, S., and Phillips, C. J., Interspecific comparison of immunohistochemical localization of retinal neurotransmitters in four species of bats, *Brain Behav. Evol.*, 30, 160, 1987.

89. Tamaki, H., Yoshio, S., and Yamashina, S., Development of cellular polarity and tight junctions in parotid acinar cells of postnatal rat, *J. Electron Microsc.*, 38, 340, 1989.

90. Tandler, B. and MacCallum, D., Ultrastructure and histochemistry of the submandibular gland of the European hedgehog, *Erinaceous europaeus* L. Z. I. acinar cells, *J. Ultrastruct. Res.*, 39, 186, 1972.

91. Tandler, B. and Phillips, C. J., Unusual secretory granules in the accessory submandibular gland of long-winged bats, *Anat. Rec.*, 211, 192A, 1985.

92. Tandler, B. and Phillips, C. J., Giant mitochondria in the seromucous secretory cells of the accessory submandibular gland of the long-haired fruit bat, *Stenonycteris lanosus*, *Anat. Rec.*, in press.

93. Tandler, B. and Riva, A., Salivary glands, in *Human Oral Embryology and Histology*, I. A. Mjör and O. Fejerskov, Eds., Munksgaard International Publishing, Copenhagen, 1986, 243.

94. Tandler, B., Nagato, T., and Phillips, C. J., Systematic implications of comparative ultrastructure of secretory acini in the submandibular salivary gland in *Artibeus*, (Chiroptera:Phyllostomidae), *J. Mammal.*, 67, 81, 1986.

95. Tandler, B., Phillips, C. J., and Nagato, T., Parotid salivary gland ultrastructure in an omnivorous neotropical bat: evolutionary divergence at the cellular level, *Zool. Scripta*, 17, 419, 1988.

96. Tandler, B., Toyoshima, K., and Phillips, C. J., Ultrastructure of the principal and accessory submandibular glands of the common vampire bat, *Am. J. Anat.*, 1990b, 189, 303.

97. Tandler, B., Phillips, C. J., Nagato, T., and Toyoshima, K., Ultrastructural diversity in chiropteran salivary glands, in *Ultrastructure of the Extraparietal Glands of the Digestive Tract*, A. Riva and P. M. Motta, Eds., Kluwer Academic Publ., New York, 1990a, 31.

98. Tandler, B., Phillips, C. J., Toyoshima, K., and Nagato, T., Comparative studies of the striated ducts of mammalian salivary glands, in *Cells and Tissues: A Three Dimensional Approach by Modern Techniques in Microscopy*, P. M. Motta, Ed., Alan R. Liss, New York, 1989, 243.

99. Tartakoff, A. M. and Vassalli, P., Comparative studies of intracellular transport of secretory proteins, *J. Bell Biol.*, 79, 694, 1978.

100. Toh, H., Hamada, N., and Ohmori, T., Accessory parotid gland in man, *J. Fukuoka Dent. Coll.*, 11, 38, 1984.

101. Tuttle, M. D. and Ryan, M. J., Bat predation and evolution of frog vocalization in the Neotropics, *Science*, 214, 677, 1981.

102. Young, J. A. and van Lennep, E. W., *The Morphology of Salivary Glands*, Academic Press, New York, 1978.

103. Weiss, A., Gilley, R. S., Phillips, C. J., and Tandler, B., Sensory innervation in the facial projections of three phyllostomid bats, *Phyllostomus hastatus*, *Trachops cirrhosus*, and *Artibeus jamaicensis*, *Cell Tissue Res.*, submitted.
104. Williams, R. S., Mitochondrial gene expression in mammalian striated muscle — evidence that variation in gene dosage is the major regulating event, *J. Biol. Chem.*, 261, 12390, 1986.
105. Yamada, J., Campos, V. J. M., Kitamura, N., Pacheco, A. C., Yamashita, T., and Caramuschi, U., Immunocytochemical study of gastro-entero-pancreatic (GEP) endocrine cells in the vam pire bat (*Desmodus rotundus*), *Gegenbaurs Morphol. Jahrb*, 130, 845, 1984.
106. Zaar, K. and Gorgas, K., Peroxisome-endoplasmic reticulum aggregates in the duck uropygial gland, *Eur. J. Cell Biol.*, 38, 322, 1985.

Cell Biology of Salivary Protein Secretion

David Castle

INTRODUCTION

Proteins secreted into saliva have important functions with respect to maintaining the oral cavity and initiating food digestion. Most of the proteins are produced and discharged by acinar cells of the salivary gland. Three sets of major salivary glands and a host of minor salivary glands typically contribute the protein components of saliva. In this chapter much of the discussion will center on the intracellular pathways of salivary protein secretion that have been characterized in the parotid gland and will be supplemented by reference to related features in other exocrine glands in instances where more thorough study has been undertaken. In certain species, e.g., rat and rabbit, the parotid appears to contain a homogeneous population of acinar cells that forms most of the tissue mass. Consequently, analysis of secretion at both the cell and tissue levels in these systems describes, for the most part, the function of a single cell type. Because acinar cells in different glands are organized similarly, it seems reasonable to assume that the pathways and mechanisms of protein discharge that have been characterized in the parotid represent general characteristics of salivary acinar cells.

ORGANIZATION OF THE INTRACELLULAR TRANSPORT PATHWAY IN PAROTID ACINAR CELLS

Parotid acinar cells are highly differentiated and polarized cells that are committed to the production of proteins for exocrine secretion. An extensive rough endoplasmic reticulum surrounds a basally located nucleus beneath a central Golgi complex and a large population of 1-μm diameter storage granules. These organelles form the principal elements of the intracellular transport pathway which is used to process and export the secretory proteins. The granules occupy as much as one third of the cytoplasmic volume and are accumulated near the apical plasma membrane that extends as a belt-like network over 12 to 13% of the cell surface (Bloom et al., 1979). The granules undergo exocytosis by fusion of their limiting membranes with the apical plasma membrane primarily in response to stimuli that are received at nerve terminals contacting the basal surface of each acinar cell (Hand, 1971).

More than 90% of the protein synthetic activity of salivary acinar cells is devoted to the production of secretory proteins. Thus, pulse-chase labeling with radioactive amino acids followed by radioautography of the tissue will trace the passage of newly synthesized secretory proteins during intracellular transport against only a low background labeling of non-exported proteins (Castle et al., 1972; Cope and Williams, 1973; Zastrow and Castle, 1987). Salivary proteins exhibit vectorial transport from the rough endoplasmic reticulum to successive membrane-bounded compartments including the Golgi complex, condensing vacuoles, and secretion granules (Figure 1) where they are posttranslationally modified, packaged, and stored prior to discharge. Passage of proteins through each of these compartments also involves their progressive concentration and sorting from other macromolecules that use segments of the same pathway. The main route of export is selective, and the salivary products are stored and released at concentrations exceeding 300 mg protein per milliliter (Arvan et al., 1984; Zastrow and Castle, 1987).

INTRACELLULAR TRANSPORT OF SALIVARY PROTEINS WITHIN THE ENDOPLASMIC RETICULUM AND GOLGI

During biosynthesis, salivary proteins are translocated into the lumen of the rough endoplasmic reticulum and are thus effectively segregated from the cytosol for the duration of their intracellular life. This first level of macromolecular sorting is mediated by a short (~20 amino acids) signal sequence (Walter and Lingappa, 1986) that is variable in primary sequence (von Heijne, 1985) and typically present at the amino terminus of secretory protein precursors [e.g., pre-amylase (Hagenbuchle et al., 1981) and preproline-rich proteins (Ziemer et al., 1984)]. Signal sequences are mostly cleaved cotranslationally and are thus not involved further in any aspect of transport. N-linked oligosaccharides are added to Asn-X-Ser/Thr consensus sequences cotranslationally on many, but not all, salivary proteins (Kousvelari et al., 1984), and the initial exoglycosidic processing of the oligosaccharides begins prior to exit from the endoplasmic reticulum (ER) (Kornfeld and Kornfeld, 1985).

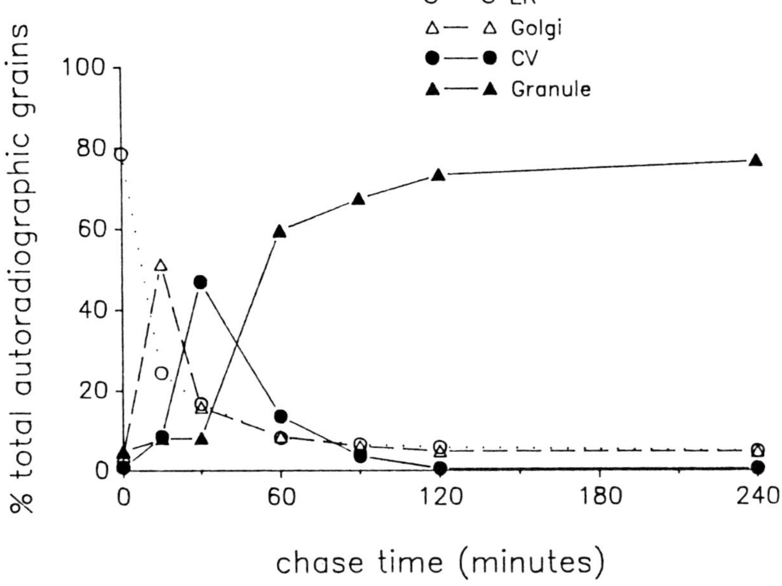

Figure 1. Intracellular transport of secretory proteins in rat parotid acinar cells analyzed by EM radioautography of isolated acini following pulse-chase labeling with [^{3}H]leucine. The distribution of radioautographic grains over cellular organelles as a function of time of chase incubation illustrates the initial labeling and rapid drainage of the endoplasmic reticulum (ER) and successive labeling of Golgi cisternae, condensing vacuoles (CV), and secretory granules. (From von Zastrow, M. and Castle, J. D., *J. Cell. Biol.*, 105, 2675, 1987. With permission.)

Protein folding and oligomerization are also initiated in the ER. These complex operations appear to be aided by binding proteins and protein disulfide isomerase, an ER luminal enzyme that catalyzes disulfide bonding (Pelham, 1989). Thus, enzymes having a number of intramolecular disulfide bonds such as α-amylase (Buisson et al., 1987) assume an active conformation within the ER (Redman and Sabatini, 1966). Moreover, a growing number of recent studies [reviewed by Hurtley and Helenius (1989)] has suggested that passage out of the ER is monitored by a "quality control" sorting process that operates to expedite transport of properly folded and oligomerized proteins and to retain misfolded polypeptides. This monitoring process may contribute to the varied rates of exit of secretory proteins from the ER (Scheele and Tartakoff, 1985; Zastrow and Castle, 1987).

Exit from the ER is widely considered to occur in a transitional region that is in close proximity to the Golgi complex (Palade, 1975). Transport to the Golgi is energy requiring, temperature sensitive, and, most likely, is mediated by vesicular shuttles (Balch, 1989). It is largely inhibited at 15°C (Saraste and Kuismanen, 1984), conditions under which smooth-surfaced membrane accumulates as tubular extensions of ER cisternae (Tartakoff, 1986). With the recent realization that exit from the ER entails selective retention of resident ER luminal and membrane proteins, the transitional elements and related vesicles are now being examined as a pre-Golgi sorting station that facilitates both vectorial transit of exported proteins and salvage/recycling of ER components (Pelham, 1989; Lippincott-Schwartz et al., 1990; Vaux et al., 1990), and possibly the degradation of misfolded or incompletely oligomerized proteins (Klausner, 1989). A role for microtubules in vesicular traffic from the ER to the Golgi in parotid acinar and other exocrine cells has been implicated from the inhibitory effects observed in the presence of colchicine in earlier studies (Patzelt et al., 1977; Busson et al., 1982). However, more recent higher-resolution studies indicate that intact microtubules are essential for the return of vesicles to the ER and that effects of microtubule disruption on ER-Golgi transport are an indirect consequence of the interference with vesicular shuttling along the retrograde pathway (Lippincott-Schwartz et al., 1990).

Numerous studies, including the cytochemical studies of parotid acinar cells (Hand and Oliver, 1984), have demonstrated that the Golgi complex is a stratified series of cisternal subcompartments that carry out a variety of post-translational modifications of secretory, lysosomal, and membrane proteins (Rothman, 1987; Farquhar and Palade, 1981; Fisher et al., 1988; Tooze et al., 1988). In particular, processing of N-linked oligosaccharides by continued exoglycosidase action and complex glycosylation appears to be organized along the *cis*-to-*trans* (entry-to-exit) Golgi axis with sequential processing activities occurring in distinct cisternae (Dunphy and Rothman, 1985; Roth and Berger, 1982; Roth et al., 1985). Traffic from one cisterna to the next is mediated by nonclathrin-coated vesicles that successively bud from one cisterna and fuse with the next (Rothman and Orci, 1990). This process occurs with very high efficiency (Rothman et al., 1984). Passage through the Golgi complex appears to be an obligatory step for all secretory proteins regardless of whether they are posttranslationally modified by Golgi enzymatic machinery, and it is clear from cytochemical and immunocytochemical studies that both glycosylated and nonglycosylated salivary proteins are found in all Golgi cisternae (Herzog and Miller, 1972; Bendayan et al., 1980; Ball et al., 1988).

As far as is known at present, transport within the Golgi complex occurs without appreciable sorting of proteins having different post-Golgi destinations. Lysosomal hydrolases are possible exceptions to this generalization, as their mannose-6-phosphate (M6P) sorting ligand is probably synthesized within *cis*-Golgi compartments (Goldberg et al., 1983) and in certain cell types (including exocrine acinar cells), the M6P receptor involved in lysosomal hydrolase sorting is also concentrated in the *cis*-Golgi (Brown and Farquhar, 1984, 1988). For other macromolecules, and possibly for lysosomal hydrolases as well (Kornfeld and Mellman, 1989), the *trans*-Golgi network has been identified as the major site of sorting (Griffiths and Simons, 1986). This reticular network, which appears to be functionally interrelated to the *trans*-most Golgi cisterna (Griffiths et al., 1989), is involved in the segregation of proteins destined for storage granules from many constitutively secreted proteins (Burgess and Kelly, 1987; Tooze et al., 1987) and is the site of sorting of membrane proteins for different domains of the plasma membrane in certain, but not all, epithelial cells (Griffiths and Simons, 1986; Bartles and Hubbard, 1988).

PATHWAYS OF SALIVARY PROTEIN SECRETION

The majority of polypeptides undergoing intracellular transport in salivary acinar cells is destined for storage granules that are characteristically released in response to secretory stimulation (Castle et al., 1972; Zastrow and Castle, 1987). In order to obtain a comprehensive picture of the mechanisms of salivary protein output, the kinetics of discharge of biosynthetically labeled polypeptides from parotid have been examined in the absence and presence of secretagogues in parallel with the radioautographic analysis of intracellular transport (Zastrow and Castle, 1987). In the absence of stimulation (autonomic inhibitors atropine, phentolamine, and propranolol were present), output of radiolabeled proteins was observed and occurred in two kinetically distinct but partially overlapping phases (Figure 2a). The early phase was initially observed at 15 min post pulse, reached a maximum at 1 to 2 h, and was estimated to be mostly complete by 7 h. The late phase was observed by 3 h, peaked at 9 to 10 h, and continued past 24 h. Less than 15% of incorporated radiolabel was discharged in the early phase while the remainder could be accounted for by the late phase. Only the late phase was stimulatable by isoproterenol. In the exocrine pancreas, where the two phases are also observed, both phases involve secretion at the apical cell surface as shown by cannulation of the pancreatic duct *in situ* (Arvan and Castle, 1987).

Examination of the radiochemical composition of early and late phase unstimulated secretion showed that it mostly contained the same proteins that were stored in parotid granules and released with stimulation. Early phase output, in contrast to late phase output, showed clear *quantitative* differences in the relative amounts of the various proteins that comprise stimulated secretion (Figure 2b). Major stored polypeptides of apparent M_r 25K and 32K were almost absent in the early phase whereas a minor 22K species released by stimulation was unusually prominent in the early phase (Zastrow and Castle, 1987). From these results, it was possible to conclude that the late phase was likely to reflect unstimulated exocytosis of secretion granules. In contrast, the early phase seemed to correspond to a pathway not previously characterized in salivary cells; unless resolved kinetically by pulse-chase labeling the smaller magnitude early phase would be obscured by the larger magnitude unstimulated exocytosis of granules.

The significant delay of early-phase unstimulated secretion well beyond passage of the wave of pulse-labeled proteins through condensing vacuoles, as estimated by radioautography (Figure 1), is inconsistent with the notion that the

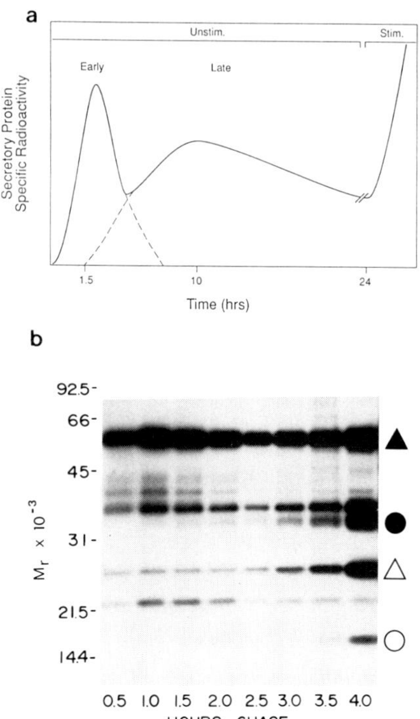

Figure 2. (a) The two phases of unstimulated secretion in acinar cells detected following pulse labeling *in vitro* with [35S]methionine. The specific radioactivity of secretory proteins released into the medium in the presence of autonomic inhibitors increases to a maximum at 1 to 2 h and subsequently declines. Following this early phase, an extended late phase is observed, initially at 3 h but continuing past 24 h. Only the late phase is stimulated by isoproterenol in parotid. (This figure is reprinted with permission of the *American Journal of Respiratory Cell and Molecular Biology* (Castle, 1990a) and summarizes observations made by Arvan and Castle (1987) and Zastrow and Castle (1987). (b) Differences in the relative radiochemical composition of early and late phase unstimulated secretion as illustrated by a fluorograph of secreted polypeptides resolved by SDS-PAGE from media collected at successive 30-min intervals of chase following pulse labeling with [35S]methionine. The composition of secretion stimulated by isoproterenol is shown for comparison in the final interval between 3.5 and 4 h. Amylase, the most prominent radiolabeled protein is marked by a solid triangle. Early phase secretion (peaking between 1 and 1.5 h) contains very little of the 32K (solid circle) and 25K (open triangle) major secretory polypeptides that are prominent components of isoproterenol-stimulated secretion but is enriched in the minor 22K polypeptide (open circle). Late phase unstimulated secretion seen between 3 and 3.5 h exhibits a relative composition that is quite similar to that obtained with stimulation. (From von Zastrow, M. and Castle, J. D., *J. Cell. Biol.*, 105, 2675, 1987. With permission.)

early phase pathway originates in the *trans*-Golgi network. Consequently, storage granule maturation became a focus for further analysis. Intermediates in parotid storage granule maturation, that were isolated by centrifugation (Zastrow and Castle, 1987; Iversen et al., 1985), proved to be unusually helpful in identifying the probable origin of early-phase secretion (Figure 3). The lowest

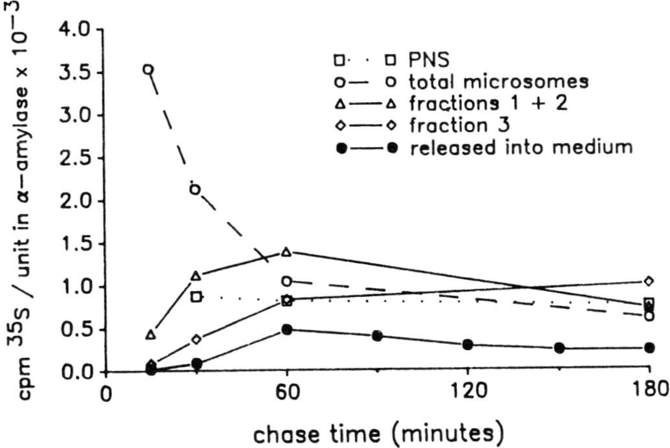

Figure 3. Comparison of intracellular transport (as judged by cell fractionation) and unstimulated secretion of pulse-labeled amylase for parotid acinar cells. The kinetics of intracellular transport are based on analysis of a total microsomal (ER + Golgi) fraction and granule subpopulations purified by gradient centrifugation in isoosmotic media (Zastrow and Castle, 1987). Drainage of labeled amylase from the total microsomal fraction occurs with a half-time of 25 min. Lower density (fractions 1 + 2) granules are subsequently labeled and appear to be precursors of higher density (fraction 3) granules that become labeled by amylase more slowly. Notably, early-phase unstimulated secretion of amylase peaks well after labeling of immature granules is observed, and the decay kinetics of unstimulated secretion are very similar to the kinetics of granule maturation. Thus, maturing granules are implicated as the origin of early-phase secretion. (From von Zastrow, M. and Castle, J. D., *J. Cell. Biol.*, 105, 2675, 1987. With permission.)

density fractions obtained were maximally labeled by radioactive secretory proteins at 30 to 60 min post pulse, and granule subfractions of progressively greater density showed increased labeling at subsequent times concomitant with decreased labeling of the lower density fractions (Zastrow and Castle, 1987). Further, the kinetics of maturation correlated quite well with the kinetics of early-phase secretion, and the granule fractions of higher buoyant density were found to be progressively depleted of the 22K secretory polypeptide that was prominent in early phase secretion (Zastrow and Castle, 1987; Castle, 1990b). Thus, it appears that vesicular shuttles originating from immature storage granules and enriched in the 22K polypeptide are the vehicles of the early pathway. The observation of frequent coated evaginations on immature granules that, by radioautography, contained radiolabeled secretory protein at a chase time corresponding to maximal early phase secretion has provided a possible structural correlate to the origin of the vesicular pathway (Zastrow and Castle, 1987).

Taken together, these studies indicate that although parotid acinar cells are highly specialized for the production of storage granules for stimulus-dependent discharge, they continuously secrete the same proteins at lower levels and by two distinct pathways in the absence of stimulation. The significance of the two unstimulated pathways, especially the early-phase route, is not known at present.

It is possible that one or both could provide essential concentrations of regulatory (Zastrow et al., 1986) or other polypeptides for oral cavity or ductal functions during periods intervening between secretory stimulation. The early-phase pathway, in particular, might also correspond to the route taken by newly synthesized membrane polypeptides of the apical surface, as nongranule pathways have been implicated in the trafficking of apical plasmalemmal components in other exocrine acinar cells (Havinga et al., 1983; Beaudoin et al., 1986).

Finally, it is important to contrast these pathways of unstimulated secretion characterized in parotid acinar cells from the constitutive pathway as originally characterized in cultured lines of endocrine cells (reviewed in Burgess and Kelly, 1987). In cultured cells, polypeptides distinct from those packaged in storage granules were shown to be secreted in the absence of stimulation, and the *trans*-Golgi network was implicated as the major site of sorting of constitutively secreted and stored proteins (Tooze et al., 1987; Tooze and Huttner, 1990). Parotid acinar cells are likely to maintain at least low levels of analogous pathways for nongranule proteins; since acinar cells are epithelial, these pathways may operate to both the apical and basal cell surfaces and involve sorting in the *trans*-Golgi network (Griffiths and Simons, 1986). However, they also contain the unstimulated pathways described above that convey the major stored secretory products to the cell surface in the absence of stimulation. The branchpoint of the latter pathways after exit from the *trans*-Golgi network in the acinar cell may reflect a heirarchical organization of secretory sorting in which the apical/basolateral "epithelial" sorting precedes the subsorting of exocrine proteins. Protein secretory pathways in salivary acinar cells are summarized diagrammatically in Figure 4.

MECHANISMS OF SALIVARY PROTEIN SORTING AND STORAGE GRANULE FORMATION

Protein Sorting

Diverging pathways to different final destinations within the cell means that sorting occurs for different classes of proteins that use the intracellular transport pathway. A number of studies examining expression of exogenous secretory proteins suggest that sorting for storage granule formation is a positive process that may rely on signals that are part of secretory protein structure. First, an exocrine protein expressed in endocrine cells and an endocrine protein expressed in exocrine cells are each sorted in the same way as endogenous secretory products (Burgess et al., 1985; Ornitz et al., 1985). Second, chimeras constructed between portions of polypeptides that are normally stored in granules and polypeptides that are either secreted by no granule pathways or are normally cytoplasmic are targeted to granules when expressed in endocrine cells (Moore and Kelly, 1986; Stoller and Shields, 1989). Finally, expression of an immuno-

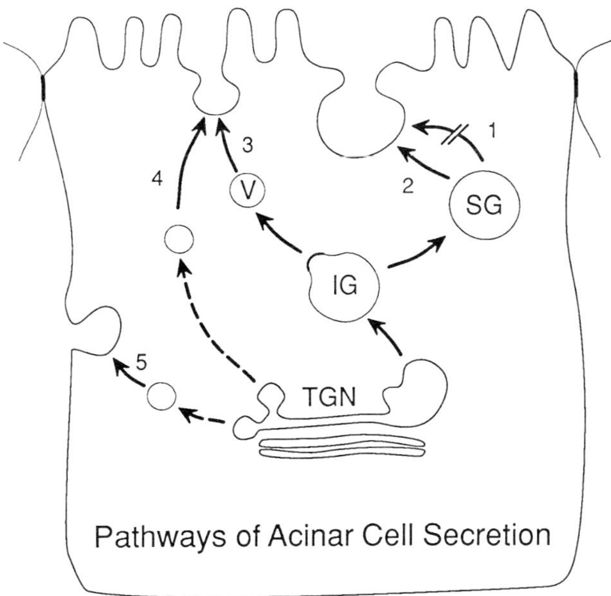

Figure 4. The pathways of secretion in salivary acinar cells. Pathway 1 represents stimulus-dependent exocytosis of storage granules (SG) while pathway 2 identifies the basal level exocytosis of granules thought to comprise the late phase of unstimulated secretion. Pathway 3 originating from immature granules (IG) is a vesicular, nongranular route representing early phase unstimulated secretion. It carries the same proteins that undergo storage in granules but in substantially different relative amounts. Pathways 4 and 5 originating in the *trans*-Golgi network (TGN) are "epithelial" pathways consisting of constitutive carriers to the apical and basolateral plasma membranes corresponding to pathways that have been identified in simple epithelia grown as polarized monolayers (Griffiths and Simons, 1986). Although these pathways have not been characterized in salivary acinar cells, they are thought to convey secretory products and membrane proteins that are not found in storage granules to the cell surface. (From Castle, J. D., *Am. J. Resp. Cell Mol. Biol.*, 2, 119, 1990a. With permission.)

globulin whose antigen is a stored secretory protein leads to storage and stimulated discharge of that immunoglobulin (Rosa et al., 1989). Positive sorting for storage granule formation could imply the presence of a sorting receptor that operates from a pregranule compartment, similar to the receptor-mediated, pH-dependent sorting of lysosomal hydrolases. In this regard, the presence of Golgi-associated proteins that exhibit pH-dependent binding to secretory proteins destined for storage granules has been described recently (Chung et al., 1989), although their function as sorting receptors has not been demonstrated. In contrast, the sorting of an immunoglobulin by binding to its antigen (as mentioned above) suggests that interactions between secretory proteins might comprise at least part of the positive sorting process involved in storage granule formation.

Interest in the possible presence of a pH-dependent receptor involved in secretory sorting has arisen from studies showing that perturbation of the pH of acidic intracellular compartments inhibits the entry of secretory proteins into forming endocrine storage granules (Moore et al., 1983; Stoller and Shields, 1989). However, cytotoxicity resulting from sustained pH manipulation may underlie some of the inhibitory effects observed (Mains and May, 1988). In parotid acinar cells, addition of the membrane-permeant weak base ammonium chloride did *not* appear to affect entry of secretory proteins into forming parotid granules, but it increased the magnitude and decreased the compositional selectivity of early-phase unstimulated secretion originating from maturing granules (Zastrow et al., 1989). These findings suggest that pH (or osmotic) perturbation mainly may affect interactions involved in the aggregation of secretory proteins for storage.

The increased magnitude and decreased selectivity of early phase secretion observed in the presence of ammonium chloride suggests that selective aggregation of content polypeptides within the forming and maturing storage granule constitutes a positive sorting process for the stimulus-regulated secretory pathway. In this case the early phase component of unstimulated secretion would contain the polypeptides that are excluded from the aggregation process, essentially as volume markers of this pathway. Sorting of polypeptides destined for storage by selective association is consistent with the interactions observed during packaging of exocrine (particularly parotid), endocrine, or neuroendocrine secretory proteins (e.g., Tandler and MacCallum, 1972; Kousvelari et al., 1982; Hashimoto et al., 1987; Fisher et al., 1988; Stoller and Shields, 1989). This process would be very efficient, especially where high concentrations of protein must undergo sorting and may complement other types of sorting processes (Kelly, 1985).

Protein Condensation

One of the important characteristics of the process of storage granule formation is the capacity to concentrate or condense secretory proteins. In the case of parotid secretion granules, biophysical measurements of the internal water space of 2 to 3 μl/mg protein translate into an internal protein concentration of 300 to 500 mg/ml (Arvan et al., 1984; Zastrow and Castle, 1987). This level has been estimated to be one or two orders of magnitude higher than in the ER (Kelly, 1985). From demonstrations that exocrine secretory proteins will condense to form intracisternal granules and apparently exclude other proteins when their exit from the ER is blocked (Tooze et al., 1989), it seems clear that selective aggregation [or "condensation sorting" (Tooze et al., 1989)] is likely to be a major factor in storage granule formation. However, packaging of secretory proteins at low osmotic activity may also involve other processes that contribute secondarily by reducing the effects of any net fixed internal charge. Moreover, the extent to which these processes occur may vary in different types of storage granules as a function of the different overall composition of secretory proteins undergoing storage.

At least three processes might contribute to secretory condensation by reducing osmotic activity during granule formation and maturation. (1) In a compartment with a substantial internal buffering capacity (Arvan et al., 1984), proton influx driven by an H^+ pump may drive cycles of electrically dissipative efflux of cations from the granule, buffering of protons through binding to the macromolecular content, and exit of water to maintain osmotic equilibrium. The possible presence of a proton pump in condensing vacuoles and immature granules that could drive such a condensation process is compatible with the observation that condensing vacuoles and immature granules in parotid (and exocrine pancreas) are relatively acidic organelles as compared to preceding Golgi and succeeding mature granule compartments (Orci et al., 1987). (2) Calcium ions may be accumulated and possibly exchanged for protons and other cations within the forming granule, thereby stabilizing condensation but decreasing intragranular acidity with progressive maturation in accord with the observations mentioned above. In this respect, a number of parotid secretory proteins bind calcium, at least weakly (Bennick, 1982), and the intragranular concentration of calcium has been estimated to be several millimolar (Wallach and Schramm, 1971). (3) Covalent modifications of secretory proteins that alter their net charge may aid in driving concentration/packaging. In parotid glands from rats that have been treated chronically with isoproterenol, the spectrum of secretory proteins that is produced and packaged in storage granules is dramatically altered and includes a very large amount of basic proline-rich proteins (Muenzer et al., 1979a; Arvan and Castle, 1986) in place of the usual proteins that are generally more acidic. Yet, the proteins are stored at about the same high concentration as in the granules from untreated animals (Arvan and Castle, 1986). Accompanying these changes which lead to an increased intragranular pH and the presence of a small fixed net positive charge (Arvan and Castle, 1986) is the amplified production of a chondroitin sulfate containing proline-rich proteoglycan (Blair, Castle, and Castle, 1991). At least part of the sulfation reaction, which would act to reduce the net positive charge by the addition of negative charges, can be shown to occur within maturing secretion granules (Blair, Castle, and Castle, 1991).

Overall, the present picture of protein sorting and condensation involved in storing parotid salivary proteins within membrane-bounded granules is envisioned to include selective associations between salivary proteins that may be based on information encoded in their primary or higher order structures and electrostatic interactions that may be maximized in a variety of ways such that net fixed intragranular charge and consequent osmotic activity are minimized.

MEMBRANE CARRIERS OF SALIVARY PROTEIN SECRETION

A mediator of protein secretion in salivary acinar cells is the membrane of the secretion granule that serves as a shuttle between the Golgi and the cell

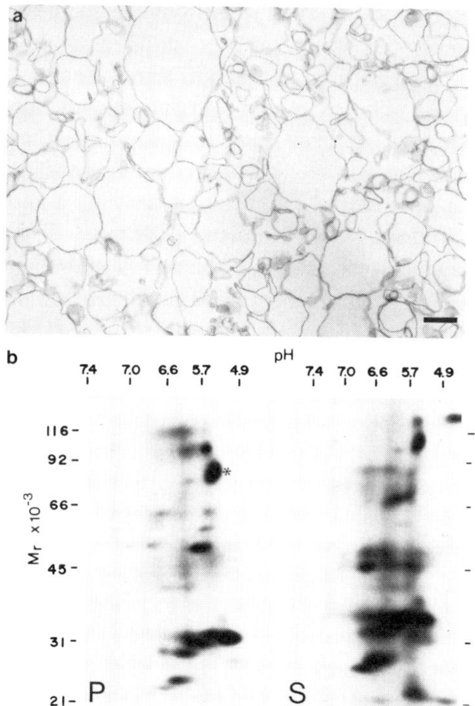

Figure 5. (a) Representative electron micrograph of granule membranes purified from rat parotid secretion granules. Comprehensive analysis of this preparation for residual contamination by secretory protein and other organelles indicates that 95% of the protein of this fraction comprises polypeptides associated with granule membranes. (From Cameron, R. S. and Castle, J. D., *J. Membrane Biol.*, 79, 127, 1984. With permission.) (b) Two-dimensional (isoelectric focusing/SDS) PAGE profiles of radioiodinated granule membrane polypeptides from parotid and submandibular salivary glands. The radioautographic patterns identify a limited number of polypeptides for each type of granule membrane, and similarities of the profiles particularly in the range of low apparent molecular weight are clear. Thus, there appears to be substantial compositional overlap between membranes that store quite different spectra of secretory products. (From Cameron R. S., Cameron, P. L., and Castle, J. C., *J. Cell Biol.*, 103, 1299, 1986. With permission.)

surface. Relatively little is known about the role of membranes in secretion. The ability to isolate highly purified preparations of storage granule membranes from salivary tissues (Figure 5a; Cameron and Castle, 1984; Cameron et al., 1986) has provided a starting point for characterizing secretion granule shuttles and dissecting their functional elements. Granule membranes are distinguished from other membranes — plasma membranes, ER membranes, mitochondrial membranes — that can be isolated from salivary acinar cells as purified fractions (Cameron and Castle, 1984; Arvan and Castle, 1982) by their low protein content (<20% by weight) and limited diversity of integral polypeptides (Figure 5b; Cameron and Castle, 1984; Cameron et al., 1986). In this membrane the polypeptide concentration and composition differs from that of *trans*-Golgi and plasma membranes (Hand and Oliver, 1984; DeCamilli et al., 1976; Cameron et al., 1986). Granule membrane proteins may be sorted, while the components of these other membranes may be excluded, during storage granule formation and during recycling following exocytosis. The combination of low protein concentration and specific composition could arise by: (1) selective association of granule membrane proteins with one another and sorting as aggregates in a lipid-rich membrane, (2) segregation in an associated clathrin–adaptin framework or, conversely, exclusion from such a framework operating to retrieve nongranule proteins from the granule, and/or (3) selective interaction with secretory proteins or crosslinking proteins between the membrane and content.

Among the polypeptides that comprise the storage granule membrane in parotid acinar cells, a subset of 10 to 15 polypeptides appear to be shared not only by storage granule membranes in other salivary glands but also in the granule membranes of the exocrine pancreas and lacrimal glands (Cameron et al., 1986). This compositional overlap extends to Golgi-derived secretory vesicles in hepatocytes (Cameron et al., 1989) and to granule membranes in granulocytes, endocrine, and neural cells (Reinhart and Neutra, 1989; Laurie et al., 1989). The set of common (or related) polypeptides may include certain of the functional elements of the secretory machinery, irrespective of the mode of secretory discharge.

One further level of analysis that eventually may aid in identifying sorting domains of membrane proteins during granule formation and recycling has involved comparison of two-dimensional peptide maps derived by enzymatic digestion of radioiodinated polypeptides for a large fraction of the entire polypeptide spectrum of the parotid storage granule membrane (Cameron et al., 1986). A common peptide "fingerprint" has been recognized that is shared by all of the membrane polypeptides examined so far but is not present in any of the secretory products that have been tested. Identical fingerprints are observed in digests of polypeptides from pancreatic zymogen granule membranes; however, the fingerprint appears to differ in the limited number of liver Golgi membrane polypeptides that have been examined. Once the primary sequences have been deduced for some of the membrane polypeptides, it should be possible to identify the primary structures of the peptides comprising the fingerprint and thereby facilitate examination of their role in secretion.

FUNCTION OF SALIVARY MEMBRANE CARRIERS IN EXOCYTOSIS AND RECYCLING

At present almost nothing is known about the actual mechanisms of membrane function in protein secretion. Salivary storage granules appear to be targeted to the apical plasma membrane, as they accumulate in the cytoplasm beneath this surface, and they appear to be released in the order that they are synthesized (Sharoni et al., 1976; Zastrow and Castle, unpublished results). Thus, granules may have apical addressing signals on their cytoplasmic surfaces [as has been implicated in secretory vesicles in other epithelial cells (Mostov and Deitcher, 1986)]; yet, they maintain an organization within the storage population, possibly indicating a sustained interaction with the cytoskeleton. Presumably, the same considerations apply to the nongranular vesicular carriers of early phase unstimulated secretion since they are directed to the apical surface of acinar cells and are released in the order of synthesis (Arvan and Castle, 1987). Under conditions of acute β-adrenergic stimulation, compound exocytosis of storage granules is observed in parotid acinar cells (Amsterdam et al., 1969). Fusion is initiated by the granules that are most proximal to the cell surface and upon discharge, these granules serve as fusion partners for granules located more deeply in the cytoplasm. Consequently, an ordered cascade of protein discharge proceeds inward to involve most or all of the storage granule population. One implication from the observation of compound exocytosis is that the storage granule membrane may contain all the necessary membrane machinery for the fusion and discharge process. Thus, in attempting to model exocytosis *in vitro*, it may be possible to devise systems that use intact secretion granules and granule membrane vesicles as fusion partners.

Following exocytosis, secretion granule membrane is selectively and efficiently reinternalized by endocytosis from the cell surface of parotid acinar cells (DeCamilli et al., 1976). The process of reinternalization appears to be calcium dependent, and in the presence of physiological concentrations of calcium in the medium, it proceeds with a half-time of 20 min (Koike and Meldolesi, 1981). The very low biosynthetic rates of parotid granule membrane polypeptides in relation to those of secretory proteins (Castle and Palade, 1978) suggests that granule membranes may be reutilized for multiple rounds of secretory discharge. Indeed, introduction of electron-dense tracers into the apical luminal space by retrograde perfusion of the parotid ducts showed that a substantial fraction of the reinternalized granule membrane returned to the Golgi complex and could be used to package new storage granules (Herzog and Farquhar, 1977). The rest of the membrane appeared to be routed to lysosomes, and the site of sorting to the different destinations has not been defined as yet. Related studies comparing the pathways taken by cationic and anionic tracers and by radiolabeled secretory proteins during reinternalization have suggested that the cargo, possibly through interactions with the membrane, may significantly influence the extent to which membrane recycling is directed to the Golgi complex or to lysosomes (Herzog and Reggio, 1980; Oliver and Hand, 1978; Romagnoli and Herzog, 1987). The pathway traced by secretory products may prove to be especially interesting, as it may indicate an affinity for storage granule membranes that is relevant to the secretory packaging process.

STIMULUS REGULATION OF SALIVARY PROTEIN SYNTHESIS, TRANSPORT, AND SECRETION

Over the past decade, the practice of categorizing secretory pathways according to their constitutive or stimulus-regulated characteristics has laid the groundwork for beginning to understand the true complexity of export pathways leading from the Golgi complex to the cell surface. Stimulus regulation has been viewed mainly in terms of processes that lead to intracellular protein storage and amplified rate of exocytosis. However, a variety of findings in parotid may prove to be particularly helpful in developing a more comprehensive view of the role of stimulus regulation in protein secretion.

Biosynthesis of Salivary Proteins

A number of studies (e.g., Sreebny et al., 1971) have shown that the rate of salivary protein synthesis in rat parotid varies with feeding and fasting. Moreover, direct examination of parotid tissue *in vitro* has demonstrated that β-adrenergic stimulation increases the rate of salivary protein synthesis, although the time course is dissociated from the amplified exocytosis of secretory products stored in granules (Lillie and Han, 1973). As well, isoproterenol increases the phosphorylation of ribosomal protein S6 in parotid, but the significance of this posttranslational modification to salivary protein synthesis is not known (Jahn et al., 1980; Freedman and Jamieson, 1982).

Glycosylation of Salivary Proteins

Isoproterenol stimulation of rat parotid tissue has been reported to increase N-linked glycosylation of selected salivary secretory proteins (Kousvelari et al., 1984). Further, at least two enzymes that are involved in the synthesis of the oligosaccharide-dolichol precursor for N-linked glycosylation appear to exhibit increased activity that is mediated by cyclic AMP in response to the β-adrenergic agonist (Banerjee et al., 1985). Whether increased glycosylation has any effects on the rates or routes of intracellular transport has not been considered to date.

Exit From the Rough Endoplasmic Reticulum

The passage of proteins from the endoplasmic reticulum to the Golgi complex is generally viewed as a constitutive process, and the rate of exit for individual proteins appears to be contingent on folding into a transport-competent conformation and, in some cases, oligomerization (Hurtley and Helenius, 1989). Egress of all secretory proteins from the ER in the exocrine pancreas appears to cease during extreme starvation or in the presence of cobalt (Palade,

1956; Tooze et al., 1989). Interestingly, treatment of rats with reserpine, which depletes sympathetic nerve termini of catecholamine neurotransmitters, appears to cause a major reduction in the exit of secretory proteins from the ER in parotid acinar cells as judged by the presence of distended ER cisternae and intracisternal granules (Muller and Roomans, 1984). This finding suggests that low levels of neural stimulation (provided by continuous neurotransmitter discharge at neuron-acinar cell termini) may be required to sustain salivary protein transport. Whether the blockade of ER-Golgi transit is a specific effect or an indirect consequence of alterations in intracellular transport (Setser et al., 1979) induced by reserpine remains to be examined in detail.

Pattern of Unstimulated Exocrine Secretion

Variations in the feeding pattern of rats prior to experimentation alters the phasic pattern of unstimulated exocrine secretion such that the early phase is either lost or substantially retarded under conditions where the granule population is partially discharged (Arvan and Castle, 1987). Thus, the relative magnitude of the early phase, or the partitioning between the two pathways, may be under physiological control. In contrast to the constitutive discharge pathway that has been characterized in endocrine cell lines (Kelly, 1985), this feature of the unstimulated early phase exocrine pathway implicates a form of regulation.

Size Variations of Forming Storage Granules Related to Secretory Stimulation

During restitution of the granule population following secretory stimulation in parotid, the initial storage granules that form are substantially smaller (<0.6 µm average diameter) than those that were present prior to stimulation (1 µm average diameter). As repopulation progresses, the average diameter of storage granules is restored and even exceeds the prestimulation size (Cope and Williams, 1981; Castle, unpublished observations). Thus, the volume of secretory storage per unit area of storage granule membrane appears to be under regulation. This property may prove to be of considerable interest when the underlying mechanisms involved in sorting of secretory and membrane proteins of storage granules are identified. Variation in the size of secretion granules that are formed after stimulation may relate to the altered pattern of early-phase unstimulated secretion that is apparently linked to secretory stimulation (discussed above) as both changes involve maturing storage granules.

Stimulus Regulation of Exocytosis

The hallmark of regulated secretion is the capacity to amplify the discharge of stored secretory proteins by stimulus-dependent increases in the rate of granule exocytosis. Amplified discharge of salivary proteins occurs primarily in response to neural stimulation, and in rat parotid β-adrenergic stimuli appear to be the principal regulators (Baum, 1987). Both cholinergic and α-adrenergic agonists will also enhance protein secretion in addition to their major role in salivary fluid and electrolyte secretion (Young et al., 1987). In general, intracellular cascades of altered phosphorylation of effector proteins are thought to couple external stimuli to amplified salivary protein secretion (Spearman and Butcher, 1989; Quissell and Tabak, 1989). Certain of the phosphoproteins that are potentially involved in the final stages of coupling are known to be associated with secretion granule membranes (Spearman et al., 1984; Marino et al., 1990). These triggering pathways of exocytosis and their phosphoprotein constituents are considered in detail elsewhere in this volume. Thus, the discussion here will be limited to potentially new insights regarding the cell biology of secretion that have arisen from delineating the pathways of discharge.

There is a general tendency to regard cAMP-mediated secretion catalyzed by β-adrenergic agonists and calcium-mediated secretion elicited by cholinergic agonists as operating at two levels (possibly synergistically) in the same stimulus-secretion cascade in parotid (Spearman and Butcher, 1989). However, the observation made by Anderson et al. (Anderson et al., 1984) that parasympathetically elicited secretion is selectively enriched in the 22K protein that was shown to be unusually prevalent in early-phase unstimulated secretion (Zastrow and Castle, 1987) now raises the possibility that sympathetic and parasympathetic stimuli may stimulate parallel pathways that preferentially access distinct pools of secretory products within acinar cells. A number of investigators have shown that cholinergic stimuli elicit the formation of cytoplasmic vacuoles from, as yet, uncharacterized membranes in proximity to the Golgi complex of acinar cells (Leslie and Putney, 1983; Schramm and Selinger, 1975). The close association of these vacuoles with the site of granule formation and their occasionally observed fusion with the cell surface (Leslie and Putney, 1983) suggest that vacuolar membrane may contribute a distinct discharge pathway. Whether this membrane functions in ion conductance in relation to stimulated electrolyte secretion is not known. Further, because the effects of cholinergic agonists on early phase secretion were not examined (Zastrow and Castle, 1987), it remains to be determined whether this pathway originating from maturing storage granules might also contain a stimulus-regulated component of discharge.

Secretory Rerouting in the Absence of Stimulation

At the other extreme from acutely amplified exocytosis in the presence of neural stimuli is the fate of storage granules in the absence of stimulation. In a variety of endocrine cells, the process of crinophagy involving granule fusion with lysosomes and degradation has been well documented as a consequence of sustained removal of secretory stimulation (e.g., Farquhar, 1971; Orci et al., 1984). In the case of parotid acinar cells, Hand has shown that prolonged starvation of rats leads to an increased incidence of storage granule degradation by a process that appears to be crinophagy (Hand, 1973). Crinophagy constitutes a switch in fusion partners for secretion granules, and at present very little is known about its regulation. It is possible that pathways leading to lysosomal fusion that normally function during endocytosis of granule membrane following exocytosis are activated prematurely during the sustained absence of stimulation. Alternatively, as the accumulated storage granule population reaches maximal levels, sorting of lysosomal and granule components along distinct post-Golgi pathways may become impaired and directly lead to degradation of secretory proteins upon activation of lysosomal hydrolases. In order to provide insight into the process of crinophagy, there is a need to define the sorting sites for membrane and content components that are directed along each pathway and to determine whether crinophagy preferentially involves newly synthesized or older secretory proteins within the storage compartment.

Overall, the process of salivary protein secretion involves a highly specialized cell type whose continuous operation aids in preserving the specialized environment of the oral cavity and whose stimulus-dependent function is closely tied to the ingestion of foodstuffs. It is clear that these processes capitalize on multiple levels of regulation, yet the studies to understand these events have only begun.

ACKNOWLEDGMENTS

The research from the author's laboratory that is reviewed in this chapter has been supported by research grants from the NIH (GM26524, DK29868, DE08941) and Biomedical Research Support Grant funds (5S07-RRo5431-26). I am very grateful to my colleagues, Drs. Anna Castle and Susan Laurie, for their critical reading of the manuscript. J.D.C. is a member of the Comprehensive Cancer Center and the Diabetes and Endocrine Research Center at the University of Virginia.

REFERENCES

1. Amsterdam, A., Ohad, I., and Schramm, M., Dynamic changes in ultrastructure of the acinar cell of the rat parotid gland during the secretory cycle, *J. Cell Biol.*, 41, 753, 1969.
2. Anderson. L. C., Garrett, J. R., Johnson, D. A., Kaufman, L., Keller, P. J., and Thulin, A., Influence of circulating catecholamines on protein secretion into rat parotid saliva during parasympathetic stimulation, *J. Physiol.*, 352, 163, 1984.
3. Arvan, P. and Castle, J. D., Plasma membrane of the rat parotid gland: preparation and partial characterization of a fraction containing the secretory surface, *Cell Biol.*, 95, 8, 1982.
4. Arvan, P. and Castle, J. D., Isolated secretion granules from parotid glands of chronically-stimulated rats possess an alkaline internal pH and inward-directed H$^+$ pump activity, *J. Cell Biol.*, 102, 1257, 1986.

5. Arvan, P. and Castle, J. D., Phasic release of newly synthesized secretory proteins in the unstimulated rat exocrine pancreas, *J. Cell Biol.*, 104, 243, 1987.

6. Arvan, P., Rudnick, G., and Castle, J. D., Ostomotic properties and internal pH of isolated rat parotid secretory granules, *J. Biol. Chem.*, 259, 13567, 1984.

7. Balch, W. E., Biochemistry of interoganelle transport, *J. Biol. Chem.*, 264, 16965, 1989.

8. Ball, W. D., Hand, A. R., and Johnson, A. O., Secretory proteins as markers for cellular phenotypes in rat salivary glands, *Dev. Biol.*, 125, 265, 1988.

9. Banerjee, D. K., Kousvelari, E. E., and Baum, B. J., Beta-adrenergic activation of glycosyltransferases in the dolichylmonophosphate-linked pathway of protein N-glycosylation, *Biochem. Biophys. Res. Comm.*, 126, 123, 1985.

10. Bartles, J. R. and Hubbard, A. L., Plasma membrane protein sorting in epithelial cells: do secretory pathway hold the key?, *Trends Biochem. Sci.*, 13, 181, 1988.

11. Baum, B. J., Regulation of salivary secretion, in *The Salivary System*, L. M. Sreebny, Ed., 123, 134, 1987.

12. Beaudoin, A. R., Groudin, G., Vachereau, A., St.-Jean, P., and Cabana, C., Detection and characterization of microvesicles in the acinar lumen and juice of unstimulated rat pancreas, *J. Histochem. Cytochem.*, 34, 1079, 1986.

13. Beaudoin, M., Roth, J., Perrelet, A., and Orci, L., Quantitative immunocytochemical localization of pancreatic secretory proteins in subcellular compartments of the rat acinar cell, *J. Histochem. Cytochem.*, 28, 149, 1980.

14. Bick, A., Salivary proline-rich proteins, *Molec. Cell. Biochem.*, 45, 83, 1982.

14a. Blair, E. A., Castle, A. M., and Castle, J. D., Proteoglycan sulfation and storage parallels storage of basic secretory proteins in exocrine cells, *Am. J. Physiol.*, 261, C897, 1991.

15. Bloom, G. D., Carlsoo, B., Danielsson, A., Gustafsson, H., and Henriksson, R., Quantitative structural analysis and the secretory behavior of the rat parotid gland after long and short term isoprenaline treatment, *Med. Biol.*, 57, 224, 1979.

16. Brown, W. J. and Farquhar, M. G., The mannose-6-phosphate receptor for lysosomal enzymes is concentrated in cis Golgi cisternae, *Cell*, 36, 295, 1984.

17. Brown, W. J. and Farquhar, M. G., The distribution of 215 kilodalton mannose-6-phosphate receptors within cis (heavy) and trans (light) Golgi subfractions varies in different cell types, *Proc. Natl. Acad. Sci. U.S.A.*, 84, 9001, 1988.

18. Buisson, G., Duiee, E., Haser, R. D., and Payan, F., Three dimensional structure of porcine pancreatic alpha-amylase at 2.9 A resolution. Role of calcium in structure and activity, *EMBO J.*, 6, 3909, 1987.

19. Burgess, T. L., Craik, C. S., and Kelly, R. B., The exocrine protein trypsinogen is targeted into the secretory granules of an endocrine cell line: studies by gene transfer, *J. Cell Biol.*, 101, 639, 1985.

20. Burgess, T. L. and Kelly, R. B., Constitutive and regulated secretion of proteins, *Annu. Rev. Cell Biol.*, 3, 243, 1987.

21. Busson-Mabilot, S., Chambaut-Guerin, A. M., Ovtracht, L., Miller, P., and Rossignol, B., Microtubules and protein secretion in rat glands: localization of short-term effects of colchicine on the secretory process, *J. Cell Biol.*, 95, 105, 1982.

22. Cameron, R. S., Arvan, P., and Castle, J. D., Secretory membranes and the exocrine storage compartment in *Handbook of Physiology Section 6: The Gastrointestinal System III, Salivary, Gastric, Pancreatic and Hepatobiliary Secretion*, J. G. Forte, Ed., American Physiological Society, New York, 107, 1989.

23. Cameron, R. S., Cameron, P. L., and Castle, J. C., A common spectrum of polypeptides occurs in secretion granule membranes of different exocrine glands, *J. Cell Biol.*, 103, 1299, 1986.

24. Cameron, R. S. and Castle. J. D., Isolation and compositional analysis of secretion granules and their membrane subfraction from the rat parotid gland, *J. Membrane Biol.*, 79, 127, 1984.

25. Castle, J. D., Sorting and secretory pathways in exocrine cells, *Am. J. Resp. Cell Mol. Biol.*, 2, 119, 1990a.

26. Castle, J. D., Multiple pathways of protein secretion in exocrine cells, *Adv. Cell Biol.*, 3, 275, 1990.

27. Castle, J. D., Jamieson, J. D., and Palade, G. E., Radioautographic analysis of the secretory process in the parotid acinar cell of the rabbit, *J. Cell Biol.*, 53, 290, 1972.

28. Castle, J. D. and Palade, G. E., Secretion granules of the rabbit parotid. Selective removal of secretory contaminants from granule membranes, *J. Cell Biol.*, 76, 323, 1978.

29. Chung, K. N., Walter, P., Aponte, G. W., and Moore, H. P. H., Molecular sorting in the secretory pathway, *Science*, 243, 192, 1989.

30. Cope, G. H. and Williams, M. A., Quantitative analyses of the constituent membranes of parotid acinar cells and of changes evident after exocytosis, *Z. Zellforsch, Mikrosk, Anat.*, 145, 311, 1973.

31. Cope, G. H. and Williams, M. A., Secretion granule formation in the rabbit parotid gland after isopdrenaline-induced secretion: sterelogical reconstruction of granule populations, *Anat. Record*, 199, 377, 1981.

32. DeCamilli, P., Peluchetti, D., and Meldolesi, J., Dynamic changes of luminal plasmalemma in stimulated parotid acinar cells. A freeze fracture study, *J. Cell Biol.*, 70, 59, 1976.

33. Dunphy, W. G. and Rothman, J. E., Compartmental organization of the Golgi stack, *Cell*, 42, 13, 1985.

34. Farquhar, M. G., Processing of secretory products by the cells of the anterior pituitary gland, *Mem. Soc. Endocrinol.*, 19, 79, 1970.

35. Farquhar, M. G. and Palade, G. E., The Golgi apparatus/complex (1954–1981) — from artifact to center stage, *J. Cell Biol.*, 91, 77, 1981.

36. Fisher, J. M., Sossin, W., Newcomb, R., and Scheller, R. H., Multiple neuropeptides derived from a common precursor are differentially packaged and transported, *Cell*, 54, 813, 1988.

37. Freedman, S. D. and Jamieson, J. D., Hormone-induced protein osphorylation. II. Localization to the ribosomal fraction from rat exocrine pancreas and parotid of a 29,000 dalton protein phosphorylated in response to secretagogues, *J. Cell Biol.*, 95, 909, 1982.

38. Goldberg, D. E., Gabel, C. A., and Kornfeld, S., Studies of the biosynthesis of the mannose-6-phosphate receptor in receptor-positive and receptor-deficient cell lines, *J. Cell Biol.*, 97, 1700, 1983.

39. Griffiths, G., Fuller, S. D., Back, R., Hollinshead, M., Pfeiffer, S., and Simons, K., The dynamic nature of the Golgi complex, *J. Cell Biol.*, 198, 277, 1989.

40. Griffiths, G. and Simons, K., The trans Golgi network: sorting at the exit site of the Golgi complex, *Science*, 234, 438, 1986.

41. Hagenbuchle, O., Tosi, M., Schibler, U., Bovey, R., Wellauer, P. K., and Young, R. A., Mouse liver and salivary gland-amylase mRNAs differ only in 5′ non-translated sequences, *Nature*, 289, 643, 1981.

42. Hand, A. R., Morphology and cytochemistry of the Golgi apparatus of the rat salivary gland acinar cells, *Am. J. Anat.*, 130, 141, 1971.

43. Hand, A. R., The effects of acute starvation on parotid acinar cells. Ultrastructural and cytochemical observations on ad libitum-fed and starved rats, *Am. J. Anat.*, 135, 71, 1973.

44. Hand, A. R. and Oliver, C., The role of GERL in the secretory process, in *Cell Biology of the Secretory Process*, P. M. Conn, Ed., Karger, Basel, Switzerland, 1984, 148.

45. Hashimoto, S., Fumagalli, G., Zanini, A., and Meldolesi, J., Sorting of three secretory proteins to distinct secretory granules in acidophilic cells of cow anterior pituitary, *J. Cell Biol.*, 105, 1579, 1987.

46. Havinga, J. R., Strous, G. J. A. M., and Poort, C., Biosynthesis of the major glycoprotein associated with zymogen-granule membranes in the pancreas, *Eur. J. Biochem.*, 133, 449, 1983.

47. Herzog, V. and Farquhar, M. G., Luminal membrane retrieved after exocytosis reaches most golgi cisternae in secretory cells, *Proc. Natl. Acad. Sci. U.S.A.*, 74, 5073, 1977.

48. Herzog, V. and Miller, F., The localization of endogeneous peroxidase in the lacrimal gland of the rat during postnatal development, *J. Cell Biol.*, 53, 662, 1972.

49. Herzog, V. and Reggio, H., Pathways of endocytosis from luminal plasma membrane in rat exocrine pancreas, *Eur. J. Cell Biol.*, 21, 141, 1980.

50. Hurtley, S. M. and Helenius, A., Protein oligomerization in the endoplasmic reticulum, *Annu. Rev. Cell Biol.*, 5, 277, 1989.

51. Iversen, J. M., Kauffman, D. L., Keller, P. J., and Robinovitch, M., Isolation and partial characterization of two populations of secretory granules from rat parotid glands, *Cell Tissue Res.*, 240, 441, 1985

52. Jahn, R., Unger, C., and Soling, H. D., Specific protein phosphorylation during stimulation of the amylase secretion by agonists or dibutyryl adenosine 3′,5′ monophosphate in the rat parotid gland, *Eur. J. Biochem.*, 112, 345, 1980.

53. Kelly, R. B., Pathways of protein secretion in eukaryotes, *Science*, 230, 25, 1985.

54. Klausner, R. D., Sorting and traffic in the central vacuolar system, *Cell*, 57, 703, 1989.

55. Koike, H. and Meldolesi, J., Post-stimulation retrieval of luminal surface membrane in parotid acinar cells is calcium-dependent, *Exp. Cell Res.*, 134, 377, 1981.

56. Kornfeld, R. and Kornfeld, S., Assembly of asparagine-linked oligosaoccharides, *Annu. Rev. Biochem.*, 54, 631, 1985.

57. Kornfeld, S. and Mellman, I., The biogenesis of lysosomes, *Annu. Rev. Cell Biol.*, 5, 483, 1989.

58. Kousvaleri, E. E., Grant, S. R., Banerjee, D. K., Newby, M. J., and Baum, B. J., Cyclic AMP mediates β-adrenergic-induced increases in N-linked protein glycosylation in rat parotid acinar cells, *Biochem. J.*, 222, 17, 1984.

59. Kousvelari, E. E., Oppenheim, F. G., and Cutler, L. S., Ultrastructural localization of salivary acidic proline-rich proteins from *Macaca fascicularis*, *J. Histochem. Cytochem.*, 30, 274, 1982.

60. Laurie, S. M., Mixon, M. B., and Castle, J. D., Identification of monoclonal antibodies against secretion granule membrane proteins of exocrine and endocrine tissues. *J. Cell Biol.*, 109, 293, 1990.

61. Leslie, B. A. and Putney, J. W., Ionic mechanisms in secretagogue induced morphological changes in rat parotid gland, *J. Cell Biol.*, 97, 1119, 1983.

62. Lillie, J. H. and Han, S. S., Secretory protein synthesis in the stimulated parotid gland. Temporal dissociation of the maximal response from secretion, *J. Cell Biol.*, 59, 708, 1973.

63. Lippincott-Schwartz, J., Donaldson, J. G., Schweizer, A., Berger, E. G., Hauri, H. P., Yuan, L. C., and Klausner, R. D., Microtubule dependent retrograde transport of proteins into the ER in the presence of brefeldin A suggests an ER recycling pathway, *Cell*, 60, 821, 1990.

64. Mains, R. E. and May, V., The role of a low pH intracellular compartment in the processing, storage and secretion of ACTH and endorphin, *J. Biol. Chem.*, 263, 7887, 1988.

65. Marino, C. R., Castle, J. D., and Gorelick, F. S., Regulated phosphorylation of secretory granule membrane proteins of the rat parotid gland, *Am. J. Physiol.*, 259, G70, 1990.

66. Moore, H. P., Gumbiner, B., and Kelly, R. B., Chloroquine diverts ACTH from a regulated to a constitutive secretory pathway in AtT-20 cells, *Nature (London)*, 302, 434, 1983.

67. Moore, H. P. H. and Kelly, R. B., Re-routing of a secretory protein by with human growth hormone sequences, *Nature*, d321, 443, 1986.

68. Mostov, K. E. and Deitcher, D. L., Polymeric immunoglobulin receptor expressed in MDCK cells transcytoses IgA, *Cell*, 46, 613, 1986.

69. Muenzer, J., Bildstein, C., Gleason, M., and Carlson, D. M., Purification of proline-rich proteins from parotid glands of isoproterenol-treated rats, *J. Biol. Chem.*, 255, 5623, 1979.

70. Muller, R. M. and Roomans, G. M., X-ray micronalysis of the rat parotid gland after chronic sympathectomimetic stimulation, *Exp. Mol. Pathol.*, 41, 363, 1984.

71. Oliver, C. and Hand, A.R., Uptake and fate of luminally administered horseradish peroxidase in resting and isoproterenol-stimulated rat parotid acinar cells, *J. Cell Biol.*, 76, 20, 1978.

72. Orci, L., Ravazzola, M., Amherdt, M., Yanaihara, C., Yanaihara, N., Halban, P., Renold, A. E., and Perrelet, A., Insulin, not C-peptide (proinsulin), is present in crinophagic bodies of the pancreatic B-cell, *J. Cell Biol.*, 98, 222, 1984.

73. Orci, L., Ravazzola, M., and Anderson, R. G. W., The condensing vacuole of exocrine cells is more acidic than the mature secretory vesicle, *Nature*, 326, 77, 1987.

74. Ornitz, D. M., Palmiter, R. D., Hammer, R. E., Brinster, R. L., Swift, G. H., and MacDonald, R. J., Specific expression of an elastase-human growth hormone fusion gene in pancreatic acinar cells of transgenic mice, *Nature*, 313, 600, 1985.

75. Palade, G. E., Intracisternal granules in the exocrine cells of the pancreas, *J. Biophys. Biochem. Cytol.*, 2, 417, 1956.

76. Palade, G. E., Intracellular aspects of the process of protein synthesis, *Science*, 189, 347, 1975.

77. Patzelt, C., Brown, D., and Jeanrenaud, B., Inhibitory effect of colchicine on amylase secretion by the rat parotid gland, *J. Cell Biol.*, 73, 578, 1977.

78. Pelham, H. R. B., Control of protein exit from the endoplasmic reticulum, *Annu. Rev. Cell Biol.*, 5, 1, 1989.

79. Quissell, D. O. and Tabak, L. A., Salivary mucin secretion, in *Handbook of Physiology Section 6: The Gastrointestinal System III, Salivary, Gastric, Pancreatic and Hepatobiliary Secretion*, J. G. Forte, Ed., American Physiological Society, New York, 1989, 79.

80. Redman, C. M. and Sabatini, D. D., Vectorial discharge of peptides released by puromycin from attached ribosomes, *Proc. Natl. Acad. Sci. U.S.A.*, 56, 608, 1966.

81. Reinhart, F. D. and Neutra, M. R., Identification of a membrane antigen in secretory granules of primate cells, *J. Cell Biol.*, 109, 294a, 1989.

82. Romagnoli, P. and Herzog, V., Reinternalization of secretory proteins during membrains recycling in rat pancreatic acinar cells, *Eur. J. Cell. Biol.*, 44, 167, 1987.

83. Rosa, P., Weiss, U., Pepperkok, R., Ansorge, W., Niehrs, C., Stelzer, E. H. K., and Huttner, W. B., An antibody against secregogranin I (chromogranin B) is packaged into secretory granules, *J. Cell Biol.*, 109, 17, 1989.

84. Roth, J. and Berger, E. G., Immunocytochemical localization of galactosyl transferase in HeLa cells. Codistribution with thiamine pyrophosphatase in *trans*-Golgi cisternae, *J. Cell Biol.*, 93, 223, 1982.

85. Rothman, J. E., Protein sorting by selective retention in the endoplasmic reticulum and Golgi stack, *Cell*, 50, 521, 1987.

86. Rothman, J. E., Miller, R. L., and Urbani, L. J., Intracompartmental transfer in the Golgi is a dissociative process: facile transfer of membrane protein between two Golgi populations, *J. Cell Biol.*, 99, 260, 1984.

87. Rothman, J. E. and Orci, L., Movement of proteins through the Golgi stack: a molecular dissection of vesicular transport, *FASEB J.*, 4, 1460, 1990.

88. Roth, J., Taatjes, D. J., Lucocq, J. M., Weinstein, J., and Paulson, J. C., Demonstration of an extensive trans-tutular network continuous with the Golgi apparatus stack that may function in glycosylation, *Cell*, 43, 287, 1985.

89. Saraste, J. and Kuismanen, E., Pre- and post-Golgi vacuoles operate in the transport of Semliki forest virus membrane glycoproteins to the cell surface, *Cell*, 38, 535, 1984.

90. Scheele, G. and Tartakoff, A., Exit of nonglycosylated secretory proteins from the rough endoplasmic reticulum is asynchronous in the exocrine pancreas, *J. Biol. Chem.*, 260, 926, 1985.

91. Schramm, M. and Selinger, Z., The functions of cyclic AMP and calcium as alternative second messengers in parotid gland and pancreas, *J. Cyclic Nucl. Res.*, 1, 181, 1975.

92. Setser, M. E., Spicer, S. S., Simson, J. A. V., and Martinez, J. R., Altered, granule discharge and amylase secretion of parotid glands in reserpine-treated rats, *Lab. Invest.*, 41, 256, 1979.

93. Sharoni, Y., Eimerl, S., and Schramm, M., Secretion of old versus new exportable protein in rat parotid slices. Control by neurotransmitters, *J. Cell Biol.*, 71, 107, 1976.

94. Spearman, T. N. and Butcher, F. R., Cellular regulation of amylase secretion in the parotid gland, in *Handbook of Physiology Section 6: The Gastrointestinal System III, Salivary, Gastric Pancreatic and Hepatobiliary Secretion*, J. G. Forte, Ed., American Physiological Society, New York, 1989, 63.

95. Spearman, T. N., Hurley, K. P., Olivas, R., Ulrich, R. G., and Butcher F. R., Subcellular localization of stimulus-affected endogenous phosphoproteins in the rat parotid gland, *J. Cell Biol.*, 99, 1354, 1984.

96. Sreebny, L. M., Johnson, D. A., and Robinovitch, M. R., Functional regulation of protein synthesis in the rat parotid gland, *J. Biol. Chem.*, 246, 3879, 1971.

97. Stoller, T. J. and Shields, D., The propeptide of preprosomatostatin mediates intracellular transport and secretion of alpha-globin from mammalian cells, *J. Cell Biol.*, 108, 1647, 1989.

98. Tandler, B. and MacCallum, D. K., Ultrastructure and histochemistry of the submandibular gland of the European hedgehog, *Erinaceous europaeus* L. *J. Ultrastruct. Res.*, 39, 186, 1972.

99. Tartakoff, A. M., Temperature and energy dependence of secretory protein transport in the exocrine pancreas, *EMBO J.*, 5, 1477, 1986.

100. Tooze, S. A. and Huttner, W. B., Cell-free protein sorting to the regulated and constitutive secretory pathways, *Cell*, 60, 837, 1990.

101. Tooze, J., Kern, H. F., Fuller, S. D., and Howell, K. E., Condensation-sorting events in the rough endoplasmic reticulum of exocrine pancreatic cells, *J. Cell Biol.*, 109, 35, 1989.

102. Tooze, J., Tooze, S. A., and Fuller, S. D., Sorting of progeny coronavirus from condensed secretory proteins at the exit of the *trans*-Golgi network of AtT20 cells, *J. Cell Biol.*, 105, 1215, 1987.

105. Tooze, S. A., Tooze, J., and Warren, G., Site of addition of N-acetyl-galactosamine to the El glycoprotein of mouse hepatitis virus – A59, *J. Cell Biol.*, 106, 1475, 1988.

104. Vaux, D., Tooze, J., and Fuller, S., Identification by anti-idiotype antibodies of an intracellular membrane protein that recognizes a mammalian endoplasmic reticulum retention signal, *Nature*, 345, 495, 1990.

105. Von Heijne, G., Signal sequences: the limits of variation, *J. Mol. Biol.*, 184, 99, 1985.

106. Wallach, D. and Schramm, M., Calcium and the exportable protein in rat parotid gland. Parallel subcellular distribution and concomitant secretion, *Eur. J. Biochem.*, 21, 483, 1971.

107. Walter, P. and Lingappa, V., Mechanisms of protein translocation across the endoplasmic reticulum membrane, *Annu. Rev. Cell Biol.*, 2, 499, 1986.

108. Young, J. A., Cook, D. I., van Lennep, E. W., and Roberts, M., Secretion of the major salivary glands, in *Physiology of the Gastrointestinal Tract*, 2nd ed., L. R. Johnson, Ed., Raven Press, NY, 1987, 733.

109. von Zastrow, M. and Castle, J. D., Protein sorting among two distinct export pathways occurs from the content of maturing exocrine storage granules, *J. Cell Biol.*, 105, 2675, 1987.

110. Zastrow, M. V., Castle, A. M., and Castle, J. D., Ammonium chloride alters secretory protein sorting within the maturing exocrine storage compartment, *J. Biol. Chem.*, 264, 6566, 1989.

111. Zastrow, M. V., Tritton, T. R., and Castle, J. D., Exocrine secretion granules contain peptide amidation activity, *Proc. Natl. Acad. Sci. U.S.A.*, 83, 3297, 1986.

112. Ziemer, M. A., Swain, W. F., Rutter, W. J., Clements, A. S., Ann, D. K., and Carlson, D. M., Nucleotide sequence analysis of a proline-rich protein cDNA and peptide homologies of rat and human proline-rich proteins, *J. Biol. Chem.*, 259, 10475, 1984.

Ion Transport Related to Fluid Secretion in Salivary Glands

R. James Turner

INTRODUCTION

In salivary glands, as in all secretory and absorptive epithelia, fluid transport is thought to be driven osmotically in response to transepithelial salt gradients (Diamond, 1979; Spring, 1983). These salt gradients are generated by ion transport systems localized to the luminal or basolateral membranes of the epithelial cells. This review will focus on recent studies which have explored the involvement of various ion transporters in fluid secretion by salivary glands. Particular emphasis will be placed on the effectiveness of combining complementary results obtained from intact glands, isolated acini and membrane vesicles in order to gain a more complete understanding of the physiology of the fluid secretory process. No attempt has been made to treat the historical aspects of this field in any detail. For these points the interested reader is referred to several excellent reviews (Young and van Lennep, 1979; Petersen, 1980; Martinez, 1987; Young et al., 1987a).

According to the two-stage hypothesis of salivary formation proposed by Thaysen et al. (Thaysen et al., 1954; Thaysen, 1960), saliva is initially formed as a nearly isotonic plasma-like "primary" secretion in the acinar lumen. This primary saliva is then modified (the "second stage" of Thaysen's hypothesis) by the removal or addition of various ions in the salivary ducts with no further secretion or absorption of water. The plasma-like ionic composition of the primary saliva has been confirmed by analyzing the electrolyte content of fluid collected from the lumen of the acinus/intercalated duct region using micropuncture techniques (Young and van Lennep, 1979; Martinez, 1987; Young et al., 1987a). In addition, microperfusion studies of excretory (extralobular) ducts have verified their ability to modify saliva by reabsorbing sodium and chloride, and secreting potassium and bicarbonate (Young and van Lennep, 1979; Martinez, 1987; Young et al., 1987a). Indeed, salivary ducts can produce steep concentration gradients of these ions between the lumen and interstitium. The hypotonicity of saliva suggests that the ducts are relatively impermeant to water and estimates of their hydraulic conductivity coefficient are extremely low (Young and van Lennep, 1979). Considerable indirect evidence indicates that these findings from excretory ducts apply also to the intralobular (striated) ducts (Young et al., 1987a), which, for technical reasons, have not been studied by microperfusion. Thus, although further experimentation is obviously required, the data currently available strongly support the contention that Thaysen's two-stage hypothesis applies, at least to a good first approximation, to most salivary glands. These include those of humans and of the commonly used laboratory animals (Young and van Lennep, 1979).

In a series of papers published in the late-1950s Lundberg (Lundberg, 1957; Lundberg, 1958) demonstrated that salivary fluid secretion in the cat sublingual gland depended on the anionic composition of the extracellular fluid. Based on this observation and electrophysiological data he proposed that salivary secretion was driven by active chloride uptake across the acinar basolateral membrane. Unfortunately, however, Lundberg's proposal fell into disfavor and for 25 years attention was mainly focused on the role of cations in the secretion of primary saliva (Young and van Lennep, 1979; Petersen, 1980). More recent studies, reviewed below, have confirmed the important role of anions in this process and demonstrated that secretion is, in fact, primarily driven by active chloride transport, albeit not of the type proposed by Lundberg.

At the same time as Lundberg carried out his studies, Burgen (Burgen, 1956) demonstrated that salivary glands lose potassium in response to parasympathetic nerve stimulation. This loss of potassium can also be evoked by direct application of muscarinic or α-adrenergic agonists to perfused glands, gland slices, or isolated acini (Putney, 1986; Martinez, 1987; Izutzu et al., 1987; Nauntofte and Dissing, 1988a). β-adrenergic agonists, which, unlike muscarinic or α-adrenergic agonists, do not induce significant fluid secretion, do not produce potassium release. An extensive series of electrophysiological and other studies has now clearly demonstrated that this potassium loss occurs via a Ca^{++} activated K^+ channel located in the basolateral membrane of the acinar cell (Petersen, 1986; Putney, 1986; Petersen and Gallacher, 1988; Dehaye et al., 1987; Nauntofte and Dissing, 1988a). In this regard, it is now well established that changes in intracellular free calcium concentrations ($[Ca^{++}]_{in}$) play a central role in the response of salivary acinar cells to agents which induce fluid secretion. The effect of neurotransmitters on calcium mobilization is reviewed in detail in Chapter 7 of this volume (Baum et al., 1992). For the purposes of the present discussion it is only important to realize that the application of these secretagogues to the acinar cell results in a rapid ($t_{1/2}$ ~0.5 s) rise in $[Ca^{++}]_{in}$. Over the next several minutes $[Ca^{++}]_{in}$ then falls to a lower value which is significantly above resting levels and which is sustained in the continued presence of the agonist.

MODELS FOR SALIVARY FLUID SECRETION

Anion Dependence of Fluid Secretion

In 1977 Silva et al. proposed a novel mechanism for salt secretion in the shark rectal gland based on the concept of active chloride secretion. This model, which is discussed in detail below, prompted researchers to re-examine the role of anions, and particularly that of chloride, in salivary fluid secretion (Case et al., 1981; Poulsen and Kristensen, 1982; Young et al., 1982; Petersen and Maruyama, 1984). Subsequent studies were performed using a variety of methodologies, however, from a didactic point of view it is instructive to first consider the results of a series of experiments carried out from 1981 to 1987 with perfused rat and rabbit submandibular glands. These studies showed that acetylcholine-induced salivary secretion is markedly reduced (~70%) when Cl^- is replaced in a Cl^-/HCO_3^--replete perfusate by a physiologically inert anion (Case et al., 1981; Case et al., 1984; Martinez and Cassity, 1985a; Novak and Young, 1986; Pirani et al., 1987), or when the loop diuretic furosemide (Poulsen et al., 1982; Martinez and Cassity, 1983; Case et al., 1984; Martinez and Cassity, 1985a; Martinez and Cassity, 1985b; Novak and Young, 1986) or its more potent and specific analogue bumetanide (Pirani et al., 1987) is present. These conditions also lead to an increase in salivary HCO_3^- concentration (Case et al., 1984; Martinez and Cassity, 1985a; Martinez and Cassity, 1985b; Pirani et al., 1987). The effect of loop diuretics was even more dramatic in HCO_3^--free media where the inhibition of fluid secretion was >95% (Martinez and Cassity, 1985a; Martinez and Cassity, 1985b; Novak and Young, 1986). In addition, salivary secretion was observed to persist in glands perfused with Cl^--free media ($[HCO_3^-] = 25$ mM), although at a reduced rate (~30% of control), and this residual secretion was blocked by carbonic anhydrase inhibitors (Case et al., 1984; Novak and Young, 1986; Pirani et al., 1987). Removal of both Cl^- and HCO_3^- from the perfusate completely eliminated acetylcholine-induced fluid secretion (Case et al., 1984; Novak and Young, 1986). Taken together these results indicate the presence of at least two components of salivary fluid secretion, one which is Cl^- dependent and sensitive to loop diuretics and another which is dependent on the presence of HCO_3^-.

An initially puzzling result from these perfusion studies was the observation that the rate of agonist-induced fluid secretion was not affected by the removal of HCO_3^- from the perfusate (Case et al., 1981; Case et al., 1982; Case et al.,

1984; Martinez and Cassity, 1985b). It was also found that carbonic anhydrase inhibitors were without effect in a Cl^-/HCO_3^--replete perfusate (Case et al., 1982; Martinez and Cassity, 1985a; Pirani et al., 1987). Both of these maneuvers did, however, result in a decrease in salivary HCO_3^- concentration (Case et al., 1982; Martinez and Cassity, 1985a; Pirani et al., 1987). A simple explanation of these observations, which is also consistent with more recent findings discussed below, is that the Cl^--dependent and HCO_3^--dependent mechanisms of salivary fluid secretion in some way compete with one another so that in the absence of HCO_3^- the Cl^--dependent mechanism is free to operate at a higher rate. In this regard, it has been demonstrated that the ability of 25 mM HCO_3^- to support salivary secretion by perfused submandibular glands is about the same as that of 25 mM Cl^- (Case et al., 1982; Novak and Young, 1984).

The models which have been proposed to account for salivary fluid secretion and the data supporting them are discussed in the following sections.

Chloride Secretion Driven by Na^+/K^+/Cl^- Cotransport

The model of Silva et al. mentioned earlier (Silva et al., 1977) proposed that salt secretion in the shark rectal gland was due to active Cl^- secretion driven by a loop diuretic sensitive cotransport system for Cl^- localized to the basolateral membrane of the secretory cell. Thus, this mechanism was thought to be a good candidate for the Cl^--dependent component of fluid secretion identified in the perfusion experiments discussed above. The applicability of this model to salivary glands has, in fact, been confirmed by a variety of studies, but before discussing the experimental evidence in favor of the model let us first consider how it functions. The model of Silva et al. as it is thought to apply to salivary glands (and to many other exocrine epithelia) is illustrated in Figure 1 (Silva et al., 1977; Petersen and Maruyama, 1984). According to this scheme, fluid secretion is the result of the concerted action of four membrane transport systems: (1) a loop diuretic sensitive Na^+/K^+/Cl^- cotransporter located in the basolateral membrane of the acinar cells (in some tissues a K^+-independent Na^+/Cl^- cotransporter may instead be involved), (2) the basolateral Ca^{++} activated K^+ channel referred to above, (3) an apical conductive pathway for Cl^-, and (4) the Na^+/K^+ ATPase. In the resting (unstimulated) state, both K^+ and Cl^- are concentrated in the acinar cell above electrochemical equilibrium, the former by Na^+/K^+ ATPase and the latter via the Na^+/K^+/Cl^- cotransporter. Stimulation of the acinar cell by secretagogues leads to a rise in $[Ca^{++}]_{in}$ (Baum et al., 1990) which in turn results in the opening of the basolateral Ca^{++} activated K^+ channel and the apical Cl^- channel. These Ca^{++} associated increases in K^+ and Cl^- conductance allow KCl to flow out of the cell resulting in an accumulation of Cl^- ions and their associated negative electrical charge in the acinar lumen. Na^+ is then thought to follow Cl^- by leaking from the interstitium through the tight junctions between the cells in order to preserve electroneutrality. The resulting

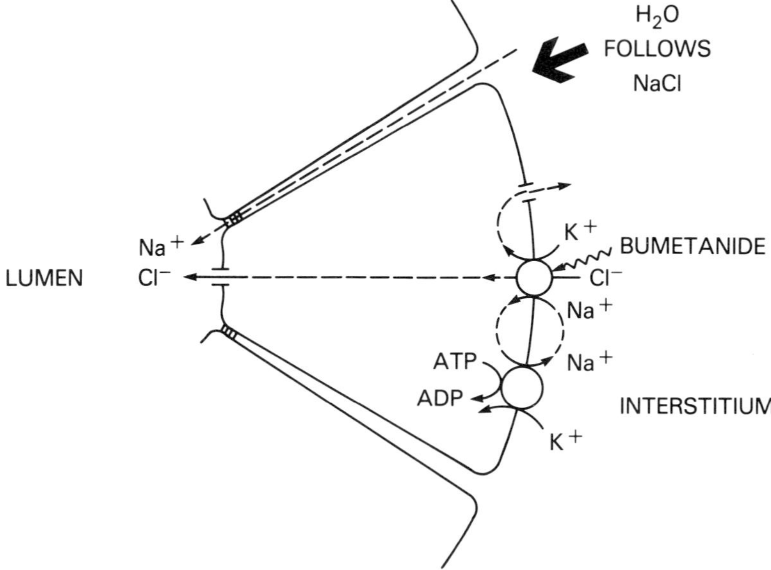

Figure 1. Model for acinar fluid secretion based on Cl⁻ entry across the basolateral membrane via a Na⁺/K⁺/Cl⁻ cotransport system and exit across the apical membrane via an agonist gated Cl⁻ channel. The loop diuretic furosemide and its more potent and specific analogue bumetanide are inhibitors of the Na⁺/K⁺/Cl⁻ cotransporter. See text for details. (Redrawn from Turner, R. J., George, J. N., and Baum, B. J., *J. Membrane Biol.*, 94, 143, 1986. With permission.)

osmotic gradient for NaCl causes a transepithelial movement of water from interstitium to lumen. In the continued presence of the agonist a net transepithelial chloride flux and a concomitant secretion of fluid is sustained as a result of Cl⁻ entry via the Na⁺/K⁺/Cl⁻ cotransporter and exit via the apical Cl⁻ channel. When the stimulus is removed, $[Ca^{++}]_{in}$ falls to resting levels, the Cl⁻ and K⁺ channels close, and the cell returns to its resting state.

Besides providing a Cl⁻-dependent, loop diuretic sensitive component of fluid secretion and incorporating the Ca⁺⁺ activated K⁺ channel known to be associated with secretory events (Petersen, 1986; Petersen and Gallacher, 1988), this model nicely accommodates many older observations concerning salivary secretion, some of which were difficult to explain using earlier models of epithelial salt and water movements based on active cation transport. In particular, it accounts for the near-isotonic plasma-like electrolyte content of the primary saliva, the anion dependence and lumen-negative potentials observed by Lundberg (Lundberg, 1957), the large venous K⁺ loss observed by Burgen (Burgen, 1956), and the requirement of extracellular Ca⁺⁺ for a sustained secretory response (Martinez and Petersen, 1971; Martinez, 1987; Putney, 1986; Baum et al., 1990). It is also consistent with the relatively high concentrations of intracellular chloride observed in salivary glands (Schneyer and Schneyer, 1960; Nauntofte and Poulsen, 1986; Lau and Case, 1988; Poulsen and Nauntofte, 1990). These concentrations (50 to 75 mM) are well above the electrochemical equilibrium value for $[Cl^-]_{in}$ (approximately 13 mM, assuming an intracellular potential of -59 mV; Petersen, 1980; Lau and Case, 1988; Nauntofte and Dissing, 1988a) but correspond well with the $[Cl^-]_{in}$ predicted for a $Na^+/K^+/Cl^-$ cotransporter operating close to thermodynamic equilibrium. Assuming a $Na^+:K^+:Cl^-$ coupling stoichiometry of 1:1:2 (see below) this value is given by the equation $[Na^+]_{in}[K^+]_{in}[Cl^-]_{in}^2 = [Na^+]_{out}[K^+]_{out}[Cl^-]_{out}^2$ (Turner, 1985). Using the intracellular concentrations for sodium and potassium determined in the rat parotid by Poulsen and Nauntofte (1990) this equation yields $[Cl^-]_{in} = 69$ mM.

In addition to the evidence for a Cl^--dependent, loop diuretic, sensitive component of fluid secretion from perfusion experiments, a number of flux studies using rat submandibular and parotid acini isolated by collagenase/ hyaluronidase digestion have provided further proof of the existence of a $Na^+/K^+/Cl^-$ cotransport system in salivary acinar cells. Dramatic inhibitions (40 to 85%) of $^{36}Cl^-$ fluxes by loop diuretics and by Na^+ or K^+ removal have been demonstrated in these preparations (Martinez and Cassity, 1985c; Kawaguchi et al., 1986; Nauntofte and Poulsen, 1986; Melvin et al., 1987; Martinez and Barker, 1989). Loop diuretic sensitive components of $^{42}K^+$ (or $^{86}Rb^+$) and $^{22}Na^+$ uptake have also been documented (Poulsen and Kristensen, 1982; Kawaguchi et al., 1986; Nauntofte and Dissing, 1988a; Soltoff et al., 1989). Moreover, direct evidence has recently been provided for a $Na^+/K^+/Cl^-$ cotransporter and its associated bumetanide binding site in basolateral membrane vesicles from the rabbit parotid (Turner et al., 1986; Turner and George, 1988a). These vesicle experiments demonstrate the presence of a loop diuretic-sensitive, KCl-dependent component of sodium flux. The properties of this transporter are consistent with a $Na^+:K^+:Cl^-$ coupling stoichiometry of 1:1:2 as observed for $Na^+/K^+/Cl^-$ cotransporters in a number of other tissues (Geck and Heinz, 1986; O'Grady et al., 1987). This coupling stoichiometry is also supported by data from oxygen consumption studies in intact salivary glands and isolated acini (Smaje et al., 1986; Poulsen and Nauntofte, 1987).

Attempts to characterize the apical Cl^- channel hypothesized in Figure 1 in vesicle studies have been frustrated by the relatively small size of this membrane (approximately 10% of the total acinar cell surface; Cope and Williams, 1973). However, patch-clamp experiments using the whole-cell configuration have provided good evidence for the presence of Ca^{++}-dependent Cl^- channels in salivary acinar cells which open in response to secretagogue stimulation (Iwatsuki et al., 1985; Cook et al., 1988; Gray, 1989) . It has also been demonstrated that Cl^- fluxes in stimulated rat parotid and submandibular acini are inhibited by Cl^- channel blockers such as diphenylamine-2-carboxylate and its analogues (Melvin et al., 1987; Martinez et al., 1987; Poulsen et al., 1988). However, definitive evidence for the localization of these Cl^- channels to the apical membrane of the acinar cells is still lacking.

In addition to the results discussed above which establish the presence of the components of the model illustrated in Figure 1 in salivary acinar cells, recent studies with intact acini have also addressed some predictions of the model and its ability to account for the observed levels of salivary fluid production. Measurements with isotopic Cl^- fluxes and Cl^--sensitive microelectrodes show a dramatic loss of acinar Cl^- content following stimulation by muscarinic (Poulsen and Kristensen, 1982; Mori et al., 1983; Martinez and Cassity, 1985c; Nauntofte and Poulsen, 1986; Melvin et al., 1987; Lau and Case, 1988; Soltoff et al., 1989) or α-adrenergic (Ambudkar et al., 1988; Martinez and Reed, 1988; Poulsen et al., 1988) agonists. A similar effect is seen in response to treatment of acini with the Ca^{++} ionophore A23187 (Martinez and Cassity, 1986, Nauntofte and Poulsen, 1986). These results strongly support the contention that Cl^- secretion is intimately involved with fluid secretion and that secretory events are mediated by $[Ca^{++}]_{in}$. A more direct demonstration of the involvement of $[Ca^{++}]_{in}$ in secretory events was recently obtained by Foskett and Melvin (1989) who simultaneously measured $[Ca^{++}]_{in}$ and cell volume in rat parotid acinar cells using optical techniques. They showed that acinar cells shrink following stimulation by the muscarinic agonist carbachol, consistent with the loss of intracellular KCl predicted by Figure 1, and that the rise in $[Ca^{++}]_{in}$ slightly precedes this shrinkage. Shrinkage could also be induced without receptor stimulation by raising $[Ca^{++}]_{in}$ with the Ca^{++} ionophore ionomycin. When carbachol induced oscillations in $[Ca^{++}]_{in}$ were observed (Berridge and Galione, 1988), corresponding oscillations in cell volume were seen. In experiments where $[Ca^{++}]_{in}$ was clamped at low values by loading the cells with a permeant Ca^{++} chelator, little rise in $[Ca^{++}]_{in}$ and no cell shrinkage was seen. Taken together with direct demonstrations of agonist-induced acinar Cl^- and K^+ loss these results provide strong evidence that $[Ca^{++}]_{in}$ is the dominant if not the only intracellular messenger involved in the acute control of the acinar K^+ and Cl^- conductance pathways.

Figure 2 shows an experiment which addresses several more specific points concerning the model in Figure 1. Here the chloride content of rat parotid acini was monitored with $^{36}Cl^-$ after the addition of carbachol at time zero. When carbachol alone was added to the preparation (Curve C), the acini showed a rapid ($t_{1/2}$ <10 s.) loss of ~50% of their intracellular Cl^- followed by a much slower ($t_{1/2}$ ~ 2 min) recovery to approximately 70% of their original chloride content. This recovery coincides with the transition between the initial and the sustained phases of fluid secretion and presumably reflects the adjustment of the cells to a new steady-state characteristic of a continuous secretory response. The recovery is blocked by the loop diuretic bumetanide (Curve B) at a concentration (0.1 mM) where it specifically inhibits $Na^+/K^+/Cl^-$ cotransport. This observation supports the model shown in Figure 1 which identifies the $Na^+/K^+/Cl^-$ cotransporter as the (only) Cl^- entry pathway. Acini which were exposed to carbachol in the presence of the anion channel blocker diphenylamine-2-carboxylate (Curve DPC) showed a blunted initial Cl^- loss and an enhanced recovery of intracellular

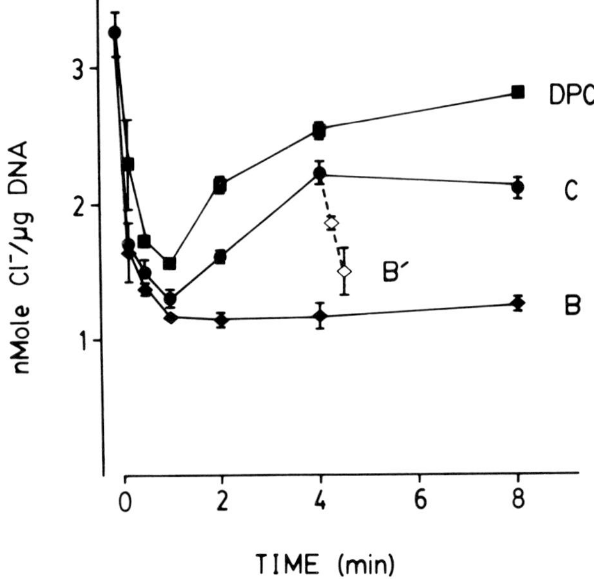

Figure 2. Response of chloride content of rat parotid acini to stimulation by the muscarinic agonist carbachol. Acini were preequilibrated with $^{36}Cl^-$ in a HCO_3^--free salt solution for 30 min at 37°C. At time zero 10 µM carbachol alone (Curve C), or with 0.1 mM bumetanide (Curve B) or 1 mM diphenylamine-2-carboxylate (Curve DPC) was added to appropriate aliquots of acini. To a fourth aliquot 10 µM carbachol was added at time zero followed by 0.1 mM bumetanide at 4 min. $^{36}C^-$ content was determined at the times indicated by a rapid filtration assay. (Reproduced from Melvin, J. E., Kawaguchi, M., Baum, B. J., and Turner, R. J., *Biochem. Biophys. Res. Commun.*, 145, 745, 1987. With permission.)

chloride content. This result is also consistent with the model which identifies the Cl^- exit pathway as an apical anion channel.

Curve B′ in Figure 2 provides a more quantitative test of the model. Here 0.1 mM bumetanide was added to the acinar preparation 4 min after the addition of carbachol. Bumetanide reverses the recovery of Cl^- content, presumably by blocking the $Na^+/K^+/Cl^-$ entry step, while allowing continued Cl^- loss via the apical channel. Using the fact that the primary saliva is secreted as an isotonic NaCl solution it is possible to convert this rate of Cl^- loss (1.4 nmol/µg DNA/ min) to a rate of fluid secretion (10 µl/mg DNA per minute). Since a rat parotid gland contains about 1.4 mg of DNA one can calculate a predicted isotonic primary salivary flow rate of 14 µl per gland per minute. This value is in excellent agreement with published results (Bodner and Baum, 1985; Abe and Dawes, 1975) for muscarinic agonist induced fluid secretion from rat parotid glands *in vivo* (approximately 8 µl per gland per minute). Thus, the results shown in Figure 2 provide both strong qualitative and quantitative support for applicability of the model of Silva et al. to fluid transport in salivary glands.

Bicarbonate Secretion Driven by the Na^+/H^+ Exchanger

As already discussed, perfusion studies with rat and rabbit submandibular glands demonstrate the presence of a HCO_3^--dependent component of fluid secretion which is sensitive to carbonic anhydrase inhibitors and persists in the absence of Cl^- (Case et al., 1984; Novak and Young, 1986; Pirani et al., 1987). This component of secretion can be dramatically blocked (~95%) by amiloride (Novak and Young, 1986), a relatively specific inhibitor of the Na^+/H^+ exchanger (Benos, 1982). These results led Young and collaborators to suggest that HCO_3^- secretion from salivary acinar cells occurs in parallel with Cl^- secretion and that this HCO_3^- secretion is driven by the Na^+/H^+ exchanger (Case et al., 1984; Novak and Young, 1986; Pirani et al., 1987). There is now considerable evidence from a variety of methodologies supporting this hypothesis.

The presence of an electroneutral Na^+/H^+ exchanger has been demonstrated in a wide variety of cell types. This transport system is thought to play a critical role in a number of physiological functions including fertilization, the regulation of intracellular pH and cell volume, the initiation of cell growth and proliferation, and the transepithelial transport of salt and acid/base equivalents (Aronson, 1985; Grinstein and Rothstein, 1986; Grinstein et al., 1989). Convincing evidence has been found for the presence of a potent Na^+/H^+ exchanger in salivary acinar cells. An amiloride sensitive component of $^{22}Na^+$ uptake into rat parotid and submandibular acini has been demonstrated (Soltoff et al., 1989; Nauntofte and Dissing, 1989; Martinez et al., 1989) and the presence of a Na^+/H^+ exchanger has been shown directly in rat and rabbit parotid acinar basolateral membrane vesicles (Manganel and Turner, 1988). In addition, it has been shown that amiloride and its analogues block the recovery of parotid and submandibular acini from an acute acid load (Melvin et al., 1988; Soltoff et al., 1989; Lau et al., 1989) providing strong evidence that the Na^+/H^+ exchanger plays an important role in pH regulation in these cells. Amiloride also dramatically reduces stimulated salivary secretion in perfused rat submandibular glands (Martinez and Cassity, 1985b; Pirani et al., 1987).

Studies with isolated rat parotid and rabbit submandibular acini strongly support the hypothesis that HCO_3^- is secreted in response to muscarinic stimulation. These preparations show a transient intracellular acidification (~0.1 pH units) in response to carbachol or acetylcholine (Nauntofte and Dissing, 1988b; Melvin et al., 1988; Lau et al., 1989; Soltoff et al., 1989; Stewart et al., 1989; Brown et al., 1989) which is blunted in a nominally HCO_3^--free medium or in the presence of carbonic anhydrase inhibitors (Nauntofte and Dissing, 1988b; Melvin et al., 1988; Lau et al., 1989; Stewart et al., 1989; Brown et al., 1989). This acidification is dramatically enhanced in the presence of amiloride (Melvin et al., 1988; Soltoff et al., 1989; Lau et al., 1989; Stewart et al., 1989). An example of this type of experiment is illustrated in Figure 3. Here rat parotid acini were preloaded with the fluorescent intracellular pH indicator BCECF. When these

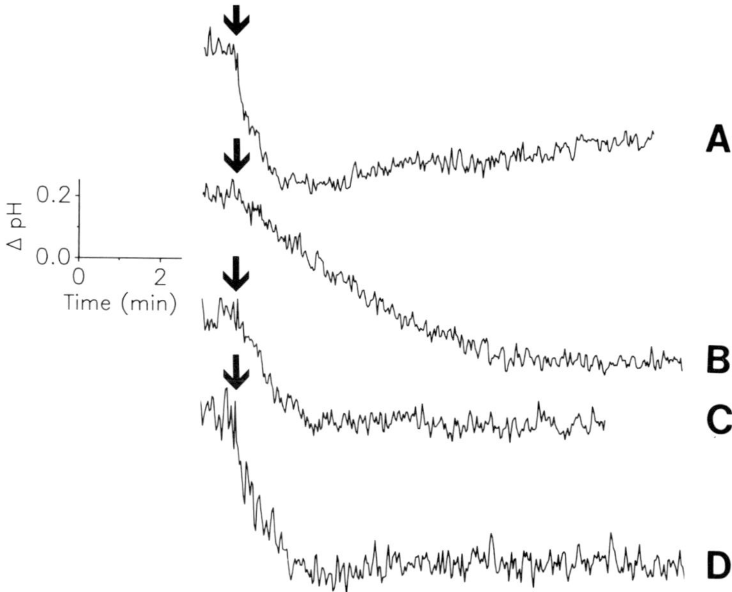

Figure 3. Carbachol-induced acidification of rat parotid acini unmasked by inhibition of the Na^+/H^+ exchanger. The effects of HCO_3^--free medium, the carbonic anhydrase inhibitor methazolamide, and the disulfonic stilbene SITS are illustrated. Rat parotid acini were preloaded with the fluorescent intracellular pH indicator BCECF and suspended in a HCO_3^--replete physiological salt solution (Trace A), in the same medium without HCO_3^- (Trace B), in HCO_3^--replete medium plus 1 mM methazolamide (Trace C) or in HCO_3^--replete medium plus 1 mM SITS (Trace D). All media contained 0.5 mM amiloride. At the times indicated by the arrows 10^{-3} M carbachol was added. The traces represent intracellular BCECF fluorescence which is a direct measure of intracellular pH (inset). Resting values for acinar pH were 7.35 ± 0.06 in HCO_3^--replete medium and 7.48 ± 0.08 in HCO_3^--free medium. (Reproduced from Melvin, J. E., Moran, A., and Turner, R. J., *J. Biol. Chem.*, 263, 19564, 1988. With permission.)

acini are stimulated with carbachol in the presence of amiloride an acidification of ~0.4 pH units occurs over a time period of ~1 min (Trace A). This acidification is carbachol dose dependent and blocked by the muscarinic antagonist atropine (Melvin et al., 1988). The rate of carbachol-induced acidification is significantly reduced in HCO_3^--free medium (Trace B) as well as in HCO_3^--replete medium in the presence of the carbonic anhydrase inhibitor methazolamide (Trace C). Since both of these maneuvers presumably reduce the available intracellular HCO_3^-, and also would normally be expected to impair rather than enhance the ability of the cells to regulate their intracellular pH, these experiments provide strong evidence that this agonist-induced acidification is largely the result of a rapid loss of intracellular HCO_3^- and that the Na^+/H^+ exchanger normally buffers the acinar cytoplasm against the resulting acid load.

One obvious candidate for this HCO_3^- exit pathway is via Cl^-/HCO_3^- exchange. The presence of such a transporter has been documented in many epithelia, including salivary acinar cells (see below). But Trace D in Figure 3

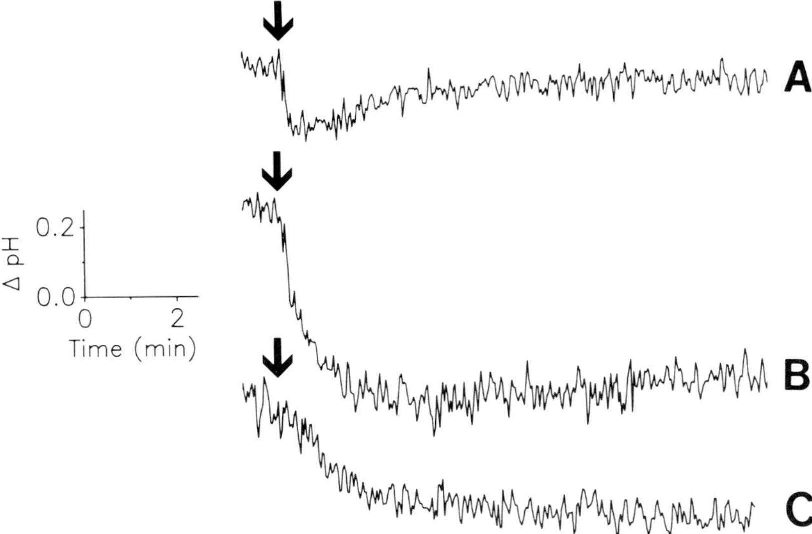

Figure 4. The effects of Cl⁻-free medium and the Cl⁻ channel blocker diphenylamine-2-carboxylate on carbachol-induced acidification of rat parotid acini. BCECF-loaded acini were suspended in a Cl⁻-free salt solution containing no additions (Trace A), 0.5 mM amiloride (Trace B), or 0.5 mM amiloride plus 1 mM diphenylamine-2-carboxylate (Trace C), then stimulated with carbachol (10^{-3} M) as indicated. All media contained 25 mM HCO_3^- with gluconate replacing chloride. The resting values for acinar pH were the same for all traces. (Reproduced from Melvin, J. E., Moran, A., and Turner, R. J., *J. Biol. Chem.*, 263, 19564, 1988. With permission.)

shows that SITS, a potent inhibitor of Cl^-/HCO_3^- exchange, does not affect the rate of carbachol induced acidification. Moreover, in Figure 4 it is shown that carbachol induced acidification persists, and, in fact, is enhanced, in Cl⁻-free medium (cf. Figure 3, Trace A and Figure 4, Trace B). The rate of acinar acidification produced by carbachol is, however, decreased by the anion channel blocker diphenylamine-2-carboxylate (Melvin et al., 1988; Brown et al., 1989) as illustrated in Trace C of Figure 4.

A model for the involvement of Na^+/H^+ exchange and HCO_3^- in salivary fluid secretion is illustrated in Figure 5 (Young et al., 1987b; Petersen and Gallacher, 1988; Melvin et al., 1988). Here CO_2 enters the acinar cell across the basolateral membrane and is converted to HCO_3^- plus a proton by intracellular carbonic anhydrase (c.a.). HCO_3^- is lost across the apical membrane via a diphenylamine-2-carboxylate sensitive pathway, presumably an anion channel, and the proton is extruded by the basolateral Na^+/H^+ exchanger. This mechanism gives rise to a net transepithelial flux of HCO_3^- in parallel with the transepithelial Cl^- flux produced by the basolateral $Na^+/K^+/Cl^-$ cotransporter and the apical Cl^- channel (Figure 1). Both anion fluxes contribute to fluid secretion in the manner outlined earlier. This model can account for virtually all of the data referenced above concerning fluid secretion by intact glands perfused with Cl⁻-free media

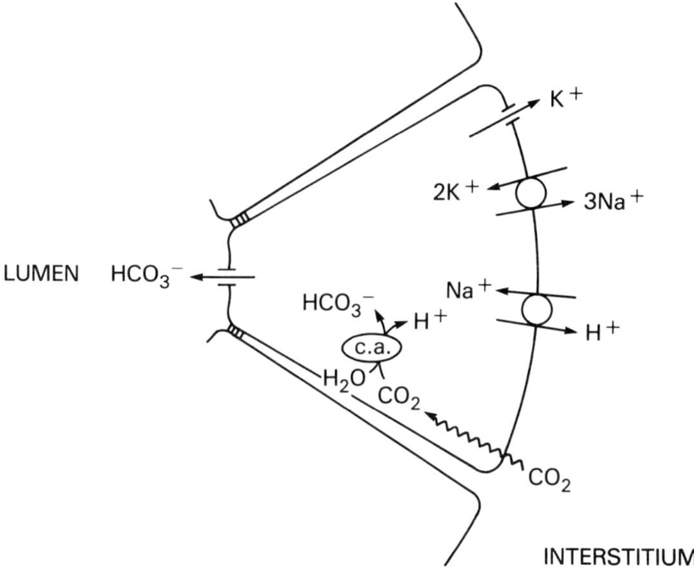

Figure 5. Model for the involvement of HCO_3^- and Na^+/H^+ exchange in salivary fluid secretion. The enzyme carbonic anhydrase (c.a.) facilitates the intracellular conversion of CO_2 to HCO_3^- plus a proton. The exit step for HCO_3^- is thought to be an apical anion channel (Melvin et al.,1988; Brown et al.,1989). See text for details. (Redrawn from Melvin, J. E., Moran, A., and Turner, R. J., *J. Biol. Chem.*, 263, 19564, 1988. With permission.)

and agonist-induced HCO_3^- loss by isolated acini. Note, however, that the presence of a diphenylamine-2-carboxylate sensitive HCO_3^- pathway in the apical membrane remains to be demonstrated. As a corollary of the above model it is worth emphasizing that the Na^+/H^+ exchanger must necessarily play a major role in intracellular pH regulation for this mode of secretion since the CO_2/HCO_3^- buffer system is obviously ineffective in a cell which is secreting HCO_3^-.

Since carbachol-induced acidification in rat parotid acini is more marked in Cl^--free medium (see above) and since both carbachol-induced acidification and Cl^- loss are inhibited by diphenylamine-2-carboxylate it is tempting to speculate that Cl^- and HCO_3^- compete for the same apical anion exit pathway (channel). Indeed, Cl^- and HCO_3^- have been shown to share channels in some cell types (Bormann et al., 1987; Reinhardt et al., 1987). However, it is possible that Cl^- and HCO_3^- have separate carbachol activated exit pathways in rat parotid acinar cells and that these anions simply compete for a common driving force, namely, the intracellular to extracellular electrical potential difference. Either of these possibilities can account for the observations discussed earlier that the rate of salivary secretion from submandibular glands perfused with Cl^-/HCO_3^--replete medium is not affected by the removal of HCO_3^- from the perfusate or by the presence of carbonic anhydrase inhibitors. In both cases one would expect increased (compensating) Cl^- loss in the absence of HCO_3^-.

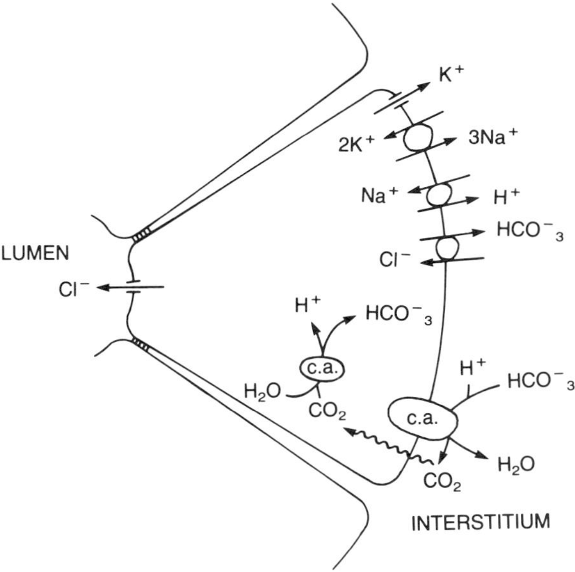

Figure 6. Model for acinar fluid secretion based on Cl⁻ entry across the basolateral membrane via a Cl⁻/HCO₃⁻ exchanger in parallel with a Na⁺/H⁺ exchanger. See text for details. (Reproduced from Turner, R. J. and George, J. N., *Am. J. Physiol.*, 254, C391, 1988b. With permission.)

Chloride Secretion Driven by Parallel Na⁺/H⁺ and Cl⁻/HCO₃⁻ Exchangers

Figure 6 shows a mechanism for epithelial chloride secretion (Turnberg et al., 1970; Warnock and Eveloff, 1982; Reuss, 1989) which has also been proposed to be relevant to salivary glands (Novak and Young, 1986; Pirani et al., 1987; Turner and George, 1988b). This model is similar in principle to the one shown in Figure 1 except that here the basolateral Na⁺/K⁺/Cl⁻ cotransporter is replaced by a basolateral Cl⁻/HCO₃⁻ exchanger in parallel with a Na⁺/H⁺ exchanger. In this model chloride is also expected to be concentrated intracellularly above electrochemical equilibrium owing to the presence of the Cl⁻/HCO₃⁻ exchanger. The condition for thermodynamic equilibrium via a Cl⁻/HCO₃⁻ antiport is given by $[Cl^-]_{in}[HCO_3^-]_{out} = [Cl^-]_{out}[HCO_3^-]_{in}$. Estimates of acinar intracellular pH range from 7.1 to 7.3 (Pirani et al., 1987; Lau et al., 1989; Nauntofte and Dissing, 1988b; Melvin et al., 1988; Stewart et al., 1989). Using a pH of 7.2 this equation yields $[Cl^-]_{in} = 73$ mM, well above the electrochemical equilibrium value for $[Cl^-]_{in}$ of 13 mM (see above). In this model, like the one shown in Figure 1, the decrease in intracellular chloride concentration resulting from secretagogue induced KCl loss leads to increased Cl⁻ entry, in this case via the Cl⁻/HCO₃⁻ exchanger. The cytoplasmic acidification resulting from the accompanying bicarbonate loss is buffered by the Na⁺/H⁺ exchanger which uses the extracellular to intracellular sodium gradient to drive protons out of the cell.

The net result is then the movement of NaCl into the cell in exchange for H_2CO_3 which in turn is recycled across the basolateral membrane as CO_2 with the catalytic assistance of carbonic anhydrase (c.a.). Note that the effectiveness of this mechanism as a transepithelial chloride pump would be enhanced by increased intracellular pH since this would result in a higher intracellular bicarbonate concentration and hence a higher driving force for chloride entry.

Good evidence has been found for the presence of a Cl^-/HCO_3^- exchanger in salivary acinar cells. The disulfonic stilbenes SITS and DIDS, which are inhibitors of Cl^-/HCO_3^- exchange (Hoffmann, 1986), have been shown to inhibit Cl^- fluxes in intact acini from the rat submandibular and parotid (Martinez and Barker, 1989; Melvin and Turner, manuscript in preparation). In perfused submandibular glands, SITS caused intracellular pH to rise (Pirani et al., 1987) as did chloride removal from the perfusate (Pirani et al., 1987). An increase in intracellular pH was also observed when isolated rabbit submandibular acini were switched to Cl^--free medium (Lau et al., 1989). These pH changes are consistent with the presence of a Cl^-/HCO_3^- exchanger since inhibition of such a transporter would lead to intracellular HCO_3^- accumulation (and a concomitant increase in intracellular pH) by blocking HCO_3^- exit, while Cl^- removal would have the same effect due to the exchange of residual intracellular Cl^- for extracellular HCO_3^-. Direct evidence for the presence of a potent Cl^-/HCO_3^- exchanger has also been obtained from rabbit acinar basolateral membrane vesicles (Turner and George, 1988b). Moreover, it was demonstrated in this preparation that comparable Cl^- fluxes could be driven via the Cl^-/HCO_3^- exchanger and the $Na^+/K^+/Cl^-$ cotransporter (for practical reasons, however, physiological ion concentrations were not employed). Evidence for an interdependence of Cl^- fluxes via the Cl^-/HCO_3^- exchanger and the $Na^+/K^+/Cl^-$ cotransporter was also found, indicating that these two transporters are present in the same vesicles and therefore in the same cells (Turner and George, 1988b).

However, despite the evidence for the presence of Cl^-/HCO_3^- exchange in salivary acinar cells, several observations lead one to question the involvement of this transporter in fluid secretion. One of these is the demonstration discussed above (Figure 3, Trace D) that it was not possible to detect a significant component of disulfonic stilbene-inhibitable HCO_3^- loss in carbachol-stimulated acini (Melvin et al., 1988; Lau et al., 1989). Also, strangely, SITS has been found to *increase* the rate of salivary secretion (20 to 30%) in rat and rabbit submandibular glands (Case et al., 1984; Novak and Young, 1986). It has been suggested that this increase is due to the increase in intracellular pH produced by this agent (Novak and Young, 1986). Indeed, increased intracellular pH would be expected to stimulate HCO_3^--dependent secretion via the mechanism illustrated in Figure 5, and the saliva formed in the presence of SITS does have increased HCO_3^- levels (Novak and Young, 1986). Also, inhibiting Cl^- entry via the Cl^-/HCO_3^- exchanger may not necessarily reduce acinar Cl^- uptake across the basolateral membrane since a compensatory uptake might be generated via

$Na^+/K^+/Cl^-$ cotransport (recall that these two Cl^- uptake mechanisms are both driven by the extracellular to intracellular Na^+ gradient, thus a reduction in activity of one would make more driving force available to the other). In support of this argument is the observation that SITS does, in fact, inhibit salivary secretion (~20%) in the perfused rabbit submandibular gland when infused in the presence of a low dose (0.1 mM) of furosemide (Novak and Young, 1986). The lack of effect of SITS on HCO_3^- loss mentioned above (Figure 3, Trace D) could also be explained by a compensating loss of HCO_3^- via the (putative) apical HCO_3^- channel when the Cl^-/HCO_3^- exchanger is blocked. These rationalizations of the curious effects of disulfonic stilbenes are also supported by calculations, discussed below, which indicate that, at least in the rat submandibular, the possible contribution of Cl^-/HCO_3^- exchange to total Cl^- secretion is small.

I feel that there are two important points to be taken from the above discussion. First, although there is reasonable evidence to indicate the involvement of Cl^-/HCO_3^- in salivary fluid secretion, this evidence cannot as yet be considered conclusive. Secondly, and perhaps more importantly, in situations like this, where multiple mechanisms may be involved in the same physiological process, one must be very careful about the design and interpretation of experiments which attempt to delineate the various participating elements.

Relative Contributions of the Three Models for Salivary Fluid Secretion

Pirani et al. (1987) have estimated the relative contributions of $Na^+/K^+/Cl^-$ cotransport, Na^+/H^+ exchange and Cl^-/HCO_3^- exchange to fluid secretion in the perfused rat submandibular gland. It is both instructive and interesting to follow their arguments. From a stimulated salivary flow rate of 140 µl/g of gland per minute and the fact that the primary saliva is secreted as an isotonic solution they calculate an acinar anion secretory rate of 21 µmol/g/min when the perfusate is Cl^-/HCO_3^- replete. The corresponding rate of salivary HCO_3^- secretion is 2 µmol/g/min. Assuming that HCO_3^- is neither secreted nor absorbed by the salivary ducts under these conditions (for which they cite evidence) this means that approximately 10% of the acinar anions secreted are HCO_3^-, the remainder being presumably Cl^-. When Cl^- is replaced by gluconate in the perfusate the secretory rate falls to 34 µl/g/min which corresponds to an anion secretory rate of 5 µmol/g/min. Since the Na^+/H^+ exchanger buffers the cytoplasm against the proton generated by intracellular HCO_3^- production (Figure 5) this must also be the rate of turnover of the exchanger. Assuming that the Na^+/H^+ exchanger transports at the same rate during gluconate and control perfusion then the difference between the rate of HCO_3^- secretion under these two conditions (3 µmol/g/min) must be due to basolateral HCO_3^- exit via the Cl^-/HCO_3^- exchanger. From these figures one can then easily calculate that the models shown in Figures 1, 5, and 6 contribute to anion secretion, and thus fluid secretion, in the ratio 16:2:3, respectively. Equivalently, as expressed by Pirani et al., the $Na^+/K^+/Cl^-$ cotransporter, the Na^+/H^+ exchanger and the Cl^-/HCO_3^- exchanger cycle at relative rates of 8:5:3. Note that the calculation of the relative contribution of the Cl^-/HCO_3^- exchanger is based on the assumption that the Na^+/H^+ exchanger operates at the same rate during gluconate and control perfusion. If this is not the case, and the Na^+/H^+ exchanger is less active when Cl^- is available for secretion, the estimate of the contribution of the Cl^-/HCO_3^- exchanger will be correspondingly reduced. Likewise, the calculation of the relative contribution of HCO_3^- secretion to fluid secretion (i.e., the model of Figure 5) is based on the assumption of little or no ductal modification of salivary HCO_3^- levels.

ACTIVATION OF THE NA$^+$/H$^+$ EXCHANGER DURING FLUID SECRETION

It has been observed that the intracellular pH of both perfused rabbit and rat submandibular glands (Stewart et al., 1989; Pirani et al., 1987) and isolated rat parotid acini (Melvin et al., 1988; Soltoff et al., 1989) rises (~0.1 pH unit) following muscarinic stimulation (this rise in intracellular pH follows the transient, HCO_3^--dependent acidification discussed earlier). In analogy with observations from a number of other cell types, it was suggested that this effect might be due to activation of the Na^+/H^+ exchanger (Pirani et al., 1987; Melvin et al., 1988; Stewart et al., 1989). Increased Na^+/H^+ exchange activity has been observed in response to a variety of stimuli in nonsalivary tissues (e.g., hormones, growth factors, fertilization, osmotic shrinkage) and this effect is thought to play an important role in the associated cellular response (Aronson, 1985; Moolenaar, 1986; Grinstein et al., 1989). It is clear from the above discussion that the Na^+/H^+ exchanger plays an important role in the fluid secretory response of salivary acinar cells and that stimulation of the exchanger and an associated increase in intracellular pH would be expected to increase fluid secretion by the HCO_3^- dependent mechanisms shown in Figures 5 and 6.

Direct evidence for the stimulation of the exchanger in rat parotid acini is provided in Figure 7. This figure shows the recovery of intracellular pH by BCECF-loaded acini following an acute acid load induced by exposure of the cells to sodium propionate. This intracellular acidification occurs as a result of the rapid permeation of propionic acid into the cytoplasm and its subsequent dissociation into propionate and protons. We have previously demonstrated that the recovery of acinar pH is markedly retarded in the presence of amiloride and thus is due to proton removal via the Na^+/H^+ exchanger (Melvin et al., 1988). Figure 7 illustrates that acini preincubated for 15 min with the muscarinic agonist carbachol (10^{-5} M) show an enhanced recovery from a propionic acid load relative to untreated controls indicating enhanced Na^+/H^+ exchange activity and thus more effective intracellular buffering in these cells (Manganel and Turner, 1989). The α-adrenergic agonist epinephrine (10^{-5} M), which like carbachol

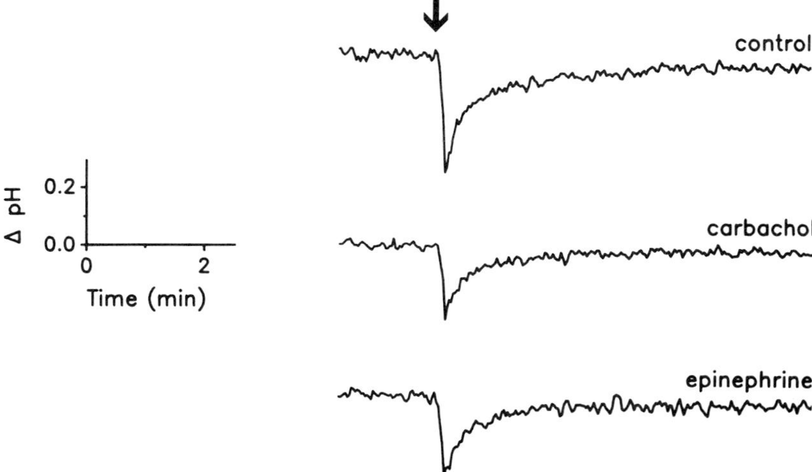

Figure 7. The recovery of treated and untreated rat parotid acini from an acute acid load. At the time indicated by the arrow BCECF-loaded acini in HCO_3^--free medium were exposed to an acute acid load by the addition of 30 mM sodium propionate. The traces represent BCECF fluorescence from untreated (control) acini and from acini pretreated with 10^{-5} M carbachol or 10^{-5} M epinephrine for 15 min. before exposure to propionate. (Reproduced from Manganel, M. and Turner, R. J.,*J. Membrane Biol.*, 111, 191, 1989. With permission.)

elicits a copious fluid secretory response (Young et al., 1987a), also renders the cells more resistant to intracellular acidification (Figure 7). This stimulation of the exchanger is also preserved in vesicles prepared from acini treated with carbachol or epinephrine (Manganel and Turner, 1989) indicating that it is due to a covalent modification of the transport protein. The time scale of this up-regulation of the exchanger ($t_{1/2}$ ~10 min) suggests that it is associated with the sustained rather than the initial phase of fluid secretion.

CONCLUDING REMARKS

From the above discussion it would seem that fluid secretion by salivary acinar cells is a rather complex process involving the interaction of a number of ion transport systems. The reason for this complexity is not yet clear. One might speculate that the relative contributions of the mechanisms illustrated in Figures 1, 5, and 6, and possibly of others not yet identified, may vary with species, gland, age, metabolic or disease state, etc. Indeed the up-regulation of the Na^+/H^+ exchanger and the associated increase in acinar intracellular pH suggests that the contribution of HCO_3^--dependent mechanisms increases at later times during the secretory response. There is also evidence for species differences both in perfused gland (Novak and Young, 1986; Pirani et al., 1987) and vesicle studies (Manganel and Turner, 1988). From an energetic point of view the model shown in Figure 1 is the more efficient since for each cycle of Na^+/K^+ ATPase six chloride ions are translocated across the epithelium while the models in Figures 5 and 6 translocate only three anions per ATP (Turner and George, 1988b). However, energetics is not the only concern here. The actual anion flux generated by each model will depend on the kinetic properties and number of the various ion carriers as well as on the magnitude of the relevant ion gradients which drive transport. In this regard it is worth mentioning that on the level of the whole cell the acinar fluid secretory response is quite dramatic (Nauntofte and Poulsen, 1986; Izutzu et al., 1987; Melvin et al., 1987; Melvin et al., 1988; Nauntofte and Dissing, 1989; Soltoff et al., 1989; Murakami et al., 1989; Foskett and Melvin, 1989) involving a marked initial loss of K^+ (~30%), Cl^- (~50%), HCO_3^- (25% per 0.1 pH unit) and H_2O (~20%) and a substantial increase in intracellular Na^+ (~400%). This is also accompanied by significant changes in intracellular pH (see above) and membrane potential (Petersen, 1980; Foskett et al., 1989). This rapid initial response is followed by a return of many of these parameters to values closer to, but still significantly different from, resting levels. A complete understanding of the contributions of the various ion transport systems discussed in this chapter to salivary fluid secretion will also require a more detailed description of these intracellular parameters.

ACKNOWLEDGMENTS

I would like to acknowledge my collaboration as well as many fruitful and stimulating discussions with Bruce J. Baum, Jean-Paul Dehaye, Janet N. George, J. Kevin Foskett, Syng Ill Lee, Michel Manganel, James E. Melvin, and Arie Moran.

REFERENCES

1. Abe, K. and Dawes, E, Circadian variations in pilocarpine and isoprenaline-induced protein secretion by rat parotid gland, *Arch. Oral Biol.*, 20, 243, 1975.
2. Ambudkar, I., Melvin, J. E., and Baum, B. J., α1-Adrenergic regulation of Cl^- and Ca^{2+} movements in rat parotid acinar cells, *Pfluegers Arch.*, 412, 75, 1988.
3. Aronson, P. S., Kinetic properties of the plasma membrane Na^+-H^+ exchanger, *Annu. Rev. Physiol.*, 47, 545, 1985.
4. Baum, B. J., Ambudkar, A. S., and Horn, V. J., Neurotransmitter control of calcium mobilization, in *The Biology of the Salivary Glands*, Vergona, K., Ed., Telford Press, Caldwell, NJ, 1992, chap. 7.
5. Benos, D. J., Amiloride: a molecular probe of sodium transport in tissues and cells, *Am. J. Physiol.*, 242, C131, 1982.
6. Berridge, M. J. and Galione, A., Cytosolic calcium oscillations, *FASEB J.*, 2, 3074. 1988.
7. Bodner, L. and Baum, B. J., Characteristics of stimulated parotid gland secretion in the aging rat, *Mech. Ageing Dev.*, 31, 337, 1985.

8. Bormann, J., Hammill, O. P., and Sakmann, B., Mechanism of anion permeation through channels gated by glycine and α-aminoisobutyric acid in mouse cultured spinal neurones, *J. Physiol.*, 385, 243, 1987.

9. Brown, P. D., Elliott, A. C., and Lau, K. R., Indirect evidence for the presence of non-specific anion channels in rabbit submandibular salivary gland acinar cells, *J. Physiol. (London)*, 414, 415, 1989.

10. Burgen, A. S. V., The secretion of potassium in saliva, *J. Physiol. (London)*, 132, 20, 1956.

11. Case, R., Hunter, M., Novak, I., and Young J. A., The anionic basis of fluid secretion by the rabbit submandibular salivary gland, *J. Physiol. (London)*, 349, 619, 1984.

12. Case, R. M., Conigrave, A. D., Favalaro, E. J., Novak, I., Thompson, C. H., and Young, J. A., The role of buffer anions and protons in secretion by the rabbit mandibular salivary gland, *J. Physiol.*, 322, 273, 1982.

13. Case, R. M., Congrave, A. D., Hunter, M., Novak, I., Thompson, C. H., and Young, J. A., Secretion of saliva by the rabbit mandibular gland *in vitro*: the role of anions, *Phil. Trans. R. Soc., Lond. B*, 296, 179, 1981.

14. Cook, D. I., Gard, G. B., Champion, M., and Young, J. A., Patch-clamp studies of the electrolyte secretory mechanism of rat mandibular gland cells stimulated with acetylcholine or isoproterenol, in *Molecular Mechanisms in Secretion*, Thorn, N. A., Treiman, M., and Petersen, O. H., Eds., Munksgaard, Copenhagen, 1988, 133.

15. Cope, G. H. and Williams, M. A., Quantitative analysis of the constituent membranes of parotid acinar cells and of the changes evident after induced exocytosis, *Z. Zellforsch Mikrosk Anat.*, 145, 311, 1973.

16. Dehaye, J.-P., Winand, J., and Christophe, J., Scorpion venom inhibits carbamylcholine-induced [86]rubidium efflux from rat parotid acini, *FEBS Lett.*, 219, 451, 1987.

17. Diamond, J. M., Osmotic water flow in leaky epithelia, *J. Membrane Biol.*, 51, 195, 1979.

18. Foskett, J. K., Gunter-Smith, P. J., Melvin, J. E., and Turner, R. J., Physiological localization of an agonist-sensitive pool of Ca^{2+} in parotid acinar cells, *Proc. Natl. Acad. Sci. U.S.A.*, 86, 167, 1989.

19. Foskett, J. K. and Melvin, J. E., Activation of salivary secretion: coupling of cell volume and $[Ca^{2+}]_i$ in single cells, *Science*, 244, 1582, 1989.

20. Geck, P. and Heinz, E., The Na-K-2Cl cotransport system, *J. Membrane Biol.*, 91, 105, 1986.

21. Gray, P. T. A., The relation of elevation of cytosolic free calcium to activation of membrane conductance in rat parotid acinar cells, *Proc. R. Soc. Lond. B*, 237, 99, 1989.

22. Grinstein, S. and Rothstein, A., Mechanisms of regulation of the Na^+/H^+ exchanger, *J. Membrane Biol.*, 90, 1, 1986.

23. Grinstein, S., Rotin, D., and Mason, M. J., Na/H exchange and growth factor-induced cytosolic pH changes. Role in cellular proliferation, *Biochim. Biophys. Acta*, 988, 73, 1989.

24. Hoffmann, E. K., Anion transport systems in the plasma membrane of vertebrate cells, *Biochim. Biophys. Acta*, 864, 1, 1986.

25. Iwatsuki, N., Maruyama, Y., Matsumoto, O., and Nishiyama, A., Activation of Ca^{2+}-dependent Cl^- and K^+ conductances in rat and mouse parotid acinar cells, *Japanese J. Physiol.*, 35, 933, 1985.

26. Izutzu, K. T., Johnson, D. E., and Goddard, M., Intracellular elemental concentrations in resting and secreting rat parotid glands, *J. Dent. Res.*, 66, 537, 1987.
27. Kawaguchi, M., Turner, R. J., and Baum, B. J., Cl^- and Rb^+ uptake in rat parotid acinar cells, *Arch. Oral Biol.*, 31, 679, 1986.
28. Lau, K. R. and Case, R. M., Evidence for apical chloride channels in rabbit mandibular salivary glands, *Pfluegers Arch.*, 411, 670, 1988.
29. Lau, K. R., Elliott, A. C., and Brown, P. D., Acetylcholine-induced intracellular acidosis in rabbit salivary gland acinar cells, *Am. J. Physiol.*, 256, C288, 1989.
30. Lundberg, A., Secretory potentials in the sublingual gland of the cat, *Acta Physiol. Scand.*, 40, 21, 1957.
31. Lundberg, A., Electrophysiology of salivary glands, *Physiol. Rev.*, 38, 21, 1958.
32. Manganel, M. and Turner, R. J., Coupled Na^+/H^+ exchange in rat parotid basolateral membrane vesicles, *J. Membrane Biol.*, 102, 247, 1988.
33. Manganel, M. and Turner, R. J., Agonist induced activation of Na/H exchange in rat parotid acinar cells exchange in rat parotid acinar cells, *J. Membrane Biol.*, 111, 191, 1989.
34. Martinez, J. R., Ion transport and water movement, *J. Dent. Res.*, 66, 638, 1987.
35. Martinez, J. R. and Barker, S., Inhibition of ^{36}Cl uptake by stilbene sulphonic acid derivatives and loop diuretics in rat submandibular salivary acini, *Arch. Oral Biol.*, 34, 535, 1989.
36. Martinez, J. R., Barker, S., and Camden, J., Amiloride inhibits ^{22}Na uptake and [^3H]QNB binding in rat submandibular cells, *Eur. J. Pharm.*, 164, 335, 1989.
37. Martinez, J. R. and Cassity, N., Effect of transport inhibitors on secretion by perfused rat submandibular gland, *Am. J. Physiol.*, 245, G711, 1983.
38. Martinez, J. R. and Cassity, N., Cl^- requirement for saliva secretion in the isolated, perfused rat submandibular gland, *Am. J. Physiol.*, 249, G464, 1985a.
39. Martinez, J. R. and Cassity, N., Effect of 4,4′-diisthiocyano-2,2″-stilbene disulphonic acid and amiloride on salivary secretion by isolated perfused rat submandibular glands, *Arch. Oral Biol.*, 30, 797, 1985.
40. Martinez, J. R. and Cassity, N., ^{36}Cl fluxes in dispersed rat submandibular acini: effects of acetylcholine and transport inhibitors, *Pfluegers Arch.*, 403, 50, 1985c.
41. Martinez, J. R. and Cassity, N., ^{36}Cl fluxes in dispersed rat submandibular acini: effects of Ca^{2+} omission and of the ionophore A23187, *Pfluegers Arch.*, 407, 615, 1986.
42. Martinez, J. R., Cassity, N., and Reed, P., An examination of functional linkage between K efflux and ^{36}Cl efflux in rat submandibular salivary gland acini *in vitro*, *Arch. Oral Biol.*, 32, 891, 1987.
43. Martinez, J. R. and Petersen, O. H., The importance of extracellular calcium for acetylcholine evoked salivary secretion, *Experientia*, 28, 167, 1971.
44. Martinez, J. R. and Reed, P., Effect of α-receptor stimulation on Cl transport by rat submandibular acini, *J. Dent. Res.*, 67, 561, 1988.
45. Melvin, J. E., Kawaguchi, M., Baum, B. J., and Turner, R. J., A muscarinic agonist-stimulated chloride efflux pathway is associated with fluid secretion in rat parotid acinar cells, *Biochem. Biophys. Res. Commun.*, 145, 754, 1987.
46. Melvin, J. E., Moran, A., and Turner, R. J., The role of HCO_3^- and Na^+/H^+ exchange in the response of rat parotid acinar cells to muscarinic stimulation, *J. Biol. Chem.*, 263, 19564, 1988.

47. Moolenaar, W. H., Effects of growth factors on intracellular pH regulation, *Annu. Rev. Physiol.*, 48, 363, 1986.

48. Mori, H., Murakami, M., Nakahari, T., and Imai, Y., Intracellular Cl⁻ activity of canine submandibular gland cells: an in vitro observation, *Jap. J. Physiol.*, 33, 869, 1983.

49. Murakami, M., Suzuki E., Miyamoto, S., Seo, Y., and Watari, H., Direct measurement of K movement by ^{39}K NMR in perfused rat mandibular salivary gland stimulated with acetylcholine, *Pfluegers Arch.*, 414, 385, 1989.

50. Nauntofte, B. and Dissing, S., K⁺ transport and membrane potentials in isolated rat parotid acini, *Am. J. Physiol.*, 255, C508, 1988a.

51. Nauntofte, B. and Dissing, S., Cholinergic-induced electrolyte transport in rat parotid acini, *Comp. Biochem. Physiol.*, 90A, 739, 1988b.

52. Nauntofte, B. and Dissing, S., Na⁺ transport in rat parotid acini, *Proc. Finn. Dent.*, Soc 85, Suppl. 1:32, Abstract, 1989.

53. Nauntofte, B. and Poulsen, J. H., Effects of Ca^{2+} and furosemide on Cl⁻ transport and O_2 uptake in rat parotid acini, *Am. J. Physiol.*, 1, C175, 1986.

54. Novak, I. and Young, J. A., The effect of transport inhibitors and anion replacement on salivary secretion, in *Secretion: Mechanisms and Control*, Case, R. M., and Lingard, J. A. Y., Eds., Manchester University Press, 1984, 81.

55. Novak, I. and Young, J. A., Two independent anion transport systems in rabbit mandibular salivary glands, *Pfluegers Arch.*, 407, 649, 1986.

56. O'Grady, S. M., Palfrey, H. C., and Field, M., Characteristics and functions of Na-K-Cl cotransport in epithelial tissues, *Am. J. Physiol.*, 253, C177, 1987.

57. Petersen, O. H., *The Electrophysiology of Gland Cells*, Academic Press, New York, 1980.

58. Petersen, O. H., Calcium activated potassium channels and fluid secretion by exocrine glands, *Am. J. Physiol.*, 251, G1, 1986.

59. Petersen, O. H. and Gallacher, D. V., Electrophysiology of pancreatic and salivary acinar cells, *Annu. Rev. Physiol.*, 50, 1988.

60. Petersen, O. H. and Maruyama, Y., Calcium-activated potassium channels and their role in secretion, *Nature (London)*, 307, 693, 1984.

61. Pirani, D., Evans, A. R., Cook, D. I., and Young, J. A., Intracellular pH in the rat mandibular salivary gland: the role of Na-H and Cl-HCO₃ antiports in secretion, *Pfluegers Arch.*, 408, 178, 1987.

62. Poulsen, J. H., Dissing, S., and Nauntofte, B., Is receptor-activated chloride transport in parotid acini controlled by the cytosolic free calcium concentration?, in *Molecular Mechanisms in Secretion*, Thorn, N. A., Treiman, M., and Petersen, O. H., Eds., Mungsgaard, Copenhagen, 1988, 152.

63. Poulsen, J. H. and Kristensen, L. O., Is stimulation induced uptake of sodium in rat parotid acinar cells mediated by a sodium/chloride co-transport system?, in *Electrolyte and Water Transport Across Gastrointestinal Epithelia*, Case, R. M., Garner, A., Turnberg, L. A., and Young, J. A., Eds., Raven Press, New York, 1982, 199.

64. Poulsen, J. H., Laugesen, L. P., and Nielsen, J. O. D., Evidence supporting that basolaterally located Na⁺-K⁺-ATPase and a cotransport system for sodium and chloride are key elements in secretion of primary saliva, in *Electrolyte and Water Transport Across Gastrointestinal Epithelia*, Case, R. M., Garner, A., Turnberg, L. A., and Young, J. A., Eds., Raven Press, New York, 1982, 157.

65. Poulsen, J. H. and Nauntofte, B., Is the stoichiometry of the parotid cotransporter 1 Na: 1 K: 2Cl:, *J. Dent. Res.*, 66, 608, 1987.

66. Poulsen, J. H. and Nauntofte, B., Role of anion transport in secretion of primary saliva, in *Epithelial Secretion of Water and Electrolytes*, J. A. Young and P. Y. D. Wong, Eds., Springer-Verlag, Berlin, 1990, 143.

67. Putney, J. W., Identification of cellular activation mechanisms associated with salivary secretion, *Annu. Rev. Physiol.*, 48, 75, 1986.

68. Reinhardt, R., Bridges, R. J., Rummel, W., and Lindemann, B., Properties of an anion selective channel from rat colonic enterocytes plasma membranes reconstituted into planar lipid bilayers, *J. Membrane Biol.*, 95, 47, 1987.

69. Reuss, L., Ion transport across gallbladder epithelium, *Physiol. Rev.*, 69, 503, 1989.

70. Schneyer, L. H. and Schneyer, C. A., Electrolyte and insulin spaces of rat salivary glands and pancreas, *Am. J. Physiol.*, 199, 649, 1960.

71. Silva, P., Stoff, J., Field, M., Fine, L., Forrest, J. N., and Epstein F. H., Mechanism of active chloride secretion by shark rectal gland: role of Na-K-ATPase in chloride transport, *Am. J. Physiol.*, 233, F298, 1977.

72. Smaje, L. H., Poulsen, J. H., and Ussing H. H., Evidence from O_2 uptake measurements for Na^+-K^+-$2Cl^-$ co-transport in the rabbit submandibular gland, *Pfluegers Arch.*, 406, 492, 1986.

73. Soltoff, S. P., McMillian, M. R., Cantley, L. C., Cragoe, E. J., and Talamo, B. R., Effects of muscarinic, alpha-adrenergic and substance P agonists and ionomycin on ion transport mechanisms in the rat parotid acinar cell, *J. Gen. Physiol.*, 93, 285, 1989.

74. Spring, K. R., Fluid transport by gallbladder epithelium, *J. Exp. Biol.*, 106, 181, 1983.

75. Stewart, M. C., Seo, Y., and Case, R. M., Intracellular pH during secretion in the perfused rabbit mandibular salivary gland measured by ^{31}P NMR spectroscopy, *Pfluegers Arch.*, 414, 200, 1989.

76. Thaysen, J. H., Handling of alkali metals by exocrine glands other than the kidney, in *The Alkali Metal Ions in Biology*, Ussing, H. H., Kruhoffer, P., Thaysen, J. H., and Thorn, N. A., Eds., Springer-Verlag, Berlin, 1960, 424.

77. Thaysen, J. H., Thorn, N. A., and Schwartz I. L., Excretion of sodium, potassium, chloride and carbon dioxide in human parotid saliva, *Am. J. Physiol.*, 178, 155, 1954.

78. Turnberg, L. A., Bieberdorf, F. A., Morawski, S. G., and Fordtran, J. S., Interrelationships of chloride, bicarbonate, sodium, and hydrogen transport in the human ileum, *J. Clin. Invest.*, 49, 547, 1970.

79. Turner, R. J., Stoichiometry of cotransport systems, *Ann. N.Y. Acad. Sci.*, 456, 10, 1985.

80. Turner, R. J. and George, J. N., Ionic dependence of bumetanide binding to the rabbit parotid Na/K/Cl cotransporter, *J. Membrane Biol.*, 102, 71, 1988a.

81. Turner, R. J. and George, J. N., Cl/HCO_3 exchange is present with Na/K/Cl cotransport in parotid acinar basolateral membranes, *Am. J. Physiol.*, 254, C391, 1988b.

82. Turner, R. J., George, J. N., and Baum, B. J., Evidence for a Na^+/K^+/Cl^- cotransport system in basolateral membrane vesicles from the rabbit parotid, *J. Membrane Biol.*, 94, 143, 1986.

83. Warnock, D. G. and Eveloff, J., NaCl entry mechanisms in the luminal membrane of the renal tubule, *Am. J. Physiol.*, 242, F561, 1982.

84. Young, J. A., Cook, D. I., Evans, L. A. R., and Pirani, D., Effects of ion transport inhibition on rat submandibular gland secretion, *J. Dent. Res.*, 66, 531, 1987b.

85. Young, J. A., Cook, D. I., Van Lennep, E. W., and Roberts, M. L., Secretion by the major salivary glands, in *Physiology of the Gastrointestinal Tract*, 2nd ed., Johnson, L., Christensen, J., Jackson, M., Jacobson, E., and Walsh, J., Eds., Raven Press, New York, 1987, 773.

86. Young, J. A., Heintze, K., Green, R., Steward, M., Poulsen, J. H., van Os, C., and Cook, D. I., Is Na/Cl cotransport the basis of transepithelial transport in absorptive and secretory epithelia?, in *Electrolyte and Water Transport Across Gastrointestinal Epithelia*, Case, R. M., Garner, A., Tunberg, L. A., and Young, J. A., Eds., Raven Press, New York, 181, 1982.

87. Young, J. A. and van Lennep, E. W., Transport in salivary and salt glands. Part I. Salivary glands, in *Membrane Transport in Biology*, Vol. 4B, *Transport Organs*, Giebish, G., Tosteson, D. C., and Ussing, H. H., Eds., Springer-Verlag, Berlin, 1979, 563.

Chapter

6

Role of Salivary Glands in Nutrition

Kenneth Etzel

INTRODUCTION

The mammalian digestive system is a complex, highly integrated system of tissues which functions to aid in the transfer and subsequent absorption of nutrients obtained from the diet. Specialized epithelial, muscle, and secretory tissue work in concert under the influence of both neuronal and hormonal stimulation to assure that the body is provided with the essential nutrients to meet its metabolic requirements. In the initial step of digestion during the process of mastication, food is mixed with saliva, chopped, emulsified, and ultimately transferred to the intestine where the bulk of nutrient absorption takes place. Although saliva and this masticatory process is not absolutely necessary for intestinal nutrient absorption, it is of fundamental importance to our appreciation of the day to day pleasures of eating. It is well documented that saliva plays an important role in the development and maintenance of a healthy oral cavity; however, when one considers the importance of saliva in oral communication, taste perception, and mastication, it is apparent that the salivary glands and their secretions are intimately involved in many daily activities which influence our overall nutritional well being. This chapter therefore will focus not only on the role that saliva plays in maintaining optimal nutritional status, but will also examine the effects of altered nutritional status on salivary function in both humans and experimental animals.

THE NUTRITIONAL IMPLICATIONS OF SALIVA IN THE ORAL CAVITY

Because of the intimate association between nutrition and the digestive system, the relationship of the diet to gastrointestinal function has been studied for years. Early observation of Beaumont (1833), Heidenhain (1878), Pavlov (1902), and Carlson (1916) were influential in unraveling the initial fundamental processes in the digestion of food. Today, the physiological and biochemical principles of digestion are being examined on the molecular level, giving us a much clearer understanding of this complex system. However, as we become more sophisticated in our studies of not only the digestive system but other areas related to nutrition, our appreciation of the more germane yet fundamental processes involved in nutrient acquisition are often overlooked. Such is the case for the initial step of digestion, mastication, which is responsible for the chopping and mixing of food in the presence of saliva. This mechanical action not only initiates the digestive process but solubilizes the dietary components which, when in contact with the taste buds, produces sensations of taste which we perceive as either pleasant or noxious. Both oral digestion and taste, therefore, require the secretions of the salivary glands for optimal function. This chapter looks at the role of saliva in mastication, taste perception, and digestion, emphasizing its fundamental yet important contribution to the overall nutritional health.

NUTRITIONAL STIMULANTS OF SALIVARY SECRETION

Salivary secretion from both the major and minor salivary glands is regulated primarily by the autonomic nervous system. Several excellent reviews outlining the physiological and pharmacological consequences of this regulation are available (Baum, 1987; Emmelin, 1987; Herrera, et al. 1988; Lavelle, 1988). In general, during feeding, fluid output is regulated primarily by the parasympathetic branch and protein output by the sympathetic branch (Spiers and Hodgson, 1976). Of particular significance to this review is the fact that both branches of the autonomic nervous system mediate the stimulation of the salivary glands produced by oral reflexes. Two major mechanisms, both responsive to dietary factors, are involved. The first responds to chewing which, via mechanoreceptors, stimulates both the sympathetic and parasympathetic branches of the autonomic nervous system (Schneyer, 1974; Gjorstrup, 1980a; Gjorstrup, 1980b; Anderson et al., 1985). The second is likewise mediated by both branches of the autonomic system but is initiated via chemoreceptors in the oral cavity (Dawes, 1984). Studies have shown that each mechanism is capable of independently stimulating salivary secretion resulting in alterations in both total volume and protein content of the secreted fluid (Newbrun, 1962; Gjorstrup,

1981). One can readily appreciate, therefore, that food which requires mastication and, at the same time, elicits a taste response, is the ideal physiological stimulant of salivary secretion.

Since the early work of Lashley (1916) numerous studies in both humans and experimental animals have confirmed the fact that the output of saliva increases in direct proportion to the degree of mastication (for a review see Johnson, 1987). In response to chewing, both the rate of fluid secretion and the organic composition are increased (Anderson et al., 1985; Anderson et al., 1981; Gjorstrup, 1980a, Gjorstrup, 1980b). Studies in experimental animals have shown that this reflex stimulation is mediated via the sympathetic and para-sympathetic nervous systems, the later being primarily responsible for increased fluid output (Gjorstrup, 1980a). Antagonists of these neuronal systems are capable of significantly reducing salivary output in both humans (Sreebny and Broich, 1987) and experimental animals (Gjorstrup, 1980a; Gjorstrup, 1977; Jirakulsomchok and Schneyer, 1979). Changes in mastication activity also result in comparable changes in parotid gland size (Schneyer and Hall, 1976; Hall and Schneyer, 1964; Hall and Schneyer, 1977; Wells and Peronace, 1967; Johnson and Sreebny, 1973a; Johnson and Sreebny, 1982; Johnson, 1981).

Chewing of food in the presence of saliva results in solubilization of chemical compounds which are capable of eliciting salivary responses via chemoreceptors located in the mouth. Differentiating between salivary responses due to mechanical stimulation vs. taste stimulation has long been the concern of investigators. Early work, taking advantage of alternating chewing patterns of the horse and mule, demonstrated that in animals with bilateral cannulae parotid saliva was greater on the chewing side than on the opposite side (for reference see Anderson et al., 1985; Gjorstrup, 1981). Similar observations in man (Lashley, 1916) showed that salivary glands were most responsive to mechanical receptors common to its own side and that parotid salivary output increased with biting force. But it was not until recently that specific mechanisms of this process were unraveled. In 1980, Gjorstrup reported that flow of saliva collected from cannulated parotid glands of conscious rabbits is approximately three times greater when animals chewed standard laboratory chow as compared to that obtained when carrots were consumed (Gjorstrup, 1980a). The increased sali-vary flow rate was significantly reduced when the pelleted diet was softened either prior to consumption or during mastication. The latter "dietary softening" was accomplished by delivering water via a cannula directly into the oral cavity. It was concluded that the greater degree of mastication required by the pelleted diet vs. the softened carrot diet was responsible for the greater salivary flow rate. This fluid secretion in response to the pelleted diet and carrots was attributed to parasympathetic activity since flow rates during consumption of either food was not altered by sympathectomy or β adrenoceptor blockade (Gjorstrup, 1980a). He observed, however, that the two kinds of food produced the same maximum output of amylase since amylase concentration of saliva in response to carrots was three to four times higher than the concentration obtained with pellets. This greater concentration of amylase was attributed to a stimulation of taste since the introduction of either sucrose or carrot juice results in increased amylase concentration in parotid saliva (Gjorstrup, 1980b). In comparing different stimuli he demonstrated that sweet stimuli produced the largest increase in amylase secretion, salty stimuli produced a small effect, and sour and bitter stimuli had a negligibly minor effect. Evidence to show that this taste stimulation is mediated by the sympathetic nervous system is given by the fact that the amylase content of saliva is significantly reduced following preganglionic sectioning of the sympathetic nerves. Even greater reduction in amylase is observed when β adrenergic blockers are administered (Gjorstrup, 1980a).

It is generally acknowledged that sour (or acid) stimulation is a strong stimulus of salivary secretion (Yamamoto, 1983). Other dietary components have been identified which, although to a lesser degree, elicit changes in volume and chemical composition of saliva. Combinations of various taste mixtures in the presence of acid produce a response which is either additive or significantly less than the response elicited by individual compounds (Feller et al., 1965). Addition of acid or sour to sucrose stimulates parotid salivary flow equal to the sum of the separate taste solution. A less than additive effect on parotid secretion was observed when the same acid solutions were added to sodium citrate.

THE EFFECT OF SALIVA ON TASTE

The presence of food in our mouths elicits a response which is perceived as either pleasant or distasteful. If pleasant, the taste sensation serves as an important psychic stimulus for continued food intake. If distasteful, further consumption of the food is terminated often resulting in the avoidance of potentially harmful compounds. Although our conscious interpretation of this acceptability of food depends upon numerous factors, the physiological mechanisms by which chemical sensations are perceived depends upon the presence of saliva in the oral cavity. Taste buds located primarily on the tongue require the chemicals which they detect, and which we ultimately sense as taste, to be solubilized. Mastication and the resultant stimulation of volume flow from the salivary gland is therefore critical to the optimal functioning of the taste buds. Individuals with dry mouths complain of reduced taste sensation (Henkin et al., 1972; Henkin, 1972). Similar alterations in taste response have been reported in surgically desalivated experimental animals (Vance, 1965; Galili, 1981; Catalanotto and Sweeney, 1972; Catalanotto and Sweeney, 1973). A recent clinical study suggests, however, that the functional integrity of the taste system does not require the presence of saliva in the mouth. Weiffenbach and co-workers demonstrated that in the chronic absence of saliva, normal taste perception is maintained (Weiffenbach et al., 1986). It appears, therefore, that the most important role that saliva plays in taste sensation is in providing a media for solubilization of chemical components rather than providing salivary components, which are proposed to be essential for maintaining the integrity of taste bud receptors (Henkin et al., 1978).

The fluid secreted from the salivary glands has characteristics which make it an ideal medium for nutrient solubilization. The ability to detect sweet, sour, salty, and bitter flavors would be difficult if salivary components which produce these sensations were present in large quantities. Fortunately, compared to dietary content, the composition of salt (Shannon, 1983; Izutsu, 1987) and sugar (Borg and Birkhed, 1988), although subject to variability, is low, as are various organic and inorganic compounds known to be responsible for sour (acidic compounds) and bitter flavors (alkaloids, glycosides, etc.) (Bartoshuk, 1985). In addition, adaptation to these low "intrinsic" concentrations of salivary electrolytes increases our sensitivity to these basic taste sensations. An excellent review of this phenomenon has recently been published (Bartoshuk, 1985).

THE ROLE OF SALIVA IN ORAL PROTECTION

In the process of mastication, forces are exerted by the teeth to bring about the initial mixing and emulsification of food. The movement of food within the oral cavity produces abrasive forces not only on the tooth structure but also on oral epithelia lining the mouth and anterior pharyngeal region. Ultimately, the food is swallowed, propelling the mass of food down the esophagus and into the stomach.

Eating is a pleasurable experience. Unless we have overt pathological disturbance in the oral cavity or the food itself is offensive to taste, the process of eating is anything but painful. This is attributable, in part, to the protective characteristics that saliva exerts on the structure of the mouth and lower gastrointestinal tract. Salivary proteins, most notably the mucin glycoproteins and basic proline-rich glycoproteins, complexed with albumin (Hatton et al., 1985), coat the surface of the food and oral tissues to provide a smooth surface over which food can easily travel (Tabak et al., 1982). Lack of these important salivary factors which make chewing and swallowing painful and difficult, significantly alter food selection, and ultimately nutrient intake.

THE EFFECT OF SALIVA ON DIGESTION

Carbohydrate Digestion

Salivary amylase is a key enzyme responsible for the hydrolytic breakdown of α 1,4 glucosidic bonds found in complex carbohydrates (Stryer, 1988). Produced in both submandibular and parotid glands, this polymorphic enzyme is biologically active only at neutral pH. Its importance in digestion has been considered to be primarily confined to initial hydrolysis of carbohydrates while food is in the mouth. Due to the high concentration and activity of this enzyme as it is secreted from the salivary glands, its overall contribution to starch digestion immediately following a meal is therefore considerable. Furthermore, studies suggest that amylase activity may continue for 15 to 30 min in the stomach if protected from gastric acid within a bolus of food (Jenkins, 1978).

The degree to which complex carbohydrates are digested by salivary amylase in the mouth is apparently related to two chewing variables. Morse and co-workers (1989) have reported that the oral digestion of carbohydrates not only is related to the time of chewing but also the degree of relaxation of the subject. In their study they observed significantly greater carbohydrate digestion, as measured by maltose production, when individuals chewed under relaxed conditions. Identical chewing patterns under stressful conditions resulted in reduced maltose production. The authors proposed that these differences in digestive function were attributed to increased parasympathetic activity of the autonomic nervous system reported to occur during relaxation (Benson et al., 1974). This increased cholinergic stimulation of the salivary glands would produce a rapid flow of saliva rich in α-amylase activity (Jenkins, 1978). Unfortunately, α-amylase activity was not directly measured in this study.

Recent work, however, has focused on the role of salivary amylase in starch hydrolysis in the duodenum (Rosenblum et al., 1988). Studies have shown that salivary amylase, which is inactivated *in vitro* at low pH, retains 56% of the initial activity after 60 min when incubated in the presence of starch (1%). This compares to only 6% of the initial activity in samples incubated without starch. Similar protection of enzyme activity from acid inactivation was provided by partially hydrolyzed starch and glucose oligosaccharides, while lactose, sucrose, and glucose were ineffective in sparing the enzyme (Rosenblum et al., 1988). These data demonstrate that salivary-type amylases are protected from the gastric environment by substrates of amylase and the hydrolytic products of starch digestion and suggest that the importance of salivary enzymes and perhaps other secretory products are not confined solely to the oral cavity. Young infants who have very low levels of pancreatic amylase tolerate moderate amounts of starches during their first months of life (Lebenthal et al., 1983). Amylase is present in the saliva of infants as well as human milk. It has been postulated that these nonpancreatic sources of amylolytic activity are responsible for the clinical tolerance of starch during the early stages of life (Heitlinger et al., 1983; Lebenthal et al., 1984). These observations and other reports (Fogel and Gray, 1973; Larkisch and Otto, 1986) suggest that salivary amylases may also significantly contribute to duodenal hydrolysis of carbohydrate in adults afflicted with diseases causing pancreatic insufficiency provided the amylolytic activity can survive passage through the stomach. Exact quantitation of the role that salivary amylase plays in duodenal digestion awaits the development of more precise techniques for the assay of the specific enzymes.

Fat Digestion

Secretions produced from the serous von Ebner glands of the tongue contain lingual lipase responsible for hydrolyzing long-chain triglycerides to glycerides and fatty acids (Hamosh and Burns, 1977). Although not secretory products of the major salivary glands (thus, the name lingual lipase), the enzyme is produced in serous lingual glands of the oral cavity and contributes to the initial phase of fat digestion prior to its arrival in the small intestine. Unlike salivary amylase, lingual lipase is enzymatically active and stable at gastric pH, permitting it to survive beyond the oral cavity and hydrolyze dietary triglycerides in the stomach and intestine (Fink et al., 1984). Under pathological conditions where pancreatic enzymes are either absent or inactivated by low intestinal pH, lingual lipase remains active as the major enzyme for intestinal fat digestion (Dutta et al., 1982).

EXTRAORAL NUTRITIONAL EFFECTS OF SALIVA

As a fluid secreted directly into the mouth, it is understandable that most of the work dealing with the role of saliva on diet and nutrition has focused on the oral cavity. Very little work, however, has investigated the role of saliva on physiological processes beyond the oral/pharyngeal region. Certainly it is acknowledged that amylase, if protected from acid degradation/inhibition by products of starch digestion, is involved in the digestion of starch after it leaves the mouth (Jenkins, 1978). Its contribution in the intestine, however, is over-shadowed by the numerous digestive enzymes which abound in the small intestine. Lingual lipase is known to be active in the lower intestine (Fink et al., 1984); however, its role in fat digestion is also considered to be minimal compared to that contributed by other pancreatic digestive enzymes.

Salivary glands have been shown to contain substances which are believed to function as secretory inhibitors of gastric acid secretion. Extracts of saliva, when injected parenterally into animals, have been shown to reduce gastric acid secretion (Code et al., 1949; Miyoshi et al., 1969; Bower et al., 1975). It is questionable, however, that these endogenous salivary factors function physi-ologically following a meal to depress gastric acid secretion or, more importantly, what role they play in effecting the digestive process which occurs in the stomach.

Several additional compounds have been identified in saliva which are believed to be involved in trace metal metabolism. Low molecular weight copper complexing substances identified in human saliva (Gollan, 1975) are thought to act to solubilize copper making it more readily available for absorption. Although various metal chelators shown to aid in the absorption of metals have been identified in secretions from other intestinal glands (Cousins, 1985), this compound is the only salivary factor which has been positively characterized to possess this absorption function. Salivary zinc-binding proteins have been identified in the rat (Etzel et al., 1984; Etzel and Koepsel, 1989); however, their role in zinc metabolism has not been characterized. Henkin and co-workers (1975) have identified a zinc-binding protein (gustin) in human saliva and suggest that this protein is an integral component of taste perception (Henkin, 1978). Citing clinical and experimental observations which report diminished taste acuity with reduced salivary gland function in man (Henkin et al., 1972) and experimental animals (Catalanotto and Sweeney, 1973; Catalanotto and Sweeney, 1972), they proposed that this salivary protein was essential for the growth and nutrition of taste buds. More recent work of Weiffenbach, working with xerostomic patients, reports that human taste perception is unimpaired in individuals with severe chronic failure of all major and minor salivary gland function (Weiffenbach et al., 1986). It is questionable, therefore, that this zinc binding protein is absolutely necessary for optimal taste sensation.

Other compounds in saliva have been identified which have been shown to bind vitamin B_{12} (Bercher et al., 1958). Although the authors speculated that this protein is in some way involved in the absorption and transport of this nutrient within the intestine, the physiological significant of this interaction has never been determined. Subsequent work on this compound is nonexistent.

Recently, considerable interest has focused upon the role of a specific group of proteins in saliva which are capable of chelating dietary compounds thought to be potentially harmful. Tannins are complex phenol compounds found in foods derived from plants. Popular sources of tannins include food such as barley, legumes, fruits, and berries as well as beverages including red wine, coffee, tea, cider, cocoa, and beer. Consumed in large amounts, tannins can cause carcinomas, liver disease, growth retardation, and other pathological problems (Mehansho et al., 1987). Studies have shown that diets rich in tannins are capable of producing dramatic increases in parotid gland weight and a 12-fold increase in the amount of proline-rich proteins secreted from the parotid gland (Mehansho et al., 1983; Mehansho et al., 1985). Proline-rich proteins are a family of proteins enriched in proline, glutamine, and glycine but lacking the aromatic and sulfur-containing amino acids (Bennick and Connell, 1971; Oppenheim et al., 1971). Although constitutive in human saliva, their synthesis in rat parotid gland is stimulated by the β-agonist, isoproterenol (Fernandez-Sorensen and Carlson, 1974; Muenzer et al., 1979a; Muenzer et al., 1979b). Specific forms of this protein not normally present in saliva are also induced by diets rich in tannins via mechanisms not yet identified. The significance of this increased synthesis of these proteins, although highly speculative, is thought to be related to their high affinity for tannins which by binding to them may reduce the toxic effects of these compounds (Mehansho et al., 1987). It has been proposed that salivary proline-rich proteins are therefore protective against the carcinogenic effects of tannins and that genetic variants of the salivary proteins may influence the incidence of esophageal cancer in different populations (Warner and Azen, 1988). Confirmation of this proposal awaits further experimental work.

EFFECT OF NUTRITIONAL STATUS ON SALIVARY GLAND FUNCTION

The effects of malnutrition on growth and development have been the basis of nutritional studies since the turn of the century. One of the classical indicators of the importance of a nutrient for mammalian systems is the ability of the specific dietary component to maintain optimal growth (increase in size) and development (increased complicity of function) when the nutrient is replaced in an otherwise nutritionally adequate diet. The development of chemically defined synthetic diets and sensitive assay systems to measure trace amounts of nutritional compounds has allowed investigators to manipulate experimental diets and, in so doing, has dramatically increased our understanding of the role that specific nutrients play in the overall process of growth and development. We now know that adequate amounts of nutrients in proper proportions are required at each critical phase of development. The salivary glands are no exception, and this section will review the role that diet and nutrition plays in the growth, development, and maintenance of the salivary glands and their secretory products.

Caloric Restriction

Studies which have investigated the effects of reduced caloric intake on salivary gland growth and function while maintaining adequate intake of protein, fat, minerals, and vitamins are essentially nonexistent. In most instances, reduced caloric intake examines the effect of not only decreased total calories consumed but more appropriately the effect of decreased intakes of proportionally the same amounts of other essential nutrients as well. Nevertheless, results of these investigations, which in essence mirror conditions often encountered by severely malnourished individuals, have shown that parotid and submandibular glands are affected by this nutritional perturbation.

In experimental animals, numerous studies have demonstrated that reduced intake of food affects the parotid gland and its secretory products (Sreebny and Johnson, 1968; Buchner and Sreebny, 1970; Johnson and Sreebny, 1973b; Alvares and Johnson, 1981; Johnson and Alvares, 1984; Johnson et al., 1984; Johansson et al., 1985). The effects of dietary restriction is dependent upon both the amount of diet withheld and the duration of the deprivation. Mild restriction of daily food intake (16%) for a period of 8 days results in increases in parotid gland weight and amylase content (Sreebny and Johnson, 1968). The increased gland weight and amylase content are attributed to the accumulated secretory products in the acinar cells, an observation confirmed by the fact that DNA and RNA content of the gland did not differ from values observed in glands of ad libidum-fed animals (Sreebny and Johnson, 1968). With a more severe restriction (approximately 33 to 50%), gland weight, and amylase content are decreased when compared to the levels observed in animals on mildly restricted diets. Once again, DNA and RNA content of the gland were not altered by the dietary changes.

In another study which investigated the effect of mild food restriction (16%) for 49 days, similar results were obtained (Buchner and Sreebny, 1970). Following an initial increase in gland weight and amylase content for the first 12 days of food restriction, total gland-weight measurements and amylase values progressively declined, albeit not below levels of ad libidum-fed controls. The increase was again attributed to increased storage products since neither cell number nor gland content of RNA were altered. After 24 days and throughout the remainder of the study, animals on restricted food intake continued to have larger parotid glands and greater amylase content than animals maintained on an ad libidum regimen. The authors attributed this increased gland weight and amylase content to normal diurnal variation in glandular secretory enzymes. In the rat, high and low levels of secretory proteins and gland weight correspond to the beginning and end, respectively, of the eating cycle (Sreebny and Johnson, 1969). It has been suggested that since animals on a restricted diet consume their food more avidly and thus more quickly than their ad libidum-fed counterparts, the secretory products of the gland have more time to accumulate during the nocturnal feeding period of the rat. If both the restricted rats and ad libidum rats are sacrificed within hours of commencement of eating, the differences in gland weight and amylase content disappear. However, compared to ad libitum-fed animals the synthetic rate of secretory proteins is increased in rats maintained on a restricted diet (Johnson and Sreebny, 1973b). It has been proposed that greater rates of food consumption (g/h) results in greater gland stimulation (Johnson, 1987). Studies to prove this theory or identify the neuronal/hormonal mechanisms which produce this stimulation have not been reported.

Severe reduction in food intake for periods greater than 4 weeks results in dramatic changes in parotid gland morphology and function. Gradual restriction of caloric intake (up to 50% of that consumed by ad libitum-fed rats) resulted in tissue changes classically associated with food restriction (Golden, 1988). Histologically, cell number and size in parotid glands of food-restricted animals are reduced compared to those measured in glands of ad libidum-fed controls (Alvares and Johnson, 1981). Reduced parotid gland weight (Johnson and Alvares, 1984) and decreased intracellular content of RNA and DNA has also been reported (Alvares and Johnson, 1981). These morphological and biochemical changes are reflected in the secretory products of the gland since salivary flow is reduced (Johnson and Alvares, 1984). The protein concentration is increased, however, and the relative proportions of specific proteins in parotid saliva are altered. Based upon densitometric analysis of SDS-polyacrylamide gel electrophoresis, it has been reported that the proline-rich proteins are increased twofold while amylase, deoxyribonuclease, and other nonspecific proteins are reduced (Johnson and Alvares, 1984). It is apparent, therefore, that the salivary glands are affected by dietary restriction and, depending upon the duration and degree of this nutritional perturbation, functional and morphological changes in the oral cavity are also likely to occur.

In contrast to the many studies in experimental models which show that overall food reduction affects salivary gland function, there are few studies which examined the effect of similar dietary restrictions in man. Protein-calorie malnutrition has been associated with reduced levels of IgA in whole saliva of young Columbian children but significant differences in other salivary factors were not observed (McMurray et al., 1977). Long-term studies of the effects of severe caloric restriction in humans is impossible for obvious ethical reasons. Nevertheless, the few studies which have been reported suggest that short-term fasting does produce changes in the composition of human saliva. Birkhed et al., (1984) observed that during successive 6-day periods of fasting (200 to 400 kcal/ day), and a lactovegetarian diet, no significant changes in secretion rate of stimulated whole saliva occurred. In contrast, Johansson and co-workers (1984) reported that in 8 healthy females, daily consumption of a 300-kcal liquid diet for 8 days resulted in a significant reduction in secretion rate and concentration of phosphate and sialic acid in stimulated whole saliva. Resting saliva in these fasted individuals had decreased concentrations of thiocyanate and lysozyme activity. In an earlier study investigating the effect of a liquid diet providing 900 kcal/day, stimulated salivary-secretion rate, protein, and amylase content were reduced (Hall et al., 1967). In this study however, the salivary effects were not observed in individuals consuming a solid diet of identical caloric content. Concerned that the effect of food restriction in man was due to reduced levels of mastication (for review see Johnson, 1987), other investigators examined the effects of a 7-day, low-calorie diet in the presence and absence of chewing (Johansson and Ericson, 1986). They reported that consumption of a low-calorie liquid diet in the absence of chewing produced a decrease in secretion rate, phosphate, calcium, lysozyme, and thiocyanate and an increase in sodium, fucose, hexosamine, and the ratio of protein to sialic acid. In patients consuming this identical diet but who also chewed paraffin wax for 10 min 5 times daily after meals, secretion rate, calcium, lysozyme, and thiocyanate were still reduced. Other measured parameters, however, were not changed by caloric restriction in the individuals who chewed paraffin wax nor did resting salivary secretions differ between the two groups. This study confirms that food deprivation, as demonstrated in experimental animals, produces reversible alterations in salivary gland function and that these changes are not solely the result of reduced motor activity stimulation.

Protein Deficiency

The effects of protein deficiency, i.e., protein reduction while maintaining adequate caloric consumption, are similar to those observed in food-restricted animals. In studies investigating the effects of protein malnutrition, energy content of all experimental diets must be maintained to avoid confounding influences of caloric deprivation. Several studies, carefully designed to only investigate the effect of reduced protein consumption, have demonstrated that protein malnutrition at different stages of development significantly affect both parotid and submandibular salivary gland function (Enwonwy, 1972; Menaker and Navia, 1973; Menaker and Navia, 1974; Watson, 1977; Johnson et al., 1977; Alvares et al., 1977; Watson and Antal, 1980; Johnson et al., 1984; Etzel et al., 1988). The degree to which these changes were manifest depended upon the severity of the protein deficiency, duration of the deprivation, and stage of development at which the nutritional deficiency was initiated.

Consumption of a protein-deficient diet (8% protein) throughout pregnancy and lactation results in reduced total net weight, protein, and DNA and RNA concentration of submandibular glands in pups born from the malnourished dams (Menaker and Navia, 1973). The effect of this protein malnutrition was reversible by administration of 10% protein supplements during the suckling period. Total stimulated volume and protein content of whole saliva from rats weaned from protein-malnourished (8% protein) dams is likewise reduced (Menaker and Navia, 1974). Reversal of these changes by protein supplementation was not examined in this study.

Continued imposition of a protein-deficient diet after weaning produces slightly different effects on salivary components. Rats whose dams were fed 8% protein diets, and subsequently at weaning maintained on 8% protein diets, had significantly decreased levels of total protein per milliliter of saliva compared to saliva of rats consuming 25% protein diets for a period of 66 days (Watson and Antal, 1980). Although amylase activity was also reduced at this time, levels of IgG and aminopeptidase in saliva were not altered. In rats that continued to consume the protein-deficient diet, total protein collected and salivary protein concentration were very similar to those values obtained from control rats fed the protein-adequate diets. The authors suggest that the initial decrease and eventual increase of protein concentration in the saliva is the result of reduced synthesis of amylase followed by an induction of amylase production by the high starch, low protein diet (Watson and Antal, 1980). Johnson and co-workers (1977) have shown that young rats fed a very low protein powdered diet (0.5%) had decreased levels of DNA, RNA, and protein content, reduced total gland weight, and significantly reduced amylase activity when compared to values obtained from rats fed a protein-adequate, powdered diet. Of particular interest in this study was the fact that unlike previous investigators no gland enlargement was observed. This inconsistency with previous studies relating parotid gland growth with protein calorie malnutrition was thought to be attributed to the lack of mastication required to consume the powdered diet used in their study.

More recent studies utilizing different protein levels and diet textures suggest, as previous studies have, that data obtained from studies relating salivary gland function to protein composition of the diet is extremely sensitive to experimental design. Johansson and Ericson (1987) show that after a period of 9 weeks rats fed a powdered 5% protein diet had lower stimulated salivary secretion rates and reduced total protein of whole saliva secreted per minute than control rats fed a 20% protein diet. Etzel and co-workers (1988), however, were unable to demonstrate differences in total protein concentration nor the volume of stimulated parotid saliva secreted between rats fed pelleted diets containing 18% protein (control) or 8% protein (deficient) for 38 days. It is apparent from these and previous studies that salivary gland function is sensitive not only to levels of dietary protein but also experimental factors including methods of saliva collection, source of saliva, diet texture, and duration of diet consumption. Clarification of the specific mechanisms responsible for these dietary-induced changes in salivary gland function will be hastened by carefully standardizing experimental design and utilizing advances in molecular biological techniques as they become available in the field of salivary research.

Vitamins and Minerals

Considering the metabolic requirements of the salivary glands which are capable of synthesizing and secreting up to 1 g of protein per 100 ml of saliva per day (Vogel, 1985), it is understandable that essential vitamins and minerals be readily available for optimal gland function. It is not surprising, therefore, that certain vitamin and mineral deficiencies have been reported to affect salivary gland function and secretion. In rats maintained on a vitamin A-deficient diet, inflammation, ductal distention, and acinar atrophy are the major histological findings in the salivary glands (Trowbridge, 1969; Anzano et al., 1980). The fact that these degenerative changes were reported in animals force fed to minimize the effects of secondary inanition suggests that vitamin A may directly affect salivary gland cells. Secretory function is also altered as indicated by decreased flow rates and protein concentration of stimulated whole saliva (Anzano et al., 1981). High doses of vitamin A in rats results in acinar degeneration and hyperplasia of rough endoplasmic reticulum and Golgi (Regezi and Rowe, 1972). Cancer of the salivary glands in experimental animals has been associated with vitamin A deficiencies (Rowe et al., 1970). In general, these pathological changes are consistent with the ascribed function of vitamin A in the maintenance of epithelial structures.

Vitamin D deficiency has been shown to reduce the volume of pilocarpine-stimulated parotid saliva in rats which were offspring of vitamin D-deficient mothers and which were maintained on a vitamin D-free diet from weaning

(Glijer et al., 1985). These salivary changes were not observed in rats which were offspring of normal mothers but were fed a vitamin D-free diet from weaning. Total amounts of amylase secreted was not affected by vitamin deficiency. Calcium secretion paralleled changes in serum calcium which was reduced to similar degrees in both groups of vitamin D-deficient rats. It was proposed, therefore, that the secretion of electrolytes and water, which in the parotid gland requires extracellular calcium, is dependent on vitamin D. It has been suggested that vitamin D is required by the parotid gland to induce the synthesis of a protein(s) which are involved in calcium metabolism (Glijer et al., 1985). This proposal is supported by *in vitro* observations demonstrating a stimulating effect of vitamin D deficiency on microsomal Ca-ATPase activity in parotid gland (Hayakawa et al., 1983) and the description of a vitamin D-dependent calcium-binding protein in rat parotid gland (Goodwin et al., 1978). Receptors for 1,25 dihydroxycholecalciferol have been isolated in acinar cells of rat parotid gland (Peterfy and Tenenhouse, 1982).

Reduced intake of minerals has been shown to influence salivary glands and their secretions. Severe (McChance, 1938; Thorn et al., 1956; Wotman et al., 1973) and moderate (Christensen et al., 1986) reductions in dietary sodium in humans produce lowered salivary sodium concentration. No changes in flow rates or protein composition were reported. In patients suffering from iron deficiency anemia, chemically stimulated salivary flow rate was significantly lower than that measured in age and body weight matched healthy male subjects (Mukherji et al., 1982). Zinc deficiency in rats produces dramatic changes in salivary gland morphology and function (Chaudhry and Meyer, 1979; Alvares and Johnson, 1981; Chaudhry et al., 1981; Johnson and Alvares, 1984). Morphological changes, characterized by distinctive election-dense secretory granules in parotid acini, were associated with decreased flow rate and alterations in protein composition of stimulated parotid saliva (Johnson and Alvares, 1984). The concentration of acidic proline-rich proteins, thought to play an important role in the maintenance of the integrity of the tooth surface, are also dramatically decreased. These changes are attributed to zinc deficiency per se, since these salivary changes were compared to those observed in rats provided with a zinc-adequate diet but restricted in amount to that consumed by the zinc-deficient animals. This comparison is crucial since zinc deficiency is associated with reduced food intake (Chesters and Quarterman, 1970). Comparisons between ad libidum-fed rats and zinc-deficient rats demonstrates that zinc deficiency also produces significant reductions in gland weight, RNA content, and cell number in parotid and submandibular gland. Protein concentration in zinc-deficient rats is increased by 86%, as is the proportion of amylase in stimulated saliva. However, since the changes in total protein concentration, gland weight, RNA and DNA content were similar in zinc-deficient and pair-fed control rats, these changes were attributed to food restriction rather than zinc deficiency per se. It is obvious, therefore, that decreased intake of zinc in the diet produces a complex sequelae of changes in salivary tissues related to both primary zinc deficiency and secondary reductions in food intake.

In contrast to studies reporting the effects of dietary deficiencies on salivary gland function are reports of changes in salivary gland morphology and function following consumption of dietary additives. The addition of pancreatin (a defatted extract of pancreas) to the diet of rats results in parotid and submandibular gland enlargement and an increase in the salivary content of basic proline-rich proteins (Mangos et al., 1969). The observed enlargement was prevented by treatment with β adrenergic antagonists suggesting that these salivary changes were mediated by β receptor activation (Brenner and Stanton, 1970). Since the salivary effects of the addition of pancreatin to the diet was not observed if the supplement was given by stomach tube, it appears that its site of action is the oral cavity (Mangos et al., 1969; Wells et al., 1965). Similar salivary changes have been reported in rats maintained on low protein diets supplemented with picolinic acid compared to rats fed diets of adequate protein without picolinic acid (Etzel et al., 1988). Picolinic acid, a compound which has been shown to enhance zinc absorption in animals fed diets of low zinc content (Evans and Johnson, 1980; Evans and Johnson, 1981), is also known to enhance β receptor function *in vitro* (Mukku et al., 1979). The fact that the addition of picolinic acid to a low protein diet resulted in both an increase in salivary protein concentration and in the proportion of basic proline-rich proteins in parotid saliva (Etzel et al., 1988) is consistent with the reported β receptor activation of this compound. It remains to be determined, however, whether picolinic acid acts via an intraoral gustatory mechanism or a systemic effect following intestinal absorption of the dietary additive.

Food Consistency

As previously discussed, mastication is a powerful stimulus of salivary secretion. The fact that both salivary volume and total amount of protein secreted in the saliva is directly proportional to the intensity of chewing suggests that food consistency does influence salivary function. In fact, numerous studies have documented the fact that both decreased and increased mastication, as required by the diet texture, dramatically affect salivary gland composition and function.

Early work of Hall and Schneyer (1964) and Sreebny and Johnson (1968) demonstrated that the parotid gland of rats maintained on liquid diets became atrophic after 7 days. No difference in exocrine function of the pancreas was observed under identical dietary conditions (Sreebny and Johnson, 1968). This parotid gland atrophy, also reported to occur in submandibular and sublingual glands (Hall and Schneyer, 1964; Wells and Peronace, 1967), is produced by both liquified chow diets (Anderson et al., 1985; Hall and Schneyer, 1964; Johnson and Sreebny, 1971) and milk diets (Hall and Schneyer, 1964). Similar changes have been observed in submandibular glands of rats which were fed by gastric intubation rather than allowed to chew and consume food naturally (Mundorft et al., 1984). Calculations of total caloric intake under experimental conditions indicate that the atrophic changes observed in the salivary glands are not the result of malnutrition (Johnson, 1982; Sreebny and Johnson, 1968; Johnson and Sreebny, 1971). Histologically, several distinct changes in the gland occur. Initially, in experimental animals there is an increase in the number and size of cytoplasm granules reflecting increased storage of secretory product (Hand and Ho, 1981). Eventually, atrophic characteristics of the parotid acinar cells begin to appear with increased amounts of lysosomal vacuoles and decreased numbers of secretory granules and the total number of cells (Hand and Ho, 1981). In fasted rats, accumulated secretory granules have been shown to fuse with lysosomal structures presumably leading to degradation of stored protein (Hand and Ball, 1988). Marked decreases in the number of both β adrenergic and muscarinic receptors are observed in parotid glands of adult rats maintained on liquid diets for periods of 7 to 21 days (Schneyer et al., 1987). In general, with prolonged consumption of liquid diets secretory volume and protein concentration of stimulated parotid saliva is reduced (Hall and Schneyer, 1964; Johnson, 1982). Except for salivary levels of deoxyribonuclease, which is increased, concentrations of amylase, ribonuclease, and proline rich proteins are decreased in response to decreased mastication (Johnson, 1982; Johnson, 1983; Johnson, 1984). These reduced levels of amylase activity in rats maintained on liquid diets have been associated with an increase in amylase-specific mRNA relative to the total mRNA pool (Zelles et al., 1989). This seemingly contradictory observation suggests that changes in salivary gland secretory function in response to dietary changes may result from alterations in translation or RNA turnover rates. Clarification of the mechanisms which regulate levels of salivary proteins under conditions of altered dietary intakes await further experimentation.

In humans, the effects of liquid diets on salivary function are similar since salivary protein concentration, amylase, proline-rich proteins, and flow rate have all been reported to be lower than those levels reported during consumption of standard diets (Hall et al., 1967; Johansson et al., 1984). Salivary pH is also reduced in response to the liquid diet (Johansson et al., 1984). Changes in salivary composition in individuals deprived of mastication for periods of less than 24 h have not been observed (Dawes, 1975).

Secretory protein content of salivary glands undergo diurnal variation in animals fed normal pelleted diets. This diurnal change consists of an increase in parotid gland weight progressing during the day followed by a gradual loss of weight commencing at the onset of feeding and continuing throughout the night. The changes in gland weight therefore are attributed to the accumulation and expulsion of secretory products during the diurnal feeding cycle (Sreebny and Johnson, 1969). Without expulsion of saliva, a process stimulated by mastication, further accumulation of secretory products is ceased, leading to an eventual atrophic gland within 4 to 6 days. Elimination of this diurnal pattern and the associated "atrophy of disuse" (Johnson and Sreebny, 1971) produced by the reduced masticatory stimulation demonstrates the importance of mastication in functional regulation and maintenance of salivary gland tissue.

Atrophy induced by consumption of liquid diets is reversible by feeding foods requiring more mastication (Johnson, 1982). Likewise, it is known that by increasing the requirement for chewing activity by adding non-nutritive bulk (cellulose) to the diet of experimental animals, gland enlargement will occur (Wells and Peronace, 1967; Johnson and Sreebny, 1973a; Schneyer and Hall, 1976; Johnson, 1981; Johnson and Sreebny, 1982). When compared to parotid glands of rats fed identical calories but pelleted without additional cellulose, rats fed the high-cellulose diet had significantly greater numbers of cells of larger size with increased RNA content. Stimulated flow rates and protein concentration of saliva are likewise increased (Johnson and Sreebny, 1982); however, no alterations in protein composition (i.e., amylase and the proline-rich proteins) were reported (Johnson and Sreebny, 1982).

Just as salivary changes were shown to be similar in humans and rats following the consumption of liquid diets, so too are the effects of increased mastication on salivary gland secretory products in humans. When the diets of a group of young boys was modified to contain foods which required more mastication, significant changes in parotid gland secretory products were observed (de Munoz et al., 1981). Salivary flow rate, amylase and protein concentration and salivary pH were all increased over values observed 45 days after dietary changes were introduced. Since caloric content of the diet was essentially identical to the diet previously consumed by the boys prior to the incorporation of the firmer foods, these changes were attributed to increased gland stimulation via mastication or stimulation of periodontal mechanoreceptors.

CONCLUSION

In the presence of friends and a comfortable surrounding, the grinding, lubricating and digestive process which goes on within us goes unnoted. The oral activities which provide us with the pleasures of eating are only seriously acknowledged when one of its components fails to contribute to the total enjoyment of a meal. Such is truly the case with saliva. In individuals who suffer from the pathological consequences of xerostomia, their daily discomfort is a neverending reminder of the importance of this oral fluid to our overall nutritional activities. William Beaumont, although acknowleging the signficiant role which saliva plays as a "preliminary to digestion" stated nonetheless that "its legitimate and only use ... is to lubricate the food and facilitate the passage of the bolus through the organs of deglutition"(Beaumont, 1833). Certainly, since his observations, numerous other functions have been ascribed to these oral secretions and yet the assignment of biological roles to specific purified proteins await the results of ongoing research. While oral protection, bacterial growth inhibition, enamel maturation, and acid neutralization remain the best documented functions of saliva, its importance in influencing nutrient intake and subsequent digestion and absorption cannot be overlooked. Intestinal secretions have long been considered to be the sole source of major enzymes and physiological ligands essential for the uptake and transport of nutrients from dietary sources. However, considering the copious amount of fluid and protein secreted during the consumption of food, and the ideal conditions under which it is mixed with dietary components, it appears incongruous not to look more closely at the role of saliva in the overall maintenance of our nutritional well being.

Hopefully, this chapter has provided convincing evidence that saliva is an important factor in nutrient acquisition and, conversely, that nutritional status is a major determinant of salivary gland function. Long-term implications of these interactions not only on oral health, but overall health in general, remain to be more accurately addressed.

REFERENCES

1. Alvares, O., Worthington, B., Johnson, D., and Sreebny, L., Sequential effects of protein-calorie malnutrition (PCM) on the ultrastructure of parotid acinar cells, *J. Dent. Res.*, 56, 240, 1977.
2. Alvares, O. F. and Johnson, D. A., Effects of zinc deficiency on rat parotid gland, *J. Oral Pathol.*, 10, 430, 1981.
3. Anderson, D. J., Hector, M. P., and Linden, R. W. A., The possible relation between mastication and parotid secretion in the rabbit, *J. Physiol. (London)*, 364, 19, 1985.
4. Anzano, M. A., Olson, J. A., and Lamb, A. J., Morphologic alterations in trachea and the salivary gland following the induction of rapid synchronous vitamin A deficiency in rats, *Am. J. Pathol.*, 98, 717, 1980.
5. Anzano, M. A., Lamb, A. J., and Olson, J. A., Impaired salivary gland secretory function following the induction of rapid, synchronous vitamin A deficiency in rats, *J. Nutr.*, 111, 496, 1981.
6. Baum, B. J., Neurotransmitter control of secretion, *J. Dent. Res.*, 66, 628, 1987.
7. Bartoshuk, L. M., *Nutrition in Oral Health and Disease*, Lea & Febiger, Philadelphia, 53, 1985.
8. Beaumont, W., *Experiments and Observations on the Gastric Juice and the Physiology of Digestion*, F. P. Allen, Plaburgh, New York, 67, 125, 1933.
9. Benson, H., Beary, J. F., and Carol, M. P., The relaxation response, *Psychiatry*, 37, 1974.
10. Bercher, R. W., Meyer, L. M., and Miller, I. F., Co^{60} Vitamin B_{12} binding capacity in normal human saliva, *Proc. Sci. Exp. Biol.*, 99, 513, 1958.
11. Birkhed, D., Heintze, U., Edwardsson, S., and Aly, K., Short term fasting and lactovegetarian diet does not affect human saliva, *Scand. J. Dent. Res.*, 92, 408, 1984.
12. Borg, A. and Birkhed, D., Secretion of glucose in human parotid saliva after carbohydrate intake, *Scand. J. Dent Res.*, 96, 551, 1988.
13. Bower, J. M., Gamble, R., Gregory, H., Gerring, E. L., and Willshire, I. R., The inhibition of gastric acid secretion by edpidermal growth factor, *Factor Experientia*, 31, 825, 1975.

14. Brenner, G. M. and Stanton, H. C., Adrenergic mechanisms responsible for submandibular salivary glandular hypertrophy in the rat, *J. Pharmacol. Exp. Ther.*, 173, 166, 1970.

15. Buchner, A. and Sreebny, L. M., Effect of prolonged food reduction on the rat parotid gland and exocrine pancreas, *J. Nutr.*, 100, 655, 1970.

16. Carlson, A. J., *The Control of Hunger in Health and Disease*, University of Chicago Press, Chicago, 1916.

17. Catalanotto, F. A. and Sweeney, E. A., The effect of surgical desalivation of the rat upon taste acuity, *Arch. Oral Biol.*, 17, 1455, 1972.

18. Catalanotto, F. A. and Sweeney, E. A., Long term effects of selective desalivation on taste acuity in rats, *Arch. Oral Biol.*, 18, 941, 1973.

19. Chaudhry, I. M. and Meyer, J., Response of submandibular gland of the rat to nutritional zinc deficiency, *J. Nutr.*, 109, 316, 1979.

20. Chaudhry, I. M., Gandor, D. W., and Gerrson, S. J., Reduction of carbonic anhydrase activity in the submandibular salivary glands of zinc deficient rats, *Arch. Oral Biol.*, 26, 399, 1981.

21. Chesters, J. K. and Quarterman, J., Effects of zinc deficiency on food intake and feeding patterns of rats, *Br. J. Nutr.*, 24, 1061, 1970.

22. Christensen, C. M., Bertino, M., Beauchamp, G. K., Navazesh, M., and Engelman, K., The influence of moderate reduction in dietary sodium on human salivary sodium concentration, *Arch. Oral Biol.*, 31, 825, 1986.

23. Code, C. F., Ratke, H. V., Livermore, G. R., and Lunberg, W., Occurrence of gastric inhibitor activity in fresh gastric and salivary mucin, *Fed. Proc.*, 8, 26, 1949.

24. Cousins, R. J., Absorption, transport and hepatic metabolism of copper and zinc: special reference to metallothionein and ceruloplasmin, *Annu. Rev. Physiol.*, 65, 238, 1985.

25. Dawes, C., The effect of food deprivation on the circadian rhythms in flow rate and protein concentration in human parotid saliva, *Arch. Oral Biol.*, 20, 547, 1975.

26. Dawes, C., Stimulus effects on protein and electrolyte concentrations in parotid saliva, *J. Physiol. (London)*, 346, 579, 1984.

27. deMunoz, B. R., Maresca, B. M., Tumilasci, O. R., and Perec, C. J., Effects of an experimental diet on parotid saliva and dental plaque pH in institutionalized children, *Arch. Oral Biol.*, 28, 575, 1981.

28. Dutta, S. K., Hamosh, M., Abrams, C. K., Hamosh, P., and Hubbard, V. S., Quantitative estimation of lingual lipase activity in the upper small intestine in adult patients with pancreatic insufficiency, *Gastroenterology*, 82, 1047, 1982.

29. Emmelin, N., Nerve interactions in salivary glands, *J. Dent. Res.*, 66, 509, 1987.

30. Enwonwu, C. O., Biochemical and morphologic changes in rat submandibular gland in experimental protein-calorie malnutrition, *Exp. Mol. Pathol.*, 16, 244, 1972.

31. Etzel, K. R., Cortez, J. E., and Johnson, D. A., Localization of zinc binding protein in rat parotid saliva, *J. Dent. Res.*, 63, 302, 1984.

32. Etzel, K. R., Cortez, J. E., and Johnson, D. A., The addition of picolinic acid to low protein diets. A word of caution, *Nutr. Res.*, 8, 1391, 1988.

33. Etzel, K. R. and Koepsel, R. R., Identification of zinc binding protein in rat saliva, *J. Dent. Res.*, 68, 354, 1989.

34. Evans, G. W. and Johnson, E. C., Zinc absorption in rats fed a low protein diet and a low protein diet supplemented with tryptophan or picolinic acid, *J. Nutr.*, 110, 1076, 1980.

35. Evans, G. W. and Johnson, E. C., Effect of iron, vitamin B_6 and picolinic acid on zinc absorption in the rat, *J. Nutr.*, 111, 68, 1981.

36. Feller, R. P., Sharon, I. M., Chauncey, H. H., and Shannon, I. L., Gustatory perception of sour, sweet, and salt mixtures using parotid gland flow rate, *J. Appl. Physiol.*, 20, 1341, 1965.

37. Fernandez-Sorensen, A. and Carlson, D. M., Isolation of a "Proline-Rich" protein from rat parotid glands following isoproterenol treatment, *Biochem. Biophys. Res. Commun.*, 60, 249, 1974.

38. Fink, C. S., Hamosh, P., and Hamosh, M., Fat digestion in the stomach: stability of lingual lipase in the gastric environment, *Pediatr. Res.*, 18, 248, 1984.

39. Fogel, M. R. and Gray, G. M., Starch hydrolysis in man: an intraluminal process not requiring membrane digestion, *J. Appl. Physiol.*, 35, 263, 1973.

40. Galili, D., Maller, D., and Brightman, V. J., The effect of desalivation by duct ligation on salivary gland extirpation on taste preference in rats, *Arch. Oral. Biol.*, 26, 853, 1981.

41. Gjorstrup, P., Effects of sympathetic nerve stimulation in the presence of a slow parasympathetic secretion in the parotid and submaxillary glands of the rabbit, *Acta Physiol. Scand.*, 101, 211, 1977.

42. Gjorstrup, P., Parotid secretion of fluid and amylase in rabbits during feeding, *J. Physiol. (London)*, 309, 101, 1980a.

43. Gjorstrup, P., Taste and chewing as stimuli for the secretion of amylase from the parotid gland of the rabbit, *Acta Physiol. Scand.*, 110, 295, 1980b.

44. Gjorstrup, P., *Advances in Physiological Sciences*, Pergamon Press, Oxford, 1981, 225.

45. Glijer, B., Peterfy, C., and Tenenhouse, A., The effect of vitamin D deficiency on secretion of saliva by rat parotid gland in vivo, *J. Physiol.*, 363, 1985.

46. Golden, M. H. N., *Nutrition in the Clinical Management of Disease*, Edward Arnold, Ltd., Baltimore, 1988, 88.

47. Gollan, J. C., Studies on the nature of amylases formed by copper with human alimentary secretions and their influence on copper absorption in the rat, *Clin. Sci. Mol. Med.*, 49, 237, 1975.

48. Goodwin, D., Noff, D., and Edelstein, S., The parotid gland: a new target for vitamin D action, *Biochem. Biophys. Acta*, 539, 249, 1978.

49. Hall, H. D. and Schneyer, C. A., Salivary gland atrophy in rat induced by liquid diet, *Proc. Soc. Exp. Biol. Med.*, 117, 789, 1964.

50. Hall, H. D., Merig, J. J., and Schneyer, C. A., Metrecal-induced changes in human saliva, *Proc. Soc. Exp Biol. Med.*, 117, 1967.

51. Hall, H. D. and Schneyer, C. A., Functional mediation of compensatory enlargement of the parotid gland, *Cell Tissue Res.*, 184, 1977.

52. Hamosh, M. and Burns, W. A., Lipolytic activity of human lingual glands, *Lab Invest.*, 37, 1977.

53. Hand, A. R. and Ho, B., Liquid diet induced alterations of rat parotid acinar cells studied by electron microscopy and enzyme cytochemistry, *Arch. Oral Biol.*, 26, 1981.

54. Hand, A. R. and Ball, W. D., Ultrastructural immunocytochemical localization of secretory proteins in autophagic vacuoles of parotid acinar cells of starved rats, *J. Oral Pathol.*, 17, 1988.

55. Hayakawa, M., Aoki, H., Terao, N., Abiko, M., and Takiguchi, H., Vitamin D-mediated decrease of Ca^{++} pump activity in the rat parotid gland, *Int. J. Biochem.*, 15, 1983.

56. Hatton, M. N., Loomis, R. E., Levine, M. J., and Tabak, L. A., Masticatory lubrication, *Biochem. J.*, 230, 1985.

57. Heidenhain, R., Ueber die pepsinbildung in den pylorusdrusen, *Arch. Ges. Physiol.*, 18, 1878.

58. Heitlinger, L. A., Lee, P. C., Dillon, W. P., and Lebenthal, E., Mammary amylase: a possible alternate pathway of carbohydrate digestion in infancy, *Pediatr. Res.*, 17, 1983.

59. Henkin, R. I., Talal, N., Larson, A., and Mattern, C. F. T., Abnormalities of taste and smell in Sjogren's syndrome, *Ann. Intern. Med.*, 76, 1972.

60. Henkin, R. I., Prevention and treatment of hypogeusia due to head and neck irradiation, *JAMA*, 200, 1972.

61. Henkin, R. I., Lippoldt, R. E., Bilstad, J., and Edelhoch, H., A zinc protein isolated from human parotid saliva, *Proc. Natl. Acad. Sci. U.S.A.*, 72, 1975.

62. Henkin, R. I., *Zinc and Copper in Clinical Medicine*, Spectrum, New York, 1978, 35.

63. Herrera, J. L., Lyons, M. F., and Johnson, L. F., Saliva: its role in health and disease, *J. Clin. Gastroenterol.*, 10, 1988.

64. Izutsu, K. T., *The Salivary System*, Sreebny, L. M., Ed., CRC Press, Boca Raton, FL, 1987, 95.

65. Jenkins, G. N., *The Physiology and Biochemistry of the Mouth*, 4th ed., Blackwell, Oxford, 1978.

66. Jirakulsomchok, D. and Schneyer, C. A., Effects on rat parotid amylase and Ca of α and β-adrenergic sympathetic stimulation, *Am. J. Physiol.*, 236, 1979.

67. Johansson, I., Ericson, T., and Steen, L., Studies on the effect of diet on saliva secretion and caries development: The effect of fasting on saliva composition of female subjects, *J. Nutr.*, 114, 1984.

68. Johansson, I., Ericson, T., Bowen, W., and Cole, M., The effect of malnutrition on caries development and saliva composition in the rat, *J. Dent. Res.*, 64, 1985.

69. Johansson, I. and Ericson, T., Effect of chewing on the secretion of salivary components during fasting, *Caries Res.*, 20, 1986.

70. Johansson, I. and Ericson, T., Saliva composition and caries development during protein deficiency and β-receptor stimulation or inhibition, *J. Oral. Path.*, 16, 1987.

71. Johnson, D. A. and Sreebny, L. M., Effect of increased mastication on the secretory process of the rat parotid gland, *Arch. Oral Biol.*, 18, 1973.

72. Johnson, D. A. and Sreebny, L. M., Effect of partial inanition on protein synthesis in rat parotid glands, *J. Dent. Res.*, 52, 1973b.

73. Johnson, D. A., Sreebny, L. M., and Enwonwu, C. O., Effect of protein-energy malnutrition and of a powdered diet on the parotid gland and pancreas of young rats, *J. Nutr.*, 107, 1977.

74. Johnson, D. A., Effect of a ground vs. a pelleted bulk diet on rat parotid gland, *Arch. Oral. Biol.*, 26, 1981.

75. Johnson, D. A., Effect of a liquid diet on the protein composition of rat parotid saliva, *J. Nutr.*, 112, 1982.

76. Johnson, D. A. and Sreebny, L. M., Effect of increasing the bulk content of the diet on the rat parotid gland and saliva, *J. Dent. Res.*, 61, 1982.

77. Johnson, D. A., Differences in basic proline-rich proteins in rat saliva following chronic isoproterenol treatment or maintenance on a liquid diet, *Arch. Oral Biol.*, 28, 1983.

78. Johnson, D. A. and Alvares, O. F., Zinc deficiency induced changes in rat parotid salivary proteins, *J. Nutr.*, 114, 1984.

79. Johnson, D. A., Changes in rat parotid salivary proteins associated with liquid diet-induced gland atrophy and isoproterenol-induced gland enlargement, *Arch Oral Biol.*, 29, 1984.

80. Johnson, D. A., Etzel, K. R., Cortez, J. E., and Girard, L. J., Changes in rat parotid salivary proteins following adrenalectomy, *J. Dent. Res.*, 63, 1984.

81. Johnson, D. A., *The Salivary System*, Sreebny, L. M., Ed., CRC Press, Boca Raton, FL, 1987, 135.

82. Johnson, D. A., Lopez, H., and Navia, J. M., Dietary-induced changes in parotid salivary proteins and dental caries, *J. Dent. Res.*, 63, 1984.

83. Lankisch, P. G. and Otto, J., Salivary asoamylase in duodenal aspirates, *Dig. Dis. Sci.*, 31, 1986.

84. Lashley, K. S., Reflex secretion of the human parotid gland, *J. Exp. Psychol.*, 1, 1916.

85. Lavelle, C. L. B., *Applied Oral Physiology*, Wright, London, 1988, 128.

86. Lebenthal, E., Lee, P. C., and Heitlinger, L. A., Impact of the development of the gastrointestinal tract on infant feeding, *J. Pediatr.*, 102, 1983.

87. Lebenthal, E. and Lee, P. C., Alternative pathways for digestion and absorption in early infancy, *J. Pediatr. Gastroenterol. Nutr.*, 3, 1984.

88. Mandel, I. D., The functions of saliva, *J. Dent. Res.*, 66, 1987.

89. Mangos, J. A., Benke, P. J., and McSherry, N., Salivary gland enlargement and functional changes during feeding of pancreatin to rats, *Pediat. Res.*, 3, 1969.

90. McCance, R. A., The effect of salt deficiency in man on the volume of the extracellular fluids, and on the composition of sweat, saliva, gastric juice and cerebrospinal fluid, *J. Physiol.*, 92, 1938.

91. McMurray, D. N., Rey, H., Casazza, L. J., and Watson, R. R., Effect of moderate malnutrition on concentrations of immunoglobulins and enzymes in tears and saliva of young Columbian children, *Am. J. Clin. Nutr.*, 30, 1977.

92. Mehansho, H., Hagerman, A., Clements, S., Butler, L., Rogler, J., and Carlson, D. M., Modulation of proline rich protein biosynthesis in rat parotid glands by sorghums with high tannin levels, *Proc. Natl. Acad. Sci. U.S.A.*, 80, 1983.

93. Mehansho, H., Clements, S., Sheares, B. T., Smith, S., and Carlson, D. M., Induction of proline-rich glycoprotein synthesis in mouse salivary glands by isoproterenol and by tannins, *J. Biol. Chem.*, 260, 1985.

94. Mehansho, H., Butler, L. G., and Carlson, D. M., Dietary tannins and salivary proline-rich proteins: interactions, induction and defense mechanisms, *Ann. Rev. Nutr.*, 7, 1987.

95. Menaker, L. and Navia, J. M., Effect of undernutrition during the perinatal period on caries development in the rat. III. Effects of undernutrition on biochemical parameters in the developing submandibular salivary gland, *J. Dent. Res.*, 52, 1973.

96. Menaker, L. and Navia, J. M., Effect of undernutrition during the perinatal period on caries development in the rat. V. Changes in whole saliva volume and protein content, *J. Dent. Res.*, 53, 1974.

97. Miyoshi, A., Moriga, M., Ohbayashi, M., and Suyama, T., Studies on a gastric secretion inhibitory substance in human saliva salivogastrone, *Arch. Intern. Med.*, 16, 1969.

98. Morse, D. R., Schacterle, G. R., Furst, L., Zaydenberg, M., and Pollack, R. L., Oral digestion of a complex-carbohydrate cereal: effects of stress and relaxation on physiological and salivary measures, *Am. J. Clin. Nutr.*, 49, 1989.

99. Muenzer, J., Bildstein, C., Gleason, M., and Carlson, D. M., Purification of proline-rich proteins from parotid glands of isoproterenol-treated rats, *J. Biol. Chem.*, 254, 1979.

100. Muenzer, J., Bildstein, C., Gleason, M., and Carlson, D. M., Properties of proline-rich proteins from parotid glands of isoproterenol-treated rats, *J. Biol. Chem.*, 254, 1979.

101. Mukherji, S., Naik, S. R., Srivastava, P. N., Choudhuri, S. N., and Chuttaini, H. K., Salivary secretions in iron deficiency anemia, *J. Oral Med.*, 37, 1982.

102. Mukku, V. R., Anderson, W. B., and Johnson, G. S., Enhancement of hormonal stimulation in intact cells, *J. Biol. Chem.*, 254, 1979.

103. Mundorff, S. A., Curzon, M. E. J., and Eisenberg, A. D., Comparison of essential nutrient supplement effects on rat growth and dental caries, *Caries Res.*, 18, 1984.

104. Newbrun, E., Observations on the amylase content and flow rate of human saliva following gustatory stimulation, *J. Dent. Res.*, 41, 1962.

105. Pavlov, I. P., *The Work of the Digestive Glands*, Charles Griffin, London, 1902.

106. Peterfy, C. and Tenenhouse, A., Vitamin D receptors in isolated rat parotid acinar cells, *Biochem. Biophys. Acta*, 721, 1982.

107. Regezi, J. A. and Rowe, N. H., Morphologic effects of hypervitaminosis A on rat submandibular gland, *Arch. Oral Biol.*, 17, 1972.

108. Rosa, R. and Nicolau, J., Sialic acid and hexosamine content in the sublinqual glands from rats fed a vitamin A deficient diet, *Arch. Oral Biol.*, 20, 1975.

109. Rosenblum, J. L., Irwin, C. L., and Alpers, D. H., Starch and glucose oligosaccharides protect salivary-type amylase activity at acid pH, *Am. J. Physiol.*, 254, 1988.

110. Rowe, N. H., Grammer, F. C., Watson, F. R., and Nickerson, N. H., A study of environmental influence upon salivary gland neoplasia in rats, *Cancer*, 26, 1970.

111. Schneyer, C. A., Role of sympathetic pathway in secretory activity induced in rat parotid by feeding, *Proc. Soc. Exp. Biol. Med.*, 147, 1974.

112. Schneyer, C. A. and Hall, H. D., Neurally mediated increase in mitosis and DNA of rat parotid with increase in bulk of diet, *Am. J. Physiol.*, 230, 1976.

113. Schneyer, C. A., Humphreys-Beher, M. G., and Hall, H. D., β-Adrenergic and muscarinic receptors of parotid gland following maintenance of rats on liquid diet, *Proc. Soc. Exp. Biol. Med.*, 184, 1987.

114. Shannon, I. L., *Handbook of Experimental Aspects of Oral Biochemistry*, CRC Press, Boca Raton, FL, 1983, 307.

115. Spiers, R. L. and Hodgson, C., Control of amylase secretion in the parotid gland during feeding, *Arch. Oral Biol.*, 21, 1976.

116. Sreebny, L. M. and Johnson, D. A., Effect of food consistency and decreased food intake on rat parotid and pancreas, *Am. J. Physiol.*, 215, 1968.

117. Sreebny, L. M. and Johnson, D. A., Diurnal variation in secretory components of the rat parotid gland, *Arch. Oral. Biol.*, 14, 1969.

118. Sreebny, L. M. and Broich, G., *The Salivary System*, CRC Press, Boca Raton, FL, 1987, 179.

119. Stryer, L., *Biochemistry*, W. H. Freeman, New York, 1988, 342.

120. Tabak, L. A., Levine, M. J., Mandel, I. D., and Ellison, S. A., Role of salivary mucins in the protection of the oral cavity, *J. Oral Pathol.*, 11, 1982.

121. Thorn, N. A., Schwartz, I. L., and Thaysen, J. H., Effect of sodium restriction on secretion of sodium and potassium in human parotid saliva, *J. Appl. Physiol.*, 35, 1956.

122. Trowbridge, H. D., Salivary gland changes in vitamin A-deficient rats, *Arch. Oral Biol.*, 14, 1969.

123. Vance, W. B., Observations on the role of salivary secretion in the regulation of food and fluid intake in the white rat, *Psychol. Monogr.*, 79, 1965.

124. Vogel, R. I., *Nutrition In Oral Health and Disease*, Lea & Febiger, Philadelphia, 1985, 84.

125. Warner, T. F. and Azen, E. A., Tannins, salivary proline-rich proteins and oesophageal cancer, *Medical Hypotheses*, 26, 1988.

126. Watson, R. R., The effect of *in utero* protein malnutrition and subsequent renutrition on rat saliva and some salivary enzymes, *Br. J. Nutr.*, 38, 1977.

127. Watson, R. R. and Antal, M., Effects of moderate chronic protein deficiency on rat salivary components, *J. Nutr.*, 110, 1980.

128. Weiffenbach, J. M., Fox, P. C., and Baum, B. J., Taste and salivary function, *Proc. Natl. Acad. Sci. U.S.A.*, 83, 1986.

129. Wells, H., Peronace, A., and Stark, L., Taste receptors and sialadenotrophic action of proteolytic enzymes in rats, *Am. J. Physiol.*, 208, 1965.

130. Wells, H. and Peronace, A. A. V., Functional hypertrophy and atrophy of the salivary glands of rats, *Am. J. Physiol.*, 212, 1967.

131. Wotman, S., Baer, L., Mandel, I. D., and Laragh, J., Salivary electrolytes, renin, and aldosterone during sodium loading and depletion, *J. Appl. Physiol.*, 35, 1973.

132. Zelles, T., Humphreys-Beher, M. G., and Schneyer, C. A., Effects of liquid diet on amylase activity and mRNA levels in rat parotid gland, *Arch. Oral Biol.*, 34, 1989.

133. Yamamoto, T., *Frontiers of Oral Physiology*, Karger, Basel, London, 1983, 102.

Neurotransmitter Control of Calcium Mobilization

Bruce J. Baum, Indu S. Ambudkar, and Valerie J. Horn

HISTORICAL VIEW OF THE IMPORTANCE OF CALCIUM TO SALIVARY SECRETION

It is well established that the secretory response of mammalian salivary glands is mediated via autonomic sympathetic and parasympathetic innervation (Garrett, 1982). Following a physiological stimulus (e.g., masticatory, gustatory), a spectrum of neurotransmitters are released from these nerves and subsequently activate specific receptors on the basolateral membrane of salivary epithelial cells (Baum, 1987; Garrett, 1982; Schneyer and Emmelin, 1974). Norepinephrine, the principal neurotransmitter released from sympathetic nerves, activates both α- and β-adrenergic receptors.

Acetylcholine, the principle neurotransmitter released from parasympathetic nerves, activates muscarinic-cholinergic receptors. Parasympathetic nerves also release vasoactive intestinal peptide (VIP) and substance P. Other neurotransmitters (e.g., ATP and 5-hydroxytryptamine) have also been suggested to play a role in salivary secretion. Recently, considerable general progress has been made to understanding the biology of autonomic neurotransmitter receptors and the reader is referred to several reviews for further information (e.g., Lefkowitz and Caron, 1988; Nathanson, 1987; Kerlavage et al., 1987).

In 1963, Douglas and Poisner first demonstrated that Ca^{2+} played a critical role in the secretory response of a salivary gland to neurotransmitter stimulation. Douglas and Poisner (1963) studied the perfused cat submaxillary gland. When the perfusion fluid contained Ca^{2+} (2.2 mM $CaCl_2$), fluid secretion from the cannulated gland duct was brisk in response to both acetylcholine and norepinephrine. If, however, Ca^{2+} was omitted from the perfusion fluid, there was essentially no fluid secretion in response to these agents. This inhibitory effect was reversible, and fluid secretion was resumed when the perfusion fluid was changed from a Ca^{2+}-free back to a Ca^{2+}-containing solution. Subsequently, similar findings have been reported with other salivary gland preparations (e.g., Hunter et al., 1983; Young and Van Lennep, 1979).

A related, and very important, observation central to understanding salivary fluid secretion was made by Burgen in 1956. Burgen reported that, following electrical stimulation of autonomic nerves innervating the dog parotid and submaxillary glands, there was a large (~two to threefold) increase in the potassium concentration in the venous drainage from these glands.

Subsequently, many investigators have utilized the release of K^+ (or a radioactive substitute, $^{86}Rb^+$) from salivary gland preparations *in vitro* as an index of a fluid secretory response to autonomic neurotransmitters (e.g., Selinger et al., 1971; Putney, 1976; Ito et al., 1982). This K^+ release response was shown to be Ca^{2+}-dependent (Selinger et al., 1971; Putney, 1978). Putney (1978), using a rat parotid slice preparation, also showed that K^+ release was biphasic and that the first transient phase was independent of extracellular Ca^{2+} while the second sustained phase required the presence of extracellular Ca^{2+}. It was speculated that the first phase was transient due to its dependence on an intracellular Ca^{2+} pool of limited size (Putney, 1978). These experiments are significant in many respects, particularly in demonstrating a role for Ca^{2+} in a secretory response measured in a nonintact gland preparation (i.e., slices or dispersed cells vs. perfused glands).

An additional historical observation which is important to understanding the relationship between Ca^{2+} and salivary fluid secretion was made by Petersen (1970). He noted that the venous K^+ response, seen in a perfused cat submaxillary gland following an acetylcholine stimulus, was dependent on the presence of Cl^- in the perfusion solution. Subsequent studies with dispersed cells from rat parotid and submandibular glands have shown that α_1-adrenergic and muscarinic-cholinergic agents elicit a marked Cl^- efflux response (Martinez and Cassity, 1985; Melvin et al., 1987) which is sufficient to account (in the rat parotid) for the *in vivo* fluid secretory rate (Melvin et al., 1987). As with the K^+ release response described above, the Cl^- efflux response from these cells appears to be Ca^{2+} dependent (Ambudkar et al., 1988; Baum et al., 1990; Iwatsuki et al., 1985; Nauntofte and Poulsen, 1986).

Thus, from an historical perspective, there is a substantial body of data supporting a central role for Ca^{2+} in salivary fluid secretion and related ion (K^+, Cl^-) fluxes. It is the purpose of this chapter to describe a current (i.e., as of late 1989) view of how Ca^{2+} is handled by salivary acinar cells. Acinar cells are believed to represent the sole site of water transport in salivary glands (Young and Van Lennep, 1979). This discussion will primarily focus on the rat parotid acinar cell for which the most biochemical mechanistic information is available (Figure 1).

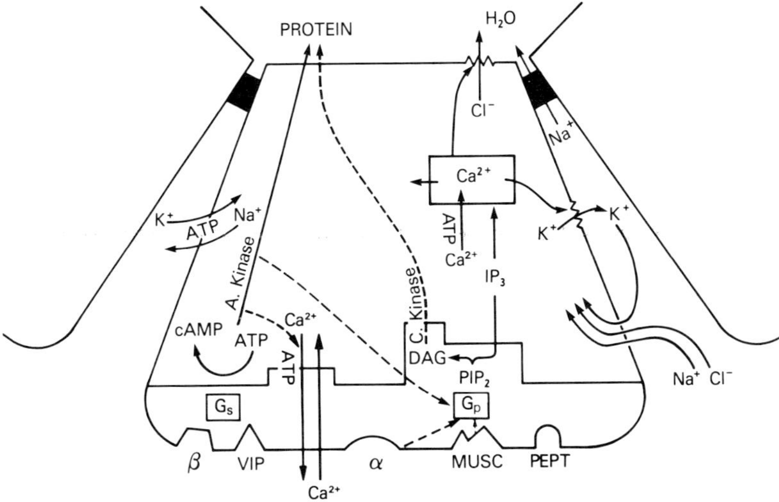

Figure 1. Model of secretory events in a rat parotid acinar cell. β = β-adrenergic receptor; VIP = vasoactive intestinal polypeptide receptor; α = α-adrenergic receptor; MUSC = muscarinic-cholinergic receptor; PEPT = substance P receptor; Gs = G protein which activates adenylate cyclase; Gp = putative G protein which activates PIP₂-specific phospholipase C; PIP₂ = phosphatidylinositol 4,5-bisphosphate; DAG = diacylglycerol; IP₃ = inositol 1,4,5 trisphosphate. Fluid secretion follows the elevation of cytosolic free Ca²⁺. See text for details.

CALCIUM HOMEOSTASIS IN EUKARYOTIC CELLS

As described above, Ca^{2+} has been long identified as a key component in the stimulation of secretory events in salivary cells. However, the intracellular steps associated with neurotransmitter stimulation of secretion have only been recently investigated. This, in great part, is due to the emergence of new, highly efficient techniques to monitor the cytosolic free calcium concentration ($[Ca^{2+}]i$) (Cubbold and Rink, 1987). These methods have allowed the measurement of $[Ca^{2+}]i$ in a number of cell types in recent years. In addition, there have been considerable advances made in understanding signal transduction events (Putney, 1986; Gilman, 1987; Berridge and Irvine, 1989). Such studies have contributed significantly towards describing the physiological basis for Ca^{2+} mobilization in cells. Although, significant progress has been made, as will be discussed below, several major areas still remain unclarified.

Reports from several laboratories have clearly established that under physiological conditions, resting [Ca]i in most cells is between 100 to 200 nM (e.g., Merritt and Rink, 1987; Ambudkar et al., 1988a; Horn et al., 1988; Muallem, 1989). Thus, there exists a gradient of Ca^{2+} across the plasma membrane which is extremely large (5000 to 10,000-fold; extracellular free Ca^{2+} being ~1.5 mM). Many studies have shown that Ca^{2+} can also be stored in in-

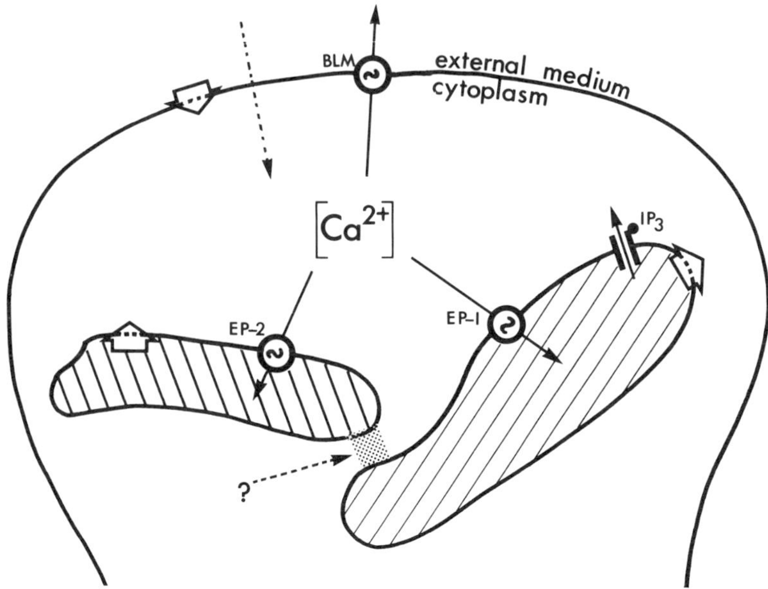

Figure 2. [Ca^{2+}]i regulation in salivary cells. The cytosolic calcium concentration ([Ca^{2+}]i) in salivary acinar cells is maintained by the concerted activities of ATP-dependent Ca^2 pumps mediating Ca^{2+} removal from cytoplasm and Ca^{2+} channels which allow Ca^{2+} entry into the cytoplasm. The Ca^{2+} pump in the basolateral plasma membrane (BLM) pumps Ca^{2+} out of the cell against its electrochemical gradient. The stimulation of this activity could account for increased ^{45}Ca^{2+} efflux from agonist-stimulated cells (Putney, 1986; Baum 1987). Calcium pumps of the endoplasmic reticulum mediate the accumulation of calcium into different intracellular pools (EP-1 and EP-2), of which EP-1 is IP$_3$-sensitive and releases Ca^{2+} via a channel activated by the binding of IP$_3$ to its receptor. Possible communication between the two pools (dotted area) has been proposed (Gill et al., 1989), though the physiological activator of such an event is as yet unknown (?). The gradients of Ca^{2+}; out $\rightarrow$ in across BLM and in $\rightarrow$ out across EP membranes are shown by the large clear arrows. Ca^{2+} entry across the plasma membrane which is stimulated by receptor activation, is shown via a dotted arrow. The proposed mechanisms for this entry pathway are discussed in Figure 4.

tracellular compartments at concentrations substantially higher than that found in the cytoplasm (Muallem, 1989). Currently, it is believed that the physiologically most important intracellular Ca^{2+} storage site is a part of the endoplasmic reticulum (Schulz et al., 1989; Gill et al., 1989) or a closely related novel organelle termed a calciosome (Volpe et al., 1988; Krause et al., 1989). It is this Ca^{2+} store which is the target pool for the rapid intracellular signal events resulting in the transient phase of K$^+$ efflux referred to above. As will be discussed below, it is this Ca^{2+} store, hereafter, referred to as ASCaP (agonist-sensitive Ca^{2+} pool), which possesses a receptor for inositol trisphosphate (IP$_3$), the second messenger which triggers the Ca^{2+} release response (Figure 2) (Spat et al., 1986; Guillemtte et al., 1987; Snyder and Suppatapone, 1989).

Eukaryotic cells in general, and salivary cells in particular, expend considerable energy in maintaining [Ca^{2+}]i at low resting levels. The [Ca^{2+}]i determined at any given time represents a net Ca^{2+} level which results from many kinetically distinct events. These include processes tending to increase $[Ca^{2+}]i$ (stimulated release from the ASCaP; gradient dependent entry across the plasma membrane) and processes tending to lower $[Ca^{2+}]i$ (transport of Ca^{2+} across the plasma membrane against its electrochemical gradient; transport into intracellular organelles). In addition, it should be recalled that cells contain many Ca^{2+} binding proteins which in effect serve to buffer intracellular $[Ca^{2+}]$.

Since $[Ca^{2+}]i$ is a physiologically crucial factor in the initial responsiveness to stimulation and in sustained regulation of fluid secretion, the salivary acinar cell must at all times tightly control the levels of cytosolic Ca^{2+}. Thus, both the magnitude and duration of the $[Ca^{2+}]i$ increase which follows stimulation can apparently be regulated by many factors such as the $[Ca^{2+}]$ in the ASCaP, the amount and the rate of generation of IP_3, the rate of turnover of IP_3, the removal of Ca^{2+} from the cytoplasm, and Ca^{+} entry into the cell. In addition, other signaling processes (e.g., the activation of protein kinase C by diacylglycerol, DAG) may influence $[Ca^{2+}]i$ (Takemura and Putney, 1989).

Stimulation of acinar cells by "classical" Ca^{2+} mobilizing agonists (e.g., agents which activate muscarinic-cholinergic and α_1-adrenergic receptors) leads to a rapid (three to fourfold) increase in $[Ca^{2+}]i$ (Merritt and Rink, 1987; Ambudkar et al., 1988a; Horn et al., 1988). This initial increase in $[Ca^{2+}]i$ is accomplished by the "downhill" movement of Ca^{2+} from the ASCaP, initiated by the binding of IP_3 to its receptor and the opening of a Ca^{2+} channel, and is followed by the entry of extracellular Ca^{2+}. Studies with parotid gland and other exocrine tissues have shown that the permeability of the cellular membrane to Ca^{2+} (i.e., Ca^{2+} entry) increases several fold during agonist stimulation and remains so for a considerable time in the continued presence of the agonist (Putney et al., 1978; Putney, 1986; Muallem, 1989; Williamson and Monck, 1989) (this will be discussed in detail below).

Two main Ca^{2+} transport systems are involved in removing (lowering) Ca^{2+} from the cytoplasm. These are the ATP-dependent Ca^{2+} pumps located in the basolateral plasma membrane (BLM pump) and in the ASCaP membrane (ER pump) (Baum and Ambudkar, 1988a). These transporters achieve the "uphill" movement of calcium by utilizing the energy released by ATP hydrolysis. The function of the Ca^{2+} pumps in the parotid acinar cell and other cells has not been physiologically established due primarily to a lack of specific agents with which to modulate these transporters. However, it has been suggested that the Ca^{2+} efflux seen in stimulated cells is mediated by the BLM pump. The ER-pump activity, in addition to decreasing $[Ca^{2+}]i$, also contributes to the refill of the ASCaP. A recent report has suggested that there may be more than one type of ER-pump in secretory cells (Thevenod et al., 1989), and c-DNA cloning of the plasma membrane pumps in other tissues has revealed the existence of more than one species of this transporter (Strehler et al., 1989). Further studies are required to assess the role of such diversity in Ca^{2+} transporters.

Knowledge of the activities of the ATP-dependent Ca^{2+} transporters has been mainly obtained from studies using isolated, enriched membrane vesicle preparations (Selinger et al., 1970; Kanagasuntheram and Teo, 1982; Immelman and Soling, 1983; Takuma et al., 1985). Recent work from our laboratory has established the kinetic parameters of the BLM and ER pumps (Ambudkar and Baum, 1988a; Baum et al., 1988). The former has a high affinity for Ca^{2+}, Kca = 100 nM, and a high velocity, Vmax = 50 nmol/mg min, which is reached at $[Ca^{2+}] > 600$ nM. The ER pump, on the other hand, has a very high affinity for Ca^{2+}, Kca = 30 nM, and a Vmax = 35nmol/mg min, which is reached at ˜100 nM Ca^{2+}. Therefore, we have suggested (Baum and Ambudkar, 1988a) that the activity of the BLM pump can be increased when $[Ca^{2+}]i$ increases following neurotransmitter stimulation, while in resting cells it is activity is at ~0.5 Vmax. The ER pump, however, can function at, or close to, Vmax levels even under resting conditions. Additionally, recent studies from our laboratory (Ambudkar and Baum, 1988a, 1988b; Baum et al., 1988), and others (Bayerdorffer et al., 1985; Thenenod and Schulz, 1988), show that ion movements (e.g., H^+, Cl^-) across the BLM and ER membranes can modulate Ca^{2+} transport activity, possibly due to charge compensatory effects. Thus, mechanisms which lead to changes in the membrane potential or intracellular pH (both of which occur following secretory stimuli) could potentially alter Ca^{2+} transport activity.

Many factors have been shown to regulate plasma membrane Ca^{2+} pumps in several tissues (for reviews see Carafoli, 1987; Muallem, 1989). These include calmodulin, protein kinases A and C, and acidic lipids. In studies with rat parotid membranes, we have shown an increase in the affinity of the BLM pump for Ca^{2+} with exogenously added calmodulin (Ambudkar et al., 1989). This occurs at $[Ca^{2+}]$ equal to that in resting cells and would result in the pump being more active at lower $[Ca^{2+}]i$. Previously, we have also shown an increase in the intracellular levels of cAMP is associated with a decrease in the affinity of the BLM pump for Ca^{2+} while increasing its Vmax (Helman et al., 1986). These effects may be due to phosphorylation of the ATP-dependent Ca^{2+} pump, as suggested from studies in the cardiac sarcolemma (Strehler at al., 1989).

While it should be an obvious corallary to the above discussion on the importance of tight regulation of $[Ca^{2+}]i$, it should be clearly noted that the deregulation of Ca^{2+} homeostasis could have serious pathological implications. For example, in liver and kidney cells, which have been widely used to study mechanisms of cellular injury, uncontrolled Ca^{2+} fluxes, i.e., that induced by certain cytotoxic agents, amplify the processes involved in cellular damage (Ambudkar et al., 1988b; Trump et al., 1980). In other cases, physiological impairments of secretory function, such as that seen in aged rat parotid gland but not the aged rat submandibular gland, appear to be the result of specific alterations in the mechanisms regulating $[Ca^{2+}]i$ (Bodner et al., 1983; Ambudkar et al., 1988c; Maki et al., 1989).

TABLE 1. Guanine Nucleotide Binding Regulatory Proteins

Name	Subunit Structure	Mr αSubunit	Toxin Sensitivity	Role	Cell Surface Receptors which Activate
T	Heterotrimeric	40 kDa	C,P	$\uparrow$ cGMP	Rhodopsin
Gs	Heterotrimeric	~45 kDa	C	$\uparrow$ cAMP	$\beta_1 + \beta_2$
Gi	Heterotrimeric	40–41 kDa	P	$\downarrow$ cAMP	M,α_2
Go	Heterotrimeric	40 kDa	P	unknown	M,α_2
Gp	Unknown	Unknown	$\pm$P	$[Ca^{2+}]i$	M,α_1, Sub P
Gz	Heterotrimeric	41 kDa	none	unknown	
ras	Monomeric	21 kDa	none	cell growth	(?)
ras-like	Monomeric	21–26 kDa	B	cell growth	(?)

Note: T, transducin; C, cholera toxin; P, pertussis toxin; B, botulinum toxin D; $\beta_1 + \beta_2$, β_1, and β_2-adrenergic receptors; M, muscarinic receptor; α_1, α_1-adrenergic receptor; α_2, α_2-adrenergic receptor; Sub P, substance P receptor. For a general review of G protein function see Casey and Gilman, 1988.

SIGNAL TRANSDUCTION AND ACTIVATION OF THE INTRACELLULAR Ca²⁺ SIGNALING SYSTEM

Many factors are involved in determining the specificity of a cellular response to an extracellular signal. The extracellular signal (e.g., a neurotransmitter or hormone; a first messenger) binds to a specific cell surface receptor. This event initiates the generation of an intracellular signal (a second messenger) which more directly mediates the physiological response. This process is called signal transduction (Rasmussen et al., 1984; Schulz et al., 1986; Baum, 1987; Exton, 1988). A large part of the specificity in signal transduction is achieved by the linkage of a cell surface receptor with a specific intracellular messenger system in a given cell type. In many cell types including salivary acinar cells (see Figure 1), substance P, α_1-adrenergic and muscarinic-cholinergic (subtypes M_1 and M_3) receptors are directly linked to Ca^{2+} signaling via the activation of a phosphatidylinositol 4,5-bisphophate (PIP$_2$)-specific phospholipase C. β_1-and β_2-Adrenergic receptors are linked to cAMP-mediated signally via the activation of adenylate cyclase. α_2-Adrenergic receptors and muscarinic receptors are linked to the inhibition of adenylate cyclase. The linkage between these receptors and the appropriate signal generating enzyme system is through different types of guanine nucleotide-binding regulatory proteins (so called G proteins; see Table 1, and for reviews see Casey and Gilman, 1988; Fain et al., 1988; Freissmuth et al., 1989; Milligan, 1988;). G proteins are heterotrimeric proteins composed of α, β, and γ subunits, the α subunit of which confers specificity. Information regarding G proteins has mainly come from studies of adenylate cyclase activation. These studies suggest that the activation of a G protein involves the

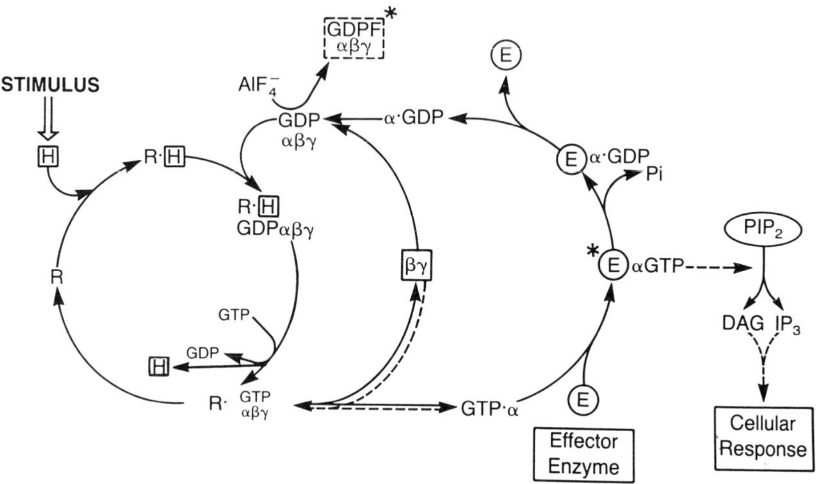

Figure 3. Schematic representation of hormone-receptor-G protein-effector interactions. Following hormone (neurotransmitter) or agonist stimulation of a cell at its specific receptor, cellular responsiveness is elicited via a signal transduction process involving a G protein and appropriate effector enzyme. Details of this process are described in the text. H = neurohormone or agonist; R = receptor; GDP = guanosine 5′-diphosphate; GTP = guanosine 5′-triphosphate; α, β, and γ = the subunits which together constitute a heterotrimeric G protein; E = effector enzyme; Pi = inorganic phosphate; PIP_2 = phosphatidylinositol 4,5-bisphosphate; IP_3 = inositol trisphosphate; DAG = diacylglycerol.

binding of the GDP-bound heterotrimer to the agonist bound receptor, the replacement of GDP by GTP, dissociation of the heterotrimer and release of the GTP bound Gα subunit. The GTP-Gα subunit is the active entity which interacts with the target effector enzyme (adenylate cyclase) resulting in second messenger generation (Figure 3).

Physiological neurotransmitters often have affinity for multiple receptor subtypes, and there may be complex interactions between different signaling systems. For example, norepinephrine released by sympathetic nerves innervating parotid acinar cells can bind to α_1-, α_2-, β_1-, and β_2-adrenergic receptors (Goodman and Gilman 1980). Thus, the specificity of a cellular response would depend on factors such as the relative binding of an agonist to the receptor and the efficacy of G protein activation. Furthermore, the factors which control the association of a given G protein with its native receptor are not known. In reconstituted vesicles, for example, G proteins will associate "promiscuously" with any available receptor (Milligan, 1988). This promiscuity is not readily evident in nature, i.e., in the parotid, specific α_2-adrenergic stimulation does not elevate $[Ca^{2+}]i$ (Ambudkar et al., 1988a). It should be readily apparent to the reader that signal transduction in general is a complex and incompletely understood process.

The Ca^{2+} signaling system can be activated in the rat parotid acinar cell by stimulation of several different receptors (Figure 1). Stimulation of α_1-adrenergic receptors and muscarinic receptors results in very similar phenomena; a rapid elevation of $[Ca^{2+}]i$ (peak response within 6 s, which is independent of extracellular Ca^{2+}) and a prolonged sustained response (of greater than 20 min duration, which is dependent on extracellular Ca^{2+}) (Merritt and Rink, 1987; Nauntofte and Dissing, 1987; Ambudkar et al., 1988a, Mertz et al., 1990). There is no apparent desensitization (induced refractoriness) of these receptors and cells are able to respond well if one dose of agonist is washed off and another dose is added. This contrasts with the stimulation of cells by substance P (Merritt and Rink, 1987; Sugiya et al., 1988). The initial Ca^{2+} response to substance P is also very rapid. However, there is a rapid desensitization of the substance P receptor and, consequently, no sustained Ca^{2+} response. In rat parotid acinar cells, β_2-adrenergic stimulation can also lead to a transient Ca^{2+} mobilization (Arkle et al., 1989; Cook et al., 1988), which is mimicked by cAMP analogs (Horn et al., 1988). Although this response is only about one-third that of classical Ca^{2+} mobilizing agonists, it suggests that in these cells there is an interaction between the two principle intracellular signaling systems. The physiological significance of this interaction remains to be determined.

Currently, it is believed that the increase in PIP_2 specific phospholipase C (PLC) activity induced following stimulation of α_1-adrenergic, muscarinic, and substance P receptors is achieved via the activation of a putative G protein, Gp (Taylor et al., 1986a; Putney et al., 1989). There is, however, no direct proof of the involvement of such a G protein in PLC activation. In fact, as of the date at which this chapter is being written (December, 1989), no such G protein has been isolated or directly characterized in any tissue. There is, however, considerable indirect evidence to support a role for a G protein in PLC activation. In membrane preparations, one can demonstrate neurotransmitter ligand-receptor-G protein interactions (and thus presume a G protein-mediated signaling) by showing alterations in the receptor affinity for a ligand (especially agonists) with the addition of GTP or GTP analogs (Ek and Nahorski, 1988; Milligan, 1988; Ott and Costa, 1989). Such alterations are due to the replacement of bound GDP on the G protein by GTP (Figure 3). A G protein with GTP bound has markedly different functional properties than a G protein with GDP bound or a G protein which has no bound guanine nucleotides. For example, the GTP-bound G protein no longer has a high affinity for the neurotransmitter receptor. Also the heterotrimeric subunits dissociate, i.e., β from α. The dissociation of the G protein from the receptor also results in both a lowered affinity of the receptor for the agonist and a free $G\alpha$ GTP subunit which is able to activate the appropriate second messenger effector enzyme (such as PLC).

Further presumptive evidence of G protein involvement in a signaling process can come from studies examining the responses of effector enzymes to guanine nucleotides. Guanine nucleotides are not freely permeable across cell membranes and therefore must be used with isolated membranes or detergent permeabilized cells. Therefore, if in such preparations the addition of GTP or an analog enhances an agonist response (e.g., stimulated PLC activation), then that receptor is presumed linked to a G-protein mediated pathway (Brass et al., 1986; Boyer et al., 1989). In rat parotid acinar cells such evidence has been presented for both the muscarinic and substance P receptors (Taylor et al., 1986b).

It is also possible to activate G proteins, independent of receptor stimulation, by using AlF_4^-. AlF_4^- is believed to act by interacting with the GDP bound to inactive G proteins and acting as a phosphate (Bigay et al., 1985; Chabre, 1989). Thus, AlF_4^- sets into play the dissociation of the G protein subunits and the resultant activation of a $G\alpha$ subunit. Therefore, if AlF_4^- appears to mimic a functional agonist response in a cell, it can be presumed that the response is G-protein mediated. In rat parotid acinar cells, AlF_4^- elicits an increase in $[Ca^{2+}]i$ by stimulating intracellular Ca^{2+} release and extracellular Ca^{2+} entry. This implies that a G protein can mediate this response (Mertz et al., 1990). The combination of data in the parotid with similar studies in other tissues, suggests that PLC activation is a G-protein-mediated process (Paris and Pouyssegur, 1987; Brock et al., 1988; Fain et al., 1988; Moriarty et al., 1988; Boyer et al., 1989). However, available data also suggest that the G protein involved is not one of the previously well characterized G proteins (Table 1). For convenience, this unknown and uncharacterized G protein has been referred to as Gp (Figure 1).

The immediate consequence of neurotransmitter-stimulated, G protein-mediated PLC activation is the cleavage of PIP_2 into 1,4,5-IP_3 and DAG (recent reviews include Fisher et al., 1989; Putney et al., 1989; Shears, 1989). In some tissues PLC activation itself is dependent on the $[Ca^{2+}]i$ concentration. However, in the parotid and in many other cell types the Ca^{2+} requirements for PLC activation, in the presence of nonhydrolyzable GTP analogs, can be met by resting free Ca^{2+} levels (i.e., ~100 nM and will not be further stimulated by Ca^{2+} until $[Ca^{2+}]i$ exceeds 1 μm) (Aub and Putney, 1985; Taylor et al., 1986a; Renard et al., 1987). Therefore PLC activation is functionally independent of $[Ca^{2+}]$ concentration throughout the physiological range of free Ca^{2+}. The substrate for PLC, PIP_2, is a minor phospholipid component of the plasma membrane. In most cells where it has been examined, PIP_2 appears to exist in a single, common, agonist-sensitive pool (Augert et al., 1989). In rat parotid acinar cells, simultaneous stimulation of α_1-adrenergic and muscarinic-cholinergic receptors does not result in an additive increase in PIP_2 hydrolysis. Such observations noted at high agonist concentrations can be explained by limitations in the size of the agonist sensitive PIP_2 pool. However, at submaximal agonist concentrations, similar results are observed, suggesting the existence of a negative interaction between these two receptor signaling pathways (Horn et al., 1990).

The DAG formed subsequent to PIP_2 cleavage activates protein kinase C (PKC). PKC can then be involved in a variety of cellular responses including, in some cells, a negative "feedback" inhibition of PLC-mediated PIP_2 hydrolysis (Carlson et al., 1989; Crouch et al., 1989). In the parotid PKC may promote amylase secretion and is involved in the desensitization of the substance P receptor (McKinney and Rubin, 1988; Sugiya et al., 1988).

MOBILIZATION OF INTRACELLULAR Ca^{2+}

As noted earlier, the IP$_3$ formed as a consequence of stimulation of Ca^{2+}-mobilizing receptors mediates the release of Ca^{2+} by binding to a specific receptor on the membrane of the ASCaP. This binding is believed to activate a Ca^{2+}-release channel by a yet unknown mechanism, allowing Ca^{2+} to enter the cytoplasm (Ferris et al., 1989; Suppattapone et al., 1988). The nature of this receptor may vary from tissue to tissue. For example, adrenal cortical cells display an IP$_3$ receptor with only a single high-affinity binding site (Guillemette et al., 1987; Rossier et al., 1989), while in Purkinje cells the IP$_3$ receptor has both high and low affinity sites with cooperative IP$_3$ binding (Meyer et al., 1988). Interestingly, antibody directed at the Purkinje cell IP$_3$ receptor shows only very limited recognition of the receptor in other cells, and localizes to the plasma membrane. This receptor may be somewhat unique (Worley et al., 1988). All of the characterized IP$_3$ receptors are very selective for 1,4,5-IP$_3$, vs. other metabolites of PIP$_2$ hydrolysis (Enyedi and Williams, 1988; Spat et al., 1986; Worley et al., 1987). An IP$_3$ receptor in rat parotid cells has been partially characterized. It appears to have only a high-affinity site and is associated with microsomal membranes (Maki et al., 1989).

The ASCaP appears to be common to all of the IP$_3$-linked receptor systems in a cell. In the rat parotid acinar cell, a number of studies have suggested that the substance P, muscarinic, α_1-adrenergic, and β_2-adrenergic (cAMP-linked) responses result in mobilizing Ca^{2+} from a common intracellular Ca^{2+} pool (Aub et al., 1982; Merritt and Rink, 1987; Ambudkar et al., 1988a). As noted above, the intracellular localization of the ASCaP is unclear. From experiments with permeabilized cells it is known that IP$_3$ can mediate Ca^{2+} release from a pool loaded in the presence of mitochondrial inhibitors and ATP. Therefore, the IP$_3$-sensitive Ca^{2+} pool is localized to a nonmitochondrial organelle. A specific, novel organelle termed a "calciosome" has been speculated to be the site of this pool (Volpe et al., 1988; Krause et al., 1989). In several cell types, membrane fractions containing IP$_3$ receptors (and also Ca^{2+}-binding proteins resembling calsequestrin) have been separated from membranes of the endoplasmic reticulum (ER) by differential centrifugation. However, in many other cell types the IP$_3$-sensitive pool copurifies with the ER fraction although it is likely to be located near the plasma membrane (Champeil et al., 1989; Dunlop and Larkins, 1988; Poggioli and Putney, 1982).

As Ca^{2+} is released from the ASCaP following a stimulus, there is an increase in [Ca^{2+}]i. This can be measured using one of a number of fluorescent Ca^{2+} probes developed over the last 5 to 7 years. The best known of these probes are quin 2 and fura 2. Both compounds are acetoxymethylesters of an avid Ca^{2+} binding agent. Being esterified, and hydrophobic, they are freely permeable across the plasma membrane. Once inside a cell, esterases hydrolyze the esters, leaving the agent trapped within the cell. When Ca^{2+} is bound to the agent, there is an increase in fluorescence which can be readily detected, calibrated, and the $[Ca^{2+}]i$ calculated (Kawanishi et al., 1989; McDonough and Button, 1989; Tsien et al., 1982). Such measurements have shown that in unstimulated parotid acinar cells resting $[Ca^{2+}]i$ averages 100 to 200 nM. After stimulation the $[Ca^{2+}]i$ rises rapidly (within ~6 s) up to a peak level of ~800 nM. This initial response appears essentially independent of extracellular calcium and is directly related to the agonist (neurotransmitter) concentration (Merritt and Rink, 1987; Ambudkar et al., 1988a; Horn et al., 1988).

The initial peak Ca^{2+} response also appears to be dependent on factors other than the limit of Ca^{2+} in the ASCaP. This becomes apparent at low levels of stimulation, where in spite of excess Ca^{2+} in the ASCaP and the opening of the IP_3-sensitive Ca^{2+}-release pathway, the initial $[Ca^{2+}]i$ peak is limited. Various suggestions have been made to explain this phenomenon. The regulation of IP_3-mediated Ca^{2+} release by $[Ca^{2+}]i$ may negatively affect IP_3 binding and Ca^{2+} release. However, since the initial peak Ca^{2+} response is limited irrespective of $[Ca^{2+}]i$, it is unlikely that $[Ca^{2+}]i$ is responsible for this limiting effect. Recently, a quantal release mechanism was described which suggested that discrete Ca^{2+} pools, varying in their content of Ca^{2+}, gave rise to the limited initial peak $[Ca^{2+}]i$ after stimulation (Muallem et al., 1989). However, this model necessitates the existence of different IP_3 sensitivities in these pools, i.e., the smallest Ca^{2+} pool would have the highest affinity for IP_3. A complex regulation, affected by the cooperative binding after four IP_3 molecules per channel, has been also proposed (Meyer et al., 1989). The recent cloning of the IP_3 sensitive Ca^{2+} release channel (Furuichi et al., 1989), should provide greater understanding of this complex and physiologically important process.

Additionally, it is possible, using fura 2, to study Ca^{2+} mobilization at a single cell level. To do this one employs a spectrofluorimeter attached to a fluorescence microscope. Interestingly, many single cells show an oscillating rather than sustained Ca^{2+} response (Berridge and Galione, 1988; Kawanishi et al., 1989; Meyer and Stryer, 1989). Single parotid acinar cells, under conditions which closely simulate the physiological situation, respond to a stimulus with a Ca^{2+} change whose magnitude and duration is proportional to the agonist concentration. Yet, under other conditions (temperatures below physiological, with a very narrow range of agonist concentrations) one sees a series of regular Ca^{2+} oscillations where the frequency (on the order of seconds), rather than the magnitude of the response, reflects the agonist concentration (Gray, 1988). The source of these oscillations is unclear. Sustained oscillations require extracellular Ca^{2+}, but the initial oscillations do not. The frequency appears to vary from individual cell to cell within a preparation, but remains as a "fingerprint" of the individual cell. These oscillations can occur in cells injected with IP_3-binding to its receptor being negatively affected by high $[Ca^{2+}]i$. Alternatively, the oscillations may reflect the periodicity in an IP_3 insensitive calcium pool fusing with the IP_3 sensitive pool, or altered rates of transport in the plasma membrane and ER membrane calcium pumps.

Whether involved in oscillations or not, there appear to be both IP_3-sensitive and IP_3-insensitive nonmitochondrial Ca^{2+} pools in most cells. In some tissues these compartments can be distinguished in the laboratory by differential centrifugation. In both microsomal membrane vesicle and permeabilized cell preparations, GTP (but not its nonhydrolyzable analogs) can release significantly more Ca^{2+} than IP_3 (Henne and Soling, 1986; Mullaney et al., 1988). This is believed to occur due to a GTP-mediated fusion of multiple intracellular vesicles, all of which contain Ca^{2+}, but only some of which can release Ca^{2+} in response to IP_3 (Comerford and Dawson, 1988). It is not clear, however, whether GTP is the physiological agent responsible for the fusion of these two pools.

Since inositol phosphates are critical to intracellular Ca^{2+} mobilization, considerable effort has been made to characterize their metabolism. Parotid acinar cells have been widely employed as models for these studies. The metabolism of $1,4,5$-IP_3 is quite complex (Balla et al., 1988; Batty et al., 1985; Daniel et al., 1987; Dixon and Hokin, 1989; Shears, 1989). Only $1,4,5$-IP_3 and $1,3,4,5$-IP_4 have known or suggested physiological roles in cell signaling. IP_4 is formed by the phosphorylation of $1,4,5$-IP_3 may more than double within the first 15 s after agonist stimulation. The amount of IP_4 generally parallels the level of $1,4,5$-IP_3. After the initial burst of $1,4,5$-IP_3 formation, its concentration decreases abruptly. In rat parotid acinar cells, the half-life of $1,4,5$-IP_3 determined after receptor inactivation by antagonist addition, is ~7.6 s (Putney et al., 1989).

The metabolism of IP_4 involves a dephosphorylation into a different IP_3 isomer, $1,3,4$-IP_3 (Shears, 1989). This isomer has no known physiological role and cannot release Ca^{2+} from the $1,4,5$-IP_3 sensitive pool. It continuously accumulates, with a short lag phase (~10 s), after activation. Since the only known source of $1,3,4$-IP_3 is the breakdown of IP_4, its accumulation can readily be used as a measure of PIP_2 hydrolysis. $1,3,4$-IP_3 is dephosphorylated sequentially into IP_2, IP_1, and inositol. LiCl can be used to block the conversion of IP_1 into inositol. This maneuver increases the accumulation of $1,3,4$-IP_3 in cells but not $1,4,5$-IP_3 or $1,3,4,5$-IP_4. Another inositol phosphate found in increased concentration after agonist stimulation is cyclic $1:2,4,5$ IP_3. This is metabolized very slowly (half-life of ~10 min) and does not appear to be active in Ca^{2+} mobilization responses in the rat parotid gland (Putney et al., 1989).

It has been demonstrated clearly in a variety of tissues, including the parotid, that PIP_2 hydrolysis is not the sole source for DAG (Martinson et al., 1989; Matozaki and Williams, 1989; and Soling et al., 1987). Phosphatidylcholine breakdown can yield DAG formation in many cells, although $1,4,5$-IP_3 and IP_4 appear to be derived solely from PIP_2 hydrolysis in all cell types studied thus far. In addition to PLC activation, agonist stimulation of cells can lead to the activation of phospholipases A2 (PLA2) and D (Axelrod et al., 1988). These are less well studied but may also involve a G protein signaling step. In some cells PLA2 may be activated by the increase in Ca^{2+} secondary to PLC activation, but in other cells it appears to be regulated independently. PLA2 activation can result in the release of an intracellular Ca^{2+} pool and subsequent elevation of $[Ca^{2+}]i$.

This appears due to a hydrolytic product, arachidonic acid, which has been shown to induce Ca^{2+} release in several cell types.

MECHANISMS OF CALCIUM ENTRY IN SALIVARY CELLS

As noted above, stimulation of salivary cells with neurotransmitter agonists results in a biphasic release of both K^+ and Cl^- (Putney, 1976; Putney, 1986). This correlates well with neurotransmitter-induced stimulation in salivary flow rate (Melvin et al., 1987). It has been accepted for a number of years that while the initial increase in the salivary flow rate is dependent on the intracellular Ca^{2+} content, sustained fluid secretion, i.e., that occurring during alimentation, required the presence (and thus entry) of extracellular Ca^{2+} (Douglas and Poisner, 1963). Similar requirements, as noted earlier, have been shown for sustained agonist stimulated K^+ and Cl^- fluxes (Selinger et al., 1973; Putney, 1986; Nauntofte and Poulson, 1986; Ambudkar et al., 1988a; Peterson and Gallacher, 1988). Thus, the entry of Ca^{2+} into salivary cells is a critical factor in the maintenance of sustained levels of fluid secretion.

There have been a number of recent reports regarding Ca^{2+} mobilization events in neurotransmitter stimulated salivary gland cells (Putney, 1986; Merritt and Rink, 1988; Horn et al., 1988; Baum and Ambudkar, 1988; Putney, 1989). The results from these studies reveal that contributions due to extracellular Ca^{2+} can be detected almost immediately following the initial Ca^{2+} response, both in determinations of [Ca]i and $^{45}Ca^{2+}$ flux. These reports also show that extracellular Ca^{2+} is required to replenish the intracellular ASCaP. Thus, agonists stimulate rates of Ca^{2+} uptake into acinar cells. This uptake can be inhibited by agents such as La^{3+}. Although these results are strongly indicative of the entry of Ca^{2+} into stimulated cells, the exact mechanism which mediates this entry is yet unknown. Since Ca^{2+} entry from the extracellular milieu will move along its electrochemical gradient, a passive type of transport can be predicted. Typically, such Ca^{2+} influx pathways would be of the "channel" type. Recently, a basal pathway of passive Ca^{2+} entry was described in hepatocytes (Bygrave, 1989), which shows properties in common with the Ca^{2+} entry during stimulation. It would be expected that regulation of the amplitude and duration of such a mechanism could occur following cell stimulation.

In excitable cells, where the most progress has been made with regard to describing Ca^{2+} entry mechanisms, voltage-operated calcium channels (VOC) are present. They are activated via a membrane depolarization event which occurs as a result of electrical stimulation (Fisher et al., 1981; Kojima et al., 1985). These VOCs, distinguished by the duration of the increased Ca^{2+} conductance in the plasma membranes, are gated (triggered open) by a change in membrane potential and are blocked specifically by a group of dihydropyridine compounds. Additional modulation of the channel activity can occur as a result of phosphorylation events (Peterson, 1988) or by G-protein activation (Hescheter et al., 1988). Thus, these channels exhibit dual levels of regulation which are of considerable relevance under physiological conditions.

In nonexcitable cells, of which the salivary acinar cell is a good example, it is clear that VOCs do not exist. Data from several studies show a lack of sensitivity of secretory events to dihydropyridine type Ca^{2+} channel antagonists. Additionally, membrane depolarization does not activate calcium entry into these cells (Morris et al., 1987). Instead, a requirement for receptor activation has been established (Putney, 1978, 1986). Thus, the terms receptor operated channel (ROC) and, more recently, second messenger-operated channel (SMOC) have been used to describe the putative type of calcium channel in these cells (Berridge and Irvine, 1989). In a number of cells (e.g., neutrophils, lymphocytes, platelets, and mast cells), electrophysiological measurements of Ca^{2+} currents have shown that Ca^{2+} entry can be induced by an effect of the second messenger IP_3 on the cytoplasmic face of the membrane (for reviews see Peterson, 1989; Penner, 1988). In secretory cells measurements of Ca^{2+} currents have not been very successful. Instead, investigators have used measurements of K^+ or Cl^- currents, which depend on intracellular Ca^{2+}, to indirectly probe Ca^{2+} movements. These studies suggest that while IP_3 alone can elicit a transient K^+ flux, for a sustained K^+ permeability, there is a requirement for both IP_3 and IP_4. These studies argue against a direct role of the receptor (ROC) in the Ca^{2+} entry process and indicate that Ca^{2+} channels in exocrine cells are likely to be of the SMOC type (Petersen, 1989; Berridge and Irvine, 1989). Although Ca^{2+} entry in secretory epithelial cells is not gated by voltage changes, several reports show that the Ca^{2+} entry is sensitive to changes in membrane potential and can be attenuated by depolarization (Merritt and Rink, 1987; Penner, 1988; He et al., 1990a,b). This again is consistent with a "channel"-like mechanism existing. In the rat parotid acinar cell, stimulation of fluid secretion results in a hyperpolarization of the basolateral membrane due to the efflux of K^+ via Ca^{2+}- and voltage-dependent K^+ channels (Petersen and Gallacher, 1988). This hyperpolarization could have a significant role in maintaining sustained Ca^{2+} entry, as has been suggested for several other nonexcitable cells. Thus, this represents a cross-activating system, where Ca^{2+} induces the efflux of potassium which in turn, by modulating the membrane potential, can regulate calcium entry.

In addition to not currently knowing the exact nature of the Ca^{2+} channel through which Ca^{2+} entry occurs, we do not understand the site at which this entry takes place, i.e., does Ca^{2+} cross the plasma membrane and directly enter the cytoplasm or does it enter an intracellular compartment. Concern for this issue mainly arose from earlier studies in which $^{86}Rb^+$ fluxes were used as a probe for Ca^{2+} movements involved in the refill of the ASCaP (Aub et al., 1982). In these studies, a transient increase in $^{86}Rb^+$ flux was observed in acinar cells stimulated with carbachol in a Ca^{2+} free medium which rapidly returned to resting levels. Thereafter, the readdition of Ca^{2+} resulted in a lower, but sustained increased $^{86}Rb^+$ flux. If the muscarinic antagonist atropine was added to cells prior to this

Ca^{2+} readdition, the ^{86}Rb$^+$ flux increase was not observed. However, if cells were allowed to remain in the Ca^{2+} replete/antagonist-containing medium for 2 min, following which a different agonist (e.g., substance P) was added, there was a substantial increase in ^{86}Rb$^+$ flux. Such results were interpreted as showing that Ca^{2+} entered the cell directly into the ASCaP or via a pathway which did not affect the bulk of [Ca^{2+}]i. Experiments with Ca^{2+}-sensitive dyes have shown that such readdition of extracellular Ca^{2+}, in the presence of an antagonist, results in only a small sustained step increase in [Ca^{2+}]i. ^{45}Ca^{2+} measurements, however, show that there is a considerable uptake of Ca^{2+} into cells during this time, also suggesting that entry is into an organelle. This appears to be the ASCaP, since Ca^{2+} can be released by exposing the cells to a second agonist (Putney, 1986). Other cells, e.g., gastric parietal and human endothelial, show similar Ca^{2+} movements. In aggregate, results suggest that Ca^{2+} can enter the ASCaP without causing significant increases in [Ca^{2+}]i (Putney, 1986; Rink and Hallam, 1989).

Two models have been proposed to account for these observations in Ca^{2+} movement. In the first model, the *cytoplasmic loading model*, shown in Figure 4.2, the entry of Ca^{2+} can occur into a restricted area in the cell, where it is not accessible to the bulk cytoplasmic Ca^{2+}, i.e., either to the Ca^{2+}-sensitive dyes or for use in activation of the K$^+$ channel. Calcium is then transported, by an ATP-dependent Ca^{2+} transporter from the cytoplasm into the ASCaP. In a recent review, Rink and Hallam (1989) have discussed experimental observations that are inconsistent with such a model. A variation of this model (see Figure 4.1) would be the entry of Ca^{2+} directly into the cytoplasm, but not into a restricted cytoplasmic location. The lack of an observed (by fluorescent dyes) increase in [Ca^{2+}]i in this model can be explained as a result of increased activity of the Ca^{2+} pumps in the plasma membrane and in the ER, thus keeping Ca^{2+} in the cytoplasm low. Indeed, studies of Ca^{2+} efflux from parotid and pancreatic acini (Putney, 1986; Muallem, 1989) have shown that there is an increase of the Ca^{2+} permeability of the plasma membrane which is dependent on the presence of agonist. Further, the rate of Ca^{2+} entry into the ER in this situation is several-fold greater than the rate of efflux from the ER. However, a detailed assessment of the rates of individual Ca^{2+} fluxes under physiological conditions is necessary to examine this model adequately. Since [Ca^{2+}]i is determined by numerous Ca^{2+} fluxes, which in turn may be regulated by other ion fluxes, such assessments would be kinetically very complicated.

The second model that has been proposed is a *direct loading model*. Based on the same studies described above, some investigators have proposed that during reloading (i.e., Ca^{2+} entry after stimulation), Ca^{2+} directly enters the ASCaP from the extracellular medium as shown in Figure 4.3. Ca^{2+} would only enter the cytosoplasm via the IP$_3$-dependent channel in the ASCaP membrane. Thus, according to this model, increases in [Ca^{2+}]i, upon readdition of Ca^{2+}, are only seen in the presence of an agonist, i.e., when the IP$_3$-activated ER Ca^{2+} channel is open. When an antagonist is added, this IP$_3$-Ca^{2+} release pathway is closed, allowing the ASCaP to reload. This model can not account for the small step increase in [Ca^{2+}]i seen upon Ca^{2+} readdition to cells in a Ca^{2+}-free medium. It

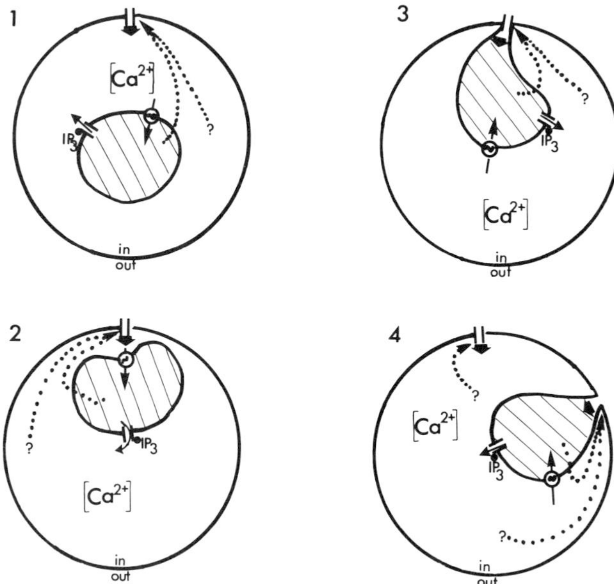

Figure 4. Possible models for Ca^{2+} entry in salivary acinar cells. Figures 4.1 and 4.2 depict *cytoplasmic loading models*. In Figure 1, Ca^{2+} enters the cytoplasm from where it is accumulated into the IP_3 sensitive Ca^{2+} pool. This model predicts that increases in the activities of the Ca^{2+} pumps compensate for the increase in $[Ca^{2+}]i$ due to influx. In the model in Figure 4.2, Ca^{2+} enters the cell, into a restricted location, not accessible directly to the bulk $[Ca^{2+}]i$. From here it is taken up by the IP_3 sensitive Ca^{2+} pool via the activity of the Ca^{2+} pump. This model predicts a highly specialized cellular architecture. In Figure 4.3, the *direct entry model* is presented. In this model, the Ca^{2+} entry is via two closely apposed membranes, BLM and ER, directly into the IP_3-sensitive Ca^{2+} pool. This model predicts a highly specialized architecture as well as a Ca^{2+} channel which can span two membrane bilayers. In Figure 4.4, the model depicts two entry pathways one directly into the pool, which could occur via mechanism as shown in 4.2 or 4.3, and a direct entry into the bulk $[Ca^{2+}]i$ pool. Dotted lines refer to the proposed regulation of the Ca^{2+} entry pathways by the status of the IP_3 sensitive Ca^{2+} pool and by other possible factors (?) in the cytoplasm, such as IP_4.

is possible that the Ca^{2+} entry involved in this step increase in $[Ca^{2+}]i$, and that involved in ASCaP refill, represent different pathways as shown in the model presented in Figure 4.4. Studies ($^{45}Ca^{2+}$ uptake) reported by Muallem with pancreatic cells (Muallem et al., 1988) show that the Ca^{2+} entry involved in reloading occurs in the presence of an antagonist and is dependent on the extracellular $[Ca^{2+}]i$. Additionally, the rate of this entry is reduced as a function of time after antagonist addition. This rate change may reflect an increase in the Ca^{2+} content of the ASCaP during refill. Studies with parotid acini reported by Takemura and Putney (1989) show that the small step increase observed in $[Ca^{2+}]i$ is dependent on the gradient of calcium across the cell membrane, but is not affected by the time of exposure to the antagonist. These results imply that Ca^{2+} entry is somewhat insensitive to receptor (second messenger) blockage.

　　A requirement for the feasibility of a direct entry model is a juxtaposition ER with plasma membranes. Morphological examination of salivary acinar cells does not show such areas. However, it is conceivable that changes in the cytoskeletal elements or the volume of cells could make such an interaction possible when the cells are stimulated. Additionally, the factors generated upon stimulation could induce such membrane interactions, e.g., the proposed role for GTP in intracellular membrane fusion (Gill et al., 1989).

An obvious complication to all of the models described above is the fact that the mechanism which initiates Ca^{2+} entry is still unclear. All studies with acinar cells reported thus far show that Ca^{2+} release from the ASCaP precedes Ca^{2+} entry, and so it has been suggested that Ca^{2+} entry is regulated by the activation of an intracellular Ca^{2+} pool. Such observations led to the "capacitative Ca^{2+} entry hypothesis", in which Ca^{2+} influx was proposed to be dependent on the extent of depletion of the IP_3-sensitive Ca^{2+} pool. However, Ca^{2+} release from this pool by itself is not sufficient to support calcium entry, since with ionophores or IP_3 sustained calcium entry is not observed. Recent studies with lacrimal gland cells suggest that IP_4 may be involved in regulating Ca^{2+} entry. Importantly, the IP_4 effect is seen only in the presence of extracellular Ca^{2+} and IP_3, a result more compatible with the direct loading model (Peterson, 1989; Berridge and Irvine, 1989). We have recently shown the activation of Ca^{2+} entry in parotid acini stimulated with AlF_4- (Mertz et al., 1990). Whether this represents a direct gating of Ca^{2+} entry by G proteins or that Ca^{2+} entry is achieved via G-protein linked PLC-hydrolysis is still to be established. A recent study, however, contradicts the involvement of second messengers in the stimulation of calcium entry. This study shows that Ca^{2+} entry can be induced in the absence of IP_3 generation by the drug thapsigargin, which apparently inhibits the ER Ca^{2+} pump somewhat specifically (Takemura et al., 1989). This condition allows Ca^{2+} to leak out of the ER. However, since there is insufficient information about the Ca^{2+} permeability of the ER membranes under physiological conditions, these studies need to be clarified further. Another point to be considered is the fact that most of the studies examining ASCaP reloading have been carried out in Ca^{2+}-free media, a situation which may alter normal Ca^{2+} homeostasis.

ROLE OF CALCIUM IN THE FLUID SECRETORY MECHANISM

It should be clear from the above discussion that a salivary acinar cell devotes considerable energy and detail into the tight control of its Ca^{2+} levels. Such regulation permits the rapid and copious fluid secretory responses which are physiologically required. The changes in $[Ca^{2+}]i$ subsequent to neurotransmitter stimulation involve the transient release of Ca^{2+} from an intracellular store and the sustained entry of Ca^{2+} from the extracellular fluid.

The combination of these events (rapid and dramatic vs. sustained, smaller changes) leads to the long-term elevation of acinar cell $[Ca^{2+}]i$, and directly results in the ion fluxes which drive fluid secretion (see Chapter 5). Specifically

(Figure 1), the elevation $[Ca^{2+}]i$ results in the opening of two ion channels in this cell, a Ca^{2+}- and voltage-activated K^+ channel (Maruyama et al., 1983; Petersen, 1986) in the basolateral membrane (the so called maxi-K^+ channel which provides the pathway for neurotransmitter-stimulated K^+ release from the cells) and a yet poorly characterized Ca^{2+}-activated Cl^- channel in the apical plasma membrane (Saito et al., 1985; Iwatsuki et al., 1985) through which neurotransmitter-stimulated Cl^- efflux is believed to take place. The specific surface localization of these two ion flux pathways provides a mechanistic explanation for the directionality (serosal to mucosal; basolateral to apical) of salivary fluid secretion. Thus, following elevation of $[Ca^{2+}]i$, the K^+ and Cl^- channels open, the K^+ channel to the interstitium and the Cl^- channel to the lumen. Potassium, which exits the cell, is utilized to help drive Cl^- into the cell against its steep electro-chemical gradient via the basolateral membrane-located loop diuretic sensitive sodium/potassium/chloride cotransporter (see Chapter 5 for details). For each Cl^- ion entering the cell, another Cl^- ion must exit. This occurs via the Cl^- exit pathway (channel) which presumably is found in the apical membrane. The negative charge in the lumen due to Cl^- release must immediately be balanced by a cation (Na^+) which is thought to leak across the tight junctions. This creates an osmotic (NaCl) gradient across the acinar epithelial cell which pulls water and results in the generation of the primary salivary fluid.

The control of $[Ca^{2+}]i$, which is central to this entire secretory signaling process, is just beginning to be clarified. Considerably more information is required to understand the effect of physiological stimuli (ie. multiple, simultaneous neurotransmitter activation events) on $[Ca^{2+}]i$ homeostasis. Further, as noted above, we currently know very little about the mechanisms of $[Ca^{2+}]i$ entry into these cells. It should also be recognized that disease occurs at a molecular level and it is likely that certain salivary hypofunctional states are a result of altered acinar cell Ca^{2+} mobilization (see Baum, 1989).

This chapter was written in late 1989. The authors suggest that the following articles may be helpful to update readers: Simon et al., *Science*, 252, 802, 1991; Dai et al., *Am. J. Physiol.*, 261, C1063, 1991; Quissell et al., *CRC Crit. Rev. Oral Biol. Med.*, 3, 83, 1992; Petersen, *J. Physiol. (London)*, 448, 1, 1992; Mertz et al., *J. Membr. Biol.*, 126, 183, 1992; and Ambudkar et al., *Mol. Cell. Biochem.*, in press.

REFERENCES

1. Ambudkar, I. S., Melvin, J. E., and Baum, B. J., α_1-Adrenergic regulation of Cl^- and Ca^{2+} movements in rat parotid acinar cells, *Pfluegers Arch.*, 412, 75, 1988.
2. Ambudkar, I. S., Smith, M. W., Phelps, P. C., Regec, A. L., and Trump B. F., Extracellular Ca^{2+}-dependent elevation in cytosolic Ca^{2+} potentiates $HgCl_2$-induced renal proximal tubular cell damage, *Toxicol. Indust. Health*, 4, 107, 1988.
3. Ambudkar, I. S. and Baum, B. J., ATP-dependent calcium transport in rat parotid basolateral membrane vesicles is modulated by membrane potential, *J. Membr. Biol.*, 102, 59, 1988a.

4. Ambudkar, I. S. and Baum, B. J., Modulation of ATP-dependent Ca^{2+} transport in rat parotid basolateral membrane vesicles by $K^+ + Cl$ flux, *Biochem. Biophys. Acta*, 941, 198, 1988b.

5. Ambudkar, I. S., Kuyatt, B. L., Roth, G. S., and Baum, B. J., Modification of ATP-dependent Ca^{2+} transport in rat parotid basolateral membrane during ageing, *Mech. Ageing Dev.*, 43, 45 1988.

6. Ambudkar, I. S., Horn, V. J., and Baum, B. J., ATP-dependent Ca^{2+} transport in the rat parotid basolateral plasma membrane is regulated by calmodulin, *Arch. Biochem. Biophys.*, 268, 576, 1989.

7. Arkle, S, Michalek, R., and Templeton, D., The relationship of intracellular free calcium activity to amylase secretion in substance P and isoprenaline-stimulated rat parotid acini, *Biochem. Pharm.*, 38, 1257, 1989.

8. Aub, D. L., McKinney, J. S., and Putney, J. W., Jr. Nature of the receptor-regulated calcium pool in the rat parotid gland, *J. Physiol.*, 331, 557, 1982.

9. Aub, D. L., Gosse, M. E., and Cote, T. E., Regulation of thyrotropin-releasing hormone receptor binding and phospholipase activation by a single GTP-binding protein, *J. Biol. Chem.*, 262, 9521, 1987.

10. Augert, G., Blackmore, P. F., and Exton, J. H., Changes in the concentration and fatty acid composition of phosphoinositides induced by hormones in hepatocytes, *J. Biol. Chem.*, 264, 2574, 1989.

11. Axelrod, J., Burch, R. M., and Jelsema, C. L., Receptor-mediated activation of phospholipase A_2 via GTP binding proteins, *Trends Neurosci.*, 11, 117, 1988.

12. Balla, T., Baukal, A. J., Guillemette, G., and Catt, K. J., Multiple pathways of inositol polyphosphate metabolism in angiotensin-stimulated adrenal glomerulosa cells, *J. Biol. Chem.*, 263, 4083, 1988.

13. Batty, I. R., Nahorski, S. R., and Irvine, R. F., Rapid formation of inositol 1,3,4,5 tetrakisphosphate following muscarinic receptor stimulation of rat cerebral cortical slices, *Biochem. J.*, 232, 211, 1985.

14. Baum, B. J., Salivary gland fluid secretion during aging, *J. Am. Geriatrics Soc.*, 37, 453, 1989.

15. Baum, B. J., Neurotransmitter control of secretion, *J. Dent. Res.*, 66, 628, 1987.

16. Baum, B. J. and Ambudkar, I. S., Regulation of calcium handling by rat parotid acinar cells, *Mol. Cell Biochem.*, 82, 67, 1988.

17. Baum, B. J., Ambudkar, I. S., Helman, J., Horn, V. J., Melvin, J. E., Mertz, L. M., and Turner, R. J., Dispersed salivary gland acinar cell preparations for use in studies of neuroreceptor coupled secretory events, *Methods Enzymol.*, 192, 26, 1990.

18. Baum, B. J., Horn, V. J., and Ambudkar, I. S., Evidence that ATP-dependent Ca^{2+} transport in rat parotid microsomal membranes requires charge compensation, *Biochem. J.*, 254, 649, 1988.

19. Bayerdorffer, E., Eckhardt, L., Haase, W., and Schulz, I., Electrogenic calcium transport in plasma membrane of rat pancreatic acinar cells, *J. Membr. Biol.*, 84, 45, 1985.

20. Berridge, M. J. and Irvine, R. F., Inositol phosphates and cell signaling, *Nature*, 341, 197, 1989.

21. Berridge, M. J., Dawson, R. M., Downes, C. P., Heslop, J. P., and Irvine, R. F., Changes in the levels of inositol phosphates after agonist-dependent hydrolysis of membrane phosphoinositides, *Biochem. J.*, 212, 473, 1983.

22. Berridge, M. J. and Galione, A., Cytosolic calcium oscillators, *FASEB J.*, 2, 3074, 1988.

23. Bigay, J., Deterre, P., Pfister, C., and Chabre, M., *FEBS Lett.*, 191, 181, 1985.

24. Bodner, L., Hoopes, M. T., Gee, M., Ito, H., Roth, G. S., and Baum, B. J., Multiple transduction mechanisms are likely involved in calcium mediated exocrine secretory events in rat parotid cells, *J. Biol. Chem.*, 258, 2774, 1983.

25. Boyer, J. L., Downes, C. P., and Harden, T. K., Kinetics of activation of phospholipase C by P_{2y} purinergic receptor agonists and guanine nucleotides, *J. Biol. Chem.*, 264, 884, 1989.

26. Boyer, J. L., Waldo, G. L., Evans, T., Northrup, J. K., Downes, C.P., and Harden, T. K., Modification of AlF_4- and receptor stimulated phospholipase C activity by G protein β subunits, *J. Biol. Chem.*, 264, 13917, 1989.

27. Brass, L. F., Laposata, M., Banja, H. S., and Rittenhouse, S. E., Regulation of the phosphoinositide hydrolysis pathway in thrombin stimulated platelets by a pertussis toxin sensitive guanine nucleotide binding protein, *J. Biol. Chem.*, 261, 16838, 1986.

28. Burch, R. M., Luini, A., and Axelrod, J., Phospholipase A_2 and phospholipase C are activated by distinct GTP-binding proteins in response to α_1-adrenergic stimulation in FRTL-5 thyroid cells, *Proc. Natl. Acad. Sci. U.S.A.*, 83, 7201, 1986.

29. Burgen, A. S. V., The secretion of potassium in saliva, *J. Physiol.*, 132, 20, 1956.

30. Bygrave, F. L., Karjalainen, A., and Altin, J. G., Passive calcium influx by plasma membrane vesicles isolated from rat liver, *Cell Calcium*, 10, 235, 1989.

31. Carafoli, E., Intracellular calcium homeostasis, *Annu. Rev. Biochem.*, 56, 395, 1987.

32. Carlson, K. E., Brass, L. F., and Manning, D. R., Thrombin and phorbol esters cause the selective phosphorylation of a G protein other than Gi in human platelets, *J. Biol. Chem.*, 264, 13298, 1989.

33. Casey, P. J. and Gilman, A. G., G protein involvement in receptor-effector coupling, *J. Biol. Chem.*, 263, 2577, 1988.

34. Chabre, M., Aluminofluoride action on G-proteins of the adenylate cyclase system is not different from that on transducin, *Biochem. J.*, 258, 931, 1989.

35. Champeil, P., Combettes, L., Berthon, B., Doucet, E., Orlowski, S., and Claret M., Fast kinetics of calcium released by myoinositol trisphosphate in permeabilized rat hepatocytes, *J. Biol. Chem.*, 264, 17665, 1989.

36. Comerford, I. G. and Dawson, A. P., The mechanism of action of GTP on Ca^{2+} efflux from rat liver microsomal vesicles, *Biochem. J.*, 249, 89, 1988.

37. Crouch, M. F. and Lapetina, E. G., Dual mechanisms of platelet hormone receptor desensitization, *J. Biol. Chem.*, 264, 584, 1989.

38. Cubbold, P. M. and Rink, T. J., Fluorescence and bioluminescence measurement of cytoplasmic free calcium, *Biochem. J.*, 248, 313, 1987.

39. Daniel, J. L., Dangelmaier, C. A., and Smith, J. B., Formation and metabolism of inositol 1,4,5 trisphosphate in human platelets, *Biochem. J.*, 264, 109, 1987.

40. Dixon, J. F. and Hokin, L. E., Kinetic analysis of the formation of inositol 1:2-cyclic phosphate in carbachol-stimulated pancreatic minilobules, *J. Biol. Chem.*, 264, 11721, 1989.

41. Douglas, W. W. and Poisner, A. M., The influence of calcium on the secretory response of the submaxillary gland to acetylcholine or to noradrenaline, *J. Physiol.*, 165, 528, 1963.

42. Dunlop, M. E. and Larkins, R. G., GTP and inositol 1,4,5-trisphosphate induced release of $^{45}Ca^{2+}$ from a membrane store colocalized with pancreatic islet-cell plasma membrane, *Biochemistry*, 253, 67, 1988.

43. Ek, B. and Nahorski, S., Muscarinic receptor coupling to inositol phospholipid metabolism in guinea pig cerebral cortex, parotid gland, and ileal smooth muscle, *Biochem. Pharm.*, 37, 4461, 1988.

44. Exton, J. H., Mechanisms of action of calcium mobilizing agonists: some variations on a young theme, *FASEB J.*, 2, 2670, 1988.

45. Farese, R. V., Rosic, N., Babischkin, J., Farese, M. G., Foster, R., and Davis, J. S., Dual activation of the inositol-trisphosphate-calcium and cyclic nucleotide intracellular signaling systems by adrenocorticotropin in rat adrenal cells, *Biochem. Biophys. Res. Commun.*, 135, 742, 1986.

46. Fargin, A, Raymond, J. R., Regan, J. W., Cotecchia, S., Lefkowitz, R. J., and Caron, M. G., Effector coupling mechanisms of the cloned 5-HT 1A receptor, *J. Biol. Chem.*, 264, 14848, 1989.

47. Ferris, C. D., Huganir, R. L., Supattapone, S., and Snyder, S. H., Purified inositol 1,4,5-trisphosphate receptor mediates calcium flux in reconstituted lipid vesicles, *Nature*, 342, 87, 1989.

48. Fisher, S. K., Holz, R. W., and Agranoff, B. W., Muscarinic receptors in chromaffin cell cultures mediate enhanced phospholipid labelling but not catecholamine secretion, *J. Neurochem.*, 37, 491, 1981.

49. Fisher, S. K., Domask, L. M., and Roland, R. M., Muscarinic receptor regulation of cytoplasmic Ca^{2+} concentration in human SK-N-SH neuroblastoma cells, *Mol. Pharmacol.*, 35, 195, 1989.

50. Furuichi, T., Uoshikawa, S., Miyawaki, A., Wade, K., Maeda, N., and Mikoshiba, K., Primary structure and functional expression of the inositol 1,4,5-trisphosphate-binding protein P400, *Nature*, 342, 32, 1989.

51. Garrett, J. K., Adventures with autonomic nerves. Perspectives in salivary glandular innervations, *Proc. R. Micr. Soc.*, 17, 242, 1982.

52. Gill, D. L., Ghosh, T. K., and Mullaney, J. M., Calcium signaling mechanism in endoplasmic reticulum activated by inositol 1,4,5-trisphosphate and GTP, *Cell Calcium*, 10, 383, 1989.

53. Gilman, A. G., G proteins: transducers of receptor-generated signals, *Annu. Rev. Biochem.*, 56, 615, 1987

54. Guillemette, G., Balla, T., Baukal, A. J., and Catt, K. J., Inositol 1,4,5 trisphosphate bind to a specific receptor and releases microsomal calcium in the anterior pituitary gland, *Proc. Natl. Acad. Sci. U.S.A.*, 84, 8185, 1987.

55. Goodman, L. and Gilman, A., *The Pharmacological Basis of Therapeutics.*, 6th ed., Macmillian, New York, 1980.

56. Gray, P. T., Oscillations of free cytosolic calcium evoked by cholinergic and catecholinergic agonists in rat parotid acinar cells, *J. Physiol.*, 406, 35, 1988.

57. He, X., Wu, X., and Baum, B. J. The effect of N-(6aminohexyl)-5-chloro-1-naphthalenesulfonamide (W-7) on muscarinic receptor-induced Ca^{2+} mobilization in a human salivary epithelial cell line, *Pfluegers Arch.*, 416, 36, 1990a.

58. He, X., Wu, X., Turner, R. J., and Baum, B. J., Evidence for two modes of Ca^{2+} entry following muscarinic stimulation of a human salivary epithelial cell line, *J. Membr. Biol.*, 115, 159, 1990b.

59. Helman, J., Roth, G. S., and Baum, B. J., α_1-Adrenergic agonist induced Ca^{2+} fluxes in rat parotid cells are not Na^+-dependent, *Biochem. J.*, 230, 313, 1985.

60. Helman, J., Kuyatt, B. L., Takuma, T., Seligman, B., and Baum, B. J., ATP-dependent calcium transport in rat parotid basolateral membrane vesicles. Modulation by agents which elevate cyclic AMP, *J. Biol. Chem.*, 261, 8919, 1986.

61. Henne, V. and Soling, H. D., Guanosine 5'-trisphosphate releases calcium from rat liver and guinea pig parotid gland endoplasmic reticulum independently of inositol 1,4,5 trisphophate, *FEBS Lett.*, 202, 267, 1986.

62. Hescheler, J., Rosenthal, W., Hinsch, K. D., Wulfern, M., Trautwein, W., and Schulz, G., Angiotensin induced stimulation of voltage-dependent Ca^{2+} currents in an adrenal cortical cell line, *EMBO J.*, 7, 619, 1988.

63. Horn, V. J., Baum, B. J., and Ambudkar, I. S., β-Adrenergic receptor stimulation induces inositol trisphosphate production and Ca^{2+} mobilization in rat parotid acinar cells, *J. Biol. Chem.*, 263, 12454, 1988.

63a. Horn, V. J., Baum, B. J., and Ambudkar, I. S., Attenuation of inositol triphosphate generation and cystolic Ca^{2+} elevation in dispersed rat parotid acini stimulated simultaneously at muscarinic and α_1-adrenergic receptors, *Biochem. Biophys. Res. Commun.*, 166, 967, 1990.

64. Hughes, A. R. and Putney, J. W., Jr., Sources of ^{3}H-labeled inositol bis and monophosphates in agonist-activated rat parotid acinar cells, *J. Biol. Chem.*, 264, 9400, 1989.

65. Hunter, M., Smith, P. A., and Case, R. M., The dependence of fluid secretion by mandibular salivary gland and pancreas on extracellular calcium, *Cell Calcium*, 4, 307, 1983.

66. Immelman, A. and Soling H. D., ATP-dependent calcium sequestration and calcium/ATP stoichiometry in isolated microsomes from guinea pig parotid glands, *FEBS Lett.*, 162, 406, 1983.

67. Ito, H., Baum, B. J., Uchida, T., Hoopes, M. T., Bodner, L., and Roth, G. S., Modulation of rat parotid cell α-adrenergic responsiveness at a step subsequent to receptor activation, *J. Biol. Chem.*, 257, 9532, 1982.

68. Iwatsuki, N., Maruyama, Y., Matsumoto, O., and Nishiyama, A., Activation of Ca^{2+}-dependent Cl^- and K^+ conductances in rat and mouse parotid acinar cells, *Jpn. J. Physiol.*, 35, 933, 1985.

69. Joseph, S. K., Rice, H. L., and Williamson, J. R., Effect of external Ca^{2+} and pH on inositol trisphosphate-mediated Ca^{2+} release from cerebellum microsomes, *Biochem. J.*, 258, 261, 1989.

70. Kanagasuntheram, P. and Teo, T. S., Parotid microsomal calcium transport, *Biochem. J.*, 208, 789, 1982.

71. Kawanishi, T., Blank, L. M., Harootunian, A. T., Smith, M. T., and Tsien, R. Y., Ca^{2+} oscillations induced by hormonal stimulation of individual fura-2 loaded hepatocytes, *J. Biol. Chem.*, 264, 12859, 1989.

72. Kaya, H., Patton, G. M., and Hung, S. L., Bradykinin induced activation of phospholipase A_2 is independent of the activation of polyphosphoinositid hydrolysing phospholipase, C, *J. Biol. Chem.*, 264, 4972, 1989.

73. Kerlavage, A. R., Fraser, C. M., and Venter, J. C., Muscarinic cholinergic receptor structure, *Trends Pharm. Sci.*, 8, 426, 1987.

74. Krause, K. H., Pittet, D., Volpe, P., Pozzan, T., Meldolesi, J., and Lew, D. P., Calciosome, a sarcoplasmic reticulum-like organelle involved in intracellular Ca^{2+} handling by non-muscle cells. Studies in human neutrophils and HL-60 cells, *Cell Calcium*, 10, 351, 1989.

75. Kojima, I., Kojima, K., and Rasmussen, M., Characteristics of angiotensin and ACTH induced calcium influx in adrenal glomerulosa cells. Evidence that angiotensin and ACTH may open a common calcium channel, *J. Biol. Chem.*, 260, 9171, 1985.

76. Lefkowitz, R. J. and Caron, M. G., Adrenergic receptors, *J. Biol. Chem.*, 263, 4993, 1988.

77. Maki, T., Kowatch, M. A., Baum, B. J., Ambudkar, I. S., and Roth, G. S., Evidence for an alteration in the microsomal Ca^{2+} release mechanism in senescent rat parotid acinar cells, *Biochem. Biophys. Acta*, 1014, 73, 1989.

78. Martinez, J. R. and Cassity, N., ^{36}Cl fluxes in dispersed rat submandibular acini: effects of acetylcholine and transport inhibitors, *Pfluegers Arch.*, 403, 50, 1985.

79. Martinson, E.A., Goldstein, D., and Brown, J. H., Muscarinic receptor activation of phosphatidylcholine hydrolysis, *J. Biol. Chem.*, 264, 14748, 1989.

80. Maruyama, Y., Gallacher, D. V., and Petersen, O. H., Voltage and Ca^{2+} activated K^+ channel in baso-lateral acinar cell membranes of mammalian salivary glands, *Nature*, 302, 827, 1983.

81. Matozaki, T. and Williams, J. A., Multiple sources of 1, 2-diacylglycerol in isolated rat pancreatic acini stimulated by cholecystokinin, *J. Biol. Chem.*, 264, 14729, 1989.

82. McDonough, P. M. and Button, D. C., Measurement of cytoplasmic calcium concentration in cell suspensions, *Cell Calcium*, 10, 171, 1989.

83. McKinney, J. S. and Rubin, R. P., Enhancement of cyclic AMP modulated salivary amylase secretion by protein kinase C activators, *Biochem. Pharm.*, 37, 4433, 1988.

84. Melvin, J. E., Kawaguchi, M., Baum, B. J., and Turner, R. J., A muscarinic agonist-stimulated chloride efflux pathway is associated with fluid secretion in rat parotid acinar cells, *Biochem. Biophys. Res. Commun.*, 145, 754, 1989.

85. Merritt, J. E. and Rink, T. J., The effects of substance P and carbachol on inositol tris- and tetrakisphosphate formation and cytosolic free calcium in rat parotid acinar cells, *J. Biol. Chem.*, 262, 14912, 1987.

86. Merritt, J. E. and Rink, T. J., Regulation of cytosolic free calcium in Fura 2 loaded rat parotid acinar cells, *J. Biol. Chem.*, 262, 17362, 1987.

87. Merritt, J. E. and Rink, T. J., Rapid increases in cytosolic free calcium in response to muscarinic stimulation of parotid acinar cells, *J. Biol. Chem.*, 262, 4958, 1987.

88. Mertz, L. M., Horn, V. J., Baum, B. J., and Ambudkar, I. S., Ca^{2+} entry in parotid acini: activation by carbachol and aluminum fluoride, *Am. J. Physiol.*, 258, C654, 1990.

89. Meyer, T. and Stryer, L., Molecular model for receptor-stimulated calcium spiking, *Proc. Natl. Acad. Sci. U.S.A.*, 85, 5051, 1988.

90. Meyer, T., Holowka, D., and Stryer, L., Highly cooperative opening of calcium channels by inositol 1,4,5-trisphosphate, *Science*, 240, 653, 1988.

91. Milligan, G., Techniques used in the identification and analysis of function of pertussis toxin-sensitive guanine nucleotide binding proteins, *Biochem. J.*, 255, 1, 1988.

92. Morris, A. P., Fuller, C. J., and Gallacher, D. V., Cholinergic receptors regulate a voltage-insensitive but Na^+-dependent calcium influx pathway in salivary acinar cells, *FEBS Lett.*, 211, 195, 1987.

93. Moriarty, T. M., Gillo, B., Carty, D. J., Premont, R.T., Landau, E. M., and Iyengar, R., β subunits of GTP binding proteins inhibit muscarinic receptor stimulation of phospholipase C, *Proc. Natl. Acad. Sci. U.S.A.*, 85, 8865, 1988.

94. Muallem, S., Schoeffield, M. D., Fimmel, C. J., and Pandol, S. J., The agonist-sensitive calcium pool in the pancreatic acinar cell. Reloading during and at the termination of stimulation, *Am. J. Physiol.*, 255, G229, 1988.

95. Muallem, S., Pandol, S. J., and Beeker, T. G., Hormone-evoked calcium release from intracellular stores is a quantal process, *J. Biol. Chem.*, 264, 205, 1988.

96. Muallem, S., Calcium transport pathways of pancreatic acinar cells, *Annu. Rev. Physiol.*, 51, 83, 1989.

97. Mullaney, J. M., Yu, M., Ghosh, T. K., and Gill, D. L., Calcium entry into the inositol 1,4,5-trisphosphate-releasable calcium pool is mediated by a GTP regulatory mechanism, *Proc. Natl. Acad. Sci. U.S.A.*, 85, 2492, 1988.

98. Nathanson, N. M., Molecular properties of the muscarinic acetylcholine receptor, *Annu. Rev. Neurosci.*, 10, 195, 1987.

99. Nauntofte, B. and Poulsen, J. M., Effects of Ca^{2+} and furosemide on Cl^- transport and O_2 uptake in rat parotid acini, *Am. J. Physiol.*, 251, C175, 1986.

100. Nauntofte, B. and Dissing, S., Stimulation induced changes in cytosolic calcium in rat parotid acini, *Am. J. Physiol.*, 253, G290, 1987.

101. Ott, S. and Costa, T., Guanine nucleotide-mediated inhibition of opoid agonist binding, *Biochem. Pharm.*, 38, 1931, 1989.

102. Paris, S. and Pouyssegur, J., Further evidence for a phospholipase C-coupled G protein in hamster fibroblasts. Induction of inositol phosphate formation by fluoroaluminate and vanadate and inhibition by pertussis toxin, *J. Biol. Chem.*, 262, 1970, 1987.

103. Penner, R., Matthews, G., and Neher, E., Regulation of calcium influx by second messengers in rat mast cells, *Nature*, 334, 449, 1988.

104. Petersen, O. H., Some factors influencing stimulation-induced release of potassium from the cat submandibular gland to fluid perfused through the gland, *J. Physiol.*, 208, 431, 1970.

105. Petersen, O. H., Calcium-activated potassium channels and fluid secretion by exocrine glands, *Am. J. Physiol.*, 251, 61, 1986.

106. Petersen, O. H., Control of potassium channels in insulin secreting cells, *ISI Atlas of Science: Biochemistry*, 1, 144, 1988.

107. Petersen, O. H. and Gallacher, D. V., Electrophysiology of pancreatic and salivary acinar cells, *Annu. Rev. Physiol.*, 50, 65, 1988.

108. Poggioli, J. and Putney, J. W., Jr., Net calcium fluxes in rat parotid acinar cells, *Pfluegers Arch.*, 392, 239, 1982.

109. Putney, J. W., Jr., Biphasic modulation of potassium release in rat parotid gland by carbachol and phenylephrine, *J. Pharmacol. Exp. Ther.*, 198, 375, 1976.

110. Putney, J. W., Jr., Role of calcium in the fade of the potassium release response in the rat parotid gland, *J. Physiol.*, 81, 383, 1978.

111. Putney, J. W., Jr., A model for receptor regulated calcium entry, *Cell Calcium*, 7, 1, 1986.

112. Putney, J. W., Jr., Identification of cellular activation mechanisms associated with salivary secretion, *Annu. Rev. Physiol.*, 48, 75, 1986.

113. Putney, J. W., Jr., Takemura, H., Hughes, A. R., Horstman, D. A., and Thastrup, O., How do inositol phosphates regulate calcium signaling, *FASEB J.*, 3, 1988, 1989.

114. Putney, J. W., Jr., Van de Walle, C. M., and Leslie, B. A., How far does phospholipase C activity depend on the cell calcium, *Biochem. J.*, 243, 391, 1978.

115. Rink, T. J. and Hallam, T. J., Calcium signaling in non-excitable cells: notes on oscillations and store refilling, *Cell Calcium*, 10, 585, 1989.

116. Rossier, M. R., Capponi, A. M., and Vallotton, M. B., The inositol 1,4,5 trisphosphate-binding site in adrenal cortical cells is distinct from the endoplasmic reticulum, *J. Biol. Chem.*, 264, 14078, 1989.

117. Saito, Y., Ozawa, T., Hayashi, H., and Nishiyama, A., Acetylcholine-induced change in intracellular Cl^- activity of mouse lacrimal acinar cells, *Pfluegers Arch.*, 405, 108, 1985.

118. Schneyer, L. H. and Emmelin, N., Salivary secretion, in *Gastrointestinal Physiology*, Jacobson, E. D. and Shambour, L. L., Eds., Butterworths, London, 1974.

119. Schulz, I., Schnefel, S., Banfic, H., and Eckhardt, L., Ca^{2+} signaling in exocrine glands in comparison to that in vascular smooth muscle cells, *Ann. N.Y. Acad. Sci.*, 488, 240, 1986.

120. Schulz, I., Thevenod, F., and Dehlinger-Kremer, M., Modulation of intracellular free Ca^{2+} concentration by IP_3-sensitive and IP_3-insensitive non-mitochondrial Ca^{2+} pools, *Cell Calcium*, 10, 325, 1989.

121. Selinger, Z., Naim, E., and Lasser, M., ATP-dependent calcium uptake by microsomal preparations from rat parotid and sub-maxillary glands, *Biochem. Biophys. Acta*, 203, 326, 1970.

122. Selinger, Z., Batzri, S., Eimerl, S., and Schramm, M., Ca^{2+} and energy requirements for K^+ release mediated by the epinephrine α-receptor in rat parotid slices, *J. Biol. Chem.*, 248, 369, 1971.

123. Shears, S. B., Parry, J. B., Tang, E. K. Y., Irvine, R. F., Michell, R. H., and Kirk, C. J., Metabolism of D-myo-inositol 1,3,4,5 tetrakisphosphate by rat liver, including the synthesis of a novel isomer of myo-inositol tetrakisphosphate, *Biochem. J.*, 246, 139, 1987.

124. Shears, S. B., Metabolism of the inositol phosphates produced upon receptor activation, *Biochem. J.*, 260, 313, 1989.

125. Snyder, S. H. and Suppatapone, S., Isolation and functional characterization of an inositol trisphosphate receptor from brain, *Cell Calcium*, 10, 337, 1989.

126. Soling, H. D., DeDomenech, E. M., Kleineki, J., and Fest, W., Early effects of β-adrenergic and muscarinic secretagogues on lipid and phospholipid metabolism in guinea pig parotid acinar cells, *J. Biol. Chem.*, 162, 11281, 1987.

127. Somlyo, A. P. and Himpens, B., Cell calcium and its regulation in smooth muscle, *FASEB J.*, 3, 2266, 1989.

128. Spat, A., Bradford, P. G., McKinney, J. S., Rubin, R. P., and Putney, J. W., Jr., A saturable receptor for ^{32}P-inositol 1,4,5 trisphosphate in hepatocytes and neutrophils, *Nature*, 319, 514, 1986.

129. Steinhelper, M. E., Fisher, R. A., Reutyak, G. E., Hanahan, D. J., and Olson, M. S., β_2-Adrenergic agonist regulation of immune aggregate and platelet-activating factor stimulated hepatic metabolism, *J. Biol. Chem.*, 264, 10976, 1989.

130. Strehler-Page, M. A., Vogel, G., and Carafoli, E., mRNAs for plasma membrane calcium pump. Isomers differing in their regulatory domain are generated by alternative splicing that involves two internal donor sites in a single exon, *Proc. Natl. Acad. Sci. U.S.A.*, 86, 6908, 1989.

131. Sugiya, H., Obie, J. F., and Putney, J. W., Jr., Two modes of regulation of the phospholipase C linked substance P receptor in rat parotid acinar cells, *Biochem. J.*, 253, 459, 1988.

132. Supattapone, S., Worley, P. F., Barban, J. M., and Snyder, S. H., Solubilization, purification and characterization of an inositol trisphosphate receptor, *J. Biol. Chem.*, 263, 1530, 1988.

133. Takemura, H. and Putney, J. W., Jr., Capacitative calcium entry in parotid acinar cells, *Biochem. J.*, 258, 409, 1989.

134. Takemura, H., Hughes, A. R., Thastrup, O., and Putney, J. W., Jr., Activation of calcium entry by the tumor promotor thapsigargin in rat parotid acinar cells. Evidence that an intracellular calcium pool, and not an inositol phosphate, regulates calcium fluxes at the plasma membrane, *J. Biol. Chem.*, 264, 12266, 1989.

135. Takuma, T., Kuyatt, B. L., and Baum B. J., Calcium transport mechanisms in basolateral membrane-enriched vesicles from rat parotid gland, *Biochem. J.*, 227, 239, 1985.

136. Taylor, C. W., Merritt, J. E., Putney, J. W., Jr., and Rubin, R. P., Effects of Ca^{2+} on phosphoinositide breakdown in exocrine pancreas, *Biochem. J.*, 238, 765, 1986a.

137. Taylor, C. W., Merritt, J. E., Putney, J. W., Jr., and Rubin, R. P., A guanine nucleotide-dependent regulatory protein couples substance P receptors to phospholipase C in rat parotid gland, *Biochem. Biophys. Res. Commun.*, 136, 362, 1986b.

138. Thevenod, F. and Schulz, I., H^+-ion dependent calcium uptake into an inositol 1,4,5-trisphophate sensitive calcium pool from rat parotid gland, *Am. J. Physiol.*, 255, G429, 1988.

139. Thevenod, F., Dehlinger-Kremer, M., Kemmer, T. P., Christian, A. L., Potter, B. V. L., and Schulz, I., Characterization of inositol 1,4,5-trisphosphate sensitive and insensitive non-mitochondrial Ca^{2+} pools in rat pancreatic acinar cells, *J. Membr. Biol.*, 109, 173, 1989.

140. Trump, B. F., Berezesky, I. K., Laicho, K. U., Osarnio, A. R., Meyner, W. J., and Smith, M. W., The role of calcium in cell injury, *Scan. Electron Microsc.*, 2, 437, 1980.

141. Tsien, R. Y., Pozzan, T., and Rink, T. J., Calcium homeostasis in intact lymphocytes. Cytoplasmic free calcium monitored with a new intracellularly trapped fluorescent indicator, *J. Cell. Biol.*, 94, 325, 1984.

142. Volpe, P., Krause, K. H., Hashimoto, S., Zorzato, F., Pozzan, T., Meldolesi, J., and Lew, D. P., Calciosome, a cytoplasmic organelle. The inositol 1,4,5-trisphosphate sensitive Ca^{2+} store of non-muscle cells, *Proc. Natl. Acad. Sci. U.S.A.*, 85, 1091, 1988.

143. Weiss, E. R., Kelleher, D. J., Woon, C. W., Spoarkar, S., Osawa, S., Heasley, L. E., and Johnson, G. L., Receptor activation of G proteins, *FASEB J.*, 2, 2841, 1988.

144. Williamson, J. R. and Monck, J. R., Hormone effects on cellular Ca^{2+} fluxes, *Annu. Rev. Physiol.*, 51, 107, 1989.

145. Worley, P. F., Barban, J. M., Suppattapone, S., Wilson, V. S., and Snyder, S. H., Characterization of inositol trisphosphate receptor binding in brain regulation by pH and calcium, *J. Biol. Chem.*, 262, 12132, 1987.

146. Worley, P. F., Baraban, J. M., Colvin, J. S., and Snyder, S. H., Inositol trisphosphate receptor localization in the brain, *Nature*, 325, 159, 1987.

147. Young, J. A. and Van Lennep, E. W., Transport in salivary and salt glands, in *Membrane Transport in Biology*, Vol. 4A, Giebisch, G., Tosteson, D. C., and Ussing, N. H., Eds., Springer-Verlag, Berlin, 1979.

Stimulus-Exocytosis Coupling Mechanism in Salivary Gland Cells

David O. Quissell

INTRODUCTION

Three anatomically distinct pairs of major salivary glands (parotid, submandibular, and sublingual) and several different minor salivary glands are responsible for the production of saliva. Salivary glands are exocrine glands and the general secretory mechanisms by which salivary acinar cells secrete are quite similar to other exocrine cells. Unlike endocrine cells, exocrine cells are involved in two different but integrated cellular processes, the production of the primary secretory fluid (i.e., water and electrolyte transport), and the secretion of high molecular weight proteins and glycoproteins, exocytosis. Both cellular processes are controlled primarily by the autonomic nervous system. In general, parasympathetic stimulation leads to the production of saliva at a rather high flow rate and with lower organic content; whereas, sympathetic stimulation leads to a much slower rate of saliva production, but of a more viscous nature (Schneyer et al., 1972).

Neurotransmitters evoke salivary gland secretion by binding to specific basolateral surface receptors; through the generation of specific intracellular second messengers, the intracellular processes involved in the production of the primary secretory fluid and exocytosis are activated. In the past, the term stimulus-secretion coupling mechanism has been used to describe the actual events that appear to be involved in both the production of the primary secretory fluid and exocytosis. Since the two separate intracellular metabolic pathways are clearly distinct, although highly integrated, I have chosen the term stimulus-exocytosis coupling mechanism, since this chapter will focus on those cellular events which appear to play a more direct role in regulating exocytosis. The emphasis of this chapter will also be on those more distal intracellular events that may play a direct role in the initiation and modulation of exocytosis. The research discussed in this chapter will focus primarily on the rat submandibular and parotid glands since most of our current understanding of the actual cellular events involved in salivary gland exocytosis has come from studies on these two tissues.

Exocytosis is a cellular process by which cells are able to release synthesized materials to an external environment. The process of release of the stored material involves the fusion of the secretory granule which surrounds the stored material with the luminal plasma membrane. In mammalian exocrine cells, the secretory material which is stored in the secretory granule is released following an appropriate cellular signal, the individual secretory granules migrate to the luminal plasma membrane, fuse with the membrane, and the secretory materials are released. The process of exocytosis in these cells is defined as regulated secretion rather than a constitutive process since exocytosis is regulated in response to external stimuli rather than at a constant rate. However, the basis of salivary gland exocytosis and its regulation (stimulus-exocytosis coupling mechanism) still constitutes a poorly understood area of cell biology.

STIMULUS

The principle external stimuli for salivary gland secretion are the various neurotransmitters which are released by both the sympathetic and parasympathetic neural pathways. Thus, the neurotransmitters are the first messengers involved in the cellular regulation of exocytosis.

Several different types of neurotransmitters have been shown to have a role in the regulation of salivary gland secretion including acetylcholine, norepinephrine, and various neuropeptides including vasoactive intestinal polypeptide (VIP), substance P, calcitonin gene-related peptide (CGRP), somatostatin, and neuropeptide Y (Baum, 1987; Ekstrom, 1987; Ekstrom et al., 1988; Wright and Leubke, 1989). The muscarinic cholinergic receptors and the α- and β-adrenergic receptors are considered to have the primary role in regulating salivary gland secretion and these receptors have been studied in the greatest detail (Baum et al., 1983). In both the rat parotid and submandibular acinar cells, the functionally important adrenergic receptors appear to be of the α_1-subtype (Ito et al., 1982) and the functionally important β_1-subtype (Au et al., 1977; Pointon and Banarjee, 1979).

The predominate receptor involved in the direct stimulation of rat parotid and submandibular gland exocytosis is the β-adrenergic receptor. In the rat submandibular gland, β-adrenergic receptor stimulation appears to be a requirement for mucin secretion (Quissell and Barzen, 1980; Quissell et al., 1981;

McPherson and Dormer, 1984). Pure cholinergic or α-adrenergic receptor stimulation will not elicit significant mucin secretion. *In vitro* studies have indicated that cholinergic or α-adrenergic receptor stimulation may have elicited a response (Fleming et al., 1987), but the magnitude of the response was only slightly above basal release. Whereas in the parotid gland of the rat and mouse, a pure cholinergic or α-adrenergic receptor-mediated response will elicit significant amylase release (Butcher and Putney, 1980), albeit the rate of amylase secretion is significantly less than that observed following a pure β-adrenergic receptor-mediated response. In addition, combined receptor stimulation has been shown to augment both parotid amylase release (Templeton, 1980) and mucin release (Quissell and Barzen, 1980).

The activation of the β-adrenergic receptor-adenylate cyclase system results in a dramatic increase in intracellular cyclic AMP (Butcher and Putney, 1980; Quissell et al., 1981). The β-adrenergic receptor-adenylate cyclase system is composed of several protein subunits, including the receptor, the catalytic subunit of adenylate cyclase and the guanine nucleotide-binding regulatory proteins, G_s-stimulating and G_i-inhibitory proteins. Secretagogues which stimulate the production of cyclic AMP activate this receptor-adenylate cyclase system (Gilman, 1984). The cyclic AMP pathway appears to be the predominate intracellular regulatory pathway which regulates rat parotid and submandibular exocytosis.

As discussed further below, both the rat parotid and submandibular glands have a cyclic AMP dependent-secretory pathway and a calcium-dependent secretory pathway which can elicit an exocytotic response. The combined stimulation of both pathways greatly augments the rate of exocytosis in both tissues, as shown in Table 1 and Figure 1. Both *in vivo* (Jones and Wilson, 1985; Asking, 1985; Argent and Arkle, 1985; Katoh and Nishiyama, 1986) and *in vitro* (Watson et al., 1985; Suzuki and Ohshika, 1985; Takemura and Ohshika, 1985; Yoshimura et al., 1985; Poat et al., 1987) studies have clearly documented this augmentation. During parasympathetic stimulation when the sympathetic neural activity is increased (Asking, 1985) amylase secretion is also enhanced. Further, during pure sympathetic neural stimulation, since norepinephrine is an equipotent α-and β-adrenergic receptor agonist, exocytosis is stimulated at a maximal rate since both intracellular Ca^{2+} and cyclic AMP levels are increased.

In some of the minor salivary glands, the rat sublingual gland, and the rat parotid gland, parasympathetic stimulation can also elicit an exocytotic response. The intracellular response following muscarinic cholinergic receptor stimulation or an α-adrenergic receptor-mediated response is an increased intracellular Ca^{2+} level (Berridge, 1984). The activation of the muscarinic cholinergic receptors leads to the activation of a membrane-bound phospholipase C which upon activation leads to the hydrolysis of phosphatidylinositol-4,5-bisphosphate resulting in the production of inositol trisphosphate and diacylglycerol. The muscarinic cholinergic receptor is coupled to phospholipase C by another GTP-binding protein (G_s). Inositol trisphosphate has been shown

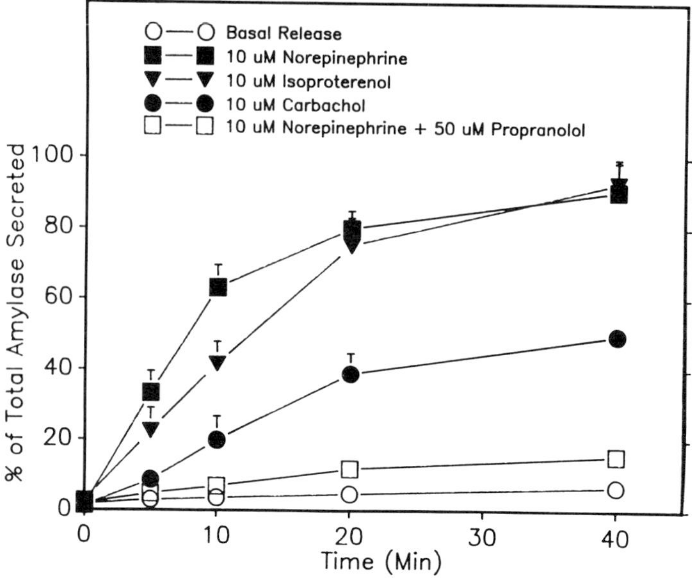

Figure 1. Effect of secretagogue stimulation on rat parotid amylase secretion. Values are the mean of at least four separate experiments and vertical bars represent ±S.E.M.

TABLE 1. Effect of Various Secretagogues on Rat Submandibular Mucin Secretion

Secretagogue	% Mucin Secretion[a]
Unstimulated	9.36
Carbamylcholine, $10^{-6}\,M$	9.96
Phenylephrine,[b] $5 \times 10^{-6}\,M$	8.87
Isoproterenol, $10^{-5}\,M$	24.60
Norepinephrine, $10^{-5}\,M$	47.00
Isoproterenol, 10^{-5}M and carbamylcholine, $10^{-5}\,M$	33.70
A23187, $10^{-6}\,M$	6.90
Bt_2cAMP, $10^{-3}\,M$	17.40
A23187, $10^{-3}\,M$ and Bt_2cAMP, $10^{-3}\,M$	31.10

Note: Values are means ±SD (<10%) of at least 4 separate experiments.

[a] Release of ^{14}C-radiolabeled trichloroacetic acid and phosphotungstic acid precipitable material after secretagogue stimulation. Extent of mucin secretion was normalized by expressing amount of radiolabeled material released as percent of total cellular precipitable ^{14}C disintegrations per minute present at the beginning of each study. (Adapted from Quissell et al. 1980, 1981.)

[b] (−)-Propranolol, 10^{-5}M, was included in the culture medium.

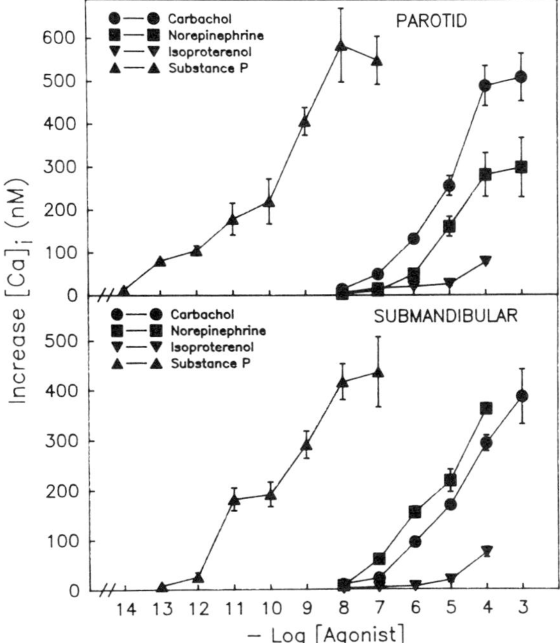

Figure 2. Calculated $[Ca^{2+}]_i$ of fura-2 loaded rat parotid and submandibular acinar cells. Intracellular $[Ca^{2}\pm]_i$ were calculated from the ratio of fluorescence (>470 nm emissions) at two excitation wavelengths, 340 and 380 nm, after correction for autofluorescence, light scatter, and fura-2 leakage. Values represent mean peak values obtained after stimulation of at least four separate experiments and vertical bars represent ±S.E.M.

to be the intracellular messenger responsible for the release of stored Ca^{2+} from the endoplasmic reticulum following receptor stimulation (Streb et al., 1983; Putney, 1986). As shown in Figure 2, many different types of secretagogues can elicit an increase in intracellular free Ca^{2+}.

Early studies had suggested that intracellular calcium may be the actual final mediator for exocytosis in salivary gland acinar cells (Putney et al., 1977), and a unifying hypothesis was set forth which indicated that the final effector of rat parotid exocytosis was calcium. This view of rat salivary gland exocytosis and its control is no longer satisfactory, for more recent studies have indicated that cyclic AMP appears to be able to mediate parotid amylase release without the elevation of cytosolic calcium (Takuma and Ichida, 1986). In permeabilized rat parotid cells, cyclic AMP significantly enhanced amylase secretion at all concentrations of free Ca^{2+} tested (Baldys-Waligorska et al., 1987). In the rat submandibular gland, the activation of the cyclic AMP pathway appears to be a prerequisite for significant mucin secretion.

The precise role of the neuropeptides in regulation of salivary secretion has yet to be determined. Substance P, substance K, VIP, and CGRP may play a more complementary role in that they help modulate or modify the response of the cells during neural stimulation (Ekstom, 1987). These neuropeptides also have the ability to change the rate of blood flow through the gland during neural stimulation. Therefore, some of the changes observed *in vivo* during neuropeptide infusion could be due to the modification of blood flow to the gland and the subsequent effects that restricted blood flow has on saliva production and composition.

EXOCYTOSIS

As in other secretory cells (Williams et al., 1989), salivary gland secretion involves the synthesis of the secretory products, packaging of the material into the secretory granules and the migration of the secretory granules to particular locations within the cell prior to release. Upon activation of the appropriate receptors and the generation of the appropriate intracellular messengers, exocytosis is initiated. The process of exocytosis is a highly specific cellular process with the secretory granule membranes fusing only to the plasma membrane of the apical surface. In 1985, Dr. Shara (Shara, 1985), presented morphological evidence which suggests that at the very onset of receptor stimulation, the initial response is not the migration of the secretory granules to the lumen but rather the apical, luminal plasma membrane develops numerous pits and tubular invaginations. Subsequently, these particular invaginations come in contact and fuse with those secretory granules close to the invaginations and the contents of the secretory granules are then discharged into the lumen. These data suggest that the luminal plasma membrane of the rat parotid acinar cells plays a significant and active role in the exocytotic process.

Unfortunately, very little is known regarding the actual molecular events involved in the fusogenic process. The precise molecular components and their regulation is not known. Morphological studies suggest that the cytoskeleton may be involved in the regulation of the secretory granule with the luminal membrane. Cytoskeletal units and actin-like filaments provide the structural elements necessary for secretory granule migration. At the same time, microfilament dissolution would have to occur to permit the juxtaposition of the secretory granule next to the luminal membrane prior to membrane fusion.

Model systems have been developed to study the exocytotic process. These systems include virus-induced cell fusion studies (Spear, 1987), *in vitro* reconstitution studies (Creutz et al., 1987), and liposome fusion to cover a few. So far no unifying mechanism has become apparent. Most studies would suggest that both cytosolic and membrane components play a major role during the actual fusogenic process (Creutz et al., 1987). In some cells, these events may be under regulation of Ca^{2+}. Both the secretory membranes and plasma membranes appear to play an important role during membrane fusion and specific lipids may also play an important regulatory role (Roos et al., 1987). Part of the difficulty with all of these various model systems is the limitation of each model system and its application to the specific exocytosis events which occurs in the intact cell. Clearly, this is an important area of cell biology research and it remains an important challenge for future research to determine the precise molecular events and participants involved in the process of exocytosis.

SIGNAL TRANSDUCTION

Cyclic AMP Pathway

The mechanism by which β-adrenergic receptor stimulation leads to exocytosis by the intracellular increase in cyclic AMP is not well understood. Cyclic AMP is thought to mediate most, if not all, of its effects by the activation of a cyclic AMP-dependent protein kinase (cA-PK). Concomitant with an increase in intracellular cyclic AMP levels following β-adrenergic receptor stimulation, rat parotid and submandibular cA-PK activity also increases, with the activation of cA-PK remaining elevated throughout the entire secretory period (Baum et al., 1981a; Spearman and Butcher, 1982; Quissell et al., 1983a). The extent of activation of cA-PK does not always correlate with the intracellular cyclic AMP levels. At the very onset of receptor stimulation, the intracellular cyclic AMP level increases dramatically. This rapid and excessive response or overshoot results in a rapid cellular response. With time after stimulation, the level of intracellular cyclic AMP can drop significantly, while the overall extent of cA-PK activation remains fairly constant (Quissell et al., 1981). Thus, when evaluating a receptor-mediated response and comparing such a response to intracellular changes in response to the stimulus, the extent of cA-PK activation is a more appropriate indicator of the cyclic AMP-dependent pathway than intracellular cyclic AMP levels. The extent of cA-PK activation and its dose-response relationship also correlates quite closely with the dose-response relationship for the rate of cellular exocytosis in both the rat parotid and submandibular glands (Baum et al., 1981b; Quissell et al., 1983a).

Two different isozymes of cA-PK have been identified and characterized in salivary glands. Both cellular types are activated by the binding of cyclic AMP to the regulatory subunits which results in the dissociation of the holoenzyme to a regulatory subunit-cyclic AMP complex and an active catalytic subunit. The free catalytic subunit is then available to phosphorylate various protein substrates. Recent studies have indicated that the subcellular distribution of cA-PK in the rat parotid and submandibular are quite similar. A cytosolic form of type I and type II cA-PK and a membrane bound of type I kinase have been demonstrated. (Baum et al., 1981a; Quissell et al., 1988, 1989b). Despite the fact that type II form of cA-PK comprises at least 65% of the total glandular activity, recent studies indicate that following β-adrenergic receptor stimulation, both the cytosolic and membrane-bound forms of type I cA-PK were activated to a far greater extent than the type II form of cA-PK (Quissell et al., 1988, 1989b). These data suggest that the type I form may play a more significant physiological role during exocytosis especially during submaximal receptor stimulation. However, given the large proportion of type II present in the cytosol, only a small increase in the activation of this form would be necessary to provide an equivalent level of active catalytic subunit. Clearly, further studies are required to more fully evaluate the importance of each kinase in the exocytolic process and to more fully understand the reason for their subcellular distribution.

Since cA-PK activation has been implicated as a major metabolic pathway in the stimulus-exocytosis coupling mechanism for both the rat parotid and submandibular glands, protein phosphorylation and/or dephosphorylation may be one of the actual molecular events by which exocytosis is regulated. Studies covering the last decade support such a possible mechanism but the problem that remains is to clearly identify and document the phosphoprotein(s) that may be directly involved in the exocytotic pathway. Stimulation of β-adrenergic receptors in rat parotid and submandibular acinar cells leads to the phosphorylation and dephosphorylation of several proteins (Jahn et al., 1980; Jahn and Söling, 1981; Baum et al., 1981b; Freedman and Jamieson, 1982a, 1982b; Quissell et al., 1983b; Spearman et al., 1984). Of these various phosphoproteins, three proteins have been identified most consistently in the rat parotid and submandibular acinar cells; 36-34 kDa (protein I), 26-24 kDa (protein II) and 22 kDa (protein III) phosphoproteins. Protein I has been identified as ribosomal protein S6 (Freedman and Jamieson, 1982b; Jahn and Söling, 1983; D.O. Quissell, unpublished data), whereas the function of the other two proteins is still quite obscure.

Protein III, from the rat parotid gland, has been purified and characterized (Thiel et al., 1988). Subcellular fractionation studies indicate that protein III is located in the endoplasmic reticulum. Because of its subcellular location (Thiel et al., 1988) and its slow rate of phosphorylation (Quissell et al., 1985) and dephosphorylation, the particular membrane-bound protein would not appear to play a direct role in the stimulus-exocytosis coupling process. However, it may still play an important physiological role during secretion, for it may be involved in the regulation of calcium sequestration by the endoplasmic reticulum during the secretory response (Plewe et al., 1984). Protein III has only one phosphorylation site, a phosphoserine, and *in vitro* studies indicate that this particular site can be phosphorylated by cA-PK. In addition, further *in vitro* studies indicate that this particular sequence, when made as a synthetic peptide, can also be phosphorylated by the catalytic subunit of cA-PK.

Of the various phosphoproteins, only the 26-kDa phosphoprotein (protein II) appears to display the appropriate rate of phosphate turnover which would be compatible with a direct role in regulating rat parotid and submandibular exocytosis (Quissell et al., 1985, 1987). This kinetic analysis is based on the assumption that since exocytosis begins rapidly after receptor stimulation, and since exocytosis stops rather abruptly following receptor blockade, those regu-

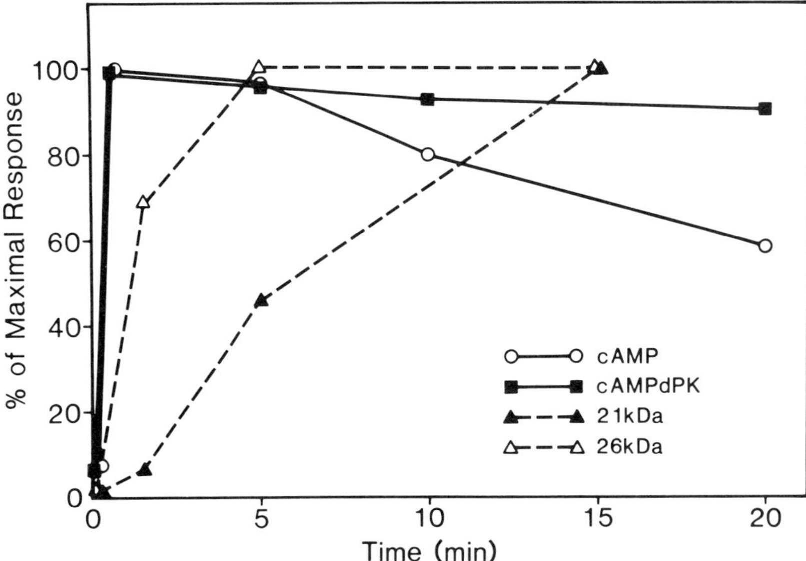

Figure 3. Time course of the effect of β-adrenergic receptor stimulation on rat submandibular cyclic AMP levels, on cellular cA-PK activity, and on endogenous protein phosphorylation of two β-adrenergic receptor-dependent, integral membrane phosphoproteins. (Adapted from Quissell et al., 1983a, 1983b, 1985. With permission.)

latory phosphoproteins involved directly in the secretory response should also demonstrate a rather rapid rate of phosphorylation and dephosphorylation. The phosphorylation and dephosphorylation rates for 21 kDa (protein III) and 36 kDa (protein I) were much slower ($t_{1/2} > 20$ min) (Figure 3). In addition, the extent of phosphorylation of the 26 kDa closely correlated with the dose-response relationship for β-adrenergic receptor-mediated secretion and with the dose-response relationship for cA-PK activation (Quissell et al., 1983a, 1985) (Figure 4).

Protein II has recently been purified from both rat parotid and submandibular acinar cells (D. O. Quissell, unpublished data) and, as yet, it is still unclear what its precise role is during exocytosis. The protein is an intrinsic membrane protein (Quissell et al., 1985) and the acceptor amino acid for phosphorylation is serine (Quissell et al., 1987). *In vitro* phosphorylation studies indicate that the 26-kDa phosphoprotein is an excellent substrate for cA-PK. The amino acid composition of the 2 26-kDa phosphoproteins are quite similar (Table 2), but recent N-terminal sequence analysis would suggest that they may not be identical or the same gene product (Table 3). However, given their general physical and chemical characteristics, their amino acid composition, and N-terminus sequence, they would appear to have a high degree of homology.

A recent report (Takuma and Ichida, 1988) suggests that cA-PK activation and cA-PK-dependent protein phosphorylation may not be directly involved in the exocytosis of rat parotid amylase. Using different types of cA-PK inhibitors

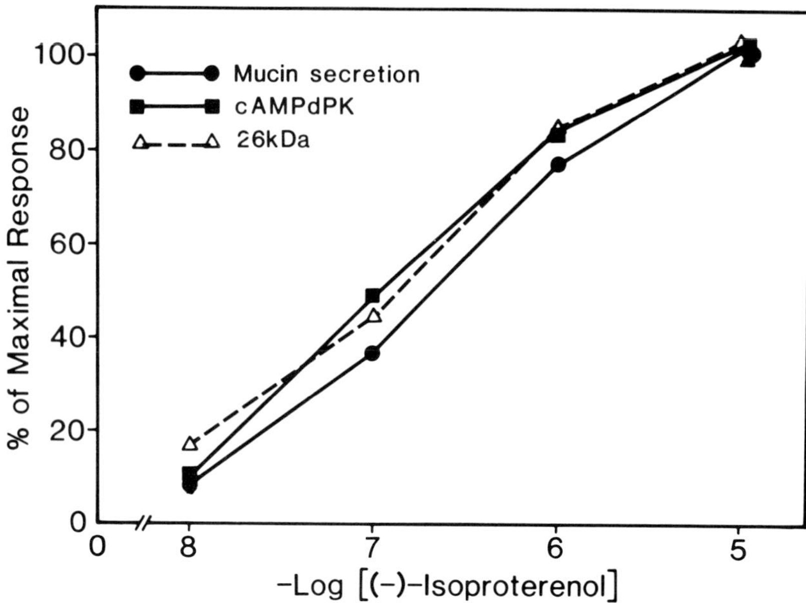

Figure 4. Concentration dependence of β-adrenergic receptor stimulation on mucin secretion, cellular cA-PK activity, and on endogenous protein phosphorylation of 26-kDa integral membrane phosphoprotein. (Adapted from Quissell et al., 1983a, 1983b, 1985. With permission.)

TABLE 2. Amino Acid Composition of 26-kDa Phosphoprotein

	Number of Amino Acids in Protein	
Amino Acid	Submandibular	Parotid
Asp	18 ± 1.6	16 ± 0.7
Glu	28 ± 0.3	28 ± 0.2
Ser	24 ± 0.4	25 ± 0.8
Gly	42 ± 0.4	41 ± 1.6
His	3 ± 0.1	3 ± 0.2
Arg	9 ± 0.3	10 ± 0.4
Thr	11 ± 0.0	12 ± 0.8
Ala	14 ± 0.3	18 ± 1.9
Pro	8 ± 0.1	9 ± 0.5
Tyr	3 ± 0.1	3 ± 0.1
Val	12 ± 0.0	15 ± 1.0
Met	1 ± 0.1	1 ± 0.2
Ile	8 ± 0.0	9 ± 0.4
Leu	18 ± 0.0	12 ± 0.2
Phe	8 ± 0.0	9 ± 0.3
Lys	9 ± 0.4	6 ± 0.2

Note: Values are mean ± S.E.M. of three separate determinations.

TABLE 3. N-Terminal Amino Acid Sequence of 26-kDa Phosphoprotein

Residue No.	Submandibular	Parotid
1	Met	Met
2	Ile	Ile
3	Ile	Ile
4	Tyr	Tyr
5	Arg	(Val)(Arg)
6	Asp	(Ala)(Asp)
7	Leu	Leu
8	Ile	Ile
9	Ser	Ser
10	His	His

Note: Obtained from three separate determinations.

and saponin-permeablized parotid cells, two of the inhibitors had only a slight effect on amylase secretion but greatly decreased the extent of phosphorylation of both the 21- and 26-kDa phosphoproteins. Another more recent report suggests that increasing levels of cyclic AMP may not be directly involved either since, when cyclic AMP phosphodiesterase was introduced into swollen submandibular acini, maximal β-adrenergic receptor stimulation could lead to enhanced mucin secretion in the absence of a significant increase in intracellular cyclic AMP (Bradbury et al., 1989). These recent reports raise a rather provocative question: if cA-PK directed phosphorylation is not directly involved, and if increased intracellular cyclic AMP is not directly involved, then what is the process by which the cyclic AMP pathway controls exocytosis, since the only other known component in the pathway is the regulatory subunit of cA-PK, and this cannot be disassociated unless cyclic AMP is bound to it? Clearly, further studies are required to resolve this issue.

In various biological systems, it is as important to terminate a biological response as it is to initiate it. Unfortunately, very little information is available regarding the identification and characterization of the various serine-specific protein phosphatases in salivary gland tissue. In the guinea-pig parotid gland, protein phosphatases of the type 1, 2A, and 2B have been identified in the cytosolic fraction with most of type 1 and 2A in the inactive state. A microsomal-associated protein phosphatase has also been identified (Mieskes and Söling, 1987) but the physiological role of each of the phosphatases in the cA-PK dependent protein phosphorylation events and their control have yet to be determined.

Calcium Pathway

The importance of calcium in the exocytotic process of the salivary glands was first indicated by the studies of Douglas and Poisner (1963). However, the precise role of calcium in the regulation of secretion is not known nor is its role in exocytosis fully understood. Recent studies have documented a calcium-dependent conductance of various ion channels as it relates to the role of calcium in regulating primary secretory fluid secretion (Cook and Young, 1989), however, very little information is currently available regarding its direct role in regulating salivary gland exocytosis. Rat parotid and submandibular acinar cells pretreated with ethylene glycol tetracetic acid to deplete their intracellular stores of calcium were greatly inhibited in their ability to secrete amylase (Butcher and Putney, 1980) or mucins following adrenergic receptor stimulation (Quissell and Barzen, 1980).

Cholinergic, α-adrenergic, and substance P receptor stimulation activates a number of calcium-dependent, intracellular responses that can lead to fluid or macromolecular secretion. In general, the evidence for these calcium-dependent processes comes from several different types of experimental approaches. Removal of extracellular calcium or the depletion of intracellular calcium stores greatly diminish or prevents fluid secretion or exocytosis. In the absence of receptor stimulation, the calcium ionophore, A23187, mimics the receptor-mediated response, and this mimic response is calcium dependent. Increases in intracellular free calcium levels following receptor stimulation also correlate with the secretory response.

In recent years, a considerable amount of information has been developed relating to the agonist-activated hydrolysis of inositol phospholipids and their subsequent role in regulating intracellular free calcium (Putney, 1986). The focus of this discussion will be on those areas in which calcium may act as a direct effector of salivary gland exocytosis. Although the mechanism of calcium entry into the salivary gland and the release of calcium from intracellular stores is actively being determined (Hughes et al., 1988), these same inositol phospholipids may have other functions in the overall sequence of events during exocytosis.

The regulation of intracellular free calcium concentrations by the parotid and submandibular acinar cells before, during, and after stimulation has not been fully characterized. Several subcellular organelles have demonstrated an ability to bind or concentrate calcium (Nyjar and Pritchard, 1971; Perec and Alonso, 1971; Selinger et al., 1970). The microsomal fraction isolated from these tissues can accumulate calcium in an ATP-dependent manner (Terman and Gunter, 1983; Watson and Siegel, 1979; Kanagasuntheram and Randle, 1976; Kanagasuntheram and Teo, 1982a; Immelmann and Söling, 1983; Hurley and Martinez, 1985, 1986), Chlorpromazine, trifluoperazine and 4,4'-di-isothiocyanostilbene-2,2'-disulfonate all inhibit calcium accumulation, but the mitochondrial transport inhibitors, ruthenium red, and carbonyl cyanide m-chlorophenylhydrazone, had no effect on calcium transport. Marker enzyme analysis suggested that a specialized region of the endoplasmic reticulum may be involved.

The actual steady-state level of free intracellular calcium levels is modulated by the release or uptake of calcium from various intracellular stores such as the endoplasmic reticulum and mitochondria, the influx of calcium from the external milieu through various plasma membrane channels, and the removal of calcium via the plasma membrane Ca^{2+}-ATPase (Merrit and Rink, 1987). Cholinergic, α-adrenergic, or substance P receptor stimulation leads to an increase in intracellular free calcium (Figure 2). The increase in free calcium is due to the release of calcium from the endoplasmic reticulum and from calcium uptake (Merrit and Rink, 1987). In recent years, a significant amount of data has developed that indicates that this particular stimulus-response coupling pathway involves inositol-1,4,5-trisphosphate (Hughes et al., 1988; Bonis et al., 1986; Henne et al., 1987). Although it has not been demonstrated in all salivary gland tissues, this particular pathway would appear to be a ubiquitous second messenger.

β-Adrenergic receptor stimulation does not appear to increase intracellular free calcium concentration at physiological levels of stimulation (Takemura, 1985; Takuma and Ichida, 1986a; Dreux et al., 1986) nor does β-adrenergic receptor stimulation facilitate calcium efflux or influx. Calcium efflux is observed only after prolonged stimulation (McPherson and Dormer, 1984; Butcher, 1980; Dreux et al., 1986).

Unlike the cyclic AMP pathway, the major problem in determining the exact role of calcium in mediating cellular exocytosis is due to the complexity of the various intracellular calcium-dependent processes. Calcium can mediate its effects through a vast number of different cellular events with some or all having a potential direct role in exocytosis. Calcium can directly bind to a large number of regulatory proteins such as calmodulin and calcimedins. It can act through a series of calcium-dependent protein kinases such as the calmodulin-dependent protein kinase and protein kinase C. It can activate or inhibit other important enzyme systems which can then lead to production of other intracellular messengers such as adenylate cyclase, phospholipases, etc. It can also stimulate phosphodiesterases that, in turn, can modulate cellular cyclic AMP levels. All of these possibilities need to be further explored before a more complete understanding of the calcium-mediated events can be obtained.

Calmodulin is widely disturbed in nature. It has been shown to be present in several different mammalian secretory tissues, including rat parotid and submandibular cells (Kanagasuntheram and Teo, 1982b; Spearman and Butcher, 1983; Arkle et al., 1986; Tojyo et al., 1987; Singh et al., 1986). Recent studies have implicated a role for calmodulin and calmodulin binding proteins in rat salivary gland exocytosis (Singh et al., 1986; Arkle et al., 1986; Toyoshima and Tandler, 1986; Piascik et al., 1986; Tojyo et al., 1987). Calmodulin clearly plays an important role in salivary gland calcium homeostatsis, and a plasma membrane calmodulin-dependent ATPase has been demonstrated. However, its precise direct role in cellular exocytosis is not understood, for it is unclear from the various studies using calmodulin antagonists if the effects observed on exocytosis are directly due to inhibition of the calmodulin-dependent secretory events or are secondary due to changes in intracellular calcium metabolism and/or nonspecific effects of these highly hydrophobic agents.

Just as protein phosphorylation may play an important role in mediating cellular exocytosis of the cyclic AMP-dependent pathway, recent studies indicate that phosphorylation may play an important role during the calcium-dependent, phospholipid-dependent protein kinase (protein kinase C) activation that occurs during receptor stimulation due to a simultaneous increase in intracellular calcium and diacylglycerol. Phorbol esters, potent activators of protein kinase C, have been shown to induce rat parotid amylase and submandibular mucin secretion (Putney et al., 1984; Takuma and Ichida, 1986b; Fleming et al., 1986). The stimulation is not calcium dependent, but when cellular calcium levels are increased via calcium ionophore, A23187, the secretory rate is augmented. Stimulation of guinea-pig parotid cells leads to translocation of protein kinase C from the cytosolic to the membrane fraction (Domenech and Söling, 1987). Phosphorylation of the ribosomal protein S6 during receptor-mediated exocytosis has been shown to be catalyzed by protein kinase C (Padel and Söling, 1985). Protein kinase C activity has been associated with rat parotid secretory granules, and two protein substrates have been identified (Dowd et al., 1987). These data suggest that cytosolic protein kinase C may phosphorylate specific granule membrane proteins, but it remains to be seen whether these phosphorylation events directly contribute to the regulation of rat parotid exocytosis. A role for protein kinase C in the exocytotic process has not been demonstrated under all experimental conditions. Using dispersed rat submandibular acinar-intercalated duct cells, we were unable to demonstrate a role for protein kinase C during exocytosis. Phorbol esters were unable to elicit a secretory response, nor were they able to augment a response using the appropriate agonists. Endogenous protein phosphorylation was unaffected by the phorbol esters, and we were unable to observe any ribosomal S6 phosphorylation (Quissell et al., 1989a). Thus, the role of this particular kinase in the exocytotic pathway may be very system dependent.

Speculation

Rat salivary gland exocytosis is regulated by two highly integrated, but separate, metabolic pathways, with the cyclic AMP dependent pathway being the most predominant one. The current hypothesis (Figure 5), based on the above-mentioned studies, suggests that there may exist within the rat parotid and submandibular acinar cell membrane fraction a "stimulus-coupled regulatory complex for exocytosis" in which the 26-kDa phosphoprotein is a subunit. The physiological function of the 26-kDa phosphoprotein is to transfer to the enzyme complex the cyclic AMP-mediated signal, and thus the 26-kDa protein is the actual regulatory site for cyclic AMP mediated exocytosis, with the extent of phosphorylation of the 26-kDa protein determining the rate of exocytosis (Quissell et al., 1985, 1987). In the rat parotid, since changes in intracellular calcium levels can also mediate an exocytotic process, there may be an additional protein present in the secretory complex, the role of which is to transfer to the

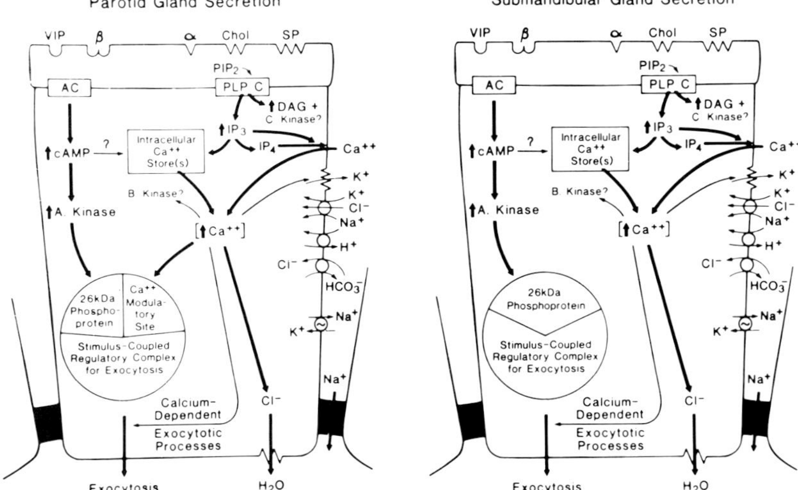

Figure 5. Current hypothesis summarizing the intracellular events that appear to be directly involved during rat parotid and submandibular exocytosis. VIP, vasocative intestinal polypeptide receptor; β, beta-adrenergic receptor; α, alpha-adrenergic receptor; Chol, cholinergic receptor; SP, substance P receptor; AC, adenylate cyclase; PIP, phosphatidylinositol 4-phosphate; PLPC, phospholipase C; DAG, diacylglycerol; IP_3, inositol 1,4,5-trisphosphate; IP_4, inositol 1,3,4,5-tetrakisphosphate.

enzyme complex the calcium-mediated signal. This protein could be calmodulin, another calcium binding protein, or a phosphoprotein.

Clearly, the idea of a secretory complex is a highly speculative one, but when one compares the complexity of the actual events involved in exocytosis, the requirements for strict metabolic control over such a process and the need for specificity as it relates to the entire process, one could easily envision that several points of metabolic control in the secretory pathway must exist. It is not unlikely that when the entire secretory pathway has been delineated, there will be several key sites or steps in which an enzyme complex, or similar protein complex, resides. Each of these steps will contain both a secretory function and regulatory function, co-residing within the same complex.

REFERENCES

1. Argent, B. E. and Arkle, S., Mechanism of action of extracellular calcium on isoprenaline-evoked amylase secretion from isolated rat parotid glands, *J. Physiol. (London)*, 369, 337, 1985.
2. Arkle, S., Pickford, P. D., Schofield, P. S., Ward, C., and Argent, B. E., Mechanism of the inhibitory effect of trifluoperazine on isoprenaline-evoked amylase secretion from isolated rat parotid glands, *Biochem. Pharmacol.*, 35, 4121, 1986.
3. Asking, B., Sympathetic stimulation of amylase secretion during a parasympathetic background activity in the rat parotid gland, *Acta Physiol. Scand.*, 124, 535, 1985.

4. Au, D. K., Malbon, C. C., and Butcher, F. R., Identification and characterization of β_1-adrenergic receptors in rat parotid membranes, *Biochim. Biophys. Acta*, 500, 361, 1977.

5. Baldys-Waglegorska, A., Pour, A., Moriarty, C. M., and Dowd, F., The effect of calcium and cyclic AMP on amylase release in digitonin-permeabilized parotid gland cells, *Biochim. Biophys. Acta*, 929, 190, 1987.

6. Baum, B. J., Neurotransmitter control of secretion, *J. Dent. Res.*, 66, 628, 1987.

7. Baum, B. J., Colpo, F. T., and Filburn, C. R., Characterization and relationship to exocrine secretion of rat parotid gland cyclic AMP-dependent protein kinase, *Arch. Oral Biol.*, 26, 333, 1981.

8. Baum, B. J., Frieberg, J. M., Ito, H., Roth, G. S., and Filburn, C. R., β-Adrenergic regulation of protein phosphorylation and its relationship to exocrine secretion in dispersed rat parotid gland acinar cell, *J. Biol. Chem.*, 256, 9731, 1981.

9. Baum, B. J., Ito, H., and Roth, G. S., *Adrenoceptors and Catecholamines Action—Part B*, Kunos, G., Ed., John Wiley & Sons, New York, 1983, 265.

10. Berridge, M. J., The molecular basis of communication within the cell, *Sci. Am.*, 253, 142, 1985.

11. Bonis, D., Giraird, F., and Rossignol, B., Inositol triphosphate induced Ca^{2+} release from rat parotid subcellular fractions, *Biol. Cell.*, 57, 71, 1986.

12. Bradbury, N. A., Dormer, R. L., and McPherson, M. A., Introduction of cyclic AMP phosphodiesterase into rat submandibular acini prevents isoproterenol-stimulated cyclic AMP rise without affecting mucin secretion, *Biochem. Biophys. Res. Commun.*, 161, 661, 1989.

13. Butcher, F. R., Regulation of calcium afflux from isolated rat parotid cells, *Biochim. Biophys. Acta*, 630, 254, 1980.

14. Butcher, F. R. and Putney, J. W. Jr., Regulation of parotid gland function by cyclic nucleotides and calcium, *Adv. Cyclic Nucleotide Res.*, 13, 215, 1980.

15. Cook, D. I. and Young, J. A., *Handbook of Physiology*, Section 6, Vol. 3, Forte, J. G., Ed., Oxford University Press, New York, 1989, 1.

16. Creutz, C. E., Zaks, W. J., Hamman, H. C., and Martin, W. H., *Cell Fusion*, Sowers, A. E., Ed., Plenum Press, New York, 1987, 45.

17. Domenech, E. M. and Söling, H. D., Effects of stimulation of muscarinic and of beta-catecholamine receptors on the intracellular distribution of protein kinase C in guinea pig exocrine glands, *Biochem. J.*, 242, 749, 1987.

18. Douglas, W. W. and Poisner, A. M., The influence of calcium on the secretory response of the submaxillary gland to acetylcholine or to noradrenaline, *J. Physiol.*, 165, 528, 1963.

19. Dowd, E., Watson, E. L., Lau, Y. S., Justin, J., Pasienuik, J., and Jacobson, K. L., Calcium-dependent protein kinase reactions associated with parotid gland secretory granule membranes, *J. Dent. Res.*, 66, 557, 1987.

20. Dreux, C., Imhoff, V., Huleux, C., Busson, S., and Rossignol, B., Forskolin, a tool for rat parotid secretion studies: ^{45}Ca afflux is not related to cAMP, *Am. J. Physiol.*, 251, C754, 1986.

21. Ekstrom, J., Neuropeptides and secretion, *J. Dent. Res.*, 66, 524, 1987.

22. Ekstrom, J., Ekman, R., Hakanson, R., Luts, A., Sundler, F., and Tobin, G., Effects of capsaicin pretreatment on neuropeptides and salivary secretion of rat parotid glands, *Br. J. Pharmacol.*, 97, 1031, 1989.

23. Fleming, N., Bilan, P.T., and Sliwinski-Lis, E., Effects of a phorbol ester and diacylglycerols on secretion of mucin and arginine esterase by rat submandibular gland cells, *Pfluegers Arch.*, 406, 6, 1986.

24. Fleming, N., Bilan, P. T., Sliwinski-Lis, E., and Carvalho, V., Muscarinic, α_1-adrenergic and peptidergic agonists stimulate phosphoinositide hydrolysis and regulate mucin secretion in rat submandibular gland cells, *Pfluegers Arch.*, 409, 416, 1987.

25. Freedman, S. D. and Jamieson, J. D., Hormone-induced protein phosphorylation. I. Relationship between secretagogue action and endogenous protein phosphorylation in intact cells from the exocrine pancreas and parotid, *J. Cell Biol.*, 95, 903, 1982a.

26. Freedman, S. D. and Jamieson, J. D., Hormone-induced protein phosphorylation. II. Localization to the ribosomal fraction from rat exocrine pancreas and parotid of a 29,000-dalton protein phosphorylated *in situ* in response to secretagogues, *J. Cell Biol.*, 95, 909, 1982b.

27. Gilman, A. G., Guanine nucleotide-binding regulatory proteins and dual control of adenylate cyclase, *J. Clin. Invest.*, 73, 1, 1984.

28. Hand, A. R., *The Salivary System*, Sreebny, L, M., Ed., CRC Press, Boca Raton, FL, 1987, 43.

29. Henne, V., Piiper, A., and Söling, H. D., Inositol 1,4,5-triphosphate and 5′-GTP induce calcium release from different intracellular pools, *FEBS Lett.*, 218, 153, 1987.

30. Hughes, A. R., Takemura, H., and Putney, J. W., Jr., Kinetics of inositol 1,4,5-triphosphate and inositol cyclic 1:2,4,5-triphosphate metabolism in intact rat parotid acinar cells, *J. Biol. Chem.*, 263, 10314, 1988.

31. Hurley, T. W. and Martinez, J. R., Characterization of the kinetic and regulatory properties of high-affinity Ca^{2+}-ATPase activity in acinar preparation of rat submandibular salivary glands, *Arch. Oral Biol.*, 30, 587, 1985.

32. Hurley, T. W. and Martinez, J. R., Characterization and localization of two forms of active Ca^{2+} transport in vesicles derived from rat submandibular glands, *Cell Calcium*, 7, 49, 1986.

33. Immelmann, A. and Söling, H. D., ATP-dependent calcium sequestration and calcium/ATP stoichiometry in isolated microsomes from guinea pig parotid glands, *FEBS Lett.*, 162, 406, 1983.

34. Ito, H., Hoopes, M. T., Baum, B. J., and Roth, G. S., K^+-release from rat parotid cells: an β_1-adrenergic mediated event, *Biochem. Pharmacol.*, 31, 567, 1982.

35. Jahn, R. and Söling, H. D., Phosphorylation of the same specific protein during amylase release evoked by β-adrenergic or cholinergic agonists in rat and mouse parotid glands, *Proc. Natl. Acad. Sci. U.S.A.*, 78, 6903, 1981.

36. Jahn, R. and Söling, H. D., Phosphorylation of the ribosomal protein S6 in response to secretagogues in the guinea pig exocrine pancreas, parotid and lacrimal gland, *FEBS Lett.*, 153, 71, 1983.

37. Jahn, R., Unger, C., and Söling, H. D., Specific protein phosphorylation during stimulation of amylase secretion by beta-agonists or dibutyryl adenosine 3′5′-monophosphate in the rat parotid gland, *Eur. J. Biochem.*, 112, 345, 1980.

38. Jones, C. J. and Wilson, S. M., The effect of autonomic agonists and nerve stimulation on protein secretion from the rat submandibular gland, *J. Physiol. (London)*, 358, 65, 1985.

39. Kanagasuntheram, P. and Randle, P. J., Calcium metabolism and amylase release in rat parotid acinar cells, *Biochem. J.*, 160, 547, 1976.

40. Kanagasuntheram, P. and Teo, T. S., Parotid microsomal Ca^{2+} transport. Subcellular localization and characterization, *Biochem. J.*, 208, 789, 1982a.

41. Kanagasuntheram, P. and Teo T. S., Calmodulin-sensitive ATP-dependent calcium transport by the rat parotid endoplasmic reticulum, *FEBS Lett.*, 141, 233, 1982b.

42. Katoh, K. and Nishiyama, A., Effects of cholinergic and adrenergic nerve stimulation on amylase release from superfused small segment of rat parotid gland, *Jpn. J. Physiol.*, 36, 305, 1986.

43. McPherson, M. A. and Dormer, R. L., Mucin release and calcium fluxes in isolated rat submandibular acini, *Biochem. J.*, 224, 473, 1984.

44. Merrit, J. E. and Rink, T. J., Regulation of cytosolic free calcium in fura-2 loaded rat parotid acinar cells, *J. Biol. Chem.*, 262, 17362, 1987.

45. Mieskes, G. and Söling, H. D., Protein phosphatases of the guinea-pig parotid gland, *Eur. J. Biochem.*, 167, 377, 1987.

46. Nyjar, M. A. and Pritchard, E. T., Calcium binding by a plasma membrane fraction isolated from rat submandibular gland, *Biochim. Biophys. Acta*, 323, 391, 1971.

47. Padel, U. and Söling, H. D., Phosphorylation of the ribosomal protein S6 during agonist-induced exocytosis in exocrine glands is catalyzed by calcium-phospholipid-dependent protein kinase C), *Eur. J. Biochem.*, 151, 1, 1985.

48. Perec, C. J. and Alonso, G. J., The effect of denervation on ATP dependent calcium uptake by microsomes of the submaxillary gland of rats, *Experientia*, 27, 897, 1971.

49. Piascik, M. T., Babich, M., Jacobson, K. L., and Watson E. L., Calmodulin activation and calcium regulation of parotid gland adenylate cyclase, *Am. J. Physiol.*, 250, C642, 1986.

50. Plewe, G., Jahn, R., Immelmann, A., Bode, C., and Söling, H. D., Specific phosphorylation of a protein in calcium accumulating endoplasmic reticulum from rat parotid glands following stimulation by agonists involving cAMP as second messenger, *FEBS Lett.*, 166, 96, 1984.

51. Poat, J. A., Cripps, H. E., Cowburn, R., and Iversen, L. L., Synergistic interactions between forskolin, isoprenaline and substance P as secretagogues in rat parotid glands, *Eur. J. Pharmacol.*, 144, 317, 1987.

52. Pointon, S. E. and Banerjee, S. P., Alpha-and beta-adrenergic receptors of the rat salivary gland, *Biochim. Biophys. Acta*, 584, 231, 1979.

53. Putney, J. W., Jr., Identification of cellular activation mechanisms associated with salivary secretion, *Annu. Rev. Physiol.*, 48, 75, 1986.

54. Putney, J. W., Jr., Formation and actions of calcium-mobilizing messenger, inositol 1,4,5-triphosphate, *Am. J. Physiol.*, 252, G149, 1987.

55. Putney, J. W., Jr., McKinney, J. S., Aub, D. L., and Leslie, B. A., Phorbol ester-induced protein secretion in parotid gland. Relationship to the role of inositol lipid breakdown and protein kinase C activation in stimulus-secretion coupling, *Mol. Pharmacol.*, 26, 261, 1984.

56. Putney, J. W., Jr., Weiss, S. J., Leslie, B. A., and Marier, S. H., Is calcium the final mediator of exocytosis in the rat parotid gland?, *J. Pharmacol. Exp. Ther.*, 203, 144, 1977.

57. Quissell, D. O. and Barzen, K. A., Secretory response of dispersed rat submandibular cells. II. Mucin secretion, *Am. J. Physiol.*, 238, C99, 1980.

58. Quissell, D. O., Barzen, K. A., and Deisher, L. M., Evidence against a direct role for protein kinase C in rat submandibular mucin secretion, *Arch. Oral Biol.*, 34, 695, 1989a.

59. Quissell, D. O., Barzen, K. A., and Deisher, L. M., Role of cyclic AMP-dependent protein kinase activation in regulating rat submandibular mucin secretion, *Biochim. Biophys. Acta*, 762, 215, 1983a.

60. Quissell, D. O., Barzen, K. A., and Lafferty, J. L., Role of calcium and cAMP in the regulation of rat submandibular mucin secretion, *Am. J. Physiol.*, 241, C76, 1981.

61. Quissell, D. O., Deisher, L. M., and Barzen, K. A., Role of protein phosphorylation in regulating rat submandibular mucin secretion, *Am. J. Physiol.*, 245, G44, 1983b.

62. Quissell, D. O., Deisher, L. M., and Barzen, K. A., The rate-determining step in cAMP-mediated exocytosis in the rat parotid and submandibular glands appears to involve analogous 26-kDa integral membrane phosphoproteins, *Proc. Natl. Acad. Sci. U.S.A.*, 82, 3237, 1985.

63. Quissell, D. O., Deisher, L. M., and Barzen, K. A., Role of protein phosphorylation in rat salivary gland exocytosis, *J. Dent. Res.*, 66, 596, 1987.

64. Quissell, D. O., Deisher, L. M., and Barzen, K. A., Subcellular distribution and activation of rat submandibular cAMP-dependent protein kinase following β-adrenergic receptor stimulation, *Biochim. Biophys. Acta*, 969, 28, 1988.

65. Quissell, D. O., Deisher, L. M., and Barzen, K. A., Activation and distribution of rat parotid cAMP-dependent protein kinase following β-adrenergic receptor stimulation *in vitro*, *Arch. Oral Biol.*, 34, 23, 1989b.

66. Roos, D. S., Davidson, R. L., and Choppin, P. W., *Cell Fusion*, Sowers, A. E., Ed., Plenum Press, New York, 1987, 123.

67. Schneyer, L. H., Young, J. A., and Schneyer, C. A., Salivary secretions of electrolytes, *Physiol. Rev.*, 52, 720, 1972.

68. Selinger, F., Naim, E., and Lassar, M., ATP-dependent calcium uptake by microsomal preparations from rat parotid and submaxillary glands, *Biochim. Biophys. Acta*, 203, 326, 1970.

69. Shara, N., Events in the luminal membrane during the early stage of secretion in rat parotid acinar cells: an immunocytochemical study, *Biomed. Res.*, 6, 117, 1985.

70. Singh, J., Brady, R. C., Dedman, J. R., and Quissell, D. O., Subcellular distribution of calmodulin and its binding proteins within the rat submandibular gland, *Am. J. Physiol.*, 251, C403, 1986.

71. Spear, P. G., *Cell Fusion*, Sowers, A. E., Ed., Plenum Press, New York, 1987, 3.

72. Spearman, T. N. and Butcher, F. R., Rat parotid gland protein kinase activation. Relationship to enzyme secretion, *Mol. Pharmacol.*, 21, 121, 1982.

73. Spearman, T. N. and Butcher, F. R., The effect of calmodulin antagonists on amylase release from the rat parotid gland *in vitro*, *Pfluegers Arch.*, 397, 220, 1983.

74. Spearman, N. T., Hurley, K. P., Olivas, R., Ulrich, R. G., and Butcher, F. R., Subcellular location of stimulus-affected endogenous phosphoproteins in the rat parotid gland, *J. Cell Biol.*, 99, 1354, 1984.

75. Streb, H., Irvine, R. F., Berridge, N. J., and Schulz, I., Release of Ca^{2+} from a nonmitochondrial intracellular store in pancreatic acinar cells by inositol 1,4,5-trisphosphate, *Nature*, 306, 67, 1983.

76. Suzuki, Y. and Ohshika, H., Beta$_1$-adrenoceptor-mediated amylase release and cyclic AMP accumulation in rat parotid gland tissue, *Jpn. J. Pharmacol.*, 37, 212, 1985.

77. Takemura, H., Changes in cytosolic free calcium concentration in isolated rat parotid cells by cholinergic and β-adrenergic agonists, *Biochem. Biophys. Res. Commun.*, 131, 1048, 1985.

78. Takemura, H. and Ohshika, H., Contribution of extracellular and intracellular calcium to the enhanced effect of an α-adrenergic agonist on amylase release from dispersed rat parotid cells, *J. Dent. Res.*, 64, 881, 1985.

79. Takuma, T. and Ichida, T., Does cyclic AMP mobilize Ca^{2+} for amylase secretion from rat parotid cells?, *Biochim. Biophys. Acta*, 887, 113, 1986a.

80. Takuma, T. and Ichida, T., Phorbol ester stimulates amylase secretion from rat parotid cells, *FEBS Lett.*, 199, 53, 1986b.

81. Takuma, T. and Ichida, T., Amylase secretion from saponin-permeabilized parotid cells evoked by cyclic AMP, *J. Biochem.*, 103, 95, 1988.

82. Templeton, D., Augmented amylase release from rat parotid gland slices, *in vitro*, *Pfluegers Arch.*, 384, 287, 1980.

83. Terman, B. I. and Gunter, T. E., Characterization of the submandibular gland microsomal calcium transport system, *Biochim. Biophys. Acta*, 730, 151, 1983.

84. Thiel, G., Schmidt, W. E., Meyer, H. E., and Söling, H. D., Purification and characterization of a 22-kDa microsomal protein from rat parotid gland which is phosphorylated following stimulation by agonists involving cAMP as second messenger, *Eur, J. Biochem.*, 170, 643, 1988.

85. Tojyo, Y., Uchida, M., and Matsumoto, Y., Inhibitory effects of calmodulin antagonists on isoproterenol- and dibutyryl cyclic AMP-stimulated amylase release from rat parotid acinar cells, *Jpn. J. Pharmacol.*, 45, 487, 1987.

86. Watson, E. L. and Siegel, I. A., Factors affecting calcium accumulation and release in canine submandibular salivary microsomes, *Arch. Oral Biol.*, 23, 323, 1978.

87. Watson, E. L., Singh, J. C., and Jacobsen, K. L., Augmentation of cholinergic-mediated amylase release by forskolin in mouse parotid gland, *Life Sci.*, 37, 2531, 1985.

88. Williams, J. A., Burnham, D. B., and Hootman, S. R., *Handbook of Physiology*, Forte, J. G., Ed., Oxford University Press, New York, 1989, 419.

89. Wright, L. L. and Luebke, J. I., Somatostatin-, vasoactive intestinal polypeptide- and neuropeptide Y-like immunoreactivity in eye- and submandibular gland-projecting sympathetic neurons, *Brain Res.*, 494, 267, 1989.

90. Yoshimura, K., Nezu, E., and Yoneyama, T., Augmentation of isoproterenol-stimulated tissue cyclic AMP level by cholinergic agonists in rat parotid gland, *Jpn. J. Physiol.*, 35, 765, 1985.

Influence of Diseases on Salivary Glands

Karen Rossie

INTRODUCTION

Salivary glands are affected by a variety of nonneoplastic diseases which produce changes at the clinical and subclinical level. The study of salivary gland disease contributes to knowledge of the biology of normal salivary glands as well as to advances in the treatment of disease. It is important to consider salivary gland disease in planning research in normal salivary glands, because outcomes may be affected if research subjects are included having subclinical salivary gland diseases.

Salivary gland diseases range from the local to the systemic. Localized diseases include inflammatory conditions, infection, obstruction, and atrophy. Nonspecific changes in the salivary glands, such as atrophy or decreased function, can be produced by many different diseases. Systemic diseases affecting salivary glands include those with a primary effect in the salivary glands, such as Sjogren's syndrome, and other diseases which have a primary effect in other parts of the body, such as diabetes mellitus.

Most nonneoplastic diseases of the salivary glands are nonfatal, but many produce significant morbidity. An example is the extensive morbidity that can be produced by severe dryness of the mouth. Further research may lead to the development of improved palliative treatments such as better saliva substitutes. Research into the pathogenesis of salivary gland diseases may eventually lead to preventive measures and cures. Additionally, studying the effect of a systemic disease on the salivary gland may yield further knowledge about that disease in general.

Clinically, the salivary glands and saliva may be useful in the diagnosis of certain systemic and localized diseases. Salivary glands can be useful in diagnosis when a specific, disease-associated change can be identified in the gland. The minor salivary glands of the lower lip, in particular, are relatively accessible and can be removed surgically with minor consequences in comparison to the major salivary glands or internal organs. Saliva is also a potential indicator of disease and is easily collected from the salivary duct orifices in the oral cavity. The composition of saliva is frequently changed in systemic disease and local salivary gland disease. Many of these changes are nonspecific, however. The development of more sensitive detection techniques may lead to the increased use of saliva for diagnostic purposes.

This chapter provides an overview of the various ways that disease manifests in the salivary glands. The effects of certain specific systemic diseases in the salivary glands are also summarized.

LOCALIZED MANIFESTATIONS OF DISEASE IN THE SALIVARY GLANDS

Disease affects the salivary glands in a number of general ways. These effects have primarily been studied at the clinical, morphological, histological, ultrastructural, and biochemical levels. Newer techniques in immunohistochemistry, immunology, cell biology, and molecular biology are contributing to a further understanding of these diseases and their effect in the salivary glands.

Physical exam, history, biopsy, and radiographs may provide useful information in the diagnosis of salivary gland disease. Effects of disease in salivary glands evident by physical exam may include tenderness to palpation, pain during stimuli, decreased function, and changes in consistency and size. Fine needle aspiration cytology is a rapid, fast, cost-effective diagnostic technique which has gained popularity for use in salivary glands. It is used most often when a neoplasm is suspected (Young, 1989). Open biopsy of the salivary gland may be necessary to confirm a diagnosis of salivary gland disease. A variety of radiographic techniques are used (Watson, 1989). Plain radiographs may demonstrate radiopaque calculi in the glands. The ducts of the salivary glands can be examined using a radiographic technique called sialography in which the duct is cannulated and a radiopaque contrast medium is injected. The glands are then visualized using a radiograph. This technique is useful for detecting calculi, duct strictures, nonopaque obstructions, fistulae, and diverticula. Sialography can demonstrate damage to ducts and acini and may be useful in locating salivary gland masses. In scintigraphy, a radioisotope, usually technetium, is taken up by intralobular ductal epithelium and secreted into the saliva. The time-activity curve of this process can be used to obtain quantitative, physiologic information. Scintigraphy is therefore useful in evaluating changes in salivary gland function and nonneoplastic swellings. Salivary glands can be examined using computerized axial tomography (CT) with or without sialography. CT is generally better than sialography for demonstrating mass lesions. Magnetic resonance imaging gives better images of the soft tissues than CT scanning, making it particularly useful in evaluating the extent of suspected salivary gland tumors.

Salivary gland diseases usually affect the ability of the glands to function (Seifert et al., 1986). Salivary flow rate may be decreased or increased. Function can be evaluated by collecting unstimulated and stimulated saliva from the individual gland duct orifices over a limited time period. Flow rates are then compared to normal values obtained using the same methodology. Whole saliva can also be collected. Several methods of saliva collection have been described with numerous minor variations (P. Fox et al., 1985; Navazesh and Christensen, 1982; Stuchell and Mandel, 1988; Sreebny and Broich, 1987). Qualitative changes may also be seen in saliva in disease and include changes in clarity and viscosity.

At the histological level, disease in the salivary glands produces many changes. Most of these changes are not specific to any particular disease; rather, they reflect the common process of inflammation and repair in these particular organs. Histologically, disease may affect the parenchyma or interstitial tissues. In the parenchyma, changes may include inflammation, hyperplasia, hypertrophy, atrophy, degeneration, necrosis, dilated ductal or acinar lumens, inspissated mucus in the lumens, and metaplasia. In some diseases there is an increase in number or size of acinar secretory granules. The interlobular and intralobular tissues may also show histologic changes such as infiltration with fibrous tissue, adipose tissue, or other substances (Seifert et al., 1986). These substances may replace normal gland parenchyma. Even more detail is apparent at the ultrastructural level. Some ultrastructural changes which have been observed in salivary gland disease include ductal metaplasia, the appearance of cilia, and changes in secretory granules, organelles, and lumenal contents (Tandler, 1987). Immunohistochemistry reveals that the substructure of the cell and the expression of certain antigens may be altered in disease. For example, certain intermediate filaments, not normally found in salivary gland epithelial cells, have been detected in some disease states (Dardick et al., 1985). The expression of an immune-related antigen, HLA-DR, is increased in certain inflammatory conditions. (Zarbo et al., 1987). Epithelial membrane antigen expression is diminished in salivary gland inflammation and may, therefore, serve as a marker for degenerative changes in the salivary glands (Tatemoto et al., 1987).

Biochemical changes in the saliva such as changes in inorganic components, pH, and protein composition are frequently found in clinical and experimental salivary gland disease. Summaries of the changes in organic and inorganic components of saliva in many diseases have appeared in recent publications (Mandel, 1980; Arglebe, 1981). These changes may indicate disorders of gland function or the presence of disease but have generally not been useful in making specific clinical diagnoses. (Ferguson, 1980). Studies using newer, more sophisticated detection techniques may yield even more information about the effects of disease on saliva and salivary glands.

Several specific localized disease entities in salivary glands are discussed below.

Salivary Hypofunction

Reduced or absent salivary secretion may occur as a result of many disease processes. The most frequent causes of severe hypofunction are certain drugs, autoimmune diseases, and irradiation of the salivary glands (Sreebny and Broich, 1987). The hypofunction may be relatively permanent, as is presently the case with the autoimmune disease Sjogren's syndrome and damage from high doses of radiation to the salivary gland, or it may be transient or reversible as it is with most medications. The change in the oral environment caused by salivary hypofunction can lead to diseases of the teeth and oral soft tissues and dysfunction in speech, chewing, swallowing, and taste (P. Fox, 1985).

The causes of salivary hypofunction that are theoretically possible include neurological problems, disorders of production, and restriction of salivary transport (Seifert et al., 1986). Neurological possibilities include decreased incoming stimuli, disturbance in conduction of stimuli to the central nervous system, disorders in the central nervous system, and disorders of salivary gland innervation. Disorders of production refer to lack of adequate basic materials or gland parenchyma. Salivary transport can be restricted by an obstruction in the ductal system.

The term xerostomia means dry mouth but has sometimes also been used to mean salivary hypofunction. The former, more subjective meaning for xerostomia is intended when the word is used in this chapter. Not everyone with salivary hypofunction complains of xerostomia; and, conversely, salivary hypofunction cannot be demonstrated in everyone who complains of xerostomia (Sreebny and Broich, 1987). Therefore, an objective assessment of salivary function, by measuring salivary flow rates, is usually indicated before a diagnosis of hypofunction can be made. Xerostomia was found in one study, however, to be a generally valid indicator of resting salivary gland hypofunction (Sreebny and Valdini, 1988).

Diseases which have been associated with xerostomia include autoimmune disease, drugs, radiation therapy, diabetes, hypertension, dehydration, depression, graft-vs.-host disease, sarcoidosis, and hyperlipoproteinemia type V (Sreebny and Valdini, 1988; Sreebny et al., 1989). The effect of aging on salivary gland function is discussed in Chapters 18 and 19 of this book. Renal, neurologic, and skin disorders may also be associated with xerostomia and deserve further study (Sreebny et al., 1989). Increased consumption of saliva, which occurs in mouth breathing, can produce xerostomia despite normal salivary function. Additionally, changes in salivary composition may produce the feeling of dry mouth without hyposecretion (P. Fox et al., 1985). Other terms that have been used for salivary hypofunction include hyposalivation and hyposialia (Seifert et al., 1986).

Hypersalivation

Chronic hypersalivation is a less common condition than hyposalivation. Acute inflammatory conditions in the mouth and teething can cause excessive salivation (Mason and Chisholm, 1975). Other conditions which may be associated with increased salivation include mercury poisoning, acrodynia, tetanus, familial dysautonomia, esophageal reflux, and rabies (Mason and Chisholm, 1975; Stuchell and Mandel, 1975). Certain medications may also produce hypersalivation. An uncommon syndrome, currently designated idiopathic paroxysmal sialorrhea, has been described (Lieblich, 1989). The patients experience paroxysmal episodes of excessive salivation preceded by nausea or epigastric pain. These episodes occur one to two times per week and last two to five minutes each. This may be related to the hypersalivation ("water brash") that occurs with esophageal reflux (Helm et al., 1987) (see "Neurological Diseases" below).

A distinction has been made between hypersalivation and sialorrhea (drooling) (Seifert et al., 1986). Sialorrhea involves the involuntary flow of saliva from the mouth and usually does not involve hypersalivation (increased salivary flow). Sialorrhea is not an uncommon problem and is seen in neuromuscular disorders such as Parkinson's and myasthenia gravis, and in spastic children. The condition is very often caused by a disordered swallowing reflex. Whether or not there is also an excessive production of saliva is not clear. Other terms that have been used for hypersalivation include ptyalism and hypersialia (Seifert et al., 1986).

Sialadenosis

Sialadenosis is a noninflammatory and nonneoplastic disease of the salivary gland characterized by bilateral swelling, usually of the parotid gland (Seifert et al., 1986). The swelling is generally painless and independent of secretory stimuli. Diseases which have been associated with sialadenosis usually fall into one of three categories: endocrine disorders, dystrophic/metabolic diseases, and disorders of the autonomic nervous system. These diseases include diabetes mellitus, malnutrition, bulimia, alcoholism, chronic liver disease, disorders of protein metabolism, and nervous diseases. Certain antihypertensive drugs and psychotropic drugs also may produce this condition.

The swelling in sialadenosis is produced by an enlargement of the salivary gland parenchyma which is due to enlargement of the acini and possibly acinar hyperplasia (Seifert et al., 1986). Acinar diameter is two to three times that of normal acini and nuclei are displaced to the basal part of the cell. Myoepithelial cells show atrophic changes. Three histologic forms of sialadenosis can be distinguished by light microscopy: the granular form (granular appearing cytoplasm), a honeycomb form (vacuolated cytoplasm), and a mixed form. The secretory particles are increased in optical density but not in number. Histologic types do not appear to correlate to clinical types. Electron microscopy reveals structures of the autonomic nervous system are degenerated. Sialadenosis appears to be caused by a neuropathy of the autonomic nervous system which produces disordered acinar protein secretion (Seifert et al., 1986; Chilla, 1981). The saliva shows a wide variety of changes in sialadenosis including increased potassium and decreased sodium (Droese, 1981).

Atrophy

Atrophy of the salivary glands is a nonspecific change seen in many diseases of the salivary glands, particularly following inflammatory diseases. At the microscopic level, a decrease in amount of gland parenchyma is apparent. The individual salivary gland lobules are often decreased in size due to a decrease in the number of epithelial cells. The number of acini is generally decreased more than the number of ducts. The space formerly occupied by parenchyma may be filled in by fatty or fibrous tissue (Seifert et al., 1986). Clinically, atrophy appears as a decrease in gland size. There may be a corresponding decrease in function.

Inflammation

Inflammation of the salivary glands, termed sialadenitis, can be caused by obstruction, radiation, immune disorders, and bacterial, viral, or mycotic infections. Pathogens can reach salivary glands by ascending the ductal system or via the blood. Factors which may contribute to sialadentis include immune-compromised states and local predisposing factors such as decreased or abnormal secretions or ductal obstruction (Seifert et al., 1986). Acute and chronic forms of sialadenitis are seen.

Acute sialadenitis most often affects the parotid gland (Seifert et al., 1986; Mason and Chisholm, 1975). It is characterized clinically by a tender swelling of the gland, sometimes with edema or redness of the overlying skin. There may be a low grade fever, malaise, and elevated white blood cell count. Digital pressure on the duct may produce a purulent discharge. Acute sialadenitis is often caused by decreased salivary flow which allows bacteria from the mouth to ascend the salivary ducts and infect the glands and tends to occur in debilitated and dehydrated patients. The microorganisms most frequently responsible include streptococci and *Staphylococcus aureus* (Lamey et al., 1987). Prevention of dehydration and antibiotics helped to decrease the incidence of this condition. Infection with antibiotic resistant strains of *S. aureus* are sometimes seen (Mason and Chisholm, 1975). Postoperative sialadenitis is a variant of acute sialadenitis seen following major surgery, particularly abdominal surgery, and also usually affects the parotid. The pathogenetic mechanism for postoperative sialadenitis may also include vascular toxic factors and proteolytic enzymes (Seifert et al., 1986). Histologically, acute sialadenitis is characterized in the early phases by swelling and desquamation of ductal epithelium, inflammatory cells in ducts, bacteria, periductal inflammatory infiltrate, and edema. Later the acini are lost, ducts degenerate, and the inflammatory infiltrate spreads with microabscess formation (Seifert et al., 1986). Treatment consists of correcting contributing factors and administering antibiotics determined by culture and sensitivity testing (Seifert et al., 1986).

Chronic sialadenitis is characterized clinically by recurrent, tender swelling of the affected salivary gland (Seifert et al., 1986). Attacks may be intermittent. Eventually, the gland may atrophy and become firm. Salivary flow rate may decrease which can contribute to ascending infections and repeated acute exacerbations. Obstruction is sometimes a causative factor. The parotid gland is affected more often when obstruction is not a factor. One possibility is that the bacterial aggregating ability of the high molecular weight of mucins may be a protective mechanism (Mandel, 1980). Histologic changes of chronic sialadenitis vary with duration. They include acinar atrophy, dilation of ducts, and metaplasia of ductal epithelium. Lost parenchyma may be replaced by fibrosis. The inflammatory infiltrate is often concentrated around ducts, blood vessels, and in interstitial tissues. There may be inspissated mucus in the ducts. Chronic sialadenitis is sometimes seen in minor salivary glands underlying inflammatory conditions of the surface mucosa. Surgical removal of the gland is sometimes necessary if chronic sialadenitis does not respond to antibiotics (Seifert et al., 1986). A variant of recurrent sialadenitis is seen in children. Congenital duct abnormalities may be a contributing factor (Morgan et al., 1989). Chronic sialadenitis in children sometimes resolves spontaneously at puberty. Immune sialadenitis and specific viral infections producing sialadenitis are discussed later in this chapter.

The composition of saliva is changed in salivary gland inflammation although the changes are not specific to inflammatory disease. In the parotid, high sodium, albumin, and lactoferrin and low phosphate characterize the acute phase (Mandel, 1984). Other changes include decreased enzyme and protein levels, increased muramidase and kallikrein, elevated magnesium, decreased zinc, and several additional protein bands. IgA, muramidase, and protein are increased with acute flare ups (Arglebe, 1981).

Obstructive Sialadenitis

Obstruction can produce recurrent, painful swelling of the involved salivary gland, particularly during stimuli, due to blockage of the flow of saliva. The obstruction is sometimes ejected spontaneously, but may have to be surgically removed. Men are affected more often than women; adults more than children. The obstructions occur most often within ducts, particularly larger ducts, and less often within the gland parenchyma. The submandibular gland is affected more often than the parotid. This may be due to the higher content of mucus which is more viscous than serous saliva. Other factors include more alkaline secretions of the submandibular gland with higher calcium and phosphate ion concentrations, narrow ductal opening, and the structure of the duct which ascends from the gland (A. Johnson, 1989). Sublingual glands and minor salivary glands are infrequently affected. A change in the electrolyte composition of the saliva may be responsible for initiating the obstruction (Seifert et al., 1986). The change leads to increased viscosity which leads to mucous retention and the formation of a gel of organic material in the terminal ducts which eventually calcifies forming a sialolith. Increased pH and increased mucin concentration also favor the mineralization of the organic matrix and enzymes may play a role in the calcification process. An alternate theory that appears to be less likely is that the inorganic components are precipitated first and then surrounded by organic substances (Seifert et al., 1986).

Obstructions may or may not be calcified. Noncalcified obstructions are seen more frequently in the parotid (A. Johnson, 1989). Salivary gland calculi, called sialoliths, can often be visualized on a radiograph. Sialoliths often show concentric lamellae consisting of alternating layers of organic and inorganic components. The organic constituents include precipitated mucus and cell debris and the inorganic components are mainly phosphates of calcium (Seifert et al., 1986). A recent, in-depth analysis of the composition of sialoliths found the organic phase was most often composed of glycoproteins, mucoids, mucopolysaccharides, cholesterol, uric acid, and xanthine. The list of crystalline components included hydroxyl- and carbonate-apatite, whitlockite, brushite, and calcium octophosphate. More rarely, calcium carbonate, struvite, and weddellite were found (Strubel and Rzepka-Glinder, 1989). Sialography may reveal the obstruction and dilate ducts. With persistent obstruction, histologic changes in the salivary gland include those typical of chronic inflammation plus inspissated mucin in ducts and dilation of ducts (Seifert et al., 1986). Granulomatous inflammation may also be seen (Van Der Walt and Leake, 1987).

Mucous Extravasation and Retention Phenomenon

In mucous extravasation phenomenon, mucus is spilled from an injured salivary duct into the adjacent soft tissues. The resulting pseudocyst is often called a mucocele. Mucoceles are not uncommon and frequently occur as a result of trauma to the area. The salivary glands of the lower lip are the most common location but any minor or major salivary gland can be affected. Clinically, there is a swelling that appears translucent and bluish if it is superficial or is covered with normal appearing mucosa if it is deep. Histologically, there is a space filled with mucus surrounded by a wall of condensed fibroblasts and granulation tissue. Inflammatory cells infiltrate the wall and are found within the mucoid contents of the pseudocyst. Mucous retention cysts are collections of mucus lined by epithelium resembling ductal epithelium. They may arise in an obstructed duct. A ranula is a mucous retention or extravasation phenomenon that occurs in the floor of the mouth. The ranula derives its name from the resemblance of the swelling to a frog's belly (Seifert et al., 1986). An uncommon plunging form extends into the neck. Mucous extravasation and retention phenomenon are treated surgically if they persist.

Radiation Sialadenitis

Ionizing radiation is damaging to salivary gland tissues. Radiation therapy is used clinically in the treatment of certain diseases, particularly malignancies, and salivary glands are sometimes included in the radiation field. The extent of damage depends on the dose actually delivered to the salivary gland, the type of radiation, and the time period over which it is delivered. The amount of damage is directly proportional to the dose and can be permanent with doses as low as 40 to 65 Gy (Seifert et al., 1986). The parotid gland is most affected because the serous acinar cells are very sensitive. Mucous glycoproteins may provide some protection for the submandibular gland (Mandel, 1984). Stimulated and unstimulated salivary flow rates are decreased due to loss of secretory parenchyma. Decreased flow is apparent as early as after the first treatment and decreases to 40% of normal after 1 week. Eventually, flow may be as low as 5 to 15% of normal (Sreebny and Broich, 1987). Some function may return 8 months later, however, permanent dryness usually results (Siefert et al., 1986). The physical properties of whole saliva are also changed due to a relative decrease in the serous component and increase in the mucous component, making it viscous and foamy (Sreebny and Broich, 1987). Irradiation damages nuclear and intracytoplasmic organelle membranes. Research comparing the effects of radiation in animals to *in vitro* cell preparations suggests that acinar damage may be secondary to a primary effect on blood vessels (Bodner et al., 1984). Other research suggests the secretory granules may be responsible for the increased radiosensitivity of serous cells because depleting the cells of these granules decreases radiation induced damage. Heavy metals are present in the secretory granules and may contribute to the pathogenesis of the damage by catalyzing lipid oxidation (Abok et al., 1984).

The early changes seen in the salivary glands histologically following irradiation occur in the serous acini and consist of vacuole formation, degranulation, and later necrosis. The blood vessels walls are edematous with swollen endothelial cells. Mucous cells are affected less than serous cells. Ducts are affected to a lesser extent than acini. Inflammatory cells eventually infiltrate the parenchyma. Later, the acini are lost and the ducts are dilated and show metaplasia. The salivary gland lobules become atrophic with intervening fibrotic tissue (Seifert et al., 1986; Sreebny and Broich, 1987). Salivary biochemical composition is also affected by irradiation of the salivary glands (Mandel, 1980).

Metaplasia

Squamous metaplasia is seen in a number of salivary gland lesions and consists of a change to squamous epithelium from more differentiated salivary gland epithelium. Squamous metaplasia has been produced and studied in rats by arterial ligation of the submandibular and sublingual glands (Standish and Shafer, 1957; Dardick et al., 1985). Squamous metaplasia appears to develop in these rats from the acinar-intercalated duct complex in degenerating and necrotic gland parenchyma, 6 to 8 days following ligation. The cells dedifferentiate and proliferate followed by accumulation of tonofilaments and desmosome formation (Dardick et al., 1985). Immunohistochemistry shows evidence of keratinization in the epithelium. Squamous metaplasia is also seen following radiation to the salivary glands (Friedman et al., 1950), following implantation of carcinogens (Standish, 1975), in a rat coronavirus infection (Jacoby et al., 1979), and in necrotizing sialometaplasia (Abrams et al., 1973).

Necrotizing sialometaplasia is an uncommon lesion consisting of necrosis of a salivary gland and the subsequent repair process. It is probably caused by localized vascular compromise (Lynch et al., 1979; Chaudhry et al., 1985). This condition can occur in any salivary gland, but is seen most often in the minor salivary glands near the junction of the hard and soft palates. Clinically, unilateral swelling develops, usually with subsequent ulceration. The name comes from the histologic appearance which consists of necrotic acini and islands of viable epithelial cells which have undergone squamous metaplasia (Abrams et al., 1973). An inflammatory infiltrate is generally also present. The clinical and histologic appearance mimic malignancy (Abrams et al., 1973). Histologically, it could be mistaken for a squamous cell carcinoma or a mucoepidermoid carcinoma. Healing with eventual regeneration of normal salivary gland occurs within 3 to 10 weeks.

The salivary glands also undergo other metaplastic changes in disease. The lining epithelium of ducts sometimes changes to squamous epithelium in chronic inflammatory conditions. Oncocytic metaplasia, seen with increasing age, may also be seen in inflammatory disease. Oncocytes have eosinophilic granular cytoplasm due to prominent mitochondria (Seifert et al., 1986). Another metaplastic change sometimes seen in salivary glands is goblet cell metaplasia in which the cells contain mucous vacuoles (Seifert et al., 1986)

Hyperplasia

Adenomatoid hyperplasia is a rare, localized swelling of the minor salivary glands which may mimic a neoplasm (F. Brown et al., 1987; Aufdemorte et al., 1985). It consists of relatively normal appearing salivary glands. Normal appearing ducts are present which help to make the distinction between this condition and a true neoplasm. The acini may appear hypertrophied in addition to hyperplastic. Adenomatoid hyperplasia usually occurs in the minor salivary glands of the hard palate. The cause is not known as yet.

SYSTEMIC DISEASES THAT AFFECT SALIVARY GLANDS

Salivary glands may be affected by certain systemic diseases. Many of these diseases have a primary affect in other parts of the body. Diseases in the following categories can affect the salivary glands and are discussed below: endocrine, metabolic, neurological, immunologic, genetic, developmental, and specific infectious disease.

Endocrine Disease

Sialadenosis of the parotid in diabetes mellitus is the most common endocrine associated sialadenosis (Chilla, 1981). The parotid enlargement in diabetes is often found in conjunction with obesity and is not common in children (Rauch and Gorlin, 1970). Histologic findings in diabetic sialadenosis include fatty infiltration of interstitial tissues and enlarged acinar cells with clear cytoplasm (Rauch and Gorlin, 1970). Diabetes is also significantly correlated with dry mouth (Sreebny et al., 1989). Salivary flow rate studies in diabetics suggest flow rates might be decreased but offer some conflicting results (Mandel, 1980). Marginal decreases in unstimulated flow rates (Kjellman, 1970) and no decrease in stimulated flow rates (Marder et al., 1975) have been reported. Significantly decreased stimulated flow rates were reported in another study (Conner et al., 1970). The latter study did not control for medications or hypertension and included poorly controlled diabetics, however. Sialadenitis is another salivary gland disease that is more common in individuals with diabetes due to decreased defense against infection and possibly due to decreased salivary flow (Rauch and Gorlin, 1970). Saliva composition is also changed in diabetic humans (Mandel, 1980; Arglebe, 1981). Changes in saliva in diabetes have not been very useful in the diagnosis of diabetes.

Sialadenosis of the parotid has been reported with almost all endocrine disorders (Chilla, 1981). Sialadenosis has been reported with conditions associated with changes in sex hormones such as pregnancy, menopause, gynecomastia, and following ovariectomy (Mason and Chisholm, 1975). It has also been reported in hypothyroidism but is rare (Rauch and Gorlin, 1970). Sialadenosis due to disease of the pituitary gland is also rarely reported (Rauch and Gorlin, 1970). Acromegaly may produce salivary gland enlargement as part of the generalized organomegaly seen with this disease (Thomson et al., 1974). Submandibular glands are affected more often than parotid glands. The enlarged salivary glands in acromegaly apparently function normally. A syndrome of adiposity, oligomenorrhea, and parotid swelling (AOP syndrome) is a rare pituitary disorder secondary to hereditary effects in the diencephalon (Seifert et al., 1986) Histologic findings in AOP syndrome include increased size of acinar cells, vacuolated cytoplasm, and decreased granules (Rauch and Gorlin, 1970). Adrenal cortical diseases affect sodium and potassium concentration in saliva and may affect salivary flow rates (Mandel, 1980).

Hormonal influences on salivary glands in experimental animal models for endocrine dysfunction have been reviewed by Johnson (D. A. Johnson, 1987). Studies in diabetic rats provide supporting evidence that diabetes has a significant effect on the salivary glands. There is a reduction in salivary gland weight of the submandibular gland but conflicting results about whether the parotid weight is decreased. Secretory bodies, lipids, and crystalloid bodies accumulate in acinar cells. The stimulated parotid flow rate is reduced. Gland enzyme content and protein composition of the saliva is changed. Some of these changes are reversed by insulin therapy.

Studies in rats and mice show effects in the salivary glands following thyroidectomy, adrenalectomy, and hypophysectomy (D. A. Johnson, 1987). In general, salivary gland weights are decreased by these procedures and salivary flow rates are reduced. The granular convoluted tubules are decreased in size. There are compositional changes in the glands and saliva. Many of these effects are reversed by administering hormone replacement. Extra hormones given to animals with intact endocrine glands often produce effects opposite to those seen in animals with endocrine glands removed.

Metabolic Disease

Sialadenosis has been reported in association with alcoholism (Mandel and Baurmash, 1971). Parotid enlargement in alcoholism is attributed to fatty infiltration and increased acinar size. Enlargement is seen in 60 to 80% of patients with alcoholic cirrhosis and increased stimulated salivary flow rate was also found in this group (Abelson et al., 1976). There has been speculation that sialadenosis in alcoholics may actually be due to alcoholism-related malnutrition, however, since other causes of cirrhosis are not associated with sialadenosis and since sialadenosis is seen in association with malnutrition (Chilla, 1981; Mandel and Baurmash, 1971). A necropsy study of alcoholics with liver disease suggests that alcohol-related salivary gland disease is specifically found in association with alcoholic cirrhosis and not with other alcohol-related diseases (Scott et al., 1988). Histologic salivary gland changes were found in association with alcoholic cirrhosis consisting of increased adipose tissue in the parotid and submandibular glands, decreased acinar tissue in the parotid, and decreased fibrovascular tissue in the submandibular gland. These histologic changes were not seen in association with alcoholic hepatitis or alcohol-related fatty change of the liver. No increase in parotid size or acinar cell size, as had previously been reported, was detected in this study.

Sialadenosis has been reported in association with other metabolic disorders. It has been reported in association with bulimia, a condition characterized by the chronic gorging of food followed by self-induced vomiting. In one report of bulimic patients, the parotid enlargement tended to be intermittent, developing 2 to 6 days after a binge ends (Levin et al., 1980). Several patients had hypokalemic alkalosis and some had moderately elevated serum amylase. One study reported resolution of the swelling in all patients following treatment of the underlying disorder (Ogren et al., 1987). Many other altered nutritional states affected the salivary glands. These are discussed in Chapter 6 of this book. A neurological mechanism may underlie the pathogenesis of all forms of sialadenosis (Seifert et al., 1986) (see "Neurologic diseases" below).

Other metabolic causes of sialadenosis have been reported less frequently. Changes in salivary gland size and function have been reported in diseases of the exocrine pancreas (Mandel, 1980). Chronic pancreatic disease in many of these reports may have been associated with other diseases such as liver disease, malnutrition, and arterial hypertension, so the extent of the contribution of the pancreatic disease is not clear. Attempts have been made to diagnose pancreatitis by looking for changes in saliva. Amylase secretion is reduced in chronic pancreatitis and increased in acute pancreatitis; however, these changes are not specific to pancreatitis (Arglebe, 1981). There have been conflicting results for other changes in the saliva associated with pancreatic diseases. Uremic sialadenosis, sometimes seen in severe kidney disease, may produce hyposalivation (Rauch and Gorlin, 1970).

Hyperlipoproteinemia, type V, a familial disorder of lipoprotein metabolism, is associated with dry mouth and parotid enlargement (Peinertsen et al., 1980). This is due to fatty infiltration replacing normal salivary gland parenchyma. Obesity and hyperlipidemia may also be associated with fatty infiltration and enlargement of the parotid gland (Mandel, 1980).

Xerostomia has been reported in hemochromatosis, a disease characterized by excess iron in the body tissues (Takeda and Ohya, 1987). The disease may be idiopathic or caused by blood transfusion or excessive dietary factors. Salivary gland involvement has been reported in the idiopathic form and in the form secondary to transfusions. Histologically, the salivary glands show iron deposition, mild atrophy, and fibrosis.

Neurological Disease

Neurological disease could theoretically produce xerostomia through decreased peripheral stimulus, disturbance of the afferent stimulus, central disorders, and disorders of salivary gland innervation (Seifert et al., 1986). Decreased peripheral stimulus is seen with decreased mastication and produces salivary gland atrophy (D. A. Johnson, 1987). There is generally not enough effect on the salivary glands to produce xerostomia, however (Sreebny and Valdini, 1987). This is discussed in more detail in the chapter on the role of salivary glands in nutrition in this text.

Central nervous system destruction is only very rarely a cause of xerostomia (Seifert et al., 1986). A syndrome consisting of adiposity, oligomenorrhea, and parotid swelling (AOP syndrome) is a rare, hereditary pituitary disorder caused by a degenerative disease of the diencephalon (Rauch and Gorlin, 1970; Seifert et al., 1986).

Patients with endogenous depression may have decreased salivary flow rates (Sreebny and Broich, 1987). Data from unmedicated subjects utilizing current saliva collection techniques are sparse, however (Sreebny and Valdini, 1987). A summary of earlier studies found multiple reports of decreased salivary flow rates in depressed subjects, for parotid or whole saliva and stimulated or unstimulated flow rates (C. Brown, 1978).

Reflux esophagitis is associated with increased salivation in a response known as "water brash" (Helm et al., 1987). Salivary flow rates double coincidentally with the onset of heartburn. The salivary flow rates do not change, however, when the esophagus is perfused with exogenous acid. An exophagosalivary reflex has been postulated to explain how this response might be mediated (Helm et al., 1987).

Salivary incontinence is sometimes associated with neurological and muscular disorders such as cerebral palsy, Parkinson's disease, epilepsy, and myasthenia gravis (Mason and Chisholm, 1975; Seifert et al., 1986). Whether there is actually overproduction of saliva in addition to decreased muscular control making swallowing difficult in these disorders is disputed. Increased unstimulated salivary flow rates have been demonstrated in persons with cerebral palsy having salivary incontinence in at least one study (Sawczuk et al., 1989). On the other hand, decreased stimulated salivary flow rates, possibly due to oral sensory problems, were reported in a study of persons with cerebral palsy involving the head and neck (Davis, 1979).

Autonomic enervation of the salivary glands shows nerves are apparently essential in maintaining normal gland structure, particularly parasympathetic nerves (Garrett, 1984). Enervation produces limited atrophy of the salivary glands (Garrett, 1984). Parasympathectomy produces submandibular atrophy; whereas, the effect of sympathetic enervation is mild atrophy and a slight increase in gland weight due to increased blood flow. In humans, chorda tympani severing reduces the secretory capacity of the submandibular gland and is sometimes used to treat sialorrhea (Seifert et al., 1986). Parasympathectomy of the parotid has less effect.

Sialadenosis, a disorder of salivary gland parenchyma, discussed earlier in this chapter, can have an underlying neurogenic, dystrophic-metabolic, or hormonal disorder (Seifert et al., 1986). Evidence suggesting a deregulation of the autonomic innervation of the salivary acini is responsible for all forms of sialadenosis is summarized by Siefert et al. (1986). Electron microscopy reveals the postganglionic sympathetic neurones exhibit a loss of neurosecretory granules, destruction of the mitochondria, hydropic swelling of the axoplasm, and terminal axolysis. The clinical finding of polyneuropathy in diseases associated with sialadenosis such as diabetes mellitus, alcoholism, chronic liver disease, nervous diseases, and disorders of protein metabolism, also supports the hypothesis that sialadenosis is caused by a primary neuropathy of the nervous system.

Submandibular salivary flow rate may be decreased on the affected side in Bell's palsy (facial paralysis) (Mandel, 1980). There may also be decreased salivary flow in Melkersson-Rosenthal syndrome, a syndrome consisting of Bell's palsy and other neurological disorders, granulomatous inflammation producing enlargement of the lip and other oral tissues, and fissured tongue (Rauch and Gorlin, 1970). Atrophy of the parotid may also be seen in Melkersson-Rosenthal syndrome. The effect of these syndromes on the salivary glands may be secondary to disturbances of the autonomic nervous system.

Increased salivation is seen in familial dysautonomia, particularly during excitement (Rauch and Gorlin, 1970). This syndrome, characterized by excessive perspiration, red blotching of the skin, defective lacrimation, fluctuations in blood pressure, and emotional instability, is thought to be due to an inborn error in catecholamine metabolism.

Hypertension is apparently associated with reduced salivary flow, even in the absence of xenogeneic medications. An early study found increased whole stimulated salivary flow rates in hypertensive patients (Wotman et al., 1967). Hypertension was correlated with xerostomia in the study in which many patients were not taking drugs that produce dry mouth (Sreebny et al., 1989). Nonmedicated, borderline hypertension patients had decreased unstimulated whole salivary flow rates in another study (Van Hooff et al., 1984). Salivary flow rate responses, in borderline hypertensives, to drugs acting on the autonomic nervous system, suggest decreased salivary flow in this group may be due to reduced parasympathetic influence on the salivary glands (Van Hooff et al., 1984). The electrolytes in saliva are also altered in hypertension (Mandel, 1980).

Frey's syndrome consists of sweating associated with gustatory stimuli in the area supplied by the auriculotemporal or great auricular nerves (Seifert et al., 1986). Sweating of the skin overlying the parotid occurs because of inappropriate regeneration of injured autonomic nerve fibers. Gustatory sweating is often seen following surgery or trauma to the parotid.

Drugs prescribed for certain systemic diseases may affect salivary gland and salivation via a neurological effect. These include anticholinergics, sympatholytic, parasympatholytic, ganglion-blocking agents, central nervous system inhibitors, parasympathomimetics, and sympathomimetics.

Immunologic Disease

Several diseases mediated by the immune system affect the salivary glands and saliva. The autoimmune disease, Sjogren's syndrome, has its primary effect on the salivary and lacrimal glands. Other immune disorders which have their primary effect in other parts of the body, may also affect salivary glands or saliva.

Sjogren's syndrome is an autoimmune disease affecting the salivary and lacrimal glands resulting in decreased production of saliva and tears, and consequently in dry eyes and dry mouth. There may also be an associated connective tissue disease. The syndrome is seen most often in women between 30 to 65 years of age (R. Fox et al., 1984). The major glands may be firm and enlarged (Daniels, 1989). In primary Sjogren's syndrome the lacrimal and salivary glands are affected without any other connective tissue disease being involved; whereas, in secondary Sjogren's syndrome, there is a connective tissue disease in addition to the lacrimal and salivary gland disease. Connective tissue diseases which are associated with secondary Sjogren's syndrome include lupus erythematosus, rheumatoid arthritis, systemic sclerosis, and primary biliary cirrhosis (Talal, 1987).

The etiology and pathogenesis of Sjogren's syndrome are not yet completely understood. Genetic, viral, endocrine, and psychoneurologic factors may all play a role in this disease, as they may in other autoimmune diseases (Talal, 1986). There is an increased susceptibility related to HLA types DR3, DQ1, DQ2, and DRw52(MT2) (Talal, 1986). Infection with Epstein-Barr virus (EBV) has been suggested as a possible inciting factor (Venables et al., 1985; R. Fox et al., 1986). Serological support for this theory comes from the finding of higher antibody titers to viral capsid antigens in patients with Sjogren's syndrome (Yamaoka et al., 1988). Another study, however, found no significant elevation in these antibody titers (Venables, 1985). Early antigen has been detected by monoclonal antibody in salivary gland biopsies from patients with the syndrome (R. Fox, 1986). EBV DNA has been demonstrated in the salivary glands of healthy individuals and individuals with Sjogren's syndrome by *in situ* hybridization (Venables et al., 1989). An increased level of EBV DNA is demonstrated by polymerase chain reaction in salivary glands from patients with Sjogren's syndrome in comparison with normal salivary glands (Saito et al., 1989). Cytomegalovirus infection has also been suggested as an inciting factor because it infects salivary glands and can remain latent afterwards, but elevated antibodies to this virus in Sjogren's syndrome have not conclusively been demonstrated (Venables et al., 1985).

Diagnosis of Sjogren's syndrome is based on finding objective evidence of at least two of the established components of the syndrome: dry eyes, dry mouth, and a connective tissue disease (Bloch et al., 1965). Salivary flow rates are decreased, especially unstimulated flow rates. Focal sialadenitis found in a biopsy of labial salivary glands is considered the most sensitive and specific indicator for Sjogren's syndrome, if certain guidelines are followed (Daniels, 1989). The histopathologic changes consist of focal lymphocytic infiltrates. In the major salivary glands, the infiltrate may eventually efface the entire gland and there may be residual foci of epithelium called epimyoepithelial islands (Daniels, 1986). Two types of epithelial cells can be identified in these islands by immunohistochemical studies of intermediate filaments: one similar to duct epithelial cells and one similar to basal duct cells or myoepithelial cells in normal glands (Kjorell et al., 1988). The lymphocytes in the infiltrate are predominantly T cells in primary and secondary Sjogren's syndrome (Matthews et al., 1986). A study of primary Sjogren's syndrome showed that T helper cells predominate (R. Fox et al., 1986). Sialadenitis associated with Sjogren's syndrome must be differentiated from nonspecific sialadenitis which is more common. A technique to differentiate these two conditions has been developed using the proportion of each class of immunoglobulins contained in plasma cells in salivary gland histological sections (DeWilde et al., 1989).

Serum autoantibodies are common in Sjogren's syndrome but are not specific for the disease. The most common circulating autoantibodies in primary or secondary Sjogren's syndrome are antinuclear antibody, anti SS-A (Ro), anti SS-B (La), and rheumatoid factor (Harley, 1989). These may also be found in other autoimmune diseases, without Sjogren's syndrome. Other diagnostic techniques such as sialography and scintigraphy may provide additional information but findings are similarly not specific for Sjogren's syndrome. Composition of the saliva is usually only slightly altered and may be within normal limits. Among the changes in saliva are elevated secretory IgA and lactoferrin (Baum and Fox, 1987). There are several animal models for Sjogren's syndrome including experimentally induced models and certain strains of mice which spontaneously manifest autoimmune disease (Hoffman and Walker, 1987).

Histologic analysis of minor salivary glands in systemic sclerosis often shows periglandular and intraglandular fibrosis in addition to the lymphocytic infiltrate found in Sjogren's syndrome (Osial et al., 1983). Fibrosis of the salivary glands is also sometimes found in the absence of inflammatory infiltrate and may be indicative of a poor prognosis for this condition.

Sarcoidosis, a chronic systemic granulomatous disease, usually affects lymph nodes, particularly of the hilum of the lung (Lazarus, 1982). Many other organs may be affected including salivary glands which are involved in 1 to 6% of the cases (Seifert et al., 1986). Heerfordt's syndrome (Uveoparotid fever) consists of involvement of the uveal tract of the eye and parotid and lacrimal glands (Seifert et al., 1986). Cranial nerve palsies may also be present. Salivary gland sarcoidosis presents as a constant swelling of the parotid gland with minimal pain. The swelling is somewhat firm and often multinodular. Gland associated lymph nodes and minor salivary glands of the palate and lips are also often affected (Cahn et al., 1964). There may be reduced salivary flow, but this is usually not severe (Seifert et al., 1986). Stimulated salivary flow rates increase to within normal limits following steroid treatment (Chisholm et al., 1971). The saliva shows nonspecific changes including decreased amylase, increased lysozymes, and increased albumin (Seifert et al., 1986). Diagnosis is confirmed by biopsy, if possible. Histologically, there are typical noncaseating granulomas with multinuclear giant cells replacing gland parenchyma (Lazarus, 1982). Weak antisalivary duct antibodies may be detected (Chisholm et al., 1971).

Graft-vs.-host disease (GVHD) is an immune-mediated disease which may occur following bone marrow transplantation. Salivary glands are among the tissues that may be affected by GVHD, resulting in destruction of the salivary glands, which manifests clinically as xerostomia (Schubert et al., 1986). The disease occurs in an acute form soon after the transplant and a chronic form which occurs after 100 days post transplant. At least some of the xerostomia may be due to residual effects of chemotherapy or radiation which are used to ablate the patient's marrow immediately prior to transplantation (Izutsu et al., 1982). Xerostomia that is increasing after 100 days post transplant is more likely to be caused by GVHD (Schubert et al., 1986). The decrease in salivary flow rates ranges from mild to severe. Sialochemistry reveals increased salivary sodium and lysozyme levels and decreased phosphate and IgA levels (Schubert et al., 1986). The histopathologic changes of GVHD in the salivary gland consist of a periductal and interstitial mononuclear inflammatory infiltrate which may invade ductal epithelium, and necrosis of ductal epithelial cells (Sale et al., 1981). Later, acinar damage, ductal ectasia, and fibrosis may develop. It has been suggested that GVHD could serve as a model for the study of Sjogren's syndrome due to the clinical and histological similarities (Gratwohl et al., 1977). Salivary gland GVHD has been studied in a rat model for bone marrow transplantation (Rossie et al., 1988).

Decreased unstimulated salivary flow rate has been reported in oral lichen planus (Lundstrom et al., 1982). This immune-mediated disease affects the oral mucosa producing a characteristic, lace-like white pattern, atrophy or erosions. Patients in this study also had an atrophic salivary gland ductal pattern demonstrated by sialography. Acinar atrophy and lymphocytic infiltrate were seen on histological examination of labial minor salivary glands. Past involvement of the overlying mucosa with lichen planus may have contributed to these histologic changes, however.

Allergic sialadenitis is rarely seen. Clinically, there is a swelling which disappears when the allergen is removed (Seifert et al., 1986; Mason and Chisholm, 1975). Salivary gland swelling may be seen in association with other allergy related disease manifestions such as urticaria, bronchial asthma, and hay fever (Touloukian et al., 1971). A high number of eosinophils is seen in plugs of proteinaceous material that can be milked from the ducts (Touloukian et al., 1971). Allergens include food, drugs, and heavy metals (Rauch and Gorlin, 1970). Histologically, there is acinar degeneration, periductal and perivascular inflammatory infiltrate including mononuclear cells and eosinophils (Seifert et al., 1986). A case has been reported of idiopathic, chronic urticaria which resolved when an inflamed submandibular gland was removed (Chaudhary et al., 1984). The glands had a marked infiltrate of eosinophils.

Amyloidosis has been reported to affect salivary glands in rare instances (Vasquez and Teruel, 1988). Amyloidosis is a group of diseases which have in common the deposition of fibrillar material, called amyloid, in tissues eventually affecting organ function. Amyloidosis can occur in a primary form and a form secondary to neoplasia or inflammation. Salivary gland involvement has been reported in both forms. Amyloid infiltrates, destroys, and replaces gland parenchyma resulting in xerostomia. There may also be salivary gland swelling.

An unusual case of swelling due to granulomatous inflammation of sublingual or floor of mouth minor salivary glands in association with Crohn's disease has been reported recently (Schmitt et al., 1987). Crohn's disease is a granulomatous disease of the gastrointestinal tract. Involvement of the submucosa of the oral cavity is sometimes seen but clinically apparent involvement of the salivary glands in uncommon.

A syndrome of parotid swelling and xerostomia has been reported in acquired immune deficiency syndrome and the related complex (Ulirsch and Jaffe, 1987) (see "Infectious Diseases" below).

Genetic and Developmental Disease

Cystic fibrosis (mucoviscidosis) is a recessive hereditary metabolic disorder affecting exocrine glands with an incidence of 1 in 2000 live Caucasian births. There is generalized dysfunction of excretory glands producing inspissation of secretions. This affects organ function and results in morbidity and mortality (Arglebe, 1981). Serious effects of the disease include chronic obstructive lung disease with recurrent infections and pancreatic insufficiency. Salivary gland involvement in cystic fibrosis has been well documented. The mucous salivary glands demonstrate disease-related morphological changes, which are thought to be secondary to obstruction of the ducts by thick secretions (Izutsu, 1987). Histologic and ultrastructural changes include dilated acinar and ductal lumens filled with thick collections of mucus. Mucus collections may show lamellar layers and microlith formation. Star-shaped hydroxyapatite crystals are seen in the acinar cytoplasm and lumens (Seifert et al., 1986). Labial minor salivary glands show decreased salivary flow rate (Izutsu, 1987) but hypersecretion has been reported in the parotid and submandibular glands (Arglebe, 1981). Saliva in cystic fibrosis has also been extensively studied. Summaries of changes in saliva seen in cystic fibrosis have been published by Arglebe and Izutsu and include increased calcium, sodium, and protein. Serous glands do not show the morphological changes seen in the mucous glands but do have changed secretions.

Developmental defects are rare in the salivary glands and are usually seen together with other developmental defects (Mason and Chisholm, 1975; Rauch and Gorlin, 1970; Seifert et al., 1986). They include aplasia or agenesis of one or more major salivary glands, and duct atresia. Salivary glands may also be hypoplastic. Polycystic parotid glands have rarely been reported. Other developmental defects include congenital fistulas, diverticula, and accessory ducts. Ectopic salivary glands are sometimes found in the neck, middle ear, mastoid bone, or within the ramus or body of the mandible. It is not uncommon to find salivary gland tissue within lymphoid tissue. Fistulas may develop from aberrant glands lacking their own ductal system. Salivary gland tissue may be found in a depression on the lingual surface of the mandible, usually near the angle (Shafer et al., 1983). This condition, variously called developmental lingual mandibular salivary gland depression, static bone cyst, or Stafne cyst, may produce a well-defined radiolucency in a radiograph of the mandible.

Infections

The more common bacterial infections of the salivary glands were discussed in the section on inflammation. Tuberculosis rarely affects the salivary glands (Seifert et al., 1986). In the region of the salivary glands, tuberculosis is usually limited to intraglandular and periglandular lymph nodes, usually of the parotid, where granulomas with caseating necrosis are seen histologically (Van De Walt and Leake, 1987). Syphilis and gonorrhea rarely produce bilateral painless swelling of the salivary glands. Actinomycosis can produce suppurative parotitis (Seifert et al., 1986). Other rare specific infections of the salivary glands include tularemia and cat scratch fever (A. Johnson, 1989). Mycotic infections which may involve the salivary glands include blastomycosis and coccidiomycosis (A. Johnson, 1989). Certain viruses have a tissue tropism for salivary glands. The most commonly identified viral infection involving the salivary glands throughout the years has been mumps, caused by paramyxovirus (Wilfert, 1985). Mumps is also known as epidemic parotitis. Prior to the advent of a vaccine, the disease was common between the ages of 5 and 10 years. As the name suggests, the parotid gland is involved far more often than the submandibular gland. The pancreas, testicles and ovaries, and central nervous system may also be affected. After entering via the upper respiratory tract, the infection has an incubation period of 14 to 21 days. Clinically, fever, malaise, and head and neck pain are followed by edematous parotid swelling and pain. The swelling is bilateral in around 70% of the cases. There may be obstruction of salivary flow and pain if salivation is stimulated. Based on serologic testing, infection is often clinically silent. Infection generally confers lifelong immunity. The glands are infrequently biopsied. They may show an interstitial inflammatory infiltrate consisting of lymphocytes and plasma cells with destruction of acini, vacuolation of ductal epithelium, and increased mucus in ducts (Seifert et al., 1986). The saliva shows an increase in amylase.

Cytomegalovirus (CMV), a herpes virus, is regularly isolated from salivary glands of patients with CMV syndromes (Griffiths and Grundy, 1988). Dysfunction is not readily apparent in salivary glands as it is in other organs where the virus is found, such as the liver, lung, brain, and colon. Primary infection is usually asymptomatic and viral latency can occur. Immune suppression may lead to reactivation of the virus. Groups that develop CMV syndromes include neonates, infants, and persons with acquired immune deficiency syndrome, renal transplants, or bone marrow transplants. Infection can be contracted from saliva, blood, other body fluids and secretions, and infected, transplanted organs. CMV can produce a number of clinical effects including infectious mononucleosis, hepatitis, hemolytic anemia, thrombocytopenia, gastrointestinal infection, retinitis, and immunosuppression. Histologic sections of salivary glands in patients with CMV infection frequently show the markedly enlarged cells with nuclear and cytoplasmic inclusion bodies that are typical of CMV infection (Seifert et al., 1986). Serum antibody, viral culture, and histology help make a diagnosis of CMV infection. Mouse CMV infection in the salivary glands has been studied in experimental models for CMV infection (Hudson, 1979). CMV has also been reported in salivary glands in rats (Jacoby et al., 1979).

EBV is another virus in the herpes group that can establish latency (Purilo, 1987). The parotid may act as a reservoir for EBV. EBV DNA has been demonstrated in salivary glands in healthy individuals and in patients with Sjogren's syndrome by *in situ* hybridization (Venables et al., 1989) and by polymerase chain reaction (Saito et al., 1989). In adolescents and adults, EBV infection produces infectious mononucleosis characterized by fever, malaise, lymphadenopathy, pharyngitis, and hepatosplenomegaly. Parotitis has occasionally been reported in association with this syndrome, but clinically apparent salivary gland disease does not generally occur as part of this syndrome (Andersson and Sterner, 1985; Mor et al., 1982). Numerous other EBV-associated syndromes have been reported, most of which occur in immune deficient patients (Purilo, 1987).

A syndrome resembling Sjogren's syndrome has been reported in association with human immunodeficiency virus (HIV) infection (Ulirsch and Jaffe, 1987; Schiodt et al., 1989). In a study of HIV-infected individuals with subjective xerostomia, the majority were found to have parotid gland enlargement, decreased salivary flow rates, and lymphocytic infiltrate of the labial salivary glands (Schiodt et al., 1989). HIV-associated salivary gland disease is different from Sjogren's syndrome in that these patients lack circulating autoantibodies against SS-A and SS-B and the T4/T8 ratio is lower for the labial salivary gland inflammatory infiltrate (Schiodt et al., 1989). Lymph node enlargement may be responsible for some of the enlargement of the major salivary glands seen in HIV infection. Involvement of the intraparotid lymph nodes in HIV-infected individuals in association with HIV-related generalized lymphadenopathy has been reported but appears to be uncommon, at least at the clinically apparent level (Poletti et al., 1988). Histologic findings in the parotid lymph nodes are similar to those found in other sites in the body in HIV-related generalized lymphadenopathy. Hypervascular reactive lymphoid hyperplasia with cyst formation and squamous metaplasia is seen.

Many other viruses may infect salivary glands but are infrequently recorded clinically. These include influenza, measles, whooping cough, encephalomyocarditis, lymphocytic choriomeningitis, coxsackie, and ECHO viruses (Seifert et al., 1986; Mason and Chisholm, 1975). Viral excretion in saliva occurs with EBV (Gerger et al., 1972), mumps, CMV, and other viruses which are not tropic for the salivary glands such as poliomyelitis, hepatitis, rabies, rubella, and viral influenza (Siefert et al., 1986).

Sialodacryoadenitis virus (SDAV) is of particular interest to researchers using rats because it infects rat salivary and lacrimal glands and because SDAV epizootic infections in laboratory rats are fairly common (Jacoby et al., 1979). This rat coronavirus produces a nonfatal infection that lasts about 1 week. Infection during an experiment studying salivary glands could significantly alter the outcome. Clinical signs of infection include swelling in the neck in the region of the salivary glands, sneezing, production of red tears, and weight loss; however, infection is sometimes subclinical. The histopathological changes at the initial stages show inflammation and destruction of the acini and ducts. Later, islands of epithelium showing squamous metaplasia are seen followed by almost complete restoration of normal glandular architecture.

It is apparent from this overview of nonneoplastic salivary gland disease that many diseases affect the salivary glands, directly and indirectly. Some salivary gland diseases have received extensive attention in the literature, while others have had only scant information published. Some diseases with possible impact on the salivary glands have had only a few, isolated case reports or small retrospective studies. Many studies were done over 20 years ago using the techniques of the day and need to be confirmed using current methodology. Salivary flow rate determination techniques sometimes varied between studies or were not well documented. It will be important to continue to improve the standardization of these techniques. Many reports of salivary gland disease have been made at the gross or clinical, biochemical, and microscopic levels. New techniques, such as those of immunohistochemistry, immunology, cell biology, and molecular biology are being employed to provide more information about the underlying mechanisms of salivary gland disease. These studies, together with prospective, controlled clinical studies utilizing current methodology, and a reexamination of early reports are needed to better understand the relationship between disease and the salivary glands.

REFERENCES

1. Abelson, C. D., Mandel, I. D., and Mortimer, K., Salivary studies in alcoholic cirrhosis, *Oral Surg. Oral Med. Oral Pathol.*, 41, 188, 1976.
2. Abok, K., Brunk, U., Jung, B., and Ericsson, J., Morphologic and histochemical studies on the differing radiosensitivity of ductular and acinar cells of the rat submandibular glands, *Virchows Arch. [B]*, 45, 443, 1984.
3. Abrams, A. M., Melrose, R. J., and Howell, F. V., Necrotizing sialometaplasia: a disease simulating malignancy, *Cancer*, 32, 130, 1973.
4. Andersson, J. and Sterner, G., A 16-month-old boy with infectious mononucleosis, parotitis and Bell's palsy, *Acta Paediatr. Scand.*, 74, 629, 1985.
5. Arglebe, C., Biochemistry of human saliva, *Adv. Otorhinolaryngol.*, 26, 97, 1981.

6. Aufdemorte, T. B., Ramzy, I., Hold, G. R., Thomas, J. R., and Duncan, D. L., Focal adenomatoid hyperplasia of salivary glands. A differential diagnostic problem in fine needle aspiration biopsy, *Acta Cytol.*, 29, 23, 1985.

7. Baum, B. J. and Fox, P. C., Chemistry of saliva, in *Sjogren's Syndrome. Clinical and Immunological Aspects*, Talal, N., Moutsopoulos, H. M., and Kassan, S. S., Eds., Springer-Verlag, New York, 1987, 34.

8. Bloch, K. J., Buchanan, W. W., Wohl, M. J., and Bunim, J. J., Sjogren's syndrome: a clinical, pathological and serological study of sixty-two cases, *Medicine*, 44, 187, 1965.

9. Bodner, L., Kuyatt, B. L., Hand, A. R., and Baum, B. J., Rat parotid cell function *in vitro* following X irradiation *in vivo*, *Radiat. Res.*, 97, 386, 1984.

10. Brown, C. C., The parotid puzzle: a review of the literature on human salivation and its application to psychophysiology, *Psychophysiology*, 7, 66, 1970.

11. Brown, F. H., Houston, G. D., Lubow, R. M., and Sagan, M. A., Adenomatoid hyperplasia of mucous salivary glands, *J. Periodontol.*, 58, 125, 1987.

12. Cahn, L. R., Eisenbud, L., and Blake, M. N., Biopsies of normal appearing palates of patients with known sarcoidosis, *Oral Surg. Oral Med. Oral Pathol.*, 18, 342, 1964.

13. Chaudhary, T. K., Wray, B. B., Guill, M. F., Crew, T. M., and Coker, N., Chronic urticaria related to sialoadenitis, *Ann. Allergy*, 53, 410, 1984.

14. Chaudhry, A. P., Salman, L., Salman S., Saxon M., and Pierri, L. K., Necrotizing sialometaplasia of palatal minor salivary glands: a report on two cases, *J. Oral Med.*, 40, 2, 1985.

14a. Chilla, R., Sialadenosis of the salivary glands of the head, *Adv. Oto-Rhino-Laryngol.*, 26, 1, 1981.

15. Chisholm, D. M., Lyell, A., Haroon, T. S., Mason, D. K., and Beeley, J. A., Salivary gland function in sarcoidosis, *Oral Surg. Oral Med. Oral Pathol.*, 31, 766, 1971.

16. Conner, S., Iranpour, B., and Mills, J., Alteration in parotid salivary flow in diabetes mellitus, *Oral Surg. Oral Med. Oral Pathol.*, 30, 55, 1970.

17. Daniels, T. E., Salivary histopathology in diagnosis of Sjogren's syndrome, *Scand. J. Rheumatol. Suppl.*, 61, 366, 1986.

18. Daniels, T. E., Clinical assessment and diagnosis of immunologically mediated salivary gland disease in Sjogren's syndrome, *J. Autoimmunity*, 2, 529, 1989.

19. Dardick, I., Jeans, M. T. D., Sinnott, N. M., Wittkuhn, J. F., Kahn, H. J., and Baumal, R., Salivary gland components involved in the formation of squamous metaplasia, *Am. J. Pathol.*, 119, 33, 1985.

20. Davis, M. J., Parotid salivary secretion and composition in cerebral palsy, *J. Dent. Res.*, 58, 1808, 1979.

21. DeWilde, P. C. M., Kater, L., Baak, J. P. A., VanHouwelingen, J. C., Hene, R. J., and Slootweg, P. J., A new and highly sensitive immunohistologic diagnostic criterion for Sjogren's syndrome, *Arthritis Rheum.*, 32, 1214, 1989.

22. Droese, M., Cytological diagnosis of sialadenosis, sialadenitis, and parotid cysts by fine-needle aspiration biopsy, *Adv. Otorhinolaryngol.*, 26, 49, 1981.

23. Ferguson, D. B., Current diagnostic uses of saliva, *J. Dent. Res.*, 66, 420, 1987.

24. Fox, P. C., Van der Ven, P. F., Sonies, B. C., Weiffenbach, J. M., and Baum, B. J., Xerostomia: evaluation of a symptom with increasing significance, *J. Am. Dent. Assoc.*, 110, 519, 1985.

24a. Fox, R. I., Howell, F. V., Bone, R. C., and Michelson, P., Primary Sjogren's syndrome: clinical and immunopathologic features, *Sem. Arthritis Rheum.*, 14, 77, 1984.

25. Fox, R. I., Pearson, G., and Vaughan, J. H., Detection of Epstein-Barr virus-associated antigens and DNA in salivary gland biopsies from patients with Sjogren's syndrome, *J. Immunol.*, 137, 3162, 1986.

26. Friedman, M. and Hall, J. W., Radiation induced squamous cell metaplasia and hyperplasia of the normal mucous glands of the oral cavity, *Radiology*, 55, 848, 1950.

27. Garrett, J. R., Innervation of salivary glands: neurohistological and functional aspects, in *The Salivary System*, Sreebny, L. M., Ed., CRC Press, Boca Raton, FL, 1984, 70.

28. Gerger, P., Nonoyama, M., and Lucas, S., Oral excretion of Epstein-Barr virus by healthy subjects and patients with infectious mononucleosis, *Lancet*, 2, 988, 1972.

29. Gratwhol, A. A., Moutsopoulos, H. M., Chused, T. M., Akizuki, M., Wolf, R. O., Sweet, J. B., and Deisseroth, A. B., Sjogren-type syndrome after allogeneic bone marrow transplantation, *Ann. Intern Med.*, 87, 703,1977.

30. Griffiths, P. D. and Grundy, J. E., The status of CMV as a human pathogen, *Epidemiol. Infect.*, 100, 1, 1988.

31. Harley, J. B., Autoantibodies in Sjogren's syndrome, *J. Autoimmunity*, 2, 383, 1989.

32. Haubrich, J., Etiology and pathogenesis of the inflammatory diseases of the cephalic salivary glands, *Adv. Otorhinolaryngol.*, 26, 39, 1981.

33. Helm, J. F., Dodds, W. J., and Hogan, W. J., Salivary response to esophageal acid in normal subjects and patients with reflux esophagitis, *Gastroenterology*, 93, 1393, 1987.

34. Hoffmann, R. W. and Walker, S. E., Animal models of Sjogren's syndrome, in *Sjogren's Syndrome. Clinical and Immunological Aspects,* Talal, N., Moutsopoulos, H. M., and Kassan, S. S., Eds., Springer-Verlag, New York, 1987, 266.

35. Hudson, J. B., The murine cytomegalovirus as a model for the study of viral pathogenesis and persistent infections, *Arch. Virol.*, 62, 1, 1979.

36. Izutsu, K. T., Salivary electrolytes and fluid protection in health and disease, in *The Salivary System*, Sreebny, L. M., Ed., CRC Press, Boca Raton, FL, 1987, 96.

37. Izutsu, K. T., Truelove, E. L., Schubert, M. M., Ensign, W. Y., and Meadows, E. E., Does graft-versus-host disease cause hyposalivation in bone marrow transplant recipients?, *Proc. J. Dent. Res.*, 61, 263, 1982.

38. Jacoby, R. O., Bhatt, P. N., and Jonas, A. M., Viral diseases, in *The Laboratory Rat,* Baker, J. H., Lindsey, J. R., and Weisbroth, S.H., Eds., Academic Press, New York, 1979, 184.

39. Johnson, A., Inflammatory conditions of the major salivary glands, *Ear Nose Throat J.*, 68, 94, 1989.

40. Johnson, D. A., Regulation of salivary glands and their secretions by masticatory, nutritional, and hormonal factors, in *The Salivary System,* Sreebny, L. M., Ed., CRC Press, Boca Raton, FL, 1987, 135.

41. Kjellman, O., Secretion rate and buffering action of whole mixed saliva in subjects with insulin-treated diabetes mellitus, *Odontol. Rev.*, 21, 1, 1970.

42. Kjorell, U., Ostberg, Y., Vertanen, I., and Thornell, L.-E., Immunohistochemical analyses of autoimmune sialadenitis in man, *J. Oral Pathol.*, 17, 374, 1988.

43. Lamey, P. J., Boyle, M. A., MacFarlane, T. W., and Samaranayake, L. P., Acute suppurative parotitis in out-patients: microbiologic and post-treatment sialographic findings, *Oral Surg. Oral Med. Oral Pathol.*, 63, 37, 1987.
44. Lazarus, A. A., Symposium on granulomatous disorders of the head and neck, *Sarcoidosis*, 15, 621, 1982.
45. Levin, P. A., Falko, J. M., Dixon, K., Gallup, E. M., and Saunders, W., Benign parotid enlargement in bulimia, *Ann. Intern Med.*, 93, 827, 1980.
46. Lieblich, S., Episodic supersalivation (Idiopathic paroxysmal sialorrhea): description of a new clinical syndrome, *Oral Surg. Oral Med. Oral Pathol.*, 68, 159, 1989.
47. Lundstrom, I. M. C., Goran, K., Anneroth, B., and Bergstedt, H. F., Salivary gland function and changes in patients with oral lichen planus, *Scand. J. Dent. Res.*, 90, 443, 1982.
48. Lynch, D., Crago, C. A., and Martinez, M. G., Necrotizing sialometaplasia. A review of the literature and report of two additional cases, *Oral Surg. Oral Med. Oral Pathol.*, 47, 63, 1979.
49. Mandel, I. D., Sialochemistry in diseases and clinical situations affecting salivary glands, *CRC Crit. Rev. Clin. Lab. Sci.*, 12, 321, 1980.
50. Mandel, I. D., Salivary gland function, *Gerodontology*, 3, 47, 1984.
51. Mandel, L. and Baurmash, H., Parotid enlargement due to alcoholism, *J. Am. Dent. Assoc.*, 82, 369, 1971.
52. Marder, M. Z., Abelson, D. C., and Mandel, I. D., Salivary alterations in diabetes mellitus, *J. Periodontol.*, 46, 567, 1975.
53. Mason, D. K. and Chisholm, D. M., Eds., *Salivary Glands in Health and Disease*, W. B. Saunders, Philadelphia, 1975, 83.
54. Matthews, J. B., Potts, A. J. C., Hamburger, J., Scott, D. G. I., and Struthers, G., An immunohistochemical study of lymphocyte subsets and expression of glandular HLA-DR in labial salivary glands from patients with symptomatic xerostomia, rheumatoid arthritis and systemic lupus erythematosus, *Scand. J. Rheumatol Suppl.*, 61, 56, 1986.
55. Mor, R., Pitlik, S., Dux, S., and Rosenfeld J., Parotitis and pancreatitis complicating infectious mononucleosis, *Isr. J. Med. Sci.*, 18, 709, 1982,
56. Morgan, D. W., Pearman, K., Raafat, F., Oates, J., and Campbell, J., Salivary disease in childhood, *Ear Nose Throat J.*, 68, 155, 1989.
57. Navazesh, M. and Christensen, C. M., A comparison of whole mouth resting and stimulated salivary measurement procedures, *J. Dent. Res.*, 61, 1158, 1982.
58. Ogren, F. P., Huerter, J. V., Pearson, P. H., Antonson, C. W., and Moore, G. F., Transient salivary gland hypertrophy in bulimics, *Laryngoscope*, 97, 951, 1987.
59. Osial, T. A., Whiteside, T. L., Buckingham, R. B., Singh, G., Barnes, E. L., Pierce, J. M., and Rodnan, G. P., Clinical and serologic study of Sjogren's syndrome in patients with progressive systemic sclerosis, *Arthritis Rheum.*, 26, 500, 1983.
60. Peinersten, J. L., Shaefer, E. J., Brewer, H. B., and Moutsopoulos, H. M., Sicca-like syndrome in type V hyperlipoproteinemia, *Arthritis Rheum.*, 23, 114, 1980.
61. Poletti, A., Manconi, R., Volpe, R., and Carbone, A., Study of AIDS-related lymphadenopathy in the intraparotid and perisubmaxillary gland lymph nodes, *J. Oral Pathol.*, 17, 164, 1988.
62. Purilo, D. T., Epstein-Barr virus: the spectrum of its manifestations in human beings, *South Med. J.*, 80, 943, 1987.

63. Rauch, S. and Gorlin, R. J., Diseases of the salivary glands, in *Thoma's Oral Pathology*, Gorlin, R. J. and Goldman, H. M., Eds., C. V. Mosby, St. Louis, 1970, 962.

64. Rossie, K. M., Sheridan, J. F., Barthold, S. W., and Tutschka, P. J., Graft-versus-host disease and sialodacryoadenitis viral infection in bone marrow transplanted rats, *Transplantation*, 45, 1012, 1988.

65. Saito, I., Servenius, B., Compton, T., and Fox, R., Detection of Epstein-Barr virus DNA by polymerase chain reaction in blood and tissue biopsies from patients with Sjogren's syndrome, *J. Exp. Med.*, 169, 2191, 1989.

66. Sale, G. E., Shulman, H. M., Schubert, M. M., Sullivan, K. M., Kopecky, K. J., Hackman, R. C., et al., Oral and ophthalmic pathology of graft versus host disease in man: predictive value of the lip biopsy, *Hum. Pathol.*, 12, 1022, 1981.

67. Sawczuk, A., Izutsu, K., Yorkstone, K., Truelove, E., LeResche, L., and Stiefel, D., Salivary flow in cerebral palsy subjects with salivary incontinence, *J. Dent. Res.*, 68, 404, 1989.

68. Schiodt, M., Greenspan, D., Daniels, T., Nelson, J., Leggott, P. J., Wara, D. S., and Greenspan, J. S., Parotid gland enlargement and xerostomia associated with labial sialadenitis in HIV-infected patients, *J. Autoimmunity*, 2, 415, 1989.

69. Schmitt S. J., Antonioli, D. A., Jaffe B., and Peppercorn M. A., Granulomatous inflammation of minor salivary gland ducts: a new oral manifestation of Crohn's disease, *Hum. Pathol.*, 18, 405, 1987.

70. Schubert, M. M., Sullivan, K. M., and Truelove, E. L., Head and neck complications of bone marrow transplantation, in *Head and Neck Management of the Cancer Patient*, Peterson, D. E., Elias, E. G. and Sonis, S. T., Eds., Martinus Nijhoff Publishing, Boston, 1986, 401.

71. Scott, J., Burns, J., and Flower, E. A., Histological analysis of parotid and submandibular glands in chronic alcohol abuse: a necropsy study, *J. Clin. Pathol.*, 41, 837, 1988.

72. Seifert, G., Miehlke, A., Haubrich, J., and Chilla, R., *Diseases of the Salivary Glands*, Georg Thieme Verlag, Inc., New York, 1986, 51, 78, 85, 101, 110, 324.

73. Shafer, W. G., Hine, M. K., and Levy, B. M., *A Textbook of Oral Pathology*, W. B. Saunders, Philadelphia, 1983, 35.

74. Sreebny, L. M. and Broich, G., Xerostomia (Dry Mouth), in *The Salivary System*, Sreebny, L. M., Ed., CRC Press, Boca Raton, FL, 1987, 179.

75. Sreebny, L. M. and Valdini, A., Xerostomia: a neglected symptom, *Arch. Intern Med.*, 147, 1333, 1987.

76. Sreebny, L. M. and Valdini, A., Xerostomia. I. Relationship to other oral symptoms and salivary gland hypofunction, *Oral Surg. Oral Med. Oral Pathol.*, 66, 451, 1988.

77. Sreebny, L. M., Valdini, A., and Yu, A., Xerostomia. Part II: Relationship to nonoral symptoms, drugs, and diseases, *Oral Surg. Oral Med. Oral Pathol.*, 68, 419, 1989.

78. Standish, S. M. and Shafer, W. G., Serial histologic effects of rat submaxillary and sublingual salivary gland duct and blood vessel ligation, *J. Dent. Res.*, 36, 886, 1957.

79. Standish, S. M., Early histologic changes induced tumors of the submaxillary salivary glands of the rat, *Am. J. Pathol.*, 33, 671, 1975.

80. Strubel, G. and Rzepka-Glinder, V., Structure and composition of sialoliths, *J. Clin. Chem. Clin. Biochem.*, 27, 244, 1989.

81. Stuchell, R. N. and Mandel, I. D., Salivary gland dysfunction and swallowing disorders, *Otolaryngol Clin.*, 21, 649, 1988.

82. Takeda, Y. and Ohya, T., Sicca symptom in a patient with hemochromatosis: minor salivary gland biopsy for differential diagnosis, *Int. J. Oral Maxillofac. Surg.*, 16, 745, 1987.

83. Talal, N., Recent developments in the immunology of Sjogren's syndrome (auto-immune exocrinopathy), *Scand. J. Rheumatol. Suppl.*, 61, 76, 1986.

84. Talal, N., Overview Sjogren's syndrome, *J. Dent. Res.*, 66, 672, 1987.

85. Tandler, B., Salivary gland changes in disease, *J. Dent. Res.*, 66, 398, 1987.

86. Tatemoto, Y., Kumasa, S., Watanabe, Y., and Mori, M., Epithelial membrane antigen as a marker of human salivary gland acinar and ductal cell function, *Acta Histochem.*, 82, 219, 1987.

87. Thomson, J. A., McCrossan, J., and Mason, D. K., Salivary gland enlargement in acromegaly, *Clin. Endocrinol. (Oxford)*, 3, 1, 1974.

88. Touloukian, R. J., Rickert, R. R., Lange, R. C., and Spencer, R. P., The micro-vascular circulation of lymphangiomas. A study of Xe 133 clearance and pathology, *Pediatrics*, 48, 36, 1971.

89. Ulirsch, R. C. and Jaffe, E. S., Sjogren's syndrome-like illness associated with acquired immunodeficiency syndrome-related complex, *Hum. Pathol.*, 18, 1063, 1987.

90. Van Der Walt, J. D. and Leake, J., Granulomatous sialadenitis of the major salivary glands. A clinicopathological study of 57 cases, *Histopathology*, 11, 131, 1987.

91. Van Hooff, M., Van Baak, M. A., Schols, M., and Rahn, K. H., Studies of salivary flow in borderline hypertension: effects of drugs acting on structures innervated by the autonomic nervous system, *Clin. Sci.*, 66, 599, 1984.

92. Vasquez, M. and Teruel, J. L., Sicca syndrome due to amyloidosis, *J. Oral Maxillofac. Surg.*, 46, 1013, 1988.

93. Venables, P. J. W., Ross, M. G. R., Charles, P. J., Melson, R. D., Griffiths, P. D., and Maini, R. N., A seroepidemiological study of cytomegalovirus and Epstein-Barr virus in rheumatoid arthritis and sicca syndrome, *Ann. Rheum. Dis.*, 44, 742, 1985.

94. Venables, P. J. W., Teo, C. G., Baboonian, C., Griffin, B. E., and Hughes, R. A., Persistence of Epstein-Barr virus in salivary gland biopsies from healthy individuals and patients with Sjogren's syndrome, *Clin. Exp. Immunol.*, 75, 359, 1989.

95. Watson, M. G., Investigation of salivary gland disease, *Ear Nose Throat J.*, 68, 84, 1989.

96. Wilfert, C. M., Mumps (epidemic parotitis), in *Cecil Textbook of Medicine*, Wyngaarden, J. B. and Smith, L. H., Eds. W. B. Saunders, Philadelphia, 1985, 1712.

97. Wotman, S., Mandel, I. D., Thompson, R. H., Jr., and Laragh, J. H., Salivary electrolytes, urea nitrogen, uric acid and salt taste thresholds in hypertension, *J. Oral Ther. Pharmacol.*, 3, 239, 1967.

98. Yamaoka, K., Miyasaka, N., and Yamamoto, K., Possible involvement of Epstein-Barr virus in polyclonal B cell activation in Sjogren's syndrome, *Arthritis Rheum.*, 31, 1014, 1988.

99. Young, J. A., Fine needle aspiration cytology of salivary glands, *Ear Nose Throat J.*, 68, 120, 1989.

100. Zarbo, R. J., Regezi, J. A., Lloyd, R. V., Crissman, J. D., and Batsakis, J. G., HLA-DR antigens in normal, inflammatory, and neoplastic salivary glands, *Oral Surg. Oral Med. Oral Pathol.*, 64, 577, 1987.

Protooncogene Expression in Rat Salivary Gland Epithelial Cells

Eleni Kousvelari

INTRODUCTION

Signal transduction is the process by which an external stimulus, such as a neurotransmitter, binds to a specific cell surface receptor, which in turn is coupled to the generation of a second messenger, to initiate a cellular response. A variety of cell surface receptors and second messengers are involved in eliciting different cellular responses in rat parotid acinar cells *in vitro* and *in vivo* (for reviews see Butcher and Putney, 1980 and Schneyer, 1972). For example specific stimulation of β-adrenoreceptors by the β-adrenergic receptor agonist isoproterenol in rat parotid acinar cells *in vitro* increases protein production, phosphorylation, protein N-glycosylation and protein secretion. Since such stimulation increases the levels of cAMP it was thought that these cellular processes, at least *in vitro*, are mediated by cAMP-dependent mechanisms (Butcher et al., 1975; Baum et al., 1981; Kousvelari et al., 1983). It is also well accepted that chronic β-adrenergic receptor stimulation in rats results in changes such as (1) hypertrophic and hyperplastic enlargement of the parotid glands and (2) transcriptional induction and synthesis of parotid tissue-specific proline rich proteins (PRP) (Schneyer, 1972; Mechansho et al., 1985; Ann et al., 1987). Although the analysis of the immediate biochemical events that occur following the interaction of isoproterenol with the β-adrenergic receptor have largely been elucidated, very little has been known about the intracellular mechanisms that transduce these signals to the nucleus to alter processes such as protein synthesis and hyperplasia in the parotid salivary glands. Recently, it has been established that addition of cAMP in macrophages, thyroid cells, and C6 glioma cells, causes a dramatic and rapid induction of the protooncogene c-*fos* (for review see Curran 1988). These observations led to the proposition that extracellular stimuli may trigger the expression of a group of cellular immediate-early genes (protooncogenes) such as the c-*fos* family of genes (fra and fos B), the c-*jun* family of genes (jun B and jun D), c-*myc*, etc. which are part of the signal transduction cascade linking cell stimulation with changes in gene expression leading to cell proliferation and differentiation. It is the thesis of this review that protooncogenes are important in understanding salivary cell processes such as those listed above.

PROTOONCOGENES

Protooncogenes are the cellular counterparts of the viral genes discovered during the study of transforming retroviruses and have been shown to control a variety of processes associated with cell growth, proliferation, and differentiation. (Bishop, 1987; Weinberg, 1985; Varmus, 1984). These genes encode proteins that function at the different stages of signal transduction to regulate cell growth: mediation of intercellular signals, intracellular transmission of information to the nucleus, and the final translation of the signal to the nucleus to influence the induction of specific gene expression. The cognate proteins of the protooncogenes appear to fall within the following categories, depending on their functions: growth factors (sis) (Doolittle et al., 1983), growth factor receptors (erb-A and erb-B), protein kinases (src, abl, fps, mos, raf) (Hunter and Cooper, 1985; Moelling et al., 1984; Maxwell and Arlinghaus, 1985), guanine nucleotide-binding proteins (ras), and nuclear proteins which act as transcription factors (fos, jun, and myc) (Table 1). The place of the protooncogenes in the cell, in accordance with their product functions, is shown in Figure 1. Starting from the extracellular compartment, the protooncogene c-*sis* has been shown to encode the B-chain of the platelet-derived growth factor (PDGF). In the group of oncogenes which encode transmembrane receptor proteins is the erb B (Downward et al., 1984). This protooncogene encodes an epidermal growth factor receptor-like protein. Protooncogene proteins which associate with the inner surface of the cellular membrane include src, abl, and ras. The src and abl proteins have tyrosine kinase activity while ras proteins are similar to the G proteins that bind guanine nucleotides (Barbacid, 1987). The link between membrane receptors and transcription factors is thought to involve the cytoplasmic proteins encoded from raf and mos which have been shown to possess serine-threonine kinase activity (Edelman et al., 1987) and may be similar to protein kinase C. In addition proteins encoded from fos, jun and myc are found in the nucleus. The Fos and Jun proteins bind to specific DNA sequences and are thought to regulate the transcription of target genes whose products in turn will cause cellular proliferation and differentiation (for review see Curran and

**TABLE 1. The Major Protooncogenes Have Been Placed
According to Their Respective Products**

Protooncogene	Viral Source	Protein
Growth factors		
sis	SSV	PDGF-B chain
Growth factor receptors		
erb-A	AEV-ES4	Thyroid
erb-B	AEV	EGF
Protein-tyrosine kinases		
src	RSV	60 kDa
abl	AB-MLV	150 kDa
fps	FSV	98 kDa
Nuclear proteins		
(transcription factors)		
fos	FBJ-MSV	55-65 kDa
jun	ASV-17	39 kDa
myc	MC 29	50-60 kDa
Guanine nucleotide		
binding proteins		
Ha-ras	Ha-MSV	21 kDa
Ki-ras	Ki-MSV	
Protein serine/threonine		
kinases		
mos	Mo-MSV	37 kDa
raf	3611-MSV	75 kDa

Franza, 1988). These "nuclear oncogenes" will be discussed in the following
section in more detail.

PROPERTIES OF FOS, JUN, AND ABL
PROTOONCOGENES AND THEIR EXPRESSION
PATTERNS IN RAT SALIVARY EPITHELIAL CELLS

The protooncogenes have mainly been classified by either their function
and/or the intracellular localization of their respective products (Figure 1). A
small group of protooncogenes expresses proteins which do not exhibit any
established functional activity, but they are localized to the cell nucleus and are
thought to act as transactivators and regulators of RNA synthesis (Kingston et al.,
1985). Among these "nuclear" protooncogenes are the families of c-*fos* and c-
jun, and c-*myc*. In particular, the proteins of the Fos and Jun families, encoded
by the respective genes, have been shown to interact with a transcriptional
regulatory element known as the AP-1 (Activator protein-1) (Franza et al., 1988;
Chiu et al., 1988). AP-1 is the phorbol-ester inducible enhancer binding protein
which recognizes DNA sequences similar to the GCN4, the yeast transcription

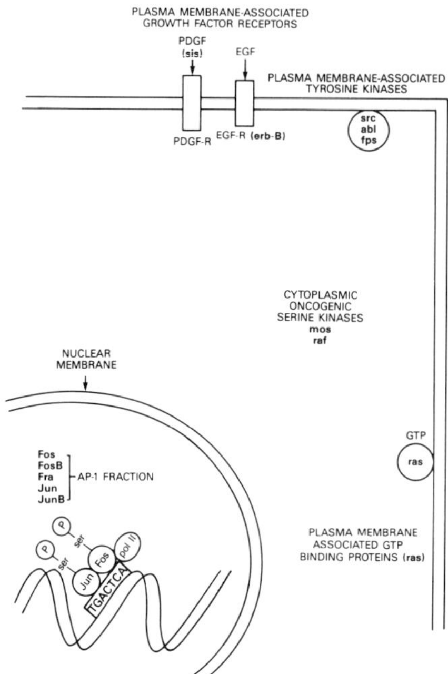

Figure 1. Schematic representation of the functional activities of some of the better-known protooncogenes and their proposed positions in the cell. The protooncogenes have been placed in accordance to their respective products. The protooncogenes sis and erb B encode proteins that act as the platelet derived-growth-factors (PDGF), or as the epidermal-growth factor receptor (EGF), respectively. The protooncogenes abl, src, and fps encode membrane associated tyrosine kinase, while the mos and raf oncogene encode cytoplasmic serine kinase. The ras oncogenes encode membrane associated CTP-binding proteins. Finally, the fos, fos B, fra, jun, and jun B protooncogenes encode the "nuclear proteins". The Fos/Jun dimer bound to AP-1 specific sequence of a gene is also illustrated in the nucleus.

factor, Sp1, and to CAAT-binding transcription factor (CTF) (Hope and Struhl, 1985; Kadonaga and Tjian, 1986). These factors activate transcription by the RNA polymerase II from a select group of promoters present in simian virus 40 (sv 40) and herpes simplex virus thymidine kinase promoters, that contain at least one properly positioned recognition site for Sp1, CTF, AP-1. The Jun protein can interact with AP-1 recognizable DNA sequences but the Fos protein requires the presence of Jun protein for DNA binding (Sassone-Corsi et al., 1988). It has been established that the Fos and Jun proteins contact each other by parallel interactions of α-helical domains containing a heptad repeat of leucines (the leucine zipper). This heterodimeric complex then binds to the AP-1 site in the DNA (Gentz et al., 1989; Turner and Tjian, 1989; Landschulz et al., 1989) to affect the regulation of gene expression. The expression of the members of the fos and jun of gene families is induced by a variety of extracellular stimuli, including, TPA, EGF, NGF, PDGF, and those using the second messengers cAMP and Ca^{++} calmodulin. Their induction is rapid (30 to 40 min) and transient. The synthesis of the c-*fos* protein, a 55-kDa phosphoprotein, follows the appearance of its mRNA and has a half-life of about 2 h. Because of the many different situations in which c-*fos* and c-*jun* gene families expression is associated, their role in these various circumstances is likely to be general (Curran, 1988). This role might be that both are necessary transcription activation factors which influence the transcription of tissue specific target genes and thus cellular proliferation and differentiation in normal and disease processes. The c-*fos* gene is the cellular homolog of the oncogene (v-fos) found in Finkel-Biskis-Jinkins murine sarcoma virus (FBJ-MSV) which induces transformation in a variety of cells. However, two alterations are required to activate the transforming potential of the c-*fos* gene; (1) a transcriptional enhancer sequence must be linked to the gene and (2) an interaction at the 3' end of the gene, which inhibits transformation, must be disrupted (Curran, 1988).

The jun B protooncogene belongs to the family of c-*jun* which is the cellular counterpart of the retrovirally transduced oncogene, v-jun, of avian sarcoma virus (ASV17). The structural differences between the viral and the cellular jun gene could account for differences in physiological activity (Vogt and Bos, 1989). The jun B gene encodes a protein similar to the 39-kDa phosphoroprotein encoded by c-*jun*.

The c-*abl* protooncogene is the normal cellular homolog of the v-abl transforming gene of the Abelson murine leukemia virus. The human c-*abl* gene has been shown to translocate from chromosome 9 to 22 in Philadelphia chromosome-positive chronic myelogenous leukemia cells. This protooncogene is a member of the tyrosine kinase family (Abelson and Rabstein, 1970; Rosenbert et al., 1975; Witte et al., 1980). The absence of obvious signal peptides or membrane spanning domains suggests that the proteins encoded by the two major c-*abl* mRNAs 6.5 and 5.3 kb belong to a group of nonreceptor kinases. One of the proteins is N-terminally myristoylated (Jackson and Baltimore, 1989) and both share sequence similarities with other nonreceptor kinase in the common protein segment upstream of the kinase domain (Ben-Neriah et al., 1986). A unique feature of the abl gene is a large C-terminal domain of 650 amino acids beyond the kinase region. Recently, it has been shown that this part of the abl may contain a partial nuclear localization signal (Van Etten et al., 1989).

The expression of c-*fos*, jun B, and c-*abl* protooncogenes was studied in rat parotid acinar cells after either *in vitro* or *in vivo* stimulation of β-adrenergic receptors by isoproterenol. The cDNA probes used in Northern blot analysis were a full-length rat c-*fos* (Curran et al., 1987), a mouse jun B (Ryseck et al., 1988) and mouse cDNAs of v-abl and clones IV and 4.4 of the c-*abl* IV cDNA (Ben-Neriah et al., 1986). The levels of c-*fos* mRNA were highly induced in rat parotid acinar cells after 60-min incubation with isoproterenol (Kousvelari et al., 1988). The product of c-*fos*, the Fos protein, followed the same kinetics of the c-*fos* mRNA and its presence was associated with the nuclei of the acinar cells and ductal elements. Treatment of rats *in vivo* with isoproterenol for 8 days causes specific changes in the parotid glands of the animal among which are hypertrophy and hyperplasia. Under these conditions c-for mRNA levels were undetectable. We, therefore, reasoned that the expression of this protooncogene does not correlate with proliferation of parotid cells. That indeed c-*fos* expression does not associate with cellular proliferation was further confirmed when parotid acinar cells were incubated with carbachol, a muscarinic agonist, which is known not to have an effect on parotid gland proliferation. The expression of c-*fos* under such stimulation was similar to that of β-adrenergic receptor stimulation (Mirels et al., 1989). Barka et al. (1986) have observed that although c-*fos* was inducible by isoproterenol in submandibular glands of 5- and 14-day-old mice, [³H]-thymidine incorporation was minimally affected. They also conclude that c-*fos* expression was not correlated with the initiation of cell division in this epithelial cell system. We have observed similar findings for c-*fos* expression in a rat submandibular duct derived cell line called RSMT-A5 (He et al., 1988). Here c-*fos* expression was highly increased, by isoproterenol and/or 8br-cAMP addition, as early as 30 min, but there was no immediate effect on cell proliferation or differentiation (Yeh et al. 1988). The jun protooncogene had a similar pattern of expression (to fos) in both rat parotid acinar cells and RSMT-A5 cells. Taken together, these results suggest that c-*fos* and jun B induction does not have an immediate effect on cell proliferation in rat salivary epithelial cells. A lack of association between expression of these protooncogenes and cell growth has also been reported for other cell types. For example, in A431 cells c-*fos* expression does not correspond with proliferation, and in NIH 3T3 c-*fos* is inducible throughout the cell cycle (Bravo et al., 1985; Bravo et al., 1986). Similarly, in normal peripheral blood lymphocytes, c-*fos* induction can occur without subsequent proliferation (Grausz et al., 1986).

It appears that in many cell types under different stimuli the expression of c-*fos* and jun B may relate to a variety of functional events other than immediate effects on cell growth and/or differentiation. Thus, c-*fos* expression may be one of the early consequences of a β-adrenergic receptor cascade of events; however, other factors, like tissue-specific genes or target genes, must be regulated to result in salivary gland cell proliferation and differentiation. The Fos protein has been demonstrated in nuclear protein complexes that modulate transcription activity in adipocytes (Distal et al., 1987). Such a function can be suggested for rat salivary epithelial cells. It is possible that the Fos/Jun complexes act as transcription activator for target genes which then will account for the long-term responses induced by chronic stimulation of β-adrenergic receptors in rat salivary glands. These suggestions are shown schematically in Figure 2.

On the other hand, the expression of the c-*abl* gene follows a different pattern of expression. This protooncogene is expressed as a 5.3-kb mRNA in the parotid and submandibular glands. Although abl expression does not change following *in vitro* isoproterenol treatment, chronic *in vivo* stimulation in rats induces the appearance of two low molecular size (1.5 and 1.3 kb) c-*abl* mRNAs. These transcripts are only detectable in the parotid glands (e.g., not submandibular glands or heart) and only with probes, v-abl and clone 4.4 of the IV c-*abl* cDNA.

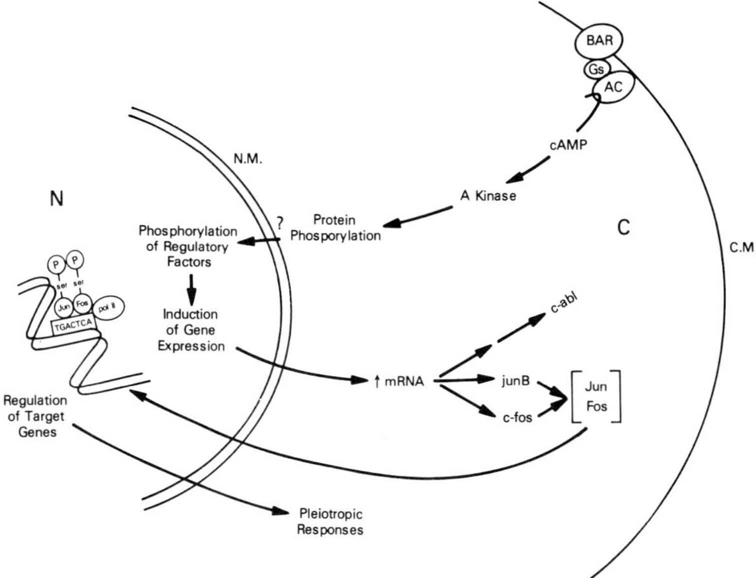

Figure 2. Signal transduction pathway of β-adrenergic receptor. Schematic representation of the events following β-adrenergic receptor stimulation in rat salivary epithelial cells. BAR (β-adrenergic receptor), Gs (CTP binding protein), CM (cell membrane), NM (nuclear membrane), N (nucleus), C (cytoplasm). The double arrow indicates the delayed induction of c-*abl* (24 h) as compared to the induction of c-*fos* and jun B (60 min).

The mechanism(s) by which the isoproterenol-inducible low molecular size mRNA transcripts are produced is unknown at present. These transcripts may have been generated from an additional promoter, differential splicing and/or premature termination of the transcribed c-*abl* mRNA. Such hypotheses are consistent with those reported by Bernards et al. (1988) and Oppi et al. (1987).

It is possible that the morphological and biochemical changes induced by isoproterenol in rat parotid glands reflect the actions of various combinations on protooncogenes and tissue-specific and/or target genes. In the cascade of events which occurs following β-adrenergic receptor stimulation, the protooncogenes fos, jun, and abl may play important roles. Studies are currently underway to delineate these events more fully.

SUMMARY AND FUTURE DIRECTIONS

Transcriptional regulatory proteins play an important role in cell growth and differentiation elicited by short-term signals that originate from either growth factors or hormones outside the cell. These signals are transmitted to the nucleus passing through receptors, transducer proteins, and cytoplasmic protein kinases. Over 40 protooncogenes have been discovered and shown to participate in the different steps of the signal transduction pathway by encoding products which act as receptors, growth factors, protein kinase, and nuclear transcription factors. Work on protooncogenes in salivary epithelial cells has just begun. Since the biological processes influenced by protooncogenes appear to be highly dependent on the cell type and/or stimulus, specific studies in salivary gland cells are important. Studies using *in situ* hybridization at different stages of salivary gland development or disease may be especially useful for answering questions such as, whether different patterns of protooncogene induction are associated with specific morphological glandular changes. The effect of chronic β-adrenergic receptor regulation on tissue-specific gene expression also needs more study. For example, the specific sequences of such genes must be identified in order to understand what transacting factors can regulate their transcription. Such studies will require well-characterized salivary acinar cell lines.

I have suggested in this review that protooncogenes may play a central role in the series of events following β-adrenergic receptor stimulation by regulating the transcription of tissue-specific and target genes which in turn will be responsible for salivary gland hypertrophy and hyperplasia. Knowledge gained by such studies may soon permit manipulations for establishing salivary acinar cells in culture and eventually allow control of growth and development in the intact organism.

ACKNOWLEDGMENT

I wish to thank Dr. Bruce J. Baum for his advice and encouragement during the writing of this manuscript.

REFERENCES

1. Abelson, H. T. and Rabstein, L. S., Lymphosarcoma: virus induced thymic independent disease in mice, *Cancer Res.*, 30, 2213, 1970.
2. Ann, D. K., Clements, S., Johnstone, E. M., and Carlson, D. M., Induction of tissue specific proline-rich protein multigene families in rat and mouse parotid glands by isoproterenol, *J. Biol. Chem.*, 262, 904, 1987.
3. Barbacid, M., ras Genes, *Annu. Rev. Biochem.*, 56, 827, 1987.
4. Barka, T., Gubits, R. M., and Van Der Noen, H. M., β-Adrenergic stimulation of c-fos gene expression in the mouse submandibular gland, *Mol. Cell Biol.*, 6, 2989, 1986.
5. Baum, B. J. and Kuyatt, B. L., Protein production and release by dispersed rat submandibular gland cells *in vitro* after adrenergic stimulation, *Life Sci.*, 2, 1143, 1981.
6. Ben-Neriah, Y., Bernards, A., Paskind, M., Daley, G. Q., and Baltimore, D., Alternative 5′ exons in c-abl mRNA, *Cell*, 44, 557, 1986.
7. Bernards, A., Paskind, M., and Baltimore, D., Four murine c-abl mRNAs arise by usage of two transcriptional promoters and alternative splicing, *Oncogene*, 2, 297, 1988.

8. Bishop, J. M., The molecular genetics of cancer, *Science*, 235, 311, 1987.

9. Bohmann, D., Bos, T. J., Admon, A., Nishimura, T., Vogt, P. K., and Tjian, R., Human proto-oncogene c-jun encodes a DNA binding protein with structural and functional properties of AP-1, *Science*, 238, 1386, 1988.

10. Bravo, R., Burckhardt, J., Curran, T., and Muller, R., Stimulation and inhibition of growth by EGF in different A431 cell clones is accompanied by the rapid induction of c-fos and c-myc protooncogenes, *EMBO J.*, 4, 1193, 1985.

11. Bravo, R., Burckhardt, J., Curran, T., and Muller, R., Expression of c-fos in NIH/3T3 cells is very low but inducible throughout the cell cycle, *EMBO J.*, 5, 695, 1986.

12. Butcher, F. R., Goldman, J. A., and Nemerovslo, M., Effects of adrenergic agents on α-amylase release and adenosine 3',5'-monophosphate accumulation in rat parotid, *Biochem. Biophys. Acta,* 392, 82, 1975.

13. Butcher, F. R. and Putney, J. W., Regulation of parotid gland function by cyclic nucleotides and calcium, *Adv. Cyclic Nucleotide Res.* 13, 215, 1980.

14. Chiu, R., Boyle, W. J., Meek, J., Smeal, T., Hunter, T., and Karin, M., The c-fos protein interacts with c-jun/AP-1 to stimulate transcription of AP-1 responsibe genes, *Cell*, 54, 541, 1988.

15. Curran, T., Gordon, M. B., Rubine, K. L., and Sambucetti, L. C., Isolation and characterization of the c-fos (Rat) cDNA and analysis of post-translational modification *in vitro*, *Oncogene*, 2, 79, 1987.

16. Curran, T., Fos, in *The Oncogene Handbook*, Reddy, E. P., Skalka, A. M., and Curran, T., Eds., Elsevier, Amsterdam, 1988, 307.

17. Curran, T. and Franza, R. B., Fos and Jun. The Ap-1 connection, *Cell*, 55, 395, 1988.

18. Downward, J., Yarden, Y., Mayers, E., et al., Close similarity of epidermal growth factor receptor and v-erb B oncogene protein sequence, *Nature*, 307, 521, 1989.

19. Doolittle, R. F., Hunkapiller, M. W., Hood, L. E., Devare, S. G., Robbins, K. C., Aaronson, S. A., and Antoniades, H. N., Simian sarcoma virus oncogene, v-sis, is derived from the gene (or genes) encoding a platelet-derived growth factor, *Science*, 221, 275, 1983.

20. Edelman, A. M., Blumenthal, D. K., and Krebs, E. G., Protein serine/threonine kinases, *Annu. Rev. Biochem.*, 56, 567, 1987.

21. Franza, B. R., Bauscher, F. J., III, Josephs, S. F., and Curran, T., Fos complex and fos related antigens recognize sequence elements that contain AP-1 sites, *Science*, 239, 1150, 1988.

22. Gentz, R., Rauscher, F. J., III, Abate, C., and Curran, T., Parallel association of fos and jun leucine zippers juxtaposes DNA binding domains, *Science*, 243, 1695, 1989.

23. Grausz, D. J., Fradelizi, D., Dautry, F., Monier, R., and Lehn, P., Modulation of c-fos and c-myc mRNA levels in normal lymmphocytes by calcium ionophore A23187 and phorbol ester, *Eur. J. Immunol.*, 16, 1217, 1986.

24. He, X., Ship, J., Wu, X., Brown, A. M., and Wellner, R. B., β-Adrenergic control of cell volume and chloride transport in an established rat submandibular cell line, *J. Cell Physiol.*, 138, 527, 1988.

25. Hope, I. A. and Struhl, K., GCN4 Protein, synthesized *in vitro*, binds to HIS3 regulatory sequences: implications for the general control of amino acid biosynthetic genes in yeast, *Cell*, 43, 177, 1985.

26. Hunter, T. and Cooper, J. A., Protein-tyrosine kinases, *Annu. Rev. Biochem.*, 54, 897, 1985.

27. Jackson, P. and Baltimore, D., N-terminal mutations activate the leukomogenic potential of the myristolylated form of c-abl, *EMBO J.*, 8, 449, 1989.

28. Kadonaga, J. T. and Tjian, R., Affinity purification of sequence-specific DNA binding proteins, *Proc. Natl. Acad. Sci. U.S.A.*, 83, 449, 1986.

29. Kingston, R. E., Baldwin, A. S., Jr., and Sharp, P. A., Transcription control by oncogenes, *Cell*, 41, 3, 1985.

30. Kousvelari, E. E., Grant, S. R., and Baum, B. J., β-Adrenergic receptor regulation of N-linked protein glycosylation in rat parotid acinar cells, *Proc. Natl. Acad. Sci. U.S.A.*, 80, 7146, 1983.

31. Kousvelari, E. E., Louis, J., Huang, L.-H., and Curran, T., Regulation of proto-oncogenes in rat parotid acinar cells *in vitro* after stimulation of β-adrenergic receptor, *Exp. Cell Res.*, 179, 194, 1988.

32. Maxwell, S. A. and Arlinghaus, R. B., Serine kinase activity associated with moloney murine sarcoma virus-124-encoded p37^{mos}, *Virology*, 143, 321, 1985.

33. Mehansho, H., Clements, S., Sheares, B.T., Smith, S., and Carlson, D. M., Induction of proline-rich glycoprotein synthesis in mouse salivary glands by isoproterenol and by tannins, *J. Biol. Chem.*, 260, 4418, 1985.

34. Mirels, L., Baum, B. M., and Kousvelari, E. E., Dissociation between c-fos gene expression and DNA synthesis in rat parotid glands, *Arch. Oral Biol.*, 34, 511, 1989.

35. Moelling, K., Heimann, B., Beimling, P., Rapp, U. R., and Sander, T., Serine and threonine-specific kinase activities of purified gag and gag-raf proteins, *Nature*, 312, 558, 1984.

36. Oppi, C., Shore, S. K., and Reddy, P. E., Nucleotide sequence of testis-derived c-abl cDNAs: implications for testis-specic transcription and abl oncogene activation, *Proc. Natl. Acad. Sci. U.S.A.*, 84, 8200, 1987.

37. Rauscher, R. J., III, Cohen, D. R., Curran, T., Bos, T. J., Vogt, P. K., Bohmann, D., Tjian, R., and Franza, B. R., Fos-associated protein p^{39} is the product of the jun proto-oncogene, *Science*, 240, 1010, 1988.

38. Rosenberg, N., Baltimore, D., and Scher, C. D., *In vitro* transformation of lymphoid cells by abelson murine leukemia virus, *Proc. Natl. Acad. Sci. U.S.A.*, 72, 1932, 1975.

39. Ryseck, R. P., Hirai, S. I., Yaniv, M., and Bravo, R., Transcriptional activation of c-jun during the Go/G1 transition in mouse fibroblasts, *Nature*, 334, 1988.

40. Sassone-Corsi, P., Lamph, W. W., Kamps, M., and Verma, I., Fos-associated cellular p^{39} is related to nuclear transcription factor AP-1, *Cell*, 54, 553, 1988.

41. Schneyer, C. A., Regulation of salivary gland size, in *Regulation of Organ and Tissue Growth*, Gross, R. J., Ed., Academic Press, New York, 1972, 211.

42. Turner, R. and Tjian, R., Leucine repeats and an adjacent DNA binding domain mediate the formation of functional c-fos-c-jun heterodimers, *Science*, 243, 1689, 1989.

43. Varmus, H. E., The molecular genetics of cellular oncogenes, *Annu. Rev. Cenet.*, 18, 553, 1984.

44. Van-Etten, R. A., Jackson, P., and Baltimore, D., The mouse Type IV c-abl gene product is a nuclear protein and activation of transforming ability is associated with the cytoplasmic localization, *Cell*, 58, 663, 1989.

45. Vogt, P. K. and Bos, T. J., The oncogene jun and nuclear signalling, *TIBS*, 14, 172, 1989.

46. Witte, O. N., Dasgupta, A., and Baltimore, D., Abelson murine leukemia virus protein is phosphorylated *in vitro* to form phosphotyrosine, *Nature*, 238, 826, 1980.

47. Yeh, C.-K., Louis, J. M., and Kousvelari, E. E., β-Adrenergic regulation of c-fos gene expression in an epithelial cell line, *FEBS Lett.*, 240, 118, 1988.

Control of Cell Growth and Proliferation

Michael G. Humphreys-Beher

INTRODUCTION

Cellular Interactions

The molecular mechanisms underlying cellular interactions regulating the control of cell proliferation and differentiation are only beginning to be understood. It is not, however, unimaginable that the overall architecture and primary constituents of the plasma membrane are intimately involved in regulating these processes. Only recently have we begun to understand, at the molecular level, which of these cell-surface molecules are involved in cell social behavior and how they operate in the context of regulating normal and abnormal cell growth responses.

Recognition

Some of the constituents of the plasma membrane, which are thought to mediate cellular interactions, most likely function by binding to specific cell surface receptors. This interaction is best exemplified by the way growth factors or specific hormones elicit alterations in cell behavior by binding to their specific receptor proteins.

Cell surface glycoproteins are capable of generating a high degree of specificity into their tertiary structure by not only their primary amino-acid sequences but through their carbohydrate structures as well. Indeed, variation of

the primary structure and oligosaccharide moieties of plasma membrane glycoproteins generate the functional diversity of cell surface proteoglycans (Ruoslahti,
1989) and blood group antigens (Feizi and Childs, 1987). The structure and
functional role of oligosaccharide moieties of proteins are interdependent, and
therefore the controlled synthesis of these structures are necessary to assure this
property of glycoproteins. Hence, developmental variability of the oligosaccharide
may provide the epitopes required for specific recognition and signaling of cell
proliferation.

Perspectives and Summary

A variety of proteins have been identified which can interact with those
oligosaccharide moieties of plasma membrane glycoproteins with a high degree
of specificity. These include lectins (Barondes, 1981; Frazier and Glaser, 1979),
glycosidases (Rauvali and Hakomori, 1981) and the glycosyltransferases (Pierce
et al., 1980). In this review I will summarize our recent advances in delineating
the molecular and biochemical role of glycosyltransferases in mediating parotid
gland acinar cell growth; specifically, the identification of β1-4 galactosyltransferase as a modulator of proliferative activity when expressed ectopically
in the plasma membrane of acinar cells. While these studies have primarily taken
place in the rat, certain parallels have been uncovered in murine developmental
studies as well as pathologies of human parotid glands in which hypertrophy and
hyperplasia have been observed. For a current review of histological development of the salivary glands, please see Redman (1987).

GLYCOSYLTRANSFERASES

The sugar moieties of glycoconjugates are attached singly or in groups by
enzymes called glycosyltransferases (for review see Schacter and Roseman,
1980). Most of the cellular glycosyltransferases are membrane bound to the
smooth endoplasmic reticulum and Golgi apparatus (Kornfeld and Kornfeld,
1985). The enzymes are specific for a particular sugar donor conjugated to a
nucleotide diphosphate (Pierce et al., 1980; Struck and Lennarz, 1980) and
specific acceptor molecule. The energy for covalent attachment of the sugar to
the oligosaccharide is provided by the phosphate bond (Schachter and Roseman,
1980)

Three basic components are involved in the reactions catalyzed by
glycosyltransferases: the enzyme, a sugar donor, and a sugar acceptor. Additionally, many of the glycosyltransferases require divalent cations for enzymatic
activity (Jentoft et al., 1976). A unique component observed in the reaction for
some galactosyltransferases is the protein α-lactalbumin (Khatra et al., 1974;

Powell et al., 1977) which alters enzyme substrate acceptor specificity (Schachter and Roseman, 1980; Brew et al., 1968; Khatra et al., 1974).

CELL SURFACE GLYCOSYLTRANSFERASES

While most glycosyltransferases have been associated with intracellular membranes, evidence continues to accumulate for the existence of plasma membrane localization of some of these enzymes (Roseman, 1970; Roth et al., 1977; Shur, 1984). Cell-surface forms of sialyltransferase (Cerven, 1977; Bernecki, 1974; Porter and Bernecki, 1975) as well as galactosyltransferase (Bosmann, 1972; Banya et al., 1988; Marchase et al., 1989) have been assayed in both normal and transformed cells. These transferases have also been found in secreted form from normal and transformed cells, in response to growth stimuli (LaMont et al., 1977; Klohs et al., 1981; Hymphreys-Beher et al., 1984), and as a consequence of uncontrolled growth (Ghanta et al., 1989; Bosmann and Hall, 1974; Ip and Dao, 1977; Davey et al., 1986).

Galactosyltransferases at the surfaces of contacting cells have also been suggested to be possible mediators of growth control. Roth and White (1972) presented evidence implicating these enzymes in contact inhibition exhibited by nontransformed tissue culture cells. In addition, the sugar donor for galactosyltransferase, UDP-galactose, has been shown to inhibit the growth of cultured cells (Roth et al., 1977; Klohs et al., 1982). Roth et al. (1982) presented data attributing the response to a galactosyltransferase present in serum added to their culture media, since heat inactivation (56°C for 30 min) eliminated this effect on cell growth.

The role of galactosyltransferase in mediating cell-social behavior has been controversial. While some evidence for the participation of the enzyme in the regulation of tissue culture cell growth has been found, *in vivo* evidence for such a role has been elusive. Indeed, increased and decreased levels of the enzyme (Patt and Grimes, 1974; Bosmann and Hall, 1974; Roth et al., 1977) as well as unique isoenzymes (Podolsky and Weiser, 1975; Ram and Munjal, 1984) have been reported for solid tumor cell growth. Such diverse studies as limb-bud generation during embryogenesis (Elmer et al., 1988), liver regeneration (Bauer et al., 1976), and salivary gland proliferation induced by β-adrenergic receptor agonist (Humphreys-Beher et al., 1984, 1987) have detected cell surface galactosyltransferase during cell proliferation. Using monospecific antibodies to galactosyltransferase, Roth, Berger and co-workers have utilized light and electron micrographic studies to localize the enzyme. While Golgi staining with this antibody was detected in all cells examined, cell surface localization was found in about half the cell types examined (Roth et al., 1977; Berger et al., 1981; Pestalozzi et al., 1982; David et al., 1984).

CELL SURFACE GALACTOSYLTRANSFERASE INDUCED BY CHRONIC ISOPROTERENOL TREATMENT

For the past several decades, isoproterenol (a β-adrenergic receptor agonist) treatment of rodent salivary glands has provided a unique model for studying cell proliferation. Chronic administration of this agent leads to hypertrophic and hyperplastic enlargement of the parotid gland (Brown-Grant, 1961; Selye et al., 1961; Schneyer, 1962; Barka, 1965). The growth promoting effects of β-agonist treatment are accompanied by increased mitotic index and polyploidy of the acinar cells (Schneyer et al., 1967). Gland hyperplasia is most pronounced during the initial phase of treatment, but is not neoplastic, for it results in limited enlargement that is partially reversible upon cessation of drug treatment (Selye et al., 1961; Schneyer, 1972). Little progress beyond these observations has been made toward delineating the events leading to the shift from quiescence to active cell growth.

The transition of rat parotid gland acinar cells to a growth state was found to be accompanied by a dramatic increase in the synthesis of β1-4 galactosyltransferase (Humphreys-Beher et al., 1984). The original observation of the increase in galactosyltransferase was in association with alterations in the oligosaccharide structures of secreted glycoprotein from the parotid gland of Wistar rats (Humphreys-Beher, 1985). A plasma membrane localization for much of the increase in β1-4 galactosyltransferase was determined by density gradient membrane fractionation (Marchase et al., 1988). Golgi enriched membranes as well as assays of several other Golgi localized glycosyltransferases failed to exhibit changes in either subcellular localization or specific activity (Table 1) (Marchase et al., 1988). The resulting specific activity for β1-4 galactosyl-transferase in the plasma membrane isolated from acinar cells following iso-proterenol treatment was 40-fold higher than control animals and nearly two thirds that of the Golgi fractions.

To further investigate the enhancement of galactosyltransferase activity seen with isoproterenol, intact parotid gland acinar cells were isolated from treated and control rats and assayed for surface enzyme activity.

Using UDP-[^{3}H]-galactose to follow the reaction, a fourfold increase of labeled galactose into endogenous acceptors was observed after isoproterenol treatment. Using the exogenous acceptor, ovalbumin, an even higher level of activity was detected. The level of increased activity could not be attributed to cell breakage and release of galactosyltransferase to the media since leakage of lactate dehydrogenase to the media was only 4% of the activity released by sonic disruption of the cells. The incorporation of label could not be attributed to internal utilization of the isotope since inclusion of excess unlabeled galactose or galactose-1-phosphate had no effect on incorporation, while excess unlabeled UDP-galactose significantly diluted macromolecular label. The appearance of cell surface galactosyltransferase in rat parotid glands occurs as early as 3 h after

TABLE 1. Glycosyltransferase Activities (nmol/min/mg protein) in Parotid Membrane Fractions

Membrane Fraction	4β-Galactosyltransferase[a]		3β-Galactosyltransferase[b]		Sialytransferase[c]		N-acetylgluco-saminyltransferase[d]	
	Control	IPR-treated	Control	IPR-treated	Control	IPR-treated	Control	IPR-treated
Total membranes	0.36 ± 0.07	1.69 ± 0.17	1.93 ± 0.14	2.06 ± 0.2	0.86 ± 0.12	0.53 ± 0.21	0.16 ± 0.05	0.12 ± 0.07
Golgi-enriched	5.41 ± 0.23	5.04 ± 0.36	4.23 ± 0.17	4.33 ± 0.15	10.24 ± 0.31	8.16 ± 0.21	2.22 ± 0.11	2.28 ± 0.10
Plasma membranes[e]	0.07 ± 0.05	3.27 ± 0.05	0.02 ± 0.01	0.02 ± 0.01	0.03 ± 0.01	0.23 ± 0.02	0.04 ± 0.02	0.03 ± 0.01

[a] Ovalbumin (10 mg/ml).

[b] Asialobovine submaxillary mucin (10 mg/ml).

[c] Asialo-fetuin (10 mg/ml).

[d] Asiasol-galacto-1-N-acetylglucosaminyl-fetuin (10 mg/ml).

[e] Enrichment of plasma membrane fractions was determined by the assay of the parotid gland acinar cell plasma membrane marker enzyme γ-glutamyltranspeptidase.[20] Plasma membranes prepared as described resulted in a tenfold enrichment in the specific activity for this enzyme over that percent in the total membrane preparation (3700 U/mg protein and 350 U/ml, respectively).

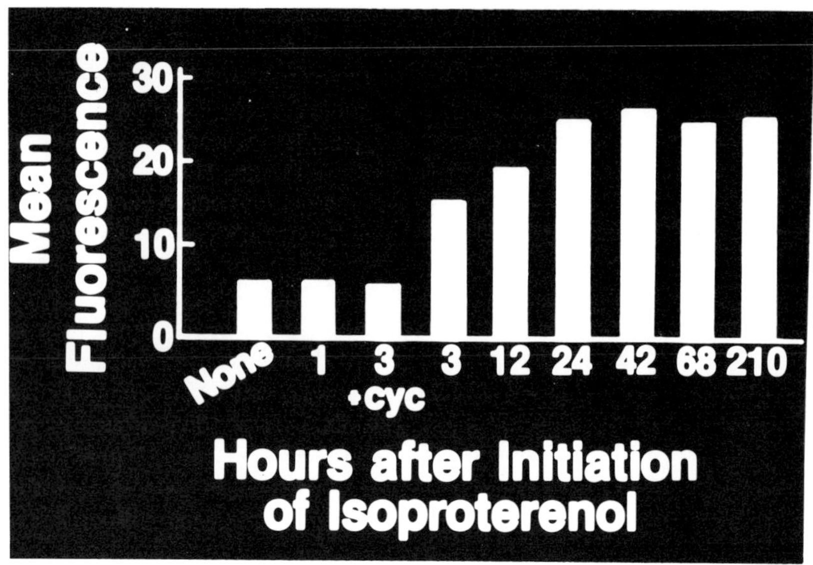

Figure 1. Levels of cell surface galactosyltransferase as a function of time after initiation of isoproterenol treatment. Quantitation was determined by fluorescence activated cell sorter utilizing antiserum to galactosyltransferase followed by reaction with an FITC-conjugated goat anti-rabbit second antibody (Marchase et al., 1988). Injections of 5 mg of isoproterenol began at 0 time and were repeated approximately every 12 h thereafter.

a single challenge of β agonist (Figure 1). A further increase is observed 12 h after treatment with isoproterenol, and with continued drug treatment increases further at 24 h before reaching a stable plateau. Pretreatment of animals with a dosage of cycloheximide sufficient to inhibit most protein synthesis (Pestka, 1971) prevents the increase in surface galactosyltransferase activity, suggesting the need for new protein biosynthesis for ectopic expression of the enzyme. Additionally, Strous et al. (1987) have provided evidence for specific phosphorylation of intracellular galactosyltransferase which is lacking in the secreted form of the enzyme, suggesting that protein phosphorylation may regulate the transferase enzyme subcellular localization.

Isolated parotid acinar cells were also assessed in these studies for surface galactosyltransferase by indirect immunofluorescence using a monospecific antibody to galactosyltransferase (Humphreys-Beher et al., 1984). Using this antiserum and a FITC-labeled secondary antibody, we obtained results on nonpermeabilized cells that correlated well with the enzyme assays and served to further confirm the cell surface expression of the enzyme (see Figure 2). Significant cellular staining was seen in isoproterenol treated cells but not in controls. These results, when coupled with the above enzyme activity assays strongly suggest that isoproterenol treatment results in a cell surface expression of galactosyltransferase with the enzyme active site oriented externally to the cell surface.

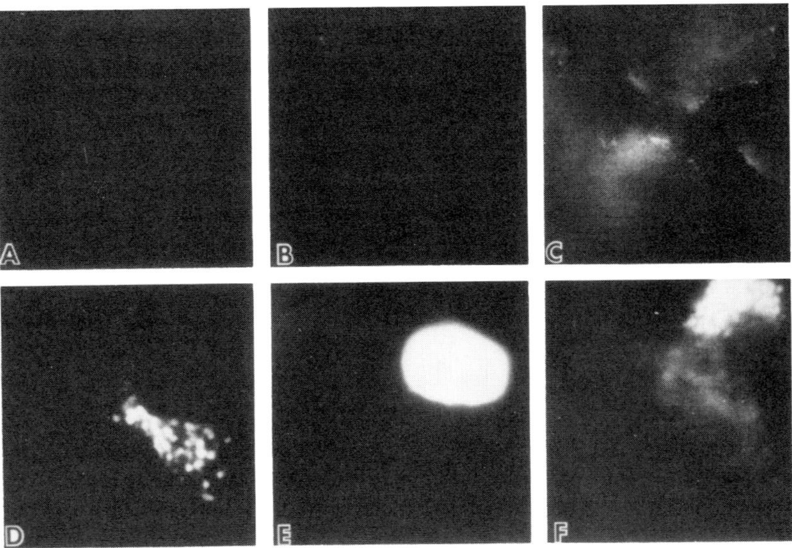

Figure 2. Immunofluorescent staining of isolated rat parotid gland acinar cells from control animals of following 5-day isoproterenol treatment. A, Intact nonpermeabilized isoproterenol-treated cells exposed to a preimmune rabbit serum; B, intact nonpermeabilized control cells exposed to antiserum for galactosyltransferase; C, isoproterenol-treated acinar cells exposed to the anti-galactosyltransferase serum; D, control acinar cells fixed and permeabilized with Triton X-100 and then exposed to antiserum to galactosyltransferase; the antibody revealed perinuclear, vesicular staining presumed to be Golgi apparatus, but no other staining in the cytoplasm nor on the cell surface; E, nuclear staining of the same cell shown in D, using Hoechst reagent; F, permeabilized isoproterenol-treated cells exposed to antiserum to galactosyltransferase. The plane of focus reveals intense vesicular, perinuclear staining presumed to be Golgi, and more diffuse staining owing (Marchase et al., 1988) the presence of antibody at the cell surface and mainly out of focus in this optical section.

IMPLICATION OF GALACTOSYLTRANSFERASE AS A MODULATOR OF PROLIFERATION

Because of the dramatic increase in cell surface galactosyltransferase in hyperplastic parotid glands and because of the possible involvement of the enzyme in growth control in other *in vitro* cell systems, it appeared possible that β1-4 galactosyltransferase might act as a modifier of the glandular growth. The specific modifier protein for galactosyltransferase, α-lactalbumin, which has proven to be an effective inhibitor of cell-surface enzyme-mediated behavior in other systems (Shur, 1984; Klohs et al., 1972), was injected in conjunction with isoproterenol to assess this potential in the rat parotid gland model.

Parotid gland enlargement as a consequence of isoproterenol treatment was found to be inhibited in a dose-dependent manner by α-lactalbumin. A control protein of similar size and charge, namely lysozyme (Shewale et al., 1984), was used as a control with no effect on isoproterenol induced gland hypertrophy (Table 2) (Humphreys-Beher et al., 1989). The specificity of α-lactalbumin for

TABLE 2. Parotid Gland Hypertrophy and 4β-Galactosyltransferase Activity

| | 4-β-Galactosyltransferase Activity[a] | | |
Treatment[b]	Gland Weight[c]	Total Membrane	Plasma Membrane
Control	0.36 ± 0.07[d]	0.14 ± 0.07[d]	0.07 ± 0.05[d]
Isoproterenol	1.69 ± 0.17	1.25 ± 0.25	3.27 ± 0.11
5μmol α-lactalbumin	0.38 ± 0.03	0.25 ± 0.90	0.06 ± 0.07
5μmol α-lactalbumin/IPR[e]	0.32 ± 0.05	1.17 ± 0.19	2.96 ± 0.25
5μmol α-lysozyme/IPR	1.63 ± 0.14	1.21 ± 0.16	2.98 ± 0.12
5μmol GDP-Man IPR	1.54 ± 0.23	1.24 ± 0.19	3.16 ± 0.08
5μmol UDP-Man IPR	0.98 ± 0.13	1.33 ± 0.12	2.91 ± 0.19
5μmol bovine serum albumin/IPR	1.67 ± 0.21	1.19 ± 0.17	3.19 ± 0.11
5μmol ovalbumin/IPR	1.05 ± 0.16	1.53 ± 0.23	3.18 ± 0.17
5μmol N-acetylglucosamine/IPR	1.64 ± 0.10	1.19 ± 0.21	3.06 ± 0.14
Pre-immune serum/IPR	1.73 ± 0.18	1.44 ± 0.21	2.89 ± 0.20
Anti-4β-galactosyltransferase serum/IPR	0.79 ± 0.19	1.37 ± 0.12	2.98 ± 0.17

[a] 4b-Galactosyltransferase activity is expressed as nanomoles of galactose/min/mg of membrane protein.
[b] All animals were provided a 5-day regimen of twice daily injections.
[c] Gland net weight determined after surgical removal and trimming of connective tissue and lymph nodes.
[d] Determinations performed on four experimental animals ± standard deviation.
[e] IPR, isoproterenol.

galactosyltransferase-expressing parotid gland acinar cells was provided by injecting ^{125}I-labeled α-lactalbumin into isoproterenol-treated and control rats. Radiolabeled protein was specifically concentrated in the parotid gland of isoproterenol-treated animals, but not in other tissue known to lack a cell-surface galactosyltransferase response upon β-adrenergic agonist challenge (Humphreys-Beher, 1989). ^{125}I-labeled α-lactalbumin showed no preferential accumulation in the parotid tissue of control rats, nor was the control protein ^{125}I-labeled lysozyme accumulated in either control or isoproterenol-treated parotid gland.

In addition to blocking gland hypertrophy, α-lactalbumin caused cessation of gland DNA synthesis as measured by ^{3}H-thymidine incorporation in the parotid gland (Table 3) (Humphreys-Beher et al., 1989). Between 36 and 41 h following the first injection of isoproterenol there was a fivefold increase in incorporation of ^{3}H-thymidine into DNA from parotid glands in the isoproterenol-treated rats compared to the DNA from glands of uninjected control animals. When glands from α-lactalbumin/isoproterenol animals were examined, no change in gland weight was detected. DNA synthesis at this time was enhanced to nearly isoproterenol-treated levels. However, by day five, synthesis of DNA was seen to have increased 35-fold with isoproterenol treatment alone while DNA synthesis in the glands of α-lactalbumin treated animals had nearly returned to basal levels. Introduction of other substrates specific for

TABLE 3. Incorporation of [³H] Thymidine into DNA

Treatment	Gland Size[a]		[³H] Thymidine Incorporated[b]	
	41 h	125 h	36 h	120 h
Control	0.32 ± 0.06	0.33 ± 0.05	125 ± 8	120 ± 10
Isoproterenol (IPR)	0.67 ± 0.07	1.67 ± 0.13	626 ± 10	4007 ± 23
5 µmol α-lactalbumin/IPR	0.34 ± 0.04	0.41 ± 0.06	551 ± 14	235 ± 8

[a] Values expressed in grams of average weight of parotid glands from two experimental animals.
[b] Values expressed as counts/min of [³H] thymidine/mg of total parotid protein.

galactosyltransferase also demonstrated a growth inhibiting effect on the parotid gland (Table 2) (Humphreys-Beher et al., 1989). These agents included the nucleotide sugar, UDP-galactose, substrate glycoprotein ovalbumin, and the monospecific antibody to galactosyltransferase.

While α-lactalbumin and other substrate macromolecules blocked gland hypertrophy and hyperplasia, isoproterenol-induced changes in gene expression of the proline-rich protein (Muenzer et al., 1979 a,b) remained unchanged. This was also true for the ectopic expression of galactosyltransferase (Table 2) (Humphreys-Beher et al., 1989). Thus it would appear that alterations in gene expression induced by isoproterenol can be separated from the cell proliferation effects *in vivo*.

SIGNAL GENERATION BY GALACTOSYLTRANSFERASE TO PROLIFERATION

Since the natural substrate of galactosyltransferase is N-acetylglucosamine, it is possible to propose that the signal for cell proliferation is generated by a conformational change either in galactosyltransferase or the corresponding membrane glycoprotein acceptor. These possibilities were evaluated in a series of experiments in which cell growth was reinitiated after priming the acinar cells by the isoproterenol/α-lactalbumin drug regimen (Humphreys-Beher, 1989). Initially, introduction of the soluble bovine galactosyltransferase was shown to restimulate proliferation. However, two conclusions can be drawn from this result. Either soluble galactosyltransferase restarts cell division by binding with α-lactalbumin, thereby freeing cell surface enzyme to interact with substrate glycoprotein to generate the transmembrane signal, or the soluble galactosyl-transferase interacts with the membrane glycoprotein substrate which generates the transmembrane signal for active cell growth. These possibilities were differentiated by introducing lectin proteins into the isoproterenol/α-lactalbu-min drug regimen. Under these circumstances acinar cell growth could be resumed by the N-acetylglucosamine specific lectin, Wheat Germ Agglutinin.

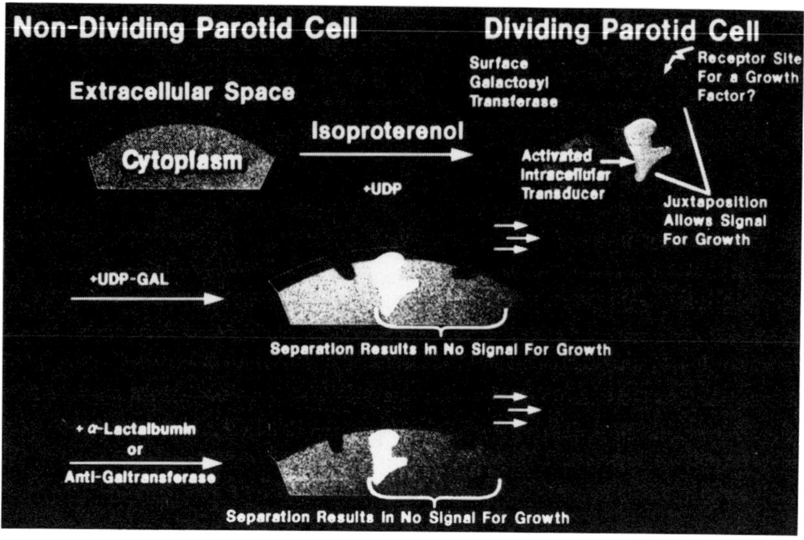

Figure 3. On the left is shown a nondividing parotid cell with little cell surface galactosyltransferase. Isoproterenol causes an increase in this enzyme as shown to the right. We hypothesize that the active site of this transferase is involved in the stable binding to the enzyme of some yet unknown cell-surface glycoprotein via a terminal N-acetylglucosamine residue of the oligosaccharide moiety. This juxtaposition is suggested to be crucial for the transduction of a signal for growth in these cells. The juxtaposition of these two surface proteins is postulated to cause a conformational change in the signal generating glycoprotein. The exogenous UDP-galactose allows for the addition of the sugar to the glycoprotein signal acceptor oligosaccharide and thus destroys the required juxtaposition. The addition of α-lactalbumin or antibody to galactosyltransferase also results in separation, in this case owing to stearic hindrance. Any of these additions would result in the inhibition of growth seen *in vivo* and *in vitro* with these agents (Humphreys-Beher et al., 1987; Ghanta et al., 1989).

The Wheat Germ Agglutinin binding could produce the same type of conformational change as the galactosyltransferase activity site binding its specific substrate glycoproteins.

Since there is no evidence for the internalization of UDP-galactose, α-lactalbumin, or antisera to galactosyltransferase, all of which inhibited the parotid gland hyperplasia, we suggest that it is the cell surface portion of this enzymatic activity that is mediating the growth inhibitory effects caused by these agents. One possible model illustrated in Figure 3 depicts β1-4 galactosyl-transferase as providing a long-term binding site for N-acetylglucosamine containing acceptor glycoproteins on the cell surface. As long as no UDP-galactose is present, catalysis cannot occur and the juxtaposition persists much as a lectin-ligand type interaction (Humphreys-Beher, 1989). This juxtaposition may be crucial for maintaining a cell surface glycoprotein, such as a growth factor receptor, in a geometry appropriate to transduction of signals for growth (Marchase et al., 1988). The presence of extracellular UDP-galactose would be expected to result in increased catalysis of acceptors by the transferase while the presence of antiserum of α-lactalbumin might be stearic hindrance, and also cause dissolution of such a juxtaposition. How such a phenomenon could result in inhibition of growth remains to be explored.

Once the signal has been generated at the cell surface by galactosyltransferase, the subsequent steps leading to activation of DNA synthesis and cell division are obscure. Several pathways of secondary messenger signaling exist in the parotid cell. Cyclic AMP generation by adenylate cyclase leads to the activation of protein kineses such as protein kinase A (Quissell et al., 1988, 1989). Protein kinase activation by phosphorylation, however, is generally considered to be involved in secretory stimulation rather than stimulation of DNA synthesis (Jahn et al., 1980; Baum et al., 1981; Quissell et al., 1985). The formation of inositol phosphates in parotid in parotid and submandibular glands occur by interaction with α-adrenergic receptors (Hawkins et al., 1986; Irvine et al., 1984). Inositol phosphate derivatives are also generated by high doses of isoproterenol and most likely result from α-adrenergic activation since the introduction of α-antagonists blocks isoproterenol-mediated inositol phosphate synthesis (unpublished observations). A third pathway of intracellular signal generated from the cell surface involves the tyrosine kineses, such as the EGF and insulin receptor (Yarden and Ullrich, 1988). Phosphorylation may play a specific role in the subcellular localization of galactosyltransferase (Strous et al., 1987).

Eventually, initiation of DNA synthesis requires the activation of nuclear genes. Most recently, Kousvelari and co-workers have identified the activation of a number of cellular protooncogenes (Yeh et al., 1988; Kousvelari et al., 1988). Induction of the c-*fos* gene has been detailed by two groups although there is no correlation with this oncogene expression and proliferation (Barka et al., 1986; Yeh et al., 1988). Further review of this topic is contained elsewhere in this book. The c-*fos* gene, involved in initiating transcription of quiescent genes, could participate in the activation of cell-division genes.

MOLECULAR MECHANISMS REGULATING CELL SURFACE GALACTOSYLTRANSFERASE

The initial observation of increased cell-surface galactosyltransferase was accompanied by evidence supporting the conclusion that there were changes in gene transcription upon isoproterenol treatment (Humphreys-Beher et al., 1984). *In vitro* translation of parotid gland RNA followed by immune precipitation of radiolabeled products identified an increase in a protein of 45,000 Da which corresponded to the reported nonglycosylated size of galactosyltransferase (Strous and Berger, 1982). Using the purported monospecific antiserum, a human cDNA library in the expression vector gt11 was screened. A positive clone was identified as a putative cDNA for galactosyltransferase (Humphreys-Beher et al., 1986). Dot-blot analysis showed increased hybridization to poly (A+)-containing RNA from isoproterenol-treated parotid glands. Northern

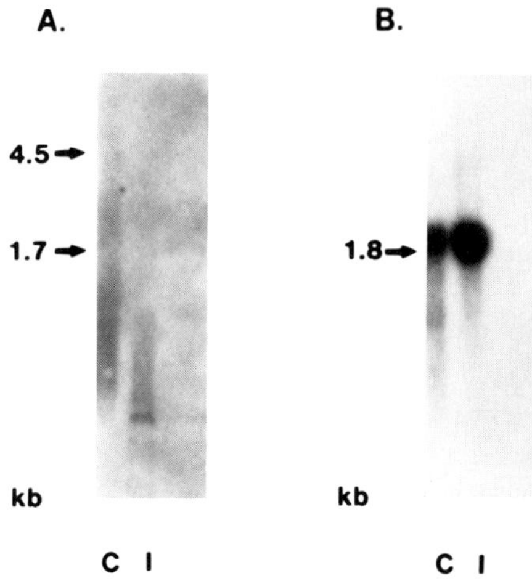

Figure 4. Northern blot analysis of parotid gland poly (A+) containing RNA. The RNA, isolated from isoproterenol-treated and control rats were separated in a one percent agarose-formaldehyde denaturing gel, transferred to nitrocellulose and hybridized to cDNA for rat galactosyltransferase (panel A) or chicken cDNA to β-tubulin (panel B). The proliferating cells of isoproterenol-treated animals demonstrate the synthesis of a unique mRNA species of approximately 4.5 to 4.7 kb. Equal quantities of RNA were loaded between the two samples based on hybridization in panel B with the tubulin probe.

analysis of this RNA showed the induction of a unique transcript of 4.5 kb upon isoproterenol treatment (Figure 4). Further studies on expression of this clone in eukaryotic cells further supported its claim to being a cDNA for galactosyltransferase (Bunnell et al., 1987). More recent studies by Lopez and Shur (1988) as well as our own DNA sequence analysis suggest this clone may in fact code for a specific galactosyltransferase modifier protein. Expression of the 1.7-kb mRNA from rat in *Escherichia coli* cells increases as intracellular galactosyltransferase activity (most likely an enzyme involved in lipopolysaccharide biosynthesis), which does not however demonstrate altered substrate specificity in the presence of α-lactalbumin nor bind to α-lactalbumin sepharose columns. As final evidence, an *E. coli* cell lysate expressing this clone failed to compete for binding to galactosyltransferase antibody in a reaction containing bovine galactosyltransferase protein. It is at this time intriguing to surmise that the galactosyltransferase modifier clone may be responsible for the ectopic expression of the enzyme to the cell plasma membrane. One possible mechanism for such occurrence could be by protein phosphorylation.

EXTRACELLULAR FACTORS STIMULATING ACINAR CELL PROLIFERATION

While isoproterenol has the most dramatic effect on inducing acinar cell proliferation, many other factors have been identified as influencing parotid gland hypertrophy and hyperplasia. Isoproterenol will also cause mitosis and enlargement of the intercalated duct cells (Hand and Ho, 1985). In addition to isoproterenol, other β-agonists also possess a growth stimulating effect (Abe and Dawes, 1980; Wells and Humphreys-Beher, 1985). Since isoproterenol mediates intracellular changes by the activation of adenylate cyclase with the subsequent increase in cAMP concentrations (Grand and Schay, 1978; Barka, 1965), treating animals with phosphodiesterase inhibitors can have the same effect on cell hyperplasia (Wells and Humphreys-Beher, 1985). Similarly, the introduction of hormones such as NGF, EGF, and thyroxine all produce parotid gland hypertrophy and hyperplasia (Schneyer and Humphreys-Beher, 1988; Humphreys-Beher et al., 1989, Gresik et al., 1981).

Dietary changes also appear to influence parotid gland hypertrophy and hyperplasia. The introduction of ground laboratory chow mixed with inert cellulose, the addition of citric acid to rat drinking water, maintenance on a liquid diet followed by solid chow, and submandibular gland ablation cause gland enlargement (Schneyer and Hall, 1975; Hall and Schneyer, 1977, 1978; Zelles et al., 1984). This enlargement as well as growth hormone appear to be mediated by β-adrenergic receptor stimulation and active participation of cell surface galactosyltransferase (Humphreys-Beher and Schneyer, 1987; Humphreys-Beher et al., 1989). Mehansho et al. (1983) identified induced salivary gland enlargement with a tannin-rich sorghum diet, although no further work has been pursued to identify mediation by cell-surface galactosyltransferase.

Documentation of salivary gland hypertrophy and hyperplasia in response to external stimuli has been noted for the submandibular gland as well. In addition to isoproterenol (Barka and Van der Noen, 1976; Fukushima and Barka, 1976; Koschel et al., 1976), partial removal of the submandibular gland (Yagil et al., 1986; Abe and Dawes, 1980) results in an increase in acinar cell mitotic indexes in both rat and mouse. Heat acclimatization of mice to temperatures of 34.5°C over a period of 4 weeks resulted in submandibular gland hypertrophy (Horowitz and Soskolne, 1978). During this period acinar cell hyperplasia was recorded during the first several days at the above temperature.

The circulating levels of hormones and growth factors appear to influence parotid gland homeostasis as it related to isoproterenol-induced hyperplasia. Studies in the mouse have indicated that submandibular-sublingual gland ablation prevents isoproterenol-induced cell division and DNA synthesis (Inoue et al., 1986). Changes in protein biosynthesis associated with isoproterenol treatment were, however, unaffected by gland ablation. Parotid gland explants could be restored to isoproterenol sensitivity by the incorporation of EGF into culture media. Similar results have been obtained with the neonatal rat in which submandibular-sublingual gland ablation drastically curtails isoproterenol-induced precocious development of the parotid gland *in vivo* (Schneyer and Humphreys-Beher, 1988; 1989). Restoration of isoproterenol-mediated DNA synthesis in these animals could be obtained by injection of NGF or EGF. Additional effects on isoproterenol-stimulated DNA synthesis of the parotid gland have been reported in the rat following removal of the parathyroid-thyroid gland complex (Gonzales et al., 1984).

EMBRYONIC AND NEONATAL ACINAR CELL PROLIFERATION

Salivary gland development in regards to galactosyltransferase expression has been addressed in a single study to date in the mouse submandibular gland (Kidd et al., 1988; Luo et al., 1987). The corresponding studies on morphological development are addressed elsewhere (Redman and Sreebny, 1971, Redman and Ball, 1978). High levels of expression of the unique 4.5-kb proliferation-associated mRNA for the galactosyltransferase modifier protein was established for proliferating epithelia for day 15 through 17 *in utero*. At this time the sublingual gland undergoes active morphogenesis from a single bud to a branched morphology (Redman and Ball, 1978). The submandibular gland on the other hand, differentiates terminal tubule cells containing several types of secretory granules (Dvorak, 1969). Low levels of hybridization with the cDNA probe were established for other stages of epithelia development as well as developing mesenchymal cells consistent with the level of enzyme necessary to function in intracellular protein glycosylation. When stimulated by chronic isoproterenol treatment epithelia cells of the adult mouse submandibular gland express the 4.5-kb mRNA for the galactosyltransferase modifier protein as well as elevated levels of enzyme activity (Kidd et al., 1988). This result is again consistent with the role of cell-surface galactosyltransferase as a mediator of normal cell and developmental proliferation of the salivary glands. In the neonatal rat, the salivary glands are at various stages of functional and morphological development. Observations on submandibular gland development indicate a temporal progression of cytodifferentiation which shows exocrine maturity is not reached until 10 to 12 weeks post-partum (Jacoby and Leeson, 1959; Cutler, 1974). Biochemical differentiation would appear to parallel cytodifferentiation as judged by the stage specific regulation of secreted proteins (Denny and Cope, 1977; Menaker and Miller, 1973; Barret and Ball, 1974). Similar parallels in the parotid gland have been found for post-partum cytodifferentiation and exocrine protein biosynthetic maturity (Ball, 1974; Robinovitch et al., 1976; Sreebny et al., 1965). Developmental regulation of cell division of the salivary glands has not been studied in detail, however. Several reports in the literature provide evidence for precocious development of the acinar cells in response to external stimuli such as isoproterenol and other hormones (Schneyer, 1972; Takuma et al., 1978; Sasaki et al., 1976). Surprisingly, the introduction of α-lactalbumin into neonatal rats also resulted in precocious development as determined by DNA and amylase synthesis of the parotid gland (Humphreys-Beher et al., 1987). This unexpected finding suggests that this protein may act as a bimodal regulator of growth, similar to that described for transforming growth factor β or cholera toxin β subunit (Roberts et al., 1985; Tucker et al., 1984). It is also worthy to note that α-lactalbumin is found naturally and in abundance in milk from mammalian species (Fitzgerald et al., 1970, 1971; Powell and Brew, 1976b; Brew et al., 1968). Galactosyltransferase has also been identified as a cell surface receptor which can interact with extracellular matrix molecules during cell differentiation and attachment (Shur, 1984; Banya et al., 1986;1988). The addition of α-lactalbumin to neonatal parotid cells may prevent certain interactions with extracellular matrix molecules which may regulate temporal differentiation. The premature loss of such interaction could be postulated to contribute to the precocious development of neonatal acinar cells exposed to α-lactalbumin.

GALACTOSYLTRANSFERASE AND PATHOLOGICAL GROWTH OF THE PAROTID GLAND

Many parallels exist between hypertrophy and hyperplasia observed in rodent models and human salivary gland pathologies. Most common occurrences of gland hypertrophy involve benign and malignant neoplastic growth (Brown et al., 1987). Parotid gland enlargement has been reported after severe malnutrition (Katsilambros, 1961; DuPlessis, 1956), excessive starch intake (Silverman and Perkins, 1966), refeeding after starvation (Watt, 1977), and metabolic disorders such as diabetes (Davidson et al., 1969). While Sjogrens syndrome and other types of allergic reactions result in parotid gland hypertrophy, enlargement in these cases is due to lymphocyte infiltration, not true salivary cell hyperplasia (Watt, 1977; Fox and Howell, 1986). The eating disorder, bulimia, also results in characteristic enlargement of the parotid gland, which on occasion necessitates surgical removal (Levin et al., 1980; Ogren et al., 1987).

A survey of galactosyltransferase enzyme activity in membrane preparations from different human salivary gland pathologies has recently been completed (Humphreys-Beher et al., 1990). As with rodent studies of acinar cell-proliferation, human parotid gland enlargement was accompanied by cell-surface galactosyltransferase. Also typical was the lack of a detectable change in any other glycosyltransferase activity. Northern blot analysis with the cDNA for the galactosyltransferase modifier protein (Humphreys-Beher et al., 1986) showed the constitutive synthesis of a 4.5-kb mRNA associated with tissue samples from parotid gland pleomorphic adenomas. Southern blot analysis of chromosomal DNA from both benign and malignant neoplasms showed a single instance of chromosomal alterations associated with the cDNA clone after digestion with two different restriction endonucleases (Humphreys-Beher et al.,

1989). The use of two restriction endonucleases in the detection of allelic variants would rule out differences due to DNA polymorphisms in the adenoid cystic carcinoma.

The direct role for galactosyltransferase in mediating acinar cell proliferation was determined by culturing neoplastic cells *in vitro*. The addition of either UDP-galactose or α-lactalbumin inhibited cell growth in a dose-dependent manner (Humphreys-Beher et al., 1989). Incubation with the control substrates, UDP-glucose or bovine serum albumin, demonstrated little effect on cell growth. Cell growth was resumed under conditions in which the media containing galactosyltransferase substrates was replaced with fresh media, indicating the effects on cell proliferation were not due to toxicity.

CONCLUSIONS AND FUTURE DIRECTIONS

In the studies summarized in this review, I have presented evidence for a direct role for the glycosyltransferase, β1-4 galactosyltransferase, in effecting parotid gland acinar cell proliferation. Normally localized to the trans-Golgi membranes of quiescent cells, the transition to active cell growth is accompanied by ectopic expression of the enzyme to the plasma membrane. Changes in gene expression for a galactosyltransferase modifier protein would appear to contribute to altering the subcellular localization of the enzyme. The mechanism for this action on galactosyltransferase remains to be resolved. Certainly, the complete sequence of the clone may help in elucidating this effect on galactosyltransferase localization. The most intriguing observation that may illuminate this mechanism of control is the phosphorylation of the Golgi form of the enzyme and dephosphorylation of secreted forms of the enzyme (Strous et al., 1987). Recent analysis of the galactosyltransferase cDNA clone of Shaper et al. (1986) suggests two mRNA species can be generated differing in length at the 5′ end by 200 nucleotides (Shaper et al., 1988). The structural implication suggests differential translation could contribute to topological distribution for a Golgi and cytoplasm localization for this glycosyltransferase. How these differing mechanisms of secretion may be reconciled with the plasma membrane localization remains for further experimentation involving *in vivo* expression of the recombinant protein.

The identification of the cell surface glycoprotein responsible for generating the cell proliferation signal has yet to be isolated. Although progress has been made in the identification of a putative protein, the induction of such a molecule in the plasma membrane may be coincidental with isoproterenol treatment (Humphreys-Beher, 1989). The identification and characterization of this glycoprotein may help delineate the intracellular signaling from the plasma membrane. Clearly, a number of pathways exist by which DNA synthesis is eventually triggered by events at the cell surface.

With cDNA clones for the galactosyltransferase modifier protein it is now possible to improve our understanding of the molecular mechanisms regulating isoproterenol-induced changes in gene regulation. Isolation and sequence analysis of the chromosomal gene will allow us to determine how isoproterenol effects transcription of the 4.5-kb mRNA species. The promotor sequences and surrounding regulatory elements may ultimately be involved in the constitutive transcription of this gene observed in human and murine neoplasms. The improvement in our knowledge of how genes involve in cell proliferation function in the salivary glands will better afford us more accurate therapies for controlling pathological conditions leading to salivary gland hypertrophy and hyperplasia.

REFERENCES

1. Abe, K. and Dawes, C., The secretion of protein and of some electrolytes in response to alpha- and beta-adrenergic agonists by rat parotid and submandibular salivary glands enlarged by chronic treatment with isoproterenol, *J. Dent. Res.*, 59, 1081, 1980.
2. Arvan, P. and Castle, J. D., Plasma membrane of the rat parotid gland: preparation and characterization of a fraction containing the secretory surface, *J. Biol. Chem.*, 95, 8, 1982.
3. Ball, W. D., Development of the rat salivary gland. IV. Amylase and ribonuclease development of the parotid and submaxillary glands, *Dev. Biol.*, 41, 267, 1974.
4. Barka, T., Induced cell proliferation: the effect of isoproterenol, *Cell Tissue Res.*, 37, 662, 1965.
5. Barka, T. and van der Noen, H., Stimulated growth of submandibular gland, *Lab. Invest.*, 35, 507, 1976.
6. Barka, T., Gubits, R. M., and van der Noen, H. M., Beta-adrenergic stimulation of c-fos gene expression in the mouse submandibular gland, *Mol. Cell Biol.*, 6, 2984, 1986.
7. Barondes, S. H., Lectins: their multiple endogenous cellular functions, *Annu. Rev. Biochem.*, 50, 207, 1981.
8. Barret, M. L. and Ball, W. D., Development of the rat salivary glands. II. The identification of a trypsin-like protease in the submaxillary gland, *Dev. Biol.*, 36, 195, 1974.
9. Baum, B. J., Freiberg, J. M., Ito, H., Roth, G. S., and Filburn, C. R., Beta-adrenergic regulation of protein phosphorylation and its relationship to exocrine secretion in dispersed rat parotid gland acinar cells, *J. Biol. Chem.*, 256, 9731, 1981.
10. Bayna, E. M., Runyan, R. B., Scully, N. F., Reichner, J., Lopez, L. C., and Shur B. D., Cell-surface galactosyl transferase as a recognition molecule during development, *Mol. Cell. Biochem.*, 72, 141, 1986.
11. Bayna, E. M., Shaper, J. H., and Shur, B. D., Temporally specific involvement of cell surface β1,4 galactosyltransferase during mouse embryo morula compaction, *Cell*, 53, 145, 1988.
12. Bauer, C. H., Hassels, B. F., and Reutter, W. G., Galactose metabolism in regenerating rat liver, *Biochem. J.*, 154, 141, 1976.
13. Berger, E. G., Mandel, T., and Schilt, U., Immunohistochemical localization of galactosyltransferase in human fibroblasts and Hela cells, *J. Histochem. Cytochem.*, 29, 364, 1981.

14. Bernecki, R. J., Plasma membrane ectoglycosyl transferase activity of L1210 murine leukemic cells, *J. Cell Physiol.*, 83, 457, 1974.

15. Bosmann, H. B. and Hall, T. C., Enzyme activity in invasive tumors of human breast and colon, *Proc. Natl. Acad. Sci. U.S.A.*, 71, 1833, 1974.

16. Bosmann, H. B., Cell surface glycosyl transferases and acceptors in normal and RNA-and DNA-virus transformed fibroblasts, *Biochem. Biophys. Res. Commun.*, 48, 523, 1972.

17. Brew, K., Vanaman, T. C., and Hill, R. L., The role of α-lactalbumin and the A protein in lactose synthetase: a unique mechanism for the control of a biological reaction, *Proc. Natl. Acad. Sci. U.S.A.*, 59, 491, 1968.

18. Brown, F. H., Houston, G. D., Lubow, R. M., and Sagan, M. A., Adenomatoid hyperplasia of mucous salivary glands, *J. Periodontol.*, 58, 125, 1987.

19. Brown-Grant, K., Enlargement of salivary gland in mice treated with isopropylnoradrenaline, *Nature* 191, 1076, 1961.

20. Bunnell, B., Humphreys-Beher, M. G., and Kidd, V. J., in *Advances in Gene Technology: The Molecular Biology of Development.*, Whalen, W., Ed., ICSU Short Reports, Cambridge University Press, Cambridge, England, 1987, 22.

21. Cerven, E., Demonstration of sialyl transfer from exogenous CMP-N-acetyl [^{14}C] neuraminic acid to cell membrane-bound acceptor molecules at the surface of intact ehrlich ascites tumor cells, *Biochem. Biophys. Acta*, 467, 72, 1977.

22. Cutler, L. S., Cytodifferentiation of the acinar cells of the rat submandibular gland, *Dev. Biol.*, 41, 31, 1974.

23. Davey, R., Harvie, R., Cahill, J., and Levi, J., Serum galactosyltransferase isoenzyme patterns of cancer patients with liver involvement, *Br. J. Cancer*, 53, 211, 1986.

24. Davidson, D., Leibel, B. S., and Berris, B., Asymptomatic parotid gland enlargement in diabetes mellitus, *Annu. Intern. Med.*, 70, 31, 1969.

25. Davis, B. W., Berger, E. G., Locher, G. W., Zeller, M., and Goldhirsch, A., Immunohistochemical localization of galactosyltransferase in the normal human ovary and fallopian tube, *J. Histochem. Cytochem.*, 30, 1146, 1984.

26. Denny, P. C. and Cope, P., Age-specific differentiations during the development of the rat submandibular gland, *Differentiation*, 8, 105, 1977.

27. DuPlessis, D. J., Parotid gland enlargement in malnutrition, *S. Afr. Med. J.*, 30, 700, 1956.

28. Dvorak, M., The secretory cells of the submaxillary gland in the perinatal period of development in the rat, *Z. Zellforsch.*, 99, 346, 1969.

29. Elmer, W. A., Pennybacker, M. F., Knudson, T. B., and Kwasigroch, T. E., Alterations in cell surface galactosyltransferase activity during limb chondrogenesis in brachypod mutant mouse embryos, *Teratology*, 38, 475, 1988.

30. Feizi, T. and Childs, R. A., Carbohydrates as antigenic determinants of glycoproteins, *Biochem. J.*, 245, 1, 1987.

31. Fitzgerald, D. K., Brodbeck, U., Kiyosawa, R., Mawal, R., Colvin, B., and Ebner, K. E., α-Lactalbumin and the lactose synthetase reaction, *J. Biol. Chem.*, 245, 2103, 1970.

32. Fitzgerald, D. K., McKenzie, L., and Ebner, K. E., Galactosyltransferase activity in a variety of sources, *Biochem. Biophys. Acta*, 235, 425, 1971.

33. Fox, R. and Howell, F., Oral problems in patients with Sjogrens syndrome, *Scand. J. Rheumatol.*, 61, 194, 1986.

34. Frasier, W. and Glaser, L., Surface components and cell regulation, *Annu. Rev. Biochem.*, 48, 491, 1979.
35. Fukushima, M. and Barka, T., The effects of 5-bromodeoxyuridine and isoproterenol on the postnatal differentiation of rat submandibular gland, *Am. J. Anat.*, 147, 159, 1976.
36. Ghanta, V. K., Hiramoto, R. N., Parker, C. A., Kidd, V. J., and Humphreys-Beher, M. G., *In vitro* growth and *in vivo* growth control of transformed lymphoid cells expression plasma membrane galactosyltransferase, *Oncogene*, in press.
37. Gonzales, R. C., Whitfield, J. F., Boynton, A. L., MacManua, J. P., and Rixon, R. H., Prereplicative changes in the soluble:calmodulin of isoproterenol-activated rat parotid glands, *J. Cell Physiol.*, 118, 257, 1984.
38. Grand, R. J. and Schay, I. M., Development of secretory function in rat parotid gland, *Pediatr. Res.*, 12, 100, 1978.
39. Gresik, E. W., van der Noen, H. and Barka, T., Hypertrophic and hyperplastic effects of thyroxine on the submandibular gland of the mouse, *Anat. Rec.*, 200, 443, 1981.
40. Hall, H. D. and Schneyer, C. A., Functional mediation of compensatory enlargement of the parotid, *Cell Tissue Res.*, 184, 249, 1977.
41. Hand, A. R. and Ho, B., Mitosis and hypertrophy of intercalated duct cells and endothelial cells in the isoproterenol-treated rat parotid gland, *J. Dent. Res.*, 64, 1031, 1985.
42. Hawkins, P. T., Stephens, L., and Downes, C. P., Rapid formation of inositol 1,3,4,5-tetrakisphosphate and inositol 1,3,4-triphospate in rat parotid glands may both result indirectly from receptor-stimulated release of inositol 1,4,5-triphosphate from phosphoinositol 4,5-bisphosphate, *Biochem. J.*, 238, 507, 1986.
43. Hawkins, P. T., Berrie, C. P., Morris A. J., and Downes, C. P., Inositol 1,2-cyclic 4,5-trisphosphate is not a product of muscarinic receptor-stimulated phosphatidyl inositol 4,5-bisphosphate hydrolysis in rat parotid glands, *Biochem. J.*, 243, 211, 1987.
44. Horowitz, M. and Soskolne, W. A., Cellular dynamics of rat submaxillary gland during heat acclimatization, *J. Appl. Physiol.*, 44, 21, 1978.
45. Hully, J. R., Benton, H. P., and Alison, M. R., Isoprenaline-induced cell proliferation in mouse salivary glands: the effect of castration, *Virchows. Arch.*, 47, 95, 1984.
46. Humphreys-Beher, M. G., Imell, M., Jentoft, N., Gleason, M., and Carlson, D. M., Isolation and characterization of UDP-galactose: N-acetylglucosamine 4β-galactosyltransferase activity induced in rat parotid glands treated with isoproterenol, *J. Biol. Chem.*, 259, 5, 799, 1984.
47. Humphreys-Beher, M. G., Bunnell, B., Ledbetter, D. H., Van Tuinen, P., and Kidd, V. J., Molecular cloning and chromosomal localization of human 4β-galactosyltransferase, *Proc. Natl. Acad. Sci. U.S.A.*, 83, 8918, 1986.
48. Humphreys-Beher, M. G. and Schneyer, C. A., Cell surface expression of 4β-galactosyltransferase accompanies rat parotid gland hypertrophy induced by changes in diet, *Biochem. J.*, 246, 387, 1987.
49. Humphreys-Beher, M. G., Schneyer, C. A., Kidd, V. J., and Marchase, R. B., Isoproterenol-mediated parotid gland hypertrophy is inhibited by effectors of 4β-galactosyltransferase, *J. Biol. Chem.*, 262, 11, 706, 1987a.

50. Humphreys-Beher, M. G., Schneyer, C. A. and Zelles, T., Alpha-lactalbumin acts as a bimodal regulator of rat parotid acinar cell growth, *Biochem. Biophys. Res. Commun.*, 147, 174, 1987b.

51. Humphreys-Beher, M. G., Restoration of α-lactalbumin-inhibited rat parotid salivary gland hypertrophy and hyperplasia by agents specific for membrane glycoprotein N-acetylglucosamine, *Arch. Oral Biol.*, 34, 811, 1989.

52. Humphreys-Beher, J. B., Zelles, T., Madea, N., Purushothamm, K. R., Cassisi, N., and Schneyer, C. A., Cell surface galactosyltransferase acts as a modulator of rat and human acinar cell proliferation, *Adv. Dent. Res.*, in press.

53. Humphreys-Beher, M. G., Zelles, T., Maeda, N., Purushotham, K. R., and Schneyer, C. A., Cell-surface galactosyltransferase acts as a general modulator of rat acinar cell proliferation, *Mol. Cell. Biochem.*, in press.

54. Inoue, H., Kikuchi, K., and Nishino, M., Effects of epidermal growth factor on the synthesis of DNA and polyamine in isoproterenol-stimulated murine parotid gland, *J. Biochem., (Japan)*, 100, 605, 1986.

55. Ip, C. and Dao, T. L., Increase in serum and tissue glycosyltransferases and glycosidases in tumor-bearing rats, *Cancer Res.*, 37, 3442, 1977.

56. Irvine, R. F., Letcher, A. J., Lander D. J., and Downes, C. P., Inositol triphosphates in carbachol-stimulated rat parotid glands, *Biochem. J.*, 223, 237, 1984.

57. Jahn, R., Unger, C., and Soling, H. D., Specific protein phosphorylation during amylase release evoked by beta-adrenergic or cholinergic agonists in rat and mouse parotid glands, *Proc. Natl. Acad. Sci. U.S.A.*, 78, 6903, 1980.

58. Jentoft, N., Cheng, P. W. and Carlson, D. M., in *Glycosyltransferases and Glycoprotein Biosynthesis*, Plenum Press, New York, 1976, 343.

59. Katsilambros, L., Asymptomatic enlargement of the parotid glands, *JAMA*, 178, 513, 1961.

60. Khatra, B. S., Herries, A. G., and Brew, K., Some kinetic properties of human milk galactosyltransferase, *Eur. J. Biochem.*, 4, 537, 1974.

61. Kidd, V. J., Luo, W., Humphreys-Beher, M. G., Slavkin, H. C., and Snead, M. L., Galactoslytransferase expression in developing mouse submandibular gland, *J. Dent. Res.*, 67, 190, 623, 1988.

62. Klohs, W. D., Mastrangelo, R., and Weiser, M. M., Release of glycosyltransferase and glycosidase activities from normal and transformed cell lines, *Cancer Res.*, 41, 2611, 1981.

63. Klohs, W. D., Wilson, J. R., and Weiser, M. M., UDP-galactose inhibited of balb/3T cell growth, *Exp. Cell. Res.*, 141, 365, 1982.

64. Kornfeld, R. and Kornfeld, S., Assembly of asparagine-linked oligosaccharides, *Annu. Rev. Biochem.*, 54, 631, 1985.

65. Koschel, K. W., Hodgson, G. S., and Radley, J. M., Characteristics of the isoprenaline stimulated proliferative response of rat submaxillary gland, *Cell Tissue Kinet.*, 9, 157, 1976.

66. Kousvelari, E., Louis, J. M., Huang, L.-H., and Curran, T., Regulation of proto-oncogenes in rat parotid acinar cells *in vitro* after stimulation of β-adrenergic receptors, *Exp. Cell Res.*, 179, 194, 1988.

67. LaMont, J. T., Gammon, M. T., and Isselbacher, K. J., Cell-surface glycosyltransferases in cultured fibroblasts: increased activity and release during serum stimulation of growth, *Proc. Natl. Acad. Sci. U.S.A.*, 74, 1086, 1976.

68. Levin, P. A., Falko, J. M., Dixon, K., Gallup, E. M., and Saunders, W., Benign parotid enlargement in bulimia, *Ann. Intern. Med.*, 93, 827, 1980.

69. Lopez, L. C. and Shur, B. D., Comparison of two independent cDNA clones reported to encode β1-4 galactosyltransferase, *Biochem. Biophys. Res. Commun.*, 156, 1223, 1988.

70. Luo, W., Kidd, V. J., Humphreys-Beher, M. G., Slavkin, H. C., and Sneed, M. L., Expression of 4β-galactosyltransferase mRNA in the ectoderm-derived cells of the developing mouse submandibular gland, *J. Cell Biol.* 105, 49 (Abstr.), 265, 1987.

71. Marchase, R. B., Kidd, V. J., Rivera, A. A., and Humphreys-Beher, M. G., Cell-surface expression of 4β-galactosyltransferase accompanies rat parotid acinar cell transition to growth, *J. Cell. Biochem.*, 36, 453, 1988.

72. Mehansho, H., Hagerman, A., Clements, S., Butler, L., Rogler, J., and Carlson, D. M., Modulation of proline-rich protein biosynthesis in rat parotid glands by sorgums with high tannin levels, *Proc. Natl. Acad. Sci. U.S.A.*, 80, 3948, 1983.

73. Menaker, L. and Miller, S. A., Regulation of protein synthesis in the submandibular salivary gland of the developing rat. Ribonuclease and ribonuclease inhibitor activity, *Arch. Oral Biol.*, 18, 1303, 1973.

74. Muenzer, J., Bildstein, Gleason, M., and Carlson, D. M., Properties of proline-rich proteins from parotid glands of isoproterenol-treated rats, *J. Biol. Chem.*, 154, 5629, 1979a.

75. Muenzer, J., Bildstein, C., Gleason, M., and Carlson, D. M., Purification of proline-rich proteins from parotid glands of isoproterenol treated rats, *J. Biol. Chem.*, 254, 5623, 1979b.

76. Orgren, F. P., Hunter, J. B., Pearson, P. H., Antonson, C. W., and Moore, E. F., Transient salivary gland hypertrophy in bulimics, *Laryngoscope*, 97, 951, 1987.

77. Patt, L. M. and Grimes, W. J., Cell surface glycolipids and glycoprotein glycosyltransferases of normal and transformed cells, *J. Biol. Chem.*, 249, 4157, 1974.

78. Pestalozzi, D. M., Hess, M., and Berger, E. G., Immunohistochemical evidence for cell surface and Golgi localization of galactosyltransferase in human stomach, jejunum, liver and pancreas, *J. Histochem. Cytochem.*, 30, 1146, 1982.

79. Pestka, A., Cycloheximide and other glutarimide antibiotics, *Annu. Rev. Microbiol.*, 25, 545, 1971.

80. Pierce, M., Turley, E. A., and Roth S., Cell surface glycosyltransferase activities, *Intl. Rev. Cytol.*, 65, 2, 1980.

81. Podolsky, D. K. and Weiser, M. M., Galactosyltransferase activities in human sera: detection of a cancer-associated isoenzyme, *Biochem. Biophys. Res. Commun.*, 65, 545, 1975.

82. Porter, C. W. and Bernecki, R. J., Ultrastructural evidence for ectoglycosyltransferases, *Nature, (London)*, 256, 648, 1975.

83. Powell, J. T. and Brew, K., Metal ion activation of galactosyltransferase, *J. Biol. Chem.*, 251, 3645, 1976a.

84. Powell, J. T. and Brew, K., A comparison of the interactions of galactosyltransferase with a glycoprotein substrate (ovalbumin) and with α-lactalbumin, *J. Biol. Chem.*, 251, 3653, 1976b.

85. Powell, J. T., Jarlfors, U. and Brew, K., Enzymic characteristics of fat globule membranes from bovine colostrum and bovine milk, *J. Cell Biol.*, 72, 617, 1977.

86. Quissell, D. O., Deisher, L. M., and Barzen, K. A., The rate determining step in cAMP mediated exocytosis in the rat parotid and submandibular glands appears to involve analogous 26 kDa integral membrane phosphoproteins, *Proc. Natl. Acad. Sci. U.S.A.*, 82, 3237, 1985.

87. Quissell, D. O., Deisher, L. M., and Barzen, K. A., Subcellular distribution and activation of rat submandibular cAMP-dependent protein kinase following β-adrenergic receptor stimulation, *Biochem. Biophys. Acta*, 969, 28, 1988.

88. Quissell, D. O., Deisher, L. M., and Barzen, K. A., Activation and distribution of rat parotid cAMP-dependent protein kinase following beta-adrenergic receptor stimulation, *in vitro*, *Arch. Oral Biol.*, 34, 23, 1989.

89. Rabinovitch, M. R., Keller, P. J., Johnson, D. A., Iversen, J. M., and Kauffman, D. L., Changes in rat parotid salivary proteins induced by chronic isoproterenol administration, *J. Dent. Res.*, 56, 290, 1976.

90. Ram, B. P. and Munjal, D. D., Isolation and characterization of cancer-associated galactosyltransferase isoenzyme, *Clin. Chem.*, 30, 1656, 1984.

91. Rauvali, H. and Hakomori, S., Studies on cell adhesion and recognition. III. The occurrence of alpha-mannosidase at the fibroblast cell surface, and its possible role in cell recognition, *J. Cell Biol.*, 88, 149, 1981.

92. Redman, R. S. and Ball, W. D., Cytodifferentiation of secretory cells in the sublingual gland of the prenatal rat: a histological, histochemical and ultrastructural study, *Am. J. Anat.*, 153, 367, 1978.

93. Redman, R. S., in *The Salivary System*, Sreebny, L. M., Ed., CRC Press, Boca Raton, FL, 1987, 2.

94. Redman, R. S. and Sreebney, L. M., Morphologic and biochemical observations on the development of the rat parotid gland, *Dev. Biol.*, 25, 248, 1971.

95. Roberts, A. B., Anzano, M. A., Wakefield, L. M., Roche, N. S., Stern, D. F., and Sporn, M. B., Type β transforming factor: a bifunctional regulator of cellular growth, *Proc. Natl. Acad. Sci. U.S.A.*, 82, 119, 1985.

96. Roseman, S., The synthesis of complex carbohydrates by multiglycosyltransferase systems and their potential function in intercellular adhesion, *Chem. Phys. Lipids.*, 5, 270, 1970.

97. Roth, S. and White, D., Intercellular contact and cell-surface galactosyltransferase activity, *Proc. Natl. Acad. Sci. U.S.A.*, 69, 485, 1972.

98. Roth, S., Roelke, M., and Dorsey, J., *Growth, Kinetics and Biochemical Regulation of Normal and Malignant Cells*, Drewinko, B. and Humphrey, R. M., Eds., Williams and Wilkins, Baltimore, MD, 1977, 245.

99. Ruoslahti, E., Proteoglycans in cell regulation, *J. Biol. Chem.*, 264, 13369, 1989.

100. Sasaki, R., Mura, M., Takevchi, T., Furihata, C., Matsushima, T., and Sugimura, T., Precocious development of rat salivary glands following dexamethasone treatment, *Biochem. Biophys. Acta*, 428, 619, 1976.

101. Schachter, H. and Roseman, S., *The Biochemistry of Glycoproteins and Proteoglycans*, Lennarz, W. J., Ed., Plenum Press, New York, 1980, 85.

102. Schneyer, C. A., Salivary gland changes after isoproternol-induced enlargement, *Am. J. Physiol.*, 203, 232, 1962.

103. Schneyer, C. A., Finley, W. H., and Finley, S. A., Increased chromosome number of rat parotid cells after isoproterenol, *Proc. Soc. Exp. Biol. Med.*, 125, 722, 1967.

104. Schneyer, C. A., *Regulation of Organ and Tissue Growth*, Goss, R. J., Ed., Academic Press, New York, 1972, 211.

105. Schneyer, C. A. and Hall, H. D., Parasympathetic regulation of mitosis induced in rat parotid by dietary change, *Am. J. Physiol.*, 229, 1975.
106. Schneyer, C. A. and Humphreys-Beher, M. G., Effects of epidermal growth factor and nerve growth factor on isoproterenol-induced DNA synthesis in rat parotid and pancreas following removal of submandibular-sublingual glands, *J. Oral Pathol.*, 17, 250, 1988.
107. Schneyer, C. A., Sucanthapree, C., Scnheyer, L. H., and Jirakulsomchok, D., Total salivary calcium and amylase output of rat parotid with electrical stimulation of autonomic innervation, *Proc. Soc. Exp. Biol. Med.*, 159, 478, 1978.
108. Schneyer, C. A. and Humphreys-Beher, M. G., Inhibition of isoproterenol-induced DNA and RNA synthesis in rat parotid and pancreas following removal of submandibular-sublingual glands, *Cell Tissue Res.*, 256, 361, 1989.
109. Selye, H., Cantin, M., and Veilleux, R., Abnormal growth and sclerosis of the salivary gland induced by chronic treatment with isoproterenol, *Growth*, 25, 243, 1961.
110. Shaper, N. L., Mann, P. L., and Shaper, J. H., Bovine galactosyltransferase: identification of a clone by direct immunological screening of a cDNA expression library, *Proc. Natl. Acad. Sci. U.S.A.*, 83, 1573, 1986.
111. Shaper, T. H., Hollis, G. F., and Shaper, N. L., Evidence for two forms of murine β1-4 galactosyltransferase based on cloning studies, *Biochemie*, 70, 1683, 1988.
112. Shewale, J. A., Shinha, S. K., and Brew, K., Evolution of α-lactalbumins, *J. Biol. Chem.*, 259, 4947, 1984.
113. Shur, B. D., The receptor function of galactosyltransferase during cellular interactions, *Mol. Cell. Biochem.*, 61, 143, 1984.
114. Silverman, M. and Perkins, R. L., Bilateral parotid enlargement and starch ingestion, *Ann. Intern. Med.*, 64, 842, 1966.
115. Sreebny, L. M., Wanamaker, B. B., and Rabinovitch, M., The DNase of the rat parotid gland, *J. Dent. Res.*, 44, 463, 1965.
116. Strous, G. J. and Berger, E. G., Biosynthesis, intracellular transport, and release of the Golgi enzyme galactosyltransferase (lactose synthetase A protein) in HeLa cells, *J. Biol. Chem.*, 257, 7623, 1982.
117. Strous, G. J., Van Kerkhof, P., Fallon, R. J., and Schwartz, A. L., Golgi galactosyltransferase contains serine-linked phosphate, *Eur. J. Biochem.*, 169, 307, 1987.
118. Struck, D. K. and Lennarz, W. J., *The Biochemistry of Glycoproteins and Proteoglycans*, Lennarz, W. J., Ed., Plenum Press, New York, 35, 1980.
119. Takeda, Y., Hirose, H., and Enotomo, S., Enlargement of rat submandibular salivary gland induced by single amputation of lower incisor teeth. Histology histometric and ultrastructural studies, *J. Oral Pathol.*, 15, 327, 1986.
120. Takuma, T., Nakanishi, M., Takagi, Y., Tanemura, T., and Kumegawa, M., Precocious differentiation of mouse parotid glands and pancreas induced by hormones, *Biochem. Biophys. Acta*, 538, 376, 1978.
121. Tucker, R. F., Shipley, G. D., Moses, H. L., and Holley, R. W., Growth inhibitor from BSC-1 cells closely related to platelet Type β transforming growth factor, *Science*, 226, 705, 1984.
122. Watt, J., Benign parotid swellings: a review, *Proc. R. Soc. Med.*, 70, 483, 1977.
123. Wells, D. J. and Humphreys-Beher, M. G., Analysis of protein synthesis in rat salivary glands after chronic treatment with β-receptor agonists and phosphodiesterase inhibitors, *Biochem. Pharmacol.*, 34, 4229, 1985.

124. Yagil, C., Michaeli, Y., and Zajicek, G., Compensatory proliferative response of the rat submandibular salivary gland to unilateral extirpation, *Virchows. Arch.*, 49, 83, 1985.
125. Yarden, Y. and Ullrich, A., Growth factor receptor tyrosine kinases, *Annu. Rev. Biochem.*, 57, 443, 1988.
126. Yeh, C.-K., Louis, J. M., and Kousvelari, E. E., β-Adrenergic regulation of c-fos gene expression in an epithelial cell line, *FEBS Lett.*, 240, 118, 1988.
127. Zelles, T., Blazsek, J., Kobor, A., and Gelencser, F., A glandula parotis gasztatorikus ingerrel letrehozott megnagyobboda'sa, *Fogorvosi Szemle*, 77, 315, 1984.

DNA Synthesis and Cell Division

Paul C. Denny and Patricia A. Denny

INTRODUCTION

The capacities for DNA replication and cell division in mature salivary gland are of special interest because they provide insights into the mechanisms for maintaining long-term stability of parenchymal cell populations and ultimately, for retaining adequate function. Which mature cell types have the capacity to divide? What are the relative contributions attributable to self-proliferation and to differentiation of the progeny of stem cells? Which cell populations are considered to be expanding and which are renewing? These fundamental questions are themselves dependent upon knowledge of turnover kinetics for each cell type, features of their cell cycles, as well as proportions and characteristics of the noncycling cells.

The abilities of salivary glands to produce a hyperplastic response to a variety of stimuli and to regenerate lost parenchyma suggest that induced cell replication may become a therapy in the future for restoration of function in salivary glands. In this regard, investigations of the different types of hyperplastic responses are valuable, especially as they relate to the types of cells that respond, their long-term stability, and the overall consequences of the response.

TECHNICAL ISSUES

As we begin to explore the above questions, it becomes apparent that technical issues are a dominant factor in judging the application and in interpreting the studies that relate to these issues. For instance, while an increase in total

DNA of a salivary gland is indicative of hyperplasia, it does not reveal which cell type(s) may be involved and indeed, even if the process occurred in the parenchymal or stromal cells. Similarly, an increase in ^{3}H-thymidine incorporation provides little information about the types of cells involved. The use of labeling index (LI) or mitotic index (MI) provides a more direct approach to analyzing cell division and DNA synthesis. The LI is usually obtained by pulsing with ^{3}H-thymidine for 1 to 2 hours. Following autoradiography, the number of radiolabeled nuclei of a cell type are related to the total number of cells that were observed for that particular cell type (Chang and Barka, 1974; Klein, 1982). Though little is known of cell cycle kinetics for parenchymal cells in adults, studies of acinar cells in neonates (Klein et al., 1976) and IPR-stimulated acinar cells in adults (Radley and Hodgson, 1971a) both suggest that a pulse of longer than 2 h might extend into M phase and beyond. This would result in an elevated LI because daughter cells would be included. A further consideration is that ^{3}H-thymidine incorporation by the nucleus of a cell does not preclude subsequent mitosis or cytokinesis. Each of the following have been documented among salivary gland acinar cells: binucleate (Epperlein and Kiefer, 1979; Denny et al., 1990b), G2-delayed (Klein and Harrington, 1977; Gelfant, 1977), and tetraploid and octaploid cells (Epperlein and Kiefer, 1979). The MI is obtained by relating the number of cells of a cell type undergoing mitosis to the total number observed of that particular cell type (Schneyer and Hall, 1969). MI can also be affected by recruitment of a G2-delayed or blocked cells (Klein and Harrington, 1977), thereby providing an inaccurate reflection of the proportion cells which are actively cycling in a population. Combining aspects of both approaches, the fraction of labeled mitosis (FLM) overcomes some of the above reservations (Klein, 1979). Observation of side-by-side radiolabeled nuclei at time intervals following a pulse label also has the potential for quantifying those cells which have continued through the cycle to complete mitosis and cytokinesis. However confusing this approach, several side-by-side pairs of radiolabeled cells were noted immediately following a single 90-min pulse in the submandibular gland (Denny et al., 1990b). Several studies have employed the use of colchicine to block mitosis at metaphase (Klein and Harrington, 1977; Sans and Galanti, 1979a). This has the advantage of expanding the temporal window for observing cells that enter into mitosis. However, MI, and possibly LI, obtained following exposure to colchicine, cannot be compared with those observed in glands lacking intervention.

MI and LI provide both an indication of the proliferative activity of a particular cell type and the basis for comparison between different cell types. When combined with the relative proportions of different types of cells in a gland, the MI or LI can be used to provide a view of the relative contributions of the different cell types to the total proliferative activity of the gland. If either of the MI or LI and the relative proportions of the different cell types are known, the relative contributions by each cell type as a percentage can be calculated. An alternative is to obtain the values by direct scanning of sections (Hand and Ho,

1985) and/or structural units (Denny et al., 1990b) The major concern with this approach is whether those areas which are scanned are structurally representative of the total gland. Perhaps there are regional differences of proliferation within a gland. In the future, it may be necessary to qualify MI, LI, and relative total proliferative activity by specific locations within a gland.

LI OR MI DURING DEVELOPMENT

There is a general decline in the LI and/or MI of all cell types with (developmental) age in rat parotid and submandibular glands (Schneyer and Hall, 1969; Chang, 1973; Alvares and Sesso, 1975; Klein, 1982; Redman, 1989) as well as in mouse submandibular glands (Srinivasan and Chang, 1979). One exception is that the MI and LI of acinar cells (AC) of rat parotid may be greater at around 15 days postnatal than at earlier ages (Schneyer and Hall, 1969; Klein, 1982). There is conflicting evidence of a similar peak of activity for AC in rat submandibular gland (Chang, 1973; Alvares and Sesso, 1975; Klein, 1982). The peak of mitotic activity in rat parotid AC is in the range of 2.4 to 2.9% (Schneyer and Hall, 1969; Klein, 1982) with an LI of 24% (Klein, 1982). The highest LI of rat and mouse submandibular glands is in the range of 10 to 14% (Chang, 1973; Alvares and Sesso, 1975; Srinivasan and Chang, 1979; Klein, 1982) with an MI of 2.3% (Klein, 1982).

The LI of intercalated duct cells (ID) in rat and mouse submandibular glands is similar to the LI of AC in the early postnatal stages (Chang, 1973; Srinivasan and Chang, 1979); however, the decline in the LI for ID proceeds at a slower pace with increasing age, so that by 3 months of age the LI of mouse submandibular gland ID is more than 10 times greater than the LI for AC (Srinivasan and Chang, 1979). The decline in LI or MI for each of the cell types appears to continue into adulthood.

LI OR MI IN ADULTS

In addition to differences in techniques, comparison by different investigators of LI and MI obtained in older animals is complicated by age or weight, species, and possibly even strain differences. Furthermore, in separate studies using 6-month old rats (Long-Evans), the average MI of female parotid AC ranged between 0.001 and 0.03% (Table 1). A diurnal fluctuation in the MI was also evident in these AC (Schneyer and Hall, 1975). Over a 24-h period, the average MI range was from 0.000 to 0.1%. Incorporation of ^{3}H-thymidine into DNA in adult mouse parotid glands has a diurnal fluctuation as well (Burns et al., 1972). Since most studies are not matched as far as the majority of the above variables are concerned, comparisons between them can rarely be made with confidence.

TABLE 1. Mitotic Index and Labeling Index for Adult Salivary Gland Cells

	AC	ID	GD	SD
Mitotic Index				
Rat parotid				
Schneyer and Hall, 1969	0.03%			
Schneyer and Hall, 1975	0.03%			
Schneyer, 1973	0.001%			
Schneyer and Hall, 1976	0.03%			
Schneyer and Wilborn, 1976	0.003%			
Rat submandibular				
Takeda, 1988	0.1%			
van den Brenk et al., 1970	0.06%			
Labeling Index				
Rat parotid				
Walker and Gobe, 1987	0.3%	8.5%		1.0%
Schwartz-Arad et al., 1988				0.0%
Rat submandibular				
Zajicek et al., 1985	0.08%	1.8%	0.44%	0.08%
Barka, 1967	0.3%			
Barka, 1965	0.4%			
van den Brenk et al., 1970	0.5%			
Mouse submandibular				
Srinivasan and Chang, 1979 (3 months old)	0.15%	2.0%	0.1%	<0.1%
Srinivasan and Chang, 1979 (5 months old)	<0.1%	1.0%	<0.1%	

Note: AC = acinar cells, ID = intercalated duct cells, GD = granular duct cells, SD = striated duct cells.

A few generalizations will be presented merely to provide estimates of ranges. For instance, the range of LI for parotid and submandibular AC in young adult rat (150 to 200g) and mouse (3 months old) parotid and/or submandibular glands falls between 0.08 and 0.5% (Table 1). With one exception, LI for ID of young male rat parotid glands, female rat submandibular, and female mouse submandibular glands falls within a range of 1.2 to 2.0% (Table 1). The exception is an LI of 8.5% for parotid glands (Walker and Gobe, 1987) from rats of the same sex and weight as one included in the above range (Schwartz-Arad et al., 1988). There is no general agreement on the LI for striated duct cells (SD) ranging from 0.00 to 1.0% in age- and sex-matched rat parotid glands (Table 1). The LI for granular duct cells (GD) was 0.44% for the submandibular glands in young female rat and 0.1% for submandibular glands in young female mouse (Table 1). Studies using colchicine (Sans and Galanti, 1979a; Sans and Galanti, 1979b; Boshell and Pennington, 1980) to determine LI or MI were not included in this aspect of the survey. In the three studies which internally compared LI for

TABLE 2. Mitotic or Radiolabeled Relative Proportions of Acinar Cells and Intercalated Duct Cells in Adult Salivary Glands

Single Pulse Label	Relative Proportion (AC:ID)
Rat male parotid	
Hand and Ho, 1985	83:17 (mitoses only)
Schwartz-Arad et al., 1988	91:9
Rat female submandibular	
Zajicek et al., 1985	47:53
Mouse female submandibular	
Denny et al., 1990b	13:87
Mouse male submandibular	
Chai and Denny, (unpublished)	66:34

Multiple Pulses	
Mouse female submandibular	
2 pulses in 5 hours	
Gresik et al., 1981	50:50
4 pulses in 24 hours	
Denny et al., 1990b	
and unpublished	58:42
9 pulses in 9 days	
Denny et al., 1990b	
and unpublished	19:81

different cell types, it was evident that ID had the highest LI (Table 1). Furthermore, in every case, the LI was at least 10 times higher than the LI of AC within the same gland.

RELATIVE PROPORTIONS OF LABELED OR DIVIDING CELLS

Several studies have reported observations in terms of the relative proportions, by cell type, of cells that are dividing or are labeled. Other studies provide the relative proportions of total cells in addition to the LI, allowing for calculation of the relative proportion of radiolabeled cells. Where possible, these have been calculated and presented in Table 2. It is clear that all cell types of the adult secretory complex undergo cell division with the possible exception of mature myoepithelial cells (Cutler and Chaudhry, 1973; Redman et al., 1980). The more distal cell types, striated granular duct cells (SGD), GD and SD are usually a very low proportion of the total dividing cells, and thus are based upon too few observations to provide an accurate estimate. An exception is the GD of male mouse submandibular glands which exhibit an LI similar to AC (Srinivasan and Chang, 1979). On the other hand, substantial numbers of dividing or radiolabeled

AC and ID can be observed, providing a much better estimate of their proportions especially relative to each other.

There is fairly good agreement between studies on the young adult male (~200g) rat parotid gland which found 83% of AC dividing and 91% incorporating ^{3}H-thymidine, relative to the proportion of ID at 17 and 9%, respectively (Table 2). The labeled AC to ID proportion in young adult female rat submandibular glands was 47 to 53% (Table 2). This may be compared to the profile from adult female mice which shows a much lower proportion of AC synthesizing DNA than ID. A similar proportion of AC to ID was also observed in a female mouse following nine daily injections of ^{3}H-thymidine (Table 2). The difference in proportion between female rats and mice might be due to structural differences, i.e., the ratio of total AC to ID in the glands, as well as to different turnover rates. The relative contribution by each of these factors is not known. A striking difference can also be seen when comparing relative proportions in adult female and male mice (Table 2). While there are striking differences in the quantity of granular ducts in male and female mouse submandibular glands (Caramia, 1966), there is not an obvious connection between this and the difference in the AC to ID proportion of labeled cells.

A cautionary note on the interpretation of a change in the relative proportions of cell types is provided by one study of the young adult rat parotid gland (Hand and Ho, 1985). Following treatment with isoproterenol, the AC to ID proportion changes from 83:17%, respectively, to 97:3%, respectively. However, as noted by the authors, this shift in proportion masks the fact that the frequency of ID mitoses increased more than fourfold as a result of the treatment. A disproportionate increase of more than 25-fold in AC mitosis frequency was primarily responsible for the shift in proportions.

A disproportionate response to stimulation may also account for another set of observations. Two independent studies using young adult female mouse submandibular glands noted AC to ID proportions of 50:50%, respectively, and 58:42%, respectively (Table 2). This might be dismissed as some of the unresolved variation noted above except for the fact that the latter observation was obtained from an experiment initiated and carried out at the same time and with the same mouse cohort as those observations noted above with AC proportion of between 13 and 19%. The common difference in the protocols of the two studies with high AC proportions in adult female mice from the other studies reported in this section, is the use of multiple short-term radiolabel pulses. The first study used 2 pulses, 5 h apart (Gresik et al., 1981) and the second used 4 pulses at 8-h intervals (Denny et al., 1990b). These observations raise the possibility that traumatization associated with the injection of precursor might initiate a wave of DNA synthesis in AC which is chronicled during a subsequent pulse of ^{3}H-thymidine. Other experiments with young adult mice have demonstrated that simply touching a parotid gland during a surgical procedure stimulates a ten-fold increase in the LI and MI of AC (Sans and Galanti, 1979a; Sans and Galanti, 1979b).

TABLE 3. Distribution of Radiolabeled Nuclei by Cell-Type and Location Following a 90 Min. Pulse with ^{3}H-Thymidine

Cell Type & Location	Female[a] Number of Radio Labeled Cells	Percentage of Total	Male[b] Number of Radio Labeled	Percentage of Total Cells
GD	3	8%	25	48%
SGD	1	3%	1	2%
ID 1° stem[c]	3		1	
ID 1° primary branch	6		1	
ID 1° secondary branch	4			
ID 2° branch	15		7	
ID TOTAL	28	78%	9	17%
AC	4	11%	17	33%
TOTAL	36	100%	50	100%

[a] Pattern of duct organization and observations reprinted from Denny et al., 1990b.
[b] Chai and Denny unpublised results.
[c] 1° = first order, 2° = second order.

A more detailed comparison of submandibular glands in adult male and female mice shows an increase in the proportion of radiolabeled GD cells in the male which might be explained largely by the morphological differences between males and females (Table 3). It is also noteworthy that the majority of radiolabeled ID are found in second order intercalated ducts in both male and female glands.

DIURNAL CYCLE

The ability of salivary gland cells to respond in a hyperplastic fashion to environmental factors is on one hand very exciting because of the potential for therapeutic exploitation. On the other hand, fundamental questions such as what are basal levels of proliferation, and what are the relative roles of extrinsic and intrinsic factors in maintaining these levels, may be difficult to reliably assess. However, within this latter framework, we shall turn our attention to those conditions which might broadly be considered to fall within the range of normal physiological experience. There is a circadian rhythm in ^{3}H-thymidine uptake in young adult mouse parotid with the peak being at approximately 2100 h (Burns et al., 1972). In 15-day old rat parotid and submandibular glands, the peak of ^{3}H-thymidine uptake into DNA was at 0600 h (Klein, 1982), but the peak of parotid AC mitosis was between 1800 and 2300 h (Klein, 1979; Klein and Whitney, 1980). By contrast, the peak of mitotic activity in adult rat parotid AC was approximately 0600 h (Schneyer and Hall, 1975). Submandibular glands of adult rats show less evidence of a diurnal variation in mitotic activity exhibiting only a narrow trough of reduced activity between 0200 and 0400 h (Blumnefeld, 1942).

The coincidence of the shift from suckling to solid food during maturation with the onset of diurnal cycling of LI and MI (Klein, 1982), and the disappearance of the cycle in adults placed on a liquid diet (Schneyer and Hall, 1975), focused attention of investigators on the role of neural stimulation, presumably from chewing, in determining the level of proliferative activity in salivary glands of adults. Adult female rats show a dramatic transitory increase in AC mitoses following the change from an all-liquid diet to solid chow (Schneyer and Hall, 1975). This effect did not occur in glands that had been unilaterally parasympathectomized, but was only slightly reduced in sympathectomized glands. Contralateral glands in both cases exhibited no reduction in the mitotic response on returning to a solid chow. Thus, it appears that parasympathetic activity is the major factor in this response and that the sympathetic neural system and/or humoral agents have minimal involvement. However, it should be noted that for adult rats maintained on solid chow, sympathectomy induces a transitory elevation of MI and total DNA in parotid gland (Schneyer, 1973a).

In addition, there is evidence suggesting that parasympathetic modulation of cell proliferation in salivary glands is an adaptive mechanism. Studies that have substituted high-bulk solid diets which require more mastication than regular solid diets show even greater levels of DNA synthesis and cell proliferation (Hall and Schneyer, 1977; Johnson and Sreebny, 1982). Comparison of pelleted with ground, high-bulk diets strengthens the perceived association between mastication and cell proliferation by showing that added bulk alone has no significant effect on cell proliferation (Johnson, 1981). Other studies have indicated that dietary factors in addition to the physical consistency of food, coupled with the diurnal feeding pattern, might also be involved in maintaining the diurnal pattern of cell proliferation (Redman and Sullivan, 1981; Redman et al., 1984). It was suggested that taste might be the modulating factor in these studies (Redman, 1987).

CELL CYCLE

The length of cell cycles of parenchymal tissues of unperturbed adult salivary glands has not been measured. The cell cycle length of parotid AC from an 8-day old rat is 13.4 h (Klein et al., 1976; Klein and Harrington, 1976). At this age, G1 phase of the cell cycle is about 2.5 h, and S is between 7 and 7.5 h. G2 is 3.5 h, and M is about 20 min.

At 15 days, the cycle in rat parotid AC is 17.4 h (Klein, 1979). The basis for this difference in cycle lengths is largely due to the increase in G1 phase to approximately 6.5 h. The proportion of adult parenchymal cells that are systematically cycling regardless of the length of G1 is not known.

The S phase in adult mouse parotid AC stimulated by a single dose of isoproterenol (IPR) is 8 h (Epperlein and Kiefer, 1979), while G2 is about 3 h

(Radley and Hodgson, 1971a). Thus, S and G2 in the IPR-stimulated AC are similar in length to those phases in the young rat. G2 appears to increase to nearly 8 h following chronic treatment with IPR (van den Brenk et al., 1970). However, the latter observation was made on adult rat submandibular AC and may not be comparable to the response by parotid AC.

EXPANDING OR RENEWING PARENCHYMAL CELL POPULATIONS IN ADULTS

In 1964, C. P. Leblond introduced a system for classifying three types of cell populations based upon their turnover characteristics (Leblond, 1964). The static population exhibits no cell proliferation and a constant DNA content throughout adulthood. An expanding population maintains a low level of proliferation that contributes to an overall increase in total DNA with age. The third type the renewing population, is characterized by a high level of proliferation with only a minimal change in total DNA. Unfortunately, the salivary glands of adults do not fit precisely into any of these categories. They clearly cannot be classed as static based upon the observed mitoses among mature parenchymal cells. The relatively low levels of cell proliferation are consistent with the definition of an expanding population (Table 1); whereas, the failure to detect significant increases in total DNA with age is consistent with a renewing population (Leblond, 1964; Bogart, 1967; Schneyer and Hall, 1969; Redman and Sreebny, 1971; Kuyatt and Baum, 1981; Denny et al., 1989). One can argue that the rate of expansion is low and that the discrepancy is based on inadequate sample sizes or inherent assay variation. However, radiolabeling studies have suggested that the average life span of AC and SD in young adult male rat parotid gland and AC and GD in adult male mouse submandibular glands are all approximately 200 days (Cameron, 1970; Schwartz-Arad et al., 1988). The life span of AC and GD in young female rat submandibular glands approximate 125 and 62 days, respectively (Zajicek et al., 1985). Thus, because these average cellular life spans are considerably less than the average life span of the animal, the most appropriate classification would appear to be "slow renewing" (Cameron, 1970). Furthermore, extrapolation of the radiolabeling studies suggest that glandular DNA should exceed a doubling during the adult phase of the life cycle if the cells of the salivary glands were expanding populations (Cameron, 1970; Zajicek et al., 1985). This change in total DNA would be easy to detect by standard DNA assays.

In a renewing population, the frequency of cell loss or death should approximate the frequency of cell addition (Leblond, 1964). Evidence for cell death at these frequencies is largely circumstantial. Over a period of 23 days, the number of radiolabeled cells in adult female rat submandibular glands declined following a single 60-min pulse with ^{3}H-thymidine (Zajicek et al., 1985). Considering the relatively high LI and MI associated with ID (Table 1), the reduction in radiolabeled ID could be largely due to isotope dilution. That is, as the labeled cells divide, the radiolabel in progeny cells eventually reaches a level which is not detectable by autoradiography. On the other hand, based on the relatively low LI and MI of AC, GD and SD (Table 1), it would seem that the substantial reductions in radiolabeled nuclei observed over 23 days would be due less to isotope dilution mediated by cell division, mediated isotope dilution than to cell death (Zajicek et al., 1985).

Active cell turnover in salivary glands is also supported by the observation that the specific activity of radiolabeled DNA in female rat submandibular glands declined 63% in 27 days at a stage when there was no change in total DNA (Barka, 1965a). The significant reductions below control levels in total DNA of adult rat parotid glands following parasympathectomy alone and in combination with sympathectomy (Schneyer, 1973a; Schneyer and Hall, 1976; Hall, 1978) might also be interpreted as revealing the cell death aspect of cell turnover by selectively blocking cell proliferation. Sympathectomy alone does not produce this effect (Schneyer, 1973a; Schneyer and Hall, 1976; Hall, 1978). Adult rats placed on a liquified stock diet also show a significant reduction in parotid DNA below control levels (Johnson, 1982). However, for both of these phenomena, there is the possibility that the loss of DNA may also be due to an induced degenerative response such as that described in rat parotid when an all-liquid diet of human infant formula was substituted for solid chow (Hand and Ho, 1981). Other studies using young rats have also demonstrated reduction in levels of total DNA per gland following change to liquid and powdered diets (Hall and Schneyer, 1969; Johnson et al., 1977) or by a zinc deficient diet (Alvares and Johnson, 1981). Because these studies were performed on rats during active growth of the parotid gland, it appears that most of the effect is due to growth retardation rather than cell death.

The most direct evidence for measurable levels of AC death under normal conditions is provided by the observation of Walker and Gobe (1987). Young adult male rat parotid glands had a frequency of approximately 0.07% apoptotic AC. While it is impossible to compare this number with the LI or MI for rat parotid AC reported earlier, it was noted that because of the fleeting nature of apoptosis, finding even a small number of apoptotic cells is indicative of significant overall cell loss (Walker and Gobe, 1987).

SELF-PROLIFERATION AND STEM OR PROGENITOR CELL-MEDIATED REPLACEMENT

The mechanisms for replacement (or expansion) of parenchymal tissue in adult salivary glands fall into two categories, self-proliferation and stem or progenitor cell-mediated replacement. The latter encompasses two quite different developmental concepts. The first envisions a pleuripotential ancestor that is capable of giving rise to all of the parenchymal cell types. The second concept is of multiple progenitor cell lines each of whose developmental potential is restricted to a single parenchymal cell type. At this time there is little evidence to aid in distinguishing between these two possibilities; however, recent advances in cell culture techniques raise the possibility of directly testing the alternatives in the future (refer to Chapters 13 to 15 in this volume). It has been suggested that an understanding of the relative contributions by each of these mechanisms might provide insight into the origins of salivary gland neoplasia (Chaudhry et al., 1972; Batsakis, 1980; Dardick et al., 1990).

As presented earlier, self-proliferation is evident in most of the parenchymal cell types of salivary glands (Table 1). However, it appears that some of this activity, as detected by ^{3}H-thymidine incorporation into DNA, does not lead to daughter cells but instead to relatively stable polyploid and/or binucleate cells (Radley, 1967; Klein and Harrington, 1977; Epperlein and Kiefer, 1979; Schwartz-Arad et al., 1989; Denny et al., 1990b). Cell volumes are roughly proportional to ploidy (Epperlein and Kiefer, 1979); thus, any increase in ploidy might be considered a part of the proliferation process. However, as nothing is known of the relationship of polyploidy to longevity, one might also speculate that polyploidy may be associated with advanced cellular age.

The distribution of DNA synthesis as indicated from a single pulse of ^{3}H-thymidine does not appear to be uniform within each parenchymal cell type (Zajicek et al., 1985; Schwartz-Arad et al., 1988). AC nearest the intercalated ducts of rat submandibular and parotid glands tended to label more frequently than those more distal. Similarly, GD and SD nearest the intercalated ducts of rat submandibular and parotid gland respectively, also labeled more frequently than those further away (Zajicek et al., 1985; Schwartz-Arad et al., 1988). The presence of side-by-side pairs of labeled cells in each of the adult submandibular gland parenchymal cell types except SD, following a single pulse, and of 4 side-by-side cells following 24 h of multiple pulses also suggests the possibility that clonality might be a factor in the nonuniform labeling patterns in adult salivary glands (Denny et al., 1990b).

The idea of stem cell-derived replacement has its origins in studies of salivary gland development and of salivary gland regeneration (Hanks and Chaudhry, 1971; Chaudhry et al., 1972). Following excision of a portion of submandibular glands from young adult rats, regeneration appears to initiate from residual ducts and to pass through stages resembling those seen during development (Hanks and Chaudhry, 1971). Furthermore, it was postulated that the ducts gave rise to two cell lines, one which differentiates into striated ducts and the other gives rise to terminal tubules and, ultimately, AC.

Evidence for active stem cell replacement of parenchymal cells in normal adult rat parotid and submandibular gland comes from a series of experiments in which the distribution and location of radiolabeled cells were noted over successive time intervals following a single pulse with ^{3}H-thymidine (Zajicek et al., 1985; Schwartz-Arad et al., 1988). The pattern of labeled cells changed with time, giving the impression of a transit of cells from the intercalated ducts in each direction to acini or granular ducts and striated ducts, respectively. Cells with histological and biochemical properties that can be interpreted to be transitional betweed ID and AC or GD have been identified at structural locations consistent with a stem cell hypothesis (Caramia, 1966; Qwarnstrom and Hand, 1983; Hazen-Martin and Simson, 1987; Denny et al., 1990a).

In evaluating observations of changes in location and distribution of labeled cells with time following a single pulse, the overall loss of labeled cells with time is the most difficult to interpret. All of the parenchymal cell types of the rat submandibular gland show the loss of labeled cells with time, often after an initial increase (Zajicek et al., 1985). ID which have the highest LI also showed the most rapid loss of labeled cells. Given the relatively high LI, isotope dilution is likely to be a major factor in the loss of labeled ID. However, ID are also clearly a "renewing" population, and thus cell death must be considered. Finally, there may be loss of ID by differentation to AC or GD and SD by stem cell activity (Zajicek et al., 1985; Schwartz-Arad et al., 1988). To assess the interplay of these factors within the intercalated ducts would require a thorough understanding of their turnover rates and cell-cycle kinetics, as well as an accurate structural model.

The structural model used for the rat submandibular and parotid gland studies is a simple, nonbranching tube (Zajicek et al., 1985; Schwartz-Arad et al., 1988). This model is suited for analyzing the ID to SD or GD relationship, but is less satisfactory for the ID to AC relationship because of the often complex branching pattern of the intercalated ducts in both glands (Jacoby and Leeson, 1959; Tamarin and Sreebny, 1964). A more complex three-dimensional structural model has been developed for similar studies in the adult mouse submandibular gland (Denny et al., 1990b).

Given the multiple factors, i.e., isotope dilution, cell death, and differentiation, that could cause the apparent loss of labeled ID in salivary glands, there is as yet no accurate measure of the average life span of ID. On the other hand, using a continuous labeling approach, ID of adult male mouse pancreas have an estimated turnover of 167 days (Tsubouchi et al., 1986). However, this is not directly comparable to salivary glands because of the considerable difference in LI, approximating 0.6% per day of continuous labeling in the pancreas (Parker et al., 1974) and 1.2 to 2.0% per single pulse in submandibular and parotid glands (Table 1). Thus, it is likely that ID of salivary glands will turnover much more rapidly than pancreas ID. The complication caused by turnover for interpreting labeled cell population changes with time following a single pulse, is that, if there are adjacent cell populations with different turnover times, eventually the population with the slower turnover time will become relatively enriched with labeled nuclei, even in the absence of a mechanism for conversion of the one cell type to the other.

Cell cycle times, as well as the percentages of cells that are cycling, provide the information needed to judge the relative contribution of self-proliferation to turnover in AC and GD or SD. In the adult male mouse, three-dimensional reconstructions and the tabulation of net increases in side-by-side labeled nuclei indicate that approximately 35% of ID, 56% of AC, and 50% of GD each divide within 24 h after a single pulse with ^{3}H-thymidine (Chai and Denny, unpublished results). Coupled with the proportion shown in Table 3, a relationship can be derived which shows that for every labeled ID that divides within a 24-h period, 3 AC and 4 GD divide in the adult male mouse submandibular gland. This observation leads to the consideration that if the amount of self-replication is adequate to balance cell death in AC, GD, or SD, then stem cell contributions to these populations may be inconsequential.

CHARACTERISTICS OF ISOPROTERENOL-STIMULATED HYPERPLASIA

Of the factors that are known to induce hyperplasia in salivary glands, IPR has been studied in greatest depth. Approximately 24 h after a single injection of IPR, both the parotid and submandibular glands of adult rats and mice show a burst of ^{3}H-thymidine incorporation into DNA that may exceed 10 times that of the nontreated glands (Burns, 1978; Nelson et al., 1979; Baserga and Heffler, 1967; Barka, 1965b; Ebberlein and Kiefer, 1979; Baserga, 1966; Barka and van der Noen, 1975). The LI for AC may increase as much as 23 times and the MI 30 times following a single injection (Sans and Galanti, 1979a). Multiple injections of IPR can boost these increases higher (Barka, 1967; Barka, 1965a; van den Brenk et al., 1970; Hand and Ho, 1985). With chronic treatment, total gland DNA is also significantly elevated (Robinovitch et al., 1977; Barka and van der Noen, 1976). Furthermore, multiple injections elicit a hyperplastic response that may approach eight times in duct cells of parotid and submandibular glands as evidenced by increases in both LI and MI (Barka, 1976; Barka, 1965a; van den Brenk et al., 1970; Hand and Ho, 1985).

The magnitude of the hyperplastic response to IPR is dose dependent (Burns, 1978; Sans and Galanti, 1979a; Barka, 1965b; Baserga et al., 1969). Females appear to be more responsive than males for a given dose level (Barka, 1967; Piantanelli et al., 1978). The magnitude of the response also varies with the phase of the cell cycle during which IPR is administered (Burns et al., 1972).

In the rat parotid and submandibular glands, the ability for AC to respond to IPR is acquired between the age of 8 and 14 days (Chang and Barka, 1974; Klein et al., 1976). As adults age, the response time tends to increase and the magnitude of the response to decrease (Roth and Adelman, 1973; Adelman et al., 1972; Piantanelli et al., 1978; Baserga et al., 1969). The latter is also accompanied by changes in the dose-response curve (Roth and Adelman, 1973). A clear relationship of IPR-induced hyperplasia to aging is most reliably seen in submandibular glands of both rats and mice, but was not evident in the parotid gland of mice (Baserga et al., 1969; Piantanelli et al., 1978).

The hyperplastic response to IPR can be inhibited by a variety of metabolic inhibitors (Baserga, 1966). Of these, actinomycin D, puromycin, and cycloheximide, provide the most direct evidence that following administration of IPR, both RNA and protein synthesis are necessary before DNA synthesis can occur (Barka, 1965b; Baserga and Heffler, 1967; Sasaki et al., 1969; Novi and Baserga, 1972).

Administration of IPR to induce hyperplasia is not without anomalous side effects whose long-term consequences are unknown. IPR appears to lengthen G1 at the expense of the S phase in young animals (Klein et al., 1976). Following a single injection of IPR to induce hyperplasia, a second injection blocks cells responding to the first injection in G2 and M phases though there is no effect on DNA synthesis (Radley and Hodgson, 1971a). Use of IPR leads to a substantial increase in polyploidy (Epperlein and Kiefer, 1979; Pipkin et al., 1981; Radley, 1967). Finally, it has been suggested that some of the increased mitotic activity observed following administration of IPR is due to the recruitment of cells from G2, rather than from G1 with subsequent passage through S phase (Klein and Harrington, 1977; Gelfant, 1977; Bybee and Tuffery, 1988). The observation of substantial numbers of nonlabeled metaphase plates following injection of IPR and ^{3}H-thymidine indicates that "G2-blocked" AC might be common in young and in adult salivary glands (Gelfant, 1977).

Multiple injections of IPR give the impression of a 48-h wave of ^{3}H-thymidine incorporation (Nelson et al., 1979). However, the obvious interpretation of a 48-h cell cycle is unlikely because other studies have shown that the probability of cells dividing more than once in response to multiple injections is low (Radley and Hodgson, 1972b). This latter observation may explain, in part, why the total DNA contents of chronically stimulated glands reach new steady-state levels rather than show increases throughout the treatment period (Barka, 1965a). The LI and MI also slow to near control levels during chronic treatment (Barka, 1965a; Hand and Ho, 1985; van den Brenk et al., 1970).

On withdrawal from IPR treatment, total DNA per gland in adults returns to near control levels in about 20 days (Barka, 1965a; Domon et al., 1978). In young animals, the DNA content of chronically stimulated glands remains approximately 25% above control values following withdrawal (Barka and van der Noen, 1976). Finally, studies of changes in specific activity of adult mouse parotid gland DNA suggest that most, but not all, cells induced to divide by IPR are selectively eliminated following withdrawal (Domdon et al., 1978). However, this conclusion could not be drawn from a similar study using rat submandibular glands though evidence for reutilization of ^{3}H-thymidine or its metabolites could not be ruled out (Barka, 1965a). Thus, the general question of the longevity of cells, mitotically derived by IPR stimulation, remains to be clarified. (Refer to Chapters 10 and 11 of this volume for discussion of the roles of protooncogenes and glycosyltransferase in mediating IPR-induced hyperplasia).

OTHER FACTORS THAT STIMULATE DNA SYNTHESIS OR CELL PROLIFERATION

In addition to IPR, there are several humoral or pharmacological agents that initiate a hyperplastic response in salivary glands. Thyroxine administered to adult female mice induces increases in LI for GD and granular intercalated duct cells (GID) of about seven times and for AC and ID for two to three times (Gresik et al., 1981). Testosterone or one of its metabolites can induce an increase in LI of GD in castrated male mice which is similar in magnitude to the thyroxine effect in female mice (Chretien, 1977). However, chemically induced hypothyroidism appeared to have no detectable effect on total DNA in adult female rat submandibular gland over a 3-week period (Morgan et al., 1985). Administration of epidermal growth factor at certain times in the diurnal cycle causes a burst of ^{3}H-thymidine incorporation in young adult male mouse parotid glands in excess of four times the controls and at other times suppresses incorporation (Yeh et al., 1981). Carbamylcholine induces an approximate fivefold increase in LI and MI of adult mouse parotid glands (Sans and Galanti, 1979a), whereas pilocarpine appears to have no hyperplastic effect (Sans and Galanti, 1979a). Low energy laser treatment and thermal stress are also reported to induce hyperplasia in rodent salivary glands (McKibben and Pechersky, 1972; Takeda, 1988). Serum from cystic fibrosis patients has a greater hyperplastic effect than normal human serum (Sehneyer and Wilborn, 1976).

As discussed earlier, the parasympathetic neural system appears to mediate the hyperplastic response of salivary glands to changes in dietary consistency (Schneyer and Hall, 1975, 1976). The parasympathetic system also mediates the hyperplastic component of a compensatory response to partial desalivation in the remaining glands (Hall and Schneyer, 1978). Removal of the sympathetic neural system has no apparent effect on the compensatory increase in cell number but does effect AC size. A compensatory hyperplastic effect has been documented for parotid AC of both adult rats and mice (Hall and Schneyer, 1977; Sans and Galanti, 1979a; Sans and Galanti, 1979b), which in addition to partial desalivation, can be elicited by unilateral duct ligation (Walker and Gobe, 1987). In the latter study, no change was noted in the LI of ID but AC and SD each showed substantial increases of about tenfold. Finally, the compensatory hyperplastic increases elicited by high-bulk diet and by partial desalivation are additive as determined by ^{3}H-thymidine incorporation (Hall and Schneyer, 1977). The response to partial desalivation is abolished when rats are placed on an all-liquid diet.

REGENERATION

The regeneration of salivary gland parenchyma following extirpation or allergic sialoadenitis has features in common with development (Hanks and Chaudhry, 1971; Boshell and Pennington, 1980; Sharawy and White, 1978). Following damage repair, the first response in the regenerating area appears to be the proliferation of rudimentary ducts resembling terminal tubules (Hanks and Chaudhry, 1971; Sharawy and White, 1978). Next, lobules form and finally, AC differentiate (Hanks and Chaudry, 1971; Boshell and Pennington, 1980). The regenerating tissue also contains the "transitory" cell types, secretory terminal tubule and proacinar cells, seen perinatally during rat submandibular gland development (Sharawy and White, 1978; Yamashina and Barka, 1972). A prolonged elevation of AC mitoses has been noted in the margins of submandibular gland bordering the extirpated region (Boshell and Pennington, 1980). Repeated administration of IPR appeared to abolish most of this prolonged hyperplastic effect. Furthermore, the cut margins showed only a partial acute response to IPR although the response of other regions of the gland was similar to IPR-treated controls. The suggestion was made that IPR accelerated the AC maturation process and suppressed cell division in a manner analogous to its action during early development (Boshell and Pennington, 1980; Schneyer, 1973b).

Another approach to the study of regeneration is the use of autografts or isografts of submandibular gland. When implanted intraperitoneally or subcutaneously, submandibular gland first undergoes extensive necrosis, then proliferation and formation of lobules filled with nondifferentiated duct-like structures (Hoshino and Lin, 1970; Sharawy and O'Dell, 1979). Mature AC are either rare or entirely lacking in these grafts. Implants returned to the site of the donor submandibular gland showed much greater differentiation of AC than those grafted subcutaneously on the back, indicating a site preference (Sharawy and O'Dell, 1979). Chronic IPR treatment appears to facilitate AC differentiation as well as ductal proliferation in both autografts and isografts at neutral sites (Hoshino and Lin, 1970; O'Dell et al., 1983). By contrast, testosterone facilitates the appearance of secretory duct cells with no apparent enhancement of formation of AC (Hoshino and Lin, 1970). Sympathectomy has no effect on the autograft regenerative process at a neutral site (O'Dell et al., 1987).

SUMMARY

In summary, it must be noted that some of the fundamental questions, such as what are the turnover kinetics for each mature parenchymal cell type, the characteristics of their cell cycles, and the proportions of cells actively cycling, remain unanswered. Thus, the relative contributions by self-proliferation and by stem cell progeny to parenchymal cell populations in adults is also unresolved.

The ability of salivary glands to undergo hyperplasia, and thus the potential for therapeutic application, remains a strong rationale for continuing to explore induced hyperplasia. IPR-stimulated hyperplasia may not hold as much promise for therapeutic development as was once thought because of the apparent instability of the induced cells and the general limitation of the response to a single division. Other agents that show promise but remain to be thoroughly evaluated for their long-term effects on gland function, are epidermal growth factor, contact stimulation, carbamylcholine, and parasympathetic neural stimulation. Finally, the relationship of parenchymal proliferation to functional renewal remains to be explored in both noninduced and induced salivary glands.

ACKNOWLEDGMENTS

The unpublished studies reported herein supported by National Institutes of Health grant DE-04960.

REFERENCES

1. Adelman, R. C., Stein, G., Roth, G. S., and Englander, D., Age-dependent regulation of mammalian DNA synthesis and cell proliferation *in vivo*, *Mech. Age. Dev.*, 1, 49, 1972.
2. Alvares, E. P. and Sesso, A., Cell proliferation, differentiation and transformation in the rat submandibular gland during early postnatal growth: a quantitative and morphological study, *Arch. Histol. Jap.*, 38, 177, 1975.
3. Alvares, O. and Johnson, D., Effects of zinc deficiency on rat parotid gland, *J. Oral Path.*, 10, 430, 1981.
4. Barka, T., Induced cell proliferation: the effect of isoproterenol, *Exptl. Cell Res.*, 37, 662, 1965a.
5. Barka, T., Stimulation of DNA synthesis by isoproterenol in the salivary gland, *Exptl. Cell Res.*, 39, 355, 1965b.
6. Barka, T., Stimulation of deoxyribonucleic acid synthesis in the salivary gland by isoproterenol: the effect of sex, *Exptl. Cell Res.*, 48, 53, 1967.
7. Barka, T. and van der Noen, H., Dissociation of rat parotid gland, *Lab. Inv.*, 32, 373, 1975.
8. Barka, T. and van der Noen, H., Stimulated growth of submandibular gland, *Lab. Invest.*, 35, 507, 1976.
9. Baserga, R. and Heffler, S., Stimulation of DNA synthesis by isoproterenol and its inhibition by actinomycin D, *Exptl. Cell Res.*, 46, 571, 1967.
10. Baserga, R., Inhibition of stimulation of DNA synthesis by isoproterenol in submandibular glands of mice, *Life Sci.*, 5, 2033, 1966.
11. Baserga, R., Sasaki, T., and Whitlock, J. P., The prereplicative phase of isoproterenol-stimulated DNA synthesis, in *Biochemistry of Cell Division*, Basenger, R., Ed., Charles C Thomas, Springfield, IL, 1969, 77.
12. Batsakis, J. G., Salivary gland neoplasia: an outcome of modified morphogenesis and cytodifferentiation, *Oral Surg.*, 49, 229, 1980.

13. Blumenfeld, C. M., Normal and abnormal mitotic activity. I. comparison of periodic mitotic activity in epidermis, renal cortex and submaxillary salivary gland of the albino rat, *Arch. Pathol.*, 33, 770, 1942.

14. Bogart, B. I., The effect of aging on the histochemistry of the rat submandibular gland, *J. Gerontol.*, 22, 372, 1967.

15. Boshell, J. L. and Pennington, C., Histological observations on the effects of isoproterenol on regenerating submandibular glands of the rat, *Cell Tissue Res.*, 213, 411, 1980.

16. Burns, E. R., A chronobiological study of the effects of a single injection of isoproterenol on mitotic index and/or DNA synthesis in normal and neoplastic cell populations, *Growth*, 42, 333, 1978.

17. Burns, E. R., Scheving, L. E., and Tsai, T., Circadian rhythm in uptake of tritiated thymidine by kidney, parotid, and duodenum of isoproterenol-treated mice, *Science*, 175, 71, 1972.

18. Bybee, A. and Tuffery, A. R., The short-term effects of a single injection of isoproterenol on proliferation in the submandibular gland, parotid gland and oesophagus *in vivo*, *Cell Tissue Kinet.*, 21, 133, 1988.

19. Cameron, I. L., Cell renewal in the organs and tissues in the nongrowing adult mouse, *Tex. Rep. Biol. Med.*, 28, 203, 1970.

20. Caramia, F., Ultrastructure of mouse submaxillary gland. I. Sexual differences, *J. Ultrastruct. Res.*, 16, 505, 1966.

21. Chang, W. W. L., Cell population changes during acinus formation in the postnatal rat submandibular gland, *Anat. Rec.*, 178, 187, 1973.

22. Chang, W. W. L. and Barka, T., Stimulation of acinar cell proliferation by isoproterenol in the postnatal rat submandibular gland, *Anat. Rec.*, 178, 203, 1974.

23. Chaudhry, A. P., Montes, M., and Cutler, L. S., Structural and functional maldevelopment of salivary glands, in *Salivary Glands and Their Secretion*, Rowe, N. H., Ed., University of Michigan Press, Ann Arbor, MI, 1972, 59.

24. Chretien, M., Action of testosterone on the differentiation and secretory activity of a target organ: the submaxillary gland of the mouse, *Int. Rev. Cytol.*, 50, 333, 1977.

25. Cutler, L. S. and Chaudhry, A. P., Differentiation of the myoepithelial cells of the rat submandibular gland *in vivo* and *in vitro*: an ultrastructural study, *J. Morphol.*, 140, 343, 1973.

26. Dardick, I., Byard, R. W., and Carnegie, J. A., A review of the proliferative capacity of major salivary glands and the relationship to current concepts of neoplasia in salivary glands, *Oral Surg. Oral Med. Oral Pathol.*, 69, 53, 1990.

27. Denny, P. C., Denny, P. A., Pimprapaiporn, W., Bove, B., and Chai, Y., Cell differentiation as replacement in mouse submandibular gland, *J. Dent. Res.*, 68, 281, 1989.

28. Denny, P. C., Chai, Y., Pimprapaiporn, W., and Denny, P. A., Three-dimensional reconstruction of adult female mouse submandibular gland secretory structures, *Anat. Rec.*, 226, 489, 1990a.

29. Denny, P. C., Chai, Y., Klauser, D. K., and Denny, P. A., Three-dimensional localization of DNA synthesis in secretory elements of adult female mouse submandibular gland, *Adv. Dent. Res.*, 4, 34, 1990b.

30. Domon, M., Okano, T., Saski, T., and Nakamura, T., Regression of the mouse parotid gland induced with isoproterenol, *Cell Tissue Kinet.*, 11, 567, 1978.

31. Epperlein, H. H. and Kiefer, G., Cellular growth in the mouse parotid gland after stimulation by isoproterenol. II. Protein accumulation during the DNA synthesis phase, *Biol. Cellulaire*, 35, 31, 1979.

32. Gelfant, S., A new concept of tissue and tumor cell proliferation, *Cancer Res.*, 37, 3845, 1977.

33. Gresik, E. W., van der Noen, H., and Barka, T., Hypertrophic and hyperplastic effects of thyroxine on the submandibular gland of the mouse, *Anat. Rec.*, 200, 443, 1981.

34. Hall, H. D. and Schneyer, C. A., Physiological activity and regulation of growth of developing parotid, *Proc. Soc. Exptl. Biol. Med.*, 131, 1288, 1969.

35. Hall, H. D. and Schneyer, C. A., Functional mediation of compensatory enlargement of the parotid gland, *Cell Tissue Res.*, 184, 249, 1977.

36. Hall, H. D. and Schneyer, C. A., Neural regulation of compensatory enlargement of the parotid gland of the rat, *Cell Tissue Res.*, 187, 147, 1978.

37. Hand, A. R. and Ho, B., Liquid-diet-induced alterations of rat parotid acinar cells studied by electron microscopy and enzyme cytochemistry, *Arch. Oral Biol.*, 26, 369, 1981.

38. Hand, A. R. and Ho, B., Mitosis and hypertrophy of intercalated duct cells and endothelial cells in the isoproterenol-treated rat parotid gland, *J. Dent. Res.*, 64, 1031, 1985.

39. Hanks, C. T. and Chaudhry, A. P., Regeneration of rat submandibular gland following partial extirpation. A light and electron microscopic study, *Am. J. Anat.*, 130, 195, 1971.

40. Hazen-Martin, D. J. and Simson, J. A. V., Electron microscopic immunostaining of nerve growth factor: secretagogue stimulated submandibular glands, *Anat. Rec.*, 219, 171, 1987.

41. Hoshino, K. and Lin, C. D., Selective effects of testosterone and isoproterenol upon regenerating submandibular gland isografts in BALB/c mice, *Anat. Rec.*, 167, 489, 1970.

42. Jacoby, F. and Leeson, C. R., The post-natal development of the rat submaxillary gland, *J. Anat.*, 93, 201, 1959.

43. Johnson, D. A., Effect of a ground versus a palliated bulk diet on the rat parotid gland, *Arch. Oral Biol.*, 26, 1091, 1981.

44. Johnson, D. A., Effect of a liquid diet on the protein composition of rat parotid saliva, *J. Nutr.*, 112, 175, 1982.

45. Johnson, D. A. and Sreebny, L. M., Effect of increasing the bulk content of the diet on the rat parotid gland and saliva, *J. Dent. Res.*, 61, 691, 1982.

46. Johnson, D. A., Sreebny, L. M., and Enwonwu, C. O., Effect of protein-energy malnutrition and of a powdered diet on the parotid gland and pancreas of young rats, *J. Nutr.*, 107, 1235, 1977.

47. Klein, R. M., Alteration of neonatal rat parotid gland acinar cell proliferation by guanethidine-induced sympathectomy cell, *Tissue Kinet.*, 12, 411, 1979.

48. Klein, R. M., Acinar cell proliferation in the parotid and submandibular salivary glands of the neonatal rat, *Cell Tissue Kinet.*, 15, 187, 1982.

49. Klein, R. M. and Harrington, D. B., Acinar cell cycle of developing rat parotid gland: comparison of males and females, *J. Dent. Res.*, 55, 712, 1976.

50. Klein, R. M. and Harrington, D. B., Isoproterenol and G2 acinar cells in the developing rat parotid gland, *J. Dent. Res.*, 56, 177, 1977.

51. Klein, R. M., Harrington, D. B., and Piliero, S. J., Isoproterenol-induced changes in cell cycle kinetics of parotid gland acinar cells in 8-day-old rats, *J. Dent. Res.*, 55, 611, 1976.

52. Klein, R. and Whitney, S., Circadian rhythms in neonatal rat parotid gland acinar cell and ileal crypt cell mitosis, *Biol. Neonate*, 38, 197, 1980.

53. Kuyatt, B. L. and Baum, B. J., Characteristics of submandibular glands from young and aged rats, *J. Dent. Res.*, 60, 936, 1981.

54. Leblond, C. D., Classification of cell populations on the basis of their proliferative behavior, *Natl. Cancer Inst. Monogr.*, 14, 119, 1964.

55. McKibben, D. H. and Pechersky, J. L., Effect of thermal stress on cell proliferation of the submandibular gland and periodontal ligament of CH3-male mice, *Arch. Oral Biol.*, 17, 291, 1972.

56. Morgan, B. L. G., Kuyatt, B. L., and Fink, J., Effects of hypothyroidism on the DNA, carbohydrate, soluble protein and sialic acid contents of rat submandibular gland, *J. Oral Pathol.*, 14, 37, 1985.

57. Nelson, N. F., Brown, K. B., Douthit, D. D., Ghatan, S., and Brown, D. G., Polyamine metabolism in isoproterenol-stimulated mouse salivary glands, *J. Dent. Res.*, 58, 1644, 1979.

58. Novi, A. M. and Baserga, R., Correlation between synthesis of ribosomal RNA and stimulation of DNA synthesis in mouse salivary glands, *Lab. Invest.*, 26, 540, 1972.

59. O'Dell, N. L., Sharawy, M., and Pennington, C. B., Effects of prior culture or isoproterenol injections on the regeneration of rat submandibular gland autografts, *Anat. Rec.*, 206, 11, 1983.

60. O'Dell, N. L., Sharawy, M., Richardson, M. C., and Pennington, C. B., Regeneration of submandibular gland autografts in sympathectomized rats, *Anat. Rec.*, 218, 373, 1987.

61. Piantanelli, L., Brogli, R., Bevilacqua, P., and Fabris, N., Age-dependence of isoproterenol-induced DNA synthesis in submandibular glands of Balb/c mice, *Mech. Ageing Dev.*, 7, 163, 1978.

62. Pipkin, J. L., Hinson, W. G., Hudson, J. L., Anson, J., and Pack, L. D., The modulating effect of isoproterenol on DNA replication and protein synthesis: synthesis patterns of the HMG proteins from electrostatically sorted salivary gland nuclei during the *in vivo* cell cycle, *Biochim. Biophys. Acta*, 655, 421, 1981.

63. Qwarnstrom, E. E. and Hand, A. R., A granular cell at the acinar-intercalated duct junction of the rat submandibular gland, *Anat. Rec.*, 206, 181, 1983.

64. Radley, J. M., Changes in ploidy in the rat submaxillary gland induced by isoprenaline, *Exptl. Cell Res.*, 48, 679, 1967.

65. Radley, J. M. and Hodgson, G. S., Effect of isoprenaline on cells in different phases of the mitotic cycle, *Exptl. Cell Res.*, 69, 148, 1971a.

66. Radley, J. M. and Hodgson, G. S., Individual cell susceptibility to undergo DNA synthesis in the proliferative response to isoprenaline, *Aust. J. Exp. Biol. Med. Sci.*, 49, 487, 1971b.

67. Redman, R. S., Development of the salivary glands, in *The Salivary System*, Sreebny, L. M., Ed., CRC Press, Boca Raton, FL, 1987, 1.

68. Redman, R. S., Proliferative activity by cell type in developing rat parotid gland, *J. Dent. Res.*, 68, 956, 1989.

69. Redman, R. S., Hsiao, L. L., and Barbour, B. J., Caries-promoting diets affect weaning rat feeding patterns and parotid maturation, *J. Dent. Res.*, 63, 243, 1984.

70. Redman, R. S. and Sreebny, L. M., Proliferative behavior of differentiating cells in the developing rat parotid gland, *J. Cell Biol.*, 46, 81, 1970.

71. Redman, R. S. and Sreebny, L. M., Morphologic and biochemical observations on the development of the rat parotid gland, *Dev. Biol.*, 25, 248, 1971.

72. Redman, R. S. and Sullivan, B. R., Dietary effects on parotid glands in prematurely weaned rats, *J. Dent. Res.*, 60, 1225, 1981.

73. Redman, R., Sweney, L. R., and McLaughlin, S. T., Differentiation of myoepithelial cells in the developing rat parotid gland, *Am. J. Anat.*, 158, 299, 1980.

74. Robinovitch, M. R., Keller, P. J., Johnson, D. A., Iversen, J. M., and Kauffman, D. L., Changes in rat parotid salivary proteins induced by chronic isoproterenol administration, *J. Dent. Res.*, 56, 290, 1977.

75. Roth, G. S. and Adelman, R. C., Possible changes in tissue sensitivity in the age-dependent stimulation of DNA synthesis *in vivo*, *J. Gerontol.*, 28, 298, 1973.

76. Sans, J. and Galanti, N., Effect of some drugs and unilateral parotidectomy upon secretion and cell proliferation in the mouse parotid gland, *Cell. Mol. Biol.*, 25, 107, 1979a.

77. Sans, J. and Galanti, N., Cell proliferation in the mouse parotid gland after unilateral parotidectomy, *Cell Biol. Int. Report*, 3, 25, 1979b.

78. Sasaki, T., Litwack, G., and Baserga, R., Protein synthesis in the early prereplicative phase of isoproterenol-stimulated synthesis of deoxyribonucleic acid, *J. Biol. Chem.*, 244, 4831, 1969.

79. Schneyer, C., Mitotic activity and cell number of denervated parotid of adult rat, *Proc. Soc. Exp. Biol. Med.*, 142, 542, 1973a.

80. Schneyer, C. A., A growth-suppressive influence of L-isoproterenol on postnatally developing parotid gland of rat, *Proc. Soc. Exp. Biol. Med.*, 143, 899, 1973b.

81. Schneyer, C. A. and Hall, H. D., Growth pattern of postnatally developing rat parotid gland, *Proc. Soc. Exptl. Biol. Med.*, 130, 603, 1969.

82. Schneyer, C. A. and Hall, H. D., Parasymathetic regulation of mitosis induced in rat parotid by dietary change, *Am. J. Physiol.*, 229, 1614, 1975.

83. Schneyer, C. A. and Hall, H. D., Neurally mediated increase in mitosis and DNA of rat parotid with increase in bulk of diet, *Am. J. Physiol.*, 230, 911, 1976.

84. Schneyer, C. A. and Wilborn, W. H., Effects of cystic fibrosis serum on the rat parotid gland, *Cell Tissue Res.*, 169, 111, 1976.

85. Schwartz-Arad, D., Arber, L., Arber, N., Zajicek, G., and Michaeli, Y., The rat parotid gland — a renewing cell population, *J. Anat.*, 161, 143, 1988.

86. Schwartz-Arad, D., Michaeli, Y., and Zajicek, G., The DNA content of the submandibular ductal cells increases with their age, *J. Dent. Res.*, 68, 716, 1989.

87. Sharawy, M. and O'Dell, N. L., Regeneration of acini in submandibular gland autografts, *Anat. Rec.*, 195, 431, 1979.

88. Sharawy, M. and White, S. C., Morphometric and fine structural study of experimental autoallergic sialadenitis of rat submandibular glands, *Virchows Arch. B Cell Path.*, 28, 255, 1978.

89. Srinivasan, R. and Chang, W. W. L., The postnatal development of the submandibular gland of the mouse, *Cell Tissue Res.*, 198, 363, 1979.

90. Takeda, Y., Irradiation effect of low-energy laser on rat submandibular salivary gland, *J. Oral Pathol.*, 17, 91, 1988.

91. Tamarin, A. and Sreebny, L. M., The rat submaxillary gland: a correlative study by light and electron microscopy, *J. Morphol.*, 117, 295, 1965.

92. Tsubouchi, S., Kano, E., and Suzuki, H., Dynamic features of duct epithelial cells in the mouse pancreas as shown by radioautography following continuous ^{3}H-thymidine infusion, *Anat. Rec.*, 214, 46, 1986.
93. Van den Brenk, H. A. S., Sparrow, N., and Moore, V., Effect of X-radiation on salivary gland growth in the rat. II. Quantitative studies of sialadenotrophism induced with isopropylnoradrenaline and its modification by single doses of X-rays, *Int. J. Radiat. Biol.*, 17, 135, 1970.
94. Walker, N. I. and Gobe, G. C., Cell death and cell proliferation during atrophy of the rat parotid gland induced by duct obstruction, *J. Pathol.*, 153, 333, 1987.
95. Yamashina, S. and Barka, T., Localization of peroxidase activity in the developing submandibular gland of normal and isoproterenol-treated rats, *J. Histochem. Cytochem.*, 20, 855, 1972.
96. Yeh, Y. C., Scheving, L. A., Tsai, T. H., and Scheving, L. E., Circadian stage-dependent effects of epidermal growth factor on deoxyribonucleic acid synthesis in ten different organs of the adult male mouse, *Endocrinology*, 109, 644, 1981.
97. Zajicek, G., Yagil, C., and Michaeli, Y., The streaming submandibular gland, *Anat. Rec.*, 213, 150, 1985.

Isolation and Maintenance of Submandibular Gland Cells

Robert S. Redman and David O. Quissell

INTRODUCTION

Salivary glands are excellent models for investigating the physiology of secretion because the influence of various agents can be assessed by their effects on the flow rate and composition of the secretory product. Collection of saliva requires no invasive surgery with human beings, and is possible with no or minor surgery with laboratory animals, facilitating repeated sampling. Thus, much of our knowledege of the physiology of secretion has been derived from experimental manipulations *in vivo*. However, elucidation of many of the mechanisms involved in controlling secretion, and the synthesis and storage of secretory products, is hindered by the many layers of complexity in the intact animal. Approaches to overcoming this limitation have included micropuncture of secretory units and ducts (Martinez et al., 1966; Young and Schogel, 1966), and preparation of isolated, perfused glands or ducts (Lundberg, 1957; Nielsen and Peterson, 1972; Martinez and Cassity, 1984) and gland slices (Bdolah et al., 1964; Martinez et al., 1983). These methods remove only some of the layers of complexity, however, and also are hampered by impaired delivery of nutrients and oxygen to the parenchymal cells. At the other extreme, preparations of cell organelles and the cytosol have yielded important knowledge of the dynamics of intracellular events in salivary gland metabolism (e,g., Robinovitch et al., 1964; Castle et al., 1975). This methodology, too, has an important limitation: the requirement that the parenchymal cells be intact for many normal physiologic responses to external influences to occur.

The isolation of intact cells from pancreas (Amsterdam and Jamieson, 1972) thus was a major addition to the armamentarium of techniques with which to investigate exocrine organ metabolism and physiology. Initially, the overall integrity and responsiveness of the cells was disappointing. Considerable improvement resulted from the use of highly purified proteases and hyaluronidase to separate the cells from each other and the stroma (Amsterdam and Jamieson, 1974). Application of this method to salivary gland acinar cells soon followed (Kanamura and Barka, 1975; Mangos et al., 1975).

With further refinements, we were able to prepare cells from rat submandibular glands that seemed to be more nearly normal by most parameters than were those of previous preparations (Quissell and Redman, 1979). Morphologically, the acinar cells were enriched at the expense of the granular duct cells, as compared with specimens from whole glands. An intriguing finding was that most of the acinar cells remained grouped together in acini, and many of these had short segments of intercalated ducts attached (Figure 1). This was in sharp contrast to the reported observations on cell preparations from pancreas and parotid glands, in which individual acinar cells predominated. This suggested at least two ways in which the dispersal method we employed effected the observed improvement in secretory responsiveness of the submandibular cells. One was by preserving the original intercellular contacts by which secretory impulses appear to be communicated among acinar cells (Hammer and Sheridan, 1978). That the dispersal method was so gentle as to preserve the cell groupings also was indirect evidence that important cell membrane structures, especially neurotransmitter receptors, were better preserved. An alternative explanation for the different degrees of dispersal among these organs is suggested by the fact that the acini of the rat submandibular gland (Cutler and Chaudhry, 1973), but not those of the parotid gland (Garrett and Parsons, 1973; Redman et al., 1980) and pancreas (Pictet et al., 1972), are invested with myoepithelial cells. The significance of this possibility is diminished by the observation of only occasional myoepithelial cells on the dispersed submandibular acini (Quissell and Redman, 1979; Quissell et al., 1986), and by the preservation of intact acini when parotid and pancreatic cells are dispersed by very gentle techniques (Williams et al., 1978; Quissell et al., 1989).

Subsequently, the same or similar dispersal techniques have been utilized to explore numerous facets of submandibular acinar cell metabolism and secretory activities (Fleming et al., 1980; Quissell, 1980; Quissell and Barzen, 1980; Quissell et al., 1981, 1983; Baum and Kuyatt, 1981; McPherson and Dormer, 1984; Martinez and Cassity, 1985; Cutler et al., 1987). However, the dispersed cells survived only a few hours, precluding investigations of longer duration, especially those involving more than one cycle of secretion, synthesis and storage of salivary proteins. Thus, there was a strong incentive for finding a means of prolonging the useful life of the cells.

STRATEGY FOR THE DEVELOPMENT OF CONDITIONS TO MAINTAIN RAT SUBMANDIBULAR ACINAR CELLS IN PRIMARY CULTURE

As we began, it was clear from previous attempts by others that it would be very difficult to maintain mature salivary gland acinar cells in culture (Tapp, 1967; Mercante, 1973; Brown, 1974; Kanamura and Barka, 1976; Wigley and Franks, 1976; Lamey et al., 1982; Yang et al., 1982). Though Oliver (1980) had maintained cultures of dispersed mature acinar cells from rat parotid gland for 30 days, they failed to divide. Furthermore, when organ cultures of embryonic pancreas and salivary glands had proceeded until acinar cells became moderately differentiated, the cells then deteriorated (Borghese, 1950; Kallman and Grobstein, 1964; Lucas, 1969; Lawson, 1970; Cutler and Chaudhry, 1973; Rufo and Barka, 1976). We were encouraged by reports that, in primary cultures from other exocrine organs, notably mammary gland (Emmerman and Pitelka, 1978; Wicha et al., 1982) and pancreas (Ruoff and Hay, 1979; Oliver, 1980; Brannon et al., 1985), the cells retained many acinar features. There also was a storehouse of information on factors involved in the promotion and maintenance of differentiation of exocrine cells in general (recently reviewed by McKeehan et al., 1990), and salivary gland cells in particular, from both *in vivo* and *in vitro* experiments. Additional helpful information was available from studies of grafts of portions of rat submandibular glands (Hoshino and Lin, 1968; Sharawy and O'Dell, 1979).

For cell culture purposes, these factors could be grouped into five categories: (1) general requirements of the media (e.g., nutrients, minerals, pH) (2) gases, (3) hormones; (4) extracellular matrix, and (5) secretory stimulation. For initial attempts, the general requirements have been standardized (Ham, 1981). Hormones to be considered included epidermal growth factor (EGF), insulin, thyroxine and adrenocorticoids (Jones, 1966; Deschodt-Lanckman et al., 1974; Sasaki et al., 1976; Kumegawa et al., 1981; Dembinski et al., 1982; Logsdon and Williams, 1983).

An extracellular matrix improves survival and orientation of cultured cells by providing attachment sites for anchorage, and promotes growth and differentiation by serving as a site for the deposition and interchange of numerous molecular factors originating both from the medium and the cells. The importance of this matrix has been demonstrated in organ cultures of tooth buds (reviewed by Slavkin, 1990) and embryonic pancreas and salivary glands, in which the mesenchyme and epithelia were separated and interchanged, or the collagen in the native matrix was enzymatically removed (Grobstein, 1953; Cunha, 1972; Ball, 1974). Subsequently, many of the structural and "message" molecules have been identified and functionally characterized (Cutler and Christian, 1984; Bernfield et al., 1985; Trelstad, 1985; Durban, 1987; LeBlond and Inoue, 1989; Beck et al., 1990; Slavkin, 1990). Cholinergic and adrenergic secretory stimuli have been shown to be important to the growth and differen-

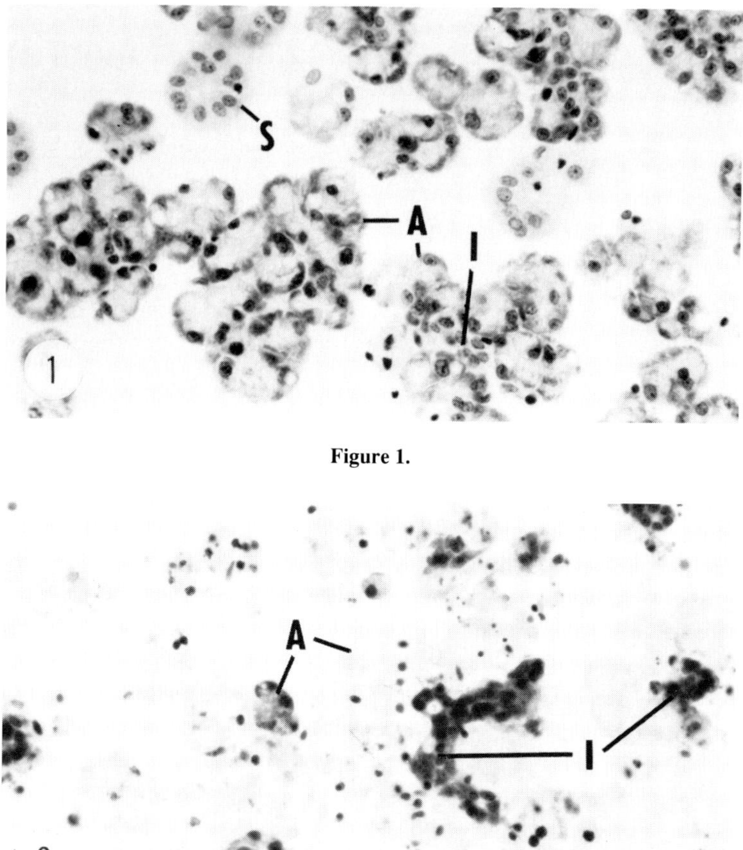

Figure 1.

Figure 2.

Figures 1 to 5. Photomicrographs of paraffin sections of dissociated rat submandibular acinar-intercalated duct complexes in the initial series of primary culture, stained with hematoxylin and eosin. A denotes acini; I, intercalated ducts; and S, striated ducts. (Figures 1 to 4 are from Quissell et al., *In Vitro Cell Devel. Biol.*, 22, 469, 1986. With permission.) (1) Freshly dissociated complexes. × 300. (2) After 2 days in culture, most acinar cells are dead but many intercalated ducts have survived. × 300. (3 and 4) After 1 day in culture at 35% oxygen, many acini seem healthy. Those with isoproterenol in the medium (Figure 4) are markedly reduced in size as compared with those without secretagogue (Figure 3). × 300. (5) After 7 days, most of the complexes have begun to degenerate, with most of the surviving cells having a flattened, squamous appearance. × 400.

tiation of salivary glands, particularly the acinar cells (Schneyer, 1972; Srinavasan et al., 1973; Srinavasan and Chang, 1977).

The overall strategy was to test combinations of modifications of the initial culture conditions until a substantial improvement in the survival of cells with

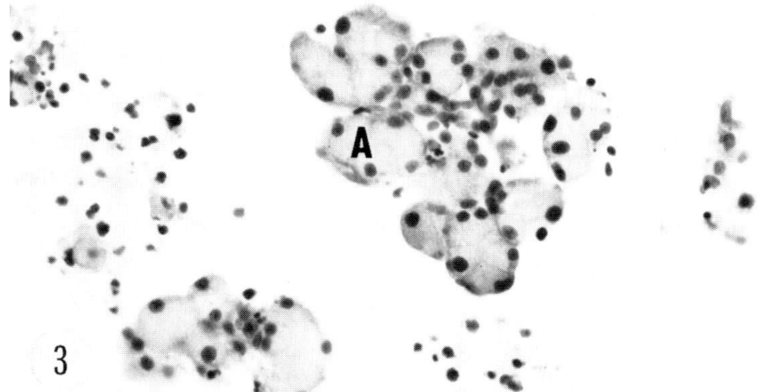

Figure 3.

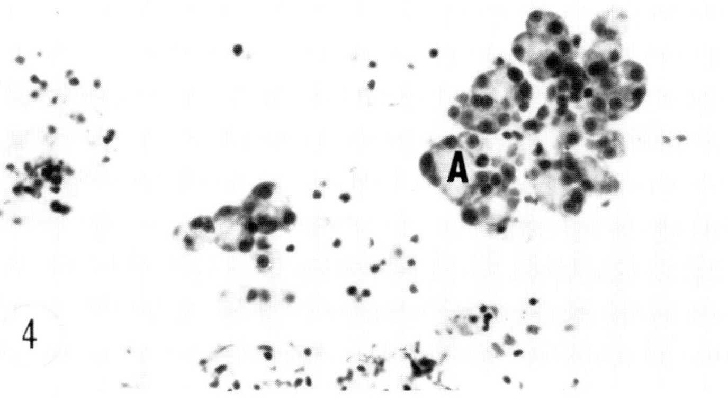

Figure 4.

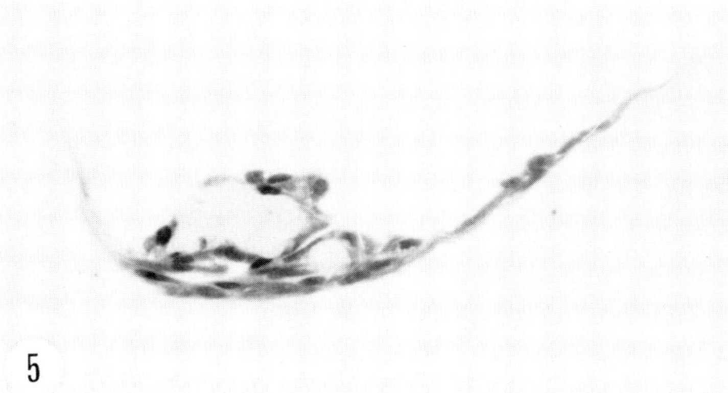

Figure 5.

recognizable acinar characteristics occurred, then withdraw the changes singly or in subsets to find out which were mainly responsible for the improvement. A screening evaluation was performed for each new set of modifications. When it appeared that a marked improvement had occurred, a detailed evaluation was undertaken on repeat cultures to characterize the improved cells at representative intervals after culture initiation. The screening evaluations relied mainly on light microscopic examination of sections of fixed, paraffin-embedded cultures, stained with hematoxylin and eosin (H & E) or special stains for mucins and other glycoproteins, specifically, periodic acid-Schiff or Alcian blue, lightly counterstained with hematoxylin (PAS-H or AB-H) (Mowry, 1963), or mucicarmine (Lauren and Sorvari, 1969). The detailed evaluations (Quissell et al., 1986; Redman et al., 1988) were performed on split or multiple cultures as required to provide representative samples for paraffin sections (a repeat of the screening examination), epoxy resin sections for high resolution light microscopy and for transmission electron microscopy (TEM), scanning electron microscopy (SEM), assays for DNA and protein, and radiolabeling estimates of the rates of synthesis of DNA, RNA, protein, and mucins, and of the secretion of mucins and other secretory proteins.

ISOLATION OF ACINAR-INTERCALATED DUCT COMPLEXES

The method for dissociating the acini and attached short segments of intercalated ducts from rat submandibular glands has been described in detail previously (Quissell and Redman, 1979). A brief synopsis is presented here to facilitate comparisons with the methods others have used for isolating acinar cells from the submandibular glands and other exocrine organs of rats and other animals. Male, Sprague-Dawley rats weighing 175 to 200 g (42 to 48 days old) were housed in a room with controlled temperature and humidity and allowed free access to a commercial pelleted diet and water. This narrow age range is important for three reasons (Quissell and Redman, 1979; Redman, 1987). (1) Maturation of the seromucous acinar cells is complete, the transition cell types of the neonatal period being scarce. (2) In older rats, the connective tissue stroma becomes more dense and tough, so that dissociating the parenchymals cells intact is more difficult. (3) The granular ducts are at an early stage of differentiation, minimizing the amounts of proteases, epidermal growth factor, and other substances with the potential for mischief in a culture system, that are liberated during the dissociation process.

ISOLATION PROCEDURE

Animals were killed by exsanguination under sodium pentobarbital anesthesia, the submandibular glands were excised, and the sublingual glands, contained in the same capsule, were removed and discarded. The submandibular glands were finely minced, suspended in dispersion medium and incubated in a gyratory shaker for 60 min at 37°C under 95% oxygen and 5% carbon dioxide. The medium was composed of Hank's salt solution (pH 7.4) containing 50 to 75 units/ml of chromatographically purified collagenase and 1.0 mg/ml type IV-hyaluronidase. The cells then were washed, filtered, washed, resuspended in the final culture medium, and incubated for another 30 min, after which aliquots were delivered to the culture dishes.

SYNOPSIS OF MODIFICATIONS IN CULTURE CONDITIONS

Details of the following initial trials of modifications were published previously (Quissell et al., 1986).

In the initial culture procedure, bovine collagen gels were prepared on ice by mixing 4 volumes of collagen stock solution with 1 volume of a 2:1 mixture of F12 and 0.34 N NaOH. Two ml of the mixture were spread evenly over the bottom of each 60-mm culture dish. The dishes were incubated at 37°C for 30 min to allow the gels to solidify. Before inoculation the dissociated acinar-intercalated duct complexes were suspended in F12 medium supplemented with 15% fetal bovine serum (FBS), 5 µg/ml insulin, 50 µg/ml Gentamycin sulfate and 1 µg/ml Amphotericin B. Four ml of a 2% cell suspension were added to each dish. Three µM of (-)-isoproterenol were added to each dish after inoculation and after each daily change of medium. Dishes were incubated at 37°C in a humidified atmosphere of 20% oxygen and 5% carbon dioxide under continuous gas flow.

Within 24 h under these conditions, most of the acinar cells and virtually all of the striated and granular duct cells were dead, but many of the intercalated duct segments survived (Figure 2). We then varied the concentrations of serum and isoproterenol, substituted normal rat serum (NRS) and rat tail collagen, and added a range of concentrations of epidermal growth factor (EGF). All had only slight effects on the survival of the acinar cells, and even in the best of these cultures the intercalated duct cells grew slowly for only a few days, then degenerated and died. Despite the report by Tapp (1967) that oxygen levels both above and below that of air (ca. 20% at sea level) were detrimental to salivary gland cells in organ culture, our next step was to increase the oxygen in the gas phase to 35%. This resulted in a dramatic improvement in the survival of the acinar cells, but not of the other cell types (Figure 3). Trials with increasing oxygen levels indicated an optimum of 85 to 90%. Under these conditions, the advantage of rat tail collagen was more noticable. Next, the collagen gel was impregnated with fibronectin, laminin, and type IV collagen, singly and in combinations. Laminin proved to be the only addition that was worthwhile. Additional manipulations in the concentrations of isoproterenol and serum, including mixtures of FBS and NRS, brought progress to where the first detailed evaluation was warranted. This was undertaken under the following culture conditions: 37°C; 90% oxygen, 5% carbon dioxide, and 5% nitrogen; 4 ml of 1% cells on 2 ml of rat tail collagen containing 5 µg/ml laminin; F12 medium supplemented with 10% FBS, 5 µg/ml insulin, 50 µg/ml Gentamycin sulfate and 1 µg/ml Amphotericin B; and 3 µM (–)-isoproterenol added initially and to each daily change of medium.

The results were as follows. Numerous viable acini and small groups of acinar cells were on the gel surface at the end of the first 24 h. The acinar cells had greatly decreased in diameter during this time (Figure 4). It was apparent that this was due mostly to the discharge of secretory granules in response to stimulation by the isoproterenol in the medium, as acinar cells in parallel cultures without isoproterenol showed no appreciable change in size. Between 2 and 4 days, most of the acinar-ductal complexes enlarged by proliferation, mostly among the acinar cells. This was accompanied by further reduction in the number of secretory granules in these cells. Though some additional enlargement occurred, increasing numbers of dead cells also accumulated in the centers of the cell groups. By 7 days, most complexes had only a few, mostly flattened cells left (Figure 5), and the cultures were discontinued at 14 days because of further deterioration.

The biochemical analysis (limited to the first 4 days of culture) gave results in parallel with the morphologic observations. One-day-old cultures were very similar to freshly prepared complexes in the secretion and synthesis of mucins, but these measures of acinar differentiation and function fell off very rapidly thereafter. Though the rates of synthesis of DNA, RNA, protein, and glycoprotein were higher at 3 and 4 days than in the fresh complexes, total protein and DNA decreased rapidly during the same period. This was interpreted as a reflection of a moderately high rate of proliferative activity being overtaken by an even greater rate of cell death.

Although we had succeeded in maintaining the acinar cells through one complete cycle of synthesis, storage, and secretion of mucin and other salivary proteins during 24 h of culture, it was clear that a vast improvement in the culture conditions still would be required to prolong this state. Details of the trials described in the following section were previously published (Redman et al., 1988).

SECOND CULTURE PROCEDURE

The next combination of modifications that was tried consisted of the addition of extracellular matrix extracts prepared by the method of Wicha et al. (1982), either from rat liver (LEM) or submandibular gland (SMEM), to the collagen gel; and the addition of retinoic acid, EGF and either dexamethasone or hydrocortisone to the medium.

Screening evaluations indicated that the complete combination promoted much improved survival of mostly plump cells, with cell groups continuing to enlarge to the end of the 11-day culture period. However, clefts and lumina were inconspicuous, and without these pathways for secretion, secretory product accumulated among the cells as eosinophilic and PAS positive, amorphous masses, albeit more slowly (4 to 7 days instead of 2 to 4 days) than in the preceding trials. Substitution and titration experiments demonstrated that neither of the extracellular matrices made a significant contribution to the improvement, the dexamethasone and hydrocortisone were interchangeable, and that a combination of retinoic acid and dexamethasone at 100 nM each was superior to either alone at any of several concentrations. Titration of the FBS revealed an optimum concentration of 1%, a tenth of that required in the previous series. Additional trials established a reduced oxygen level, 50%, as optimum. Under these conditions, the contribution of the EGF was found to be the spreading of a layer of cells across the gel surface from the original cell groups, accompanied by a general increase in the rate of cell proliferation. Most importantly, as the cell groups enlarged, they developed and maintained an organoid architecure, resembling primitive acini and small ducts, for 21 days of culture. At this point, it was decided that another major advance had been made. The modifications of the initial culture conditions in the ensuing second detailed evaluation were 1% FBS, 100 nM dexamethasone, 100 nM retinoic acid and 80 ng/ml EGF in the medium, and 50% oxygen in the gas phase.

From 2 to 22 days, the gel surface was dotted with acinar-intercalated duct complexes (Figure 6). For the first few days, the secretory granules of the acinar cells had the affinity for mucin stains and the ultrastructural substructure of the seromuocous granules of freshly isolated acini. The large number of these granules in the apical cytoplasm showed that these acinar cells were well differentiated. The number of cells containing granules and the number of granules per cell declined rapidly over the next 6 days, and after 8 days only an occasional cell contained a few small granules. Concomitantly, many cells accumulated lysosomes and autophagic vacuoles which seemed to contain the remains of unsecreted granules (Figure 7). From 4 to 22 days (end of culture) the complexes steadily enlarged, while a mostly monolayer sheet of cells spread over the surface of the gel among them (Figure 6). The cells in the sheets and at the bases of the complexes were attached to the gel by a basal lamina, and the surfaces facing the media, lumina or canaliculi were studded with small microvilli (Figures 7 and 8). Thus, although after 8 days the cytodifferentiation of the cells had become rudimentary or atrophic, most retained much of the morphodifferentiation of acini. An additional observation of note was the presence, subjacent to the basal lamina, of masses of small fibrils resembling those of type VII collagen.

The biochemical results were in general agreement with the morphologic observations. Throughout the culture period, the rate of DNA synthesis was

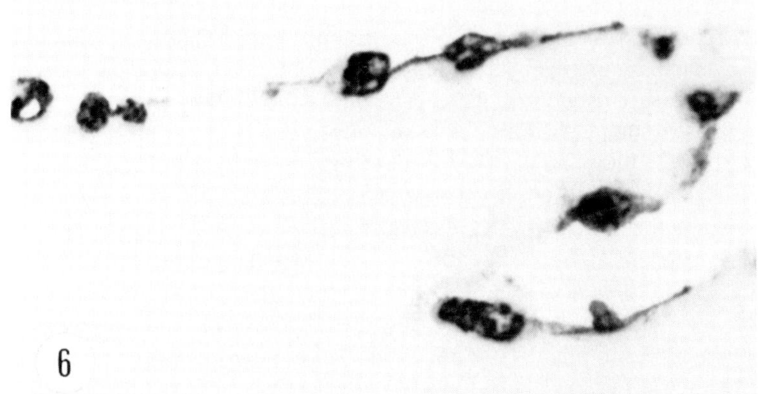

Figure 6. Survey photomicrograph of a paraffin section after 22 days under third-culture conditions. Sheets of cells are almost contiguous among the numerous complexes that have survived, attached to the gel, and enlarged severalfold. Periodic acid-Schiff and hematoxylin stain; ×65. (From Redman et al., *In Vitro Cell Dev. Biol.*, 24, 734, 1988. With permission.)

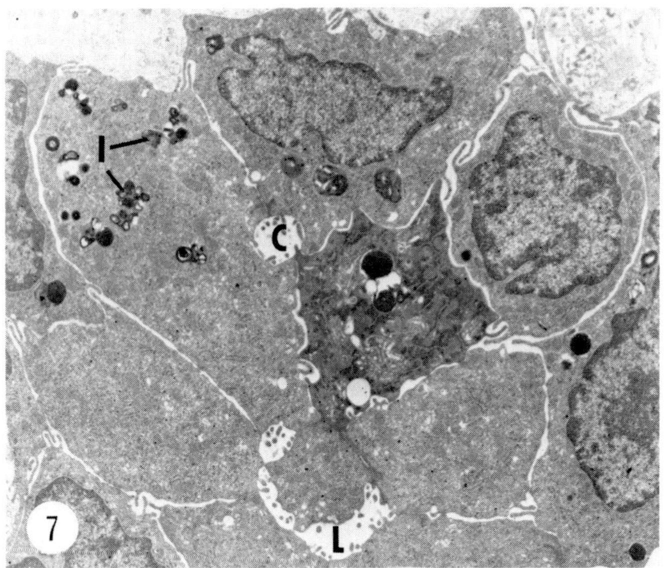

Figure 7.

Figures 7 and 8. Electron micrographs of complexes after 22 days under third culture conditions, with standard (glutaraldehyde, osmium tetroxide) fixation, embedding (epoxy resin), and stains (uranyl acetate and lead citrate). (From Redman et al., *In Vitro Cell Dev. Biol.*, 24, 734, 1988. With permission.) (7) Presence of lumina (L) and secretory canaliculi (C) in this complex indicate the retention of acinar morphodifferentiation, but acinar cytodifferentiation is barely recognizable, as there are few secretory granules and only modest amounts of rough endoplasmic reticulum. Many

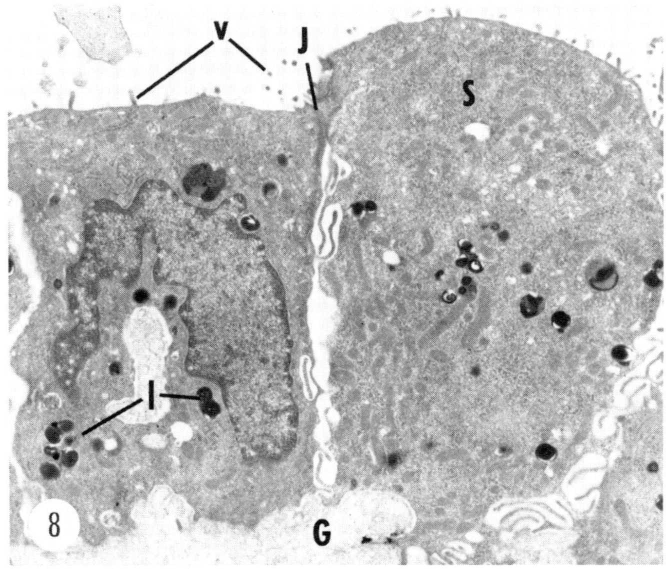

Figure 8.

Figures 7 and 8 (contd.). cells contain varying numbers of lysosomes and autophagic vacuoles (l). × 7200. (8) Cells in this sheet between complexes are polarized, with the basal side attached to the gel (G) and the side toward the medium having luminal features, e.g., microvilli (v) and tight junctions (J). Most have features of rudimentary acinar or intercalated duct cells, but a few (S) resemble striated duct cells, having basolateral folds and elongated mitochodria. Many cells in the sheets also contain a number of lysosomes and autophagic vacuoles. × 8300.

much higher than that of freshly isolated acini, peaking at almost 28-fold higher at 4 days (Figure 9). Nonetheless, the total number of cells, as judged by total DNA, declined from 50 to 20% of that of fresh complexes between 1 and 22 days. Although the secretion (Figure 10) and synthesis of mucin was sharply curtailed after only 1 day in culture, the rates of synthesis of proteins and glycoproteins in general remained close to that of fresh complexes through day 22 (Figure 11).

The results indicated that the new culture conditions greatly improved and prolonged the attachment, growth, and morphodifferentiation of the cells in the complexes, while significantly slowing the loss of most of the features of acinar cytodifferentiation, as compared to the previous conditions.

SERUM-FREE CULTURES

Innumerable substances in serum have been shown to affect growth and differentition. Thus, only in the absence of serum is it possible to unequivocally establish the requirements for specific nutrients, hormones, neurotransmitters, carrier proteins, and other substances, for various effects on a given tissue or cell

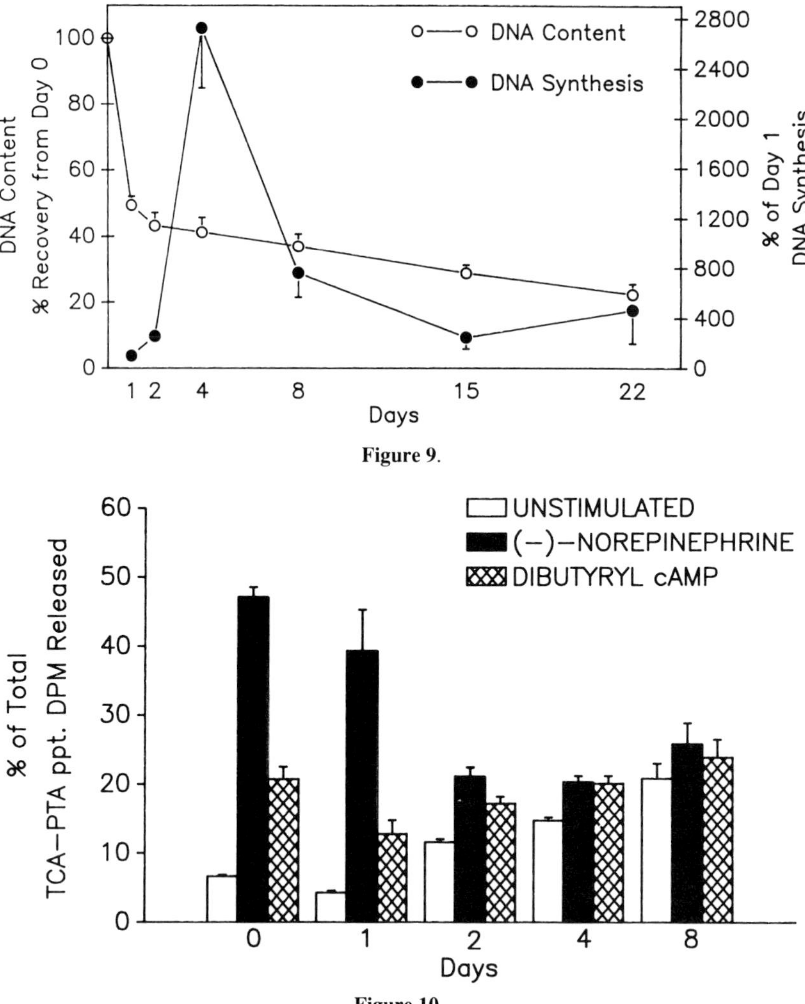

Figure 9.

Figure 10.

Figures 9 to 11. Biochemical measures of cellular activity under optimum culture conditions. Each symbol and bar represent the mean value ± SEM of five separate experiments. (From Redman et al., *In Vitro Cell Dev. Biol.*, 24, 734, 1988. With permission.) (9) Relative amounts of DNA, and DNA synthetic rate (uptake of 3H-thymidine over 24 h), as functions of time in culture. (10) Secretion of 3H-glucosamine-labeled material in response to the addition of 10 μM (–)-norepinephrine and 2 mM dibutyrl cAMP as functions of time in culture. (11) Rates of synthesis of protein (35S-methionine-labeling) and glycoprotein (3H-glucosamine-labeling) as functions of time in culture.

type in primary culture. Having progressed to where only 1% serum was required, we seemed poised to eliminate serum entirely. Thus, the major objective of the next series of experiments was to find a combination of substances that would replace serum in the medium to the extent that the acinar cells would do at least as well as they had in the preceding series.

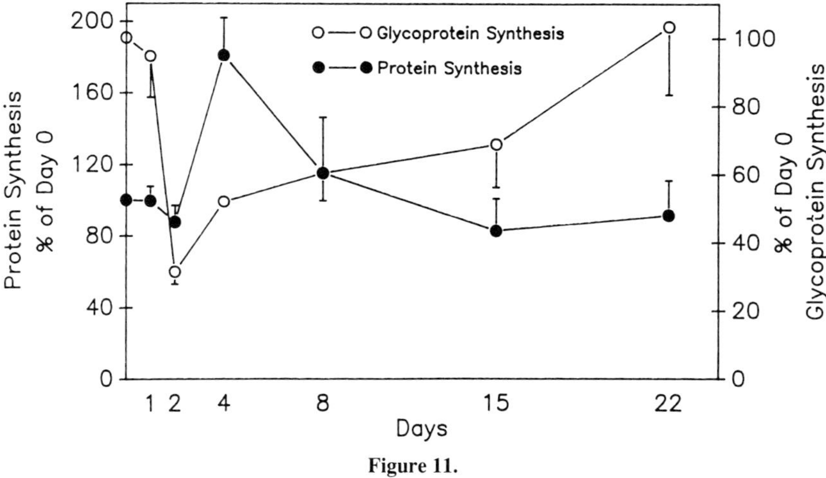

Figure 11.

The starting point was highly purified bovine serum albumin (BSA), as this protein is important not only in the maintenance of osmolality, but also as a carrier for hormones, other proteins and peptides, and lipids (Ham, 1981). Transferrin and a trace element mixture (Ham, 1981) were added to ensure that the cells would be able to utilize iron and other elements. An additional consideration was the major abnormalities in the secretory proteins synthesized by the acinar cells of both parotid and submandibular acinar cells which have been shown to be caused by even short-term stimulation with isoproterenol (Denny and Denny, 1981; Johnson, 1984; Yagil et al., 1986). Therefore, the initial experiments were aimed at comparing the effects of other, more "physiological" secretagogues, with those of isoproterenol on the maintenence of acinar differentiation by the cultured cells.

THIRD-CULTURE PROCEDURE

The initial serum-free culture conditions were 0.5% BSA, 10 nM each of dexamethasone and retinoic acid, 5 µg/ml each of insulin and transferrin, and 80 ng/ml EGF in the medium; rat tail collagen gel with 5 µg/ml laminin; and 50% oxygen. The tested secretagogues, added at several concentrations with each daily change of medium, were isoproterenol, norepinephrine, carbachol, carbachol plus norepinephrine, and a no-secretagogue control.

Carbachol gave the best results, but those with norepinephrine and carbachol plus norepinephrine also were encouraging. Carbachol at 0.3 µM was designated the new "standard" agonist. Subsequent experiments demonstrated improvements from the addition of trace elements and 0.2 nM thyroxine (T4). A retitration of the oxygenation showed that the optimum level had now shifted to 35%. At this point another detailed evaluation was undertaken.

The new culture conditions for this evaluation were 0.5% BSA; 5 μg/ml each of insulin and transferrin; 0.1 μ*M* each of dexamethasone and retinoic acid; 80 n*M* EGF; 0.2 n*M* T3; trace elements; and 35% oxygen.

Light microscopy of sections of both paraffin and epoxy resin-embedded tissues provided the following overview. There was very rapid growth of mostly plump cuboidal cells. Within 8 days the gel surface was covered with a layer several cells thick, with many lumen-like channels. This organization, reminiscent of endometrium, remained almost constant through 22 days. Both the channels and the cytoplasm of most cells contained mucins, as demonstrated by positve staining with PAS, AB, and mucicarmine. In the epoxy resin sections, many of the cells adjacent to the gel contained empty-looking secretory granules similar to those seen in freshly prepared complexes. Other cells were arranged in narrow duct-like structures. Thus, there seemed to be fairly good morpho- and cyto-differentiation of acini among the cell layers next to the gel. However, there was a problem of major importance: a cloud of dead cells on the medial surface that increased in depth with the age of the culture. This indicated that though there was rapid cell proliferation, it was matched by the rate of cell death. Further inspection revealed that most of the cells contained droplets with a green caste indicative of lipid. These were few and small in the cells next to the gel, but multiplied, enlarged and coalesced into globules in succeeding layers to the surface, with dying and dead cells being stuffed with them. It was concluded that the culture conditions were stimulating too much cellular proliferation, accompanied by some sort of metabolic derangement. The detailed evaluation was suspended until this problem could be solved. The subsequent screening evaluations required the use of epoxy resin-embedded materials, as the lipid droplets were not recognizable in paraffin sections.

There was a possibility the BSA had not been sufficiently purged of the lipids normally attracted to it in serum, and so the cells were being overloaded with lipid. However, vigorous solvent exraction of lipids from the BSA was of no benefit.

Next, different media (L 15, RPMI, Dulbecco, F12, and F12 plus Dulbecco) were tried, each with 1% FBS, 1% dialyzed FBS, and 0.5% BSA plus the dialysate from the FBS. In general, lipid droplets were not a problem in any of the media with FBS, and neither excessive cell death nor lipid droplets were eliminated by the BSA plus FBS dialysate with any of the media. Addition of pituitary extract also was of no benefit.

Growth-stimulating, as well as metabolic, effects of insulin have been demonstrated in pancreatic acinar cells in culture (Brannon et al., 1988). This suggested another plausible explanation for the problems with our cultures. Titration of the insulin proved to be very fruitful. Growth was slowed and accumulation of lipid droplets gradually decreased with decreasing concentrations to 0.05 μg/ml, below which growth and differentiation rapidly fell off (Figures 12 and 13). Increasing insulin but adding glucagon gave results similar

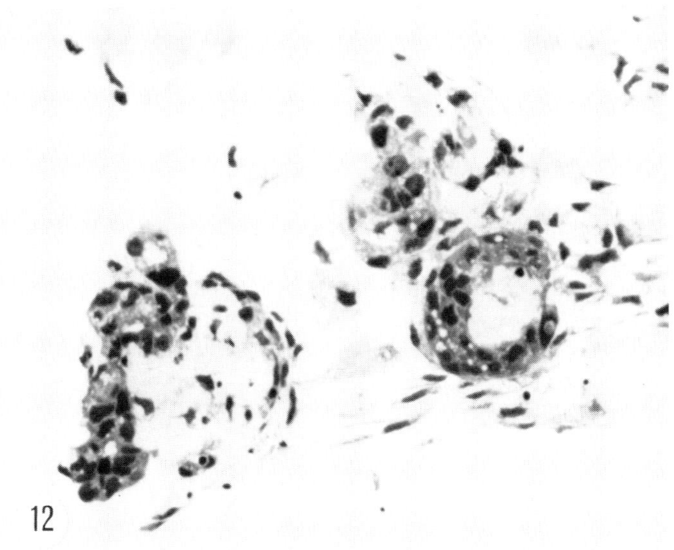

Figure 12.

Figures 12 to 15. Photomicrographs of complexes after 15 days in serum-free culture conditions. Conditions were identical except for variations in the amount of insulin and/or glucagon added before each daily change of medium. Cultures were fixed and embedded as in Figures 7 and 8. Semithin (ca. 1 μM) sections were stained with toluidine blue and azure II. (12) Addition of 0.005 μg insulin resulted in slow growth and sparse secretory granules. ×435. (13) With 0.2 μg insulin, there was rapid growth, modest acinar differentiation and widespread cell death. Many cells accumulated lipid droplets, and these frequently coalesced into globules (arrows). The excessive cell death and accumulation of lipids began at, and tended to worsen with, increasing amounts over 0.1 μg insulin. ×435. (14) With 0.05 μg insulin, or, as illustrated here and in the following figure, with 0.5 μg insulin plus 300 ng glucagon, there was slower growth, moderate acinar differentiation, much less cell death, and very few lipid droplets. With 0.5 μg insulin, the accumulations of dead cells and lipid droplets were roughly inversely proportional to the amount of glucagon added, up to 300 ng. × 435. (15) Higher magnification view of one of the complexes in Figure 14, showing numerous seromucous granules, some of which (arrows) appear to have undergone exocytosis into the lumina. × 910.

to a roughly equivalent concentration of insulin alone. Thus, 0.05 μg/ml insulin in 0.5% BSA seemed to have resolved the problem, and a similar improvement was obtained with 0.5 μg insulin plus 300 ng glucagon. Complexes enlarged slowly to 15 days, maintaining moderate acinar differentiation, with minimal cell death and lipid accumulation (Figures 14 and 15).

At this point we believe we have achieved the growth of moderately well-differentiated acini in serum-free medium. The extent to which this is so will be revealed by the ongoing detailed evaluation.

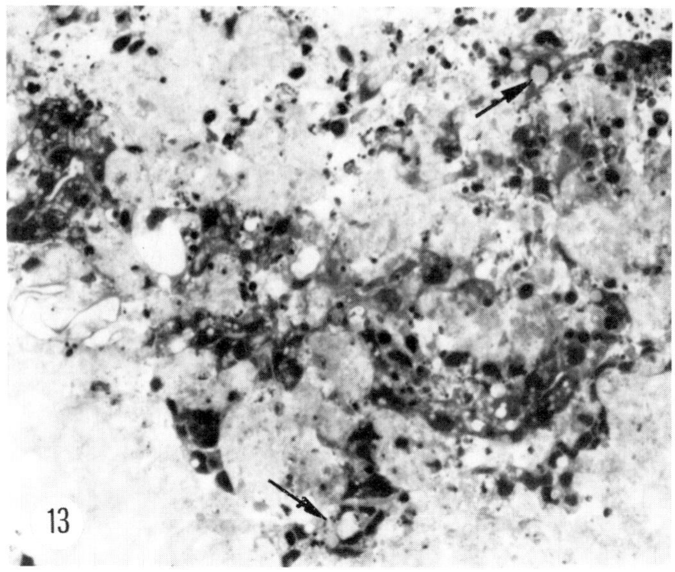

Figure 13.

FUTURE DIRECTIONS

We are encouraged by the recent successes of Oliver et al. (1987), Brannon et al. (1988), and Durban (1990). Oliver and colleagues have further refined the culture of rat parotid acinar cells to where they both proliferate and remain well differentiated, except for the loss of almost all amylase activity. However, they retain considerable salivary peroxidase activity (Kiser et al., 1990). Furthermore, these cells have remained in a moderately well-differentiated state for up to 6 months in culture (Yeh et al., 1991). Brannon and associates have progressed to the maintenance of moderately well-differentiated pancreatic acinar cells in serum-free medium, and have begun to use this system as a model for inquiries into the mechanisms of regulation of secretory protein synthesis (Justice and Brannon, 1989). Durban has developed a system for selectively culturing ductal cells from mouse submandibular glands, inducing them to differentiate into granular duct cells by adding dihydrotestosterone and T3 to the culture medium.

We plan to investigate the effects of a number of other modifications. Changes in the structure or suspension of the collagen gel to provide a unilateral flow of oxygen and nutrients to the bases of the cells may encourage better polarization and differentiation. Reexamination of other substances with effects on salivary secretion, such as VIP and substance P, seems warranted. Immortalization will be desirable, to provide cell lines with dependable characteristics. The results of early attempts by others suggest that this may be difficult, as the

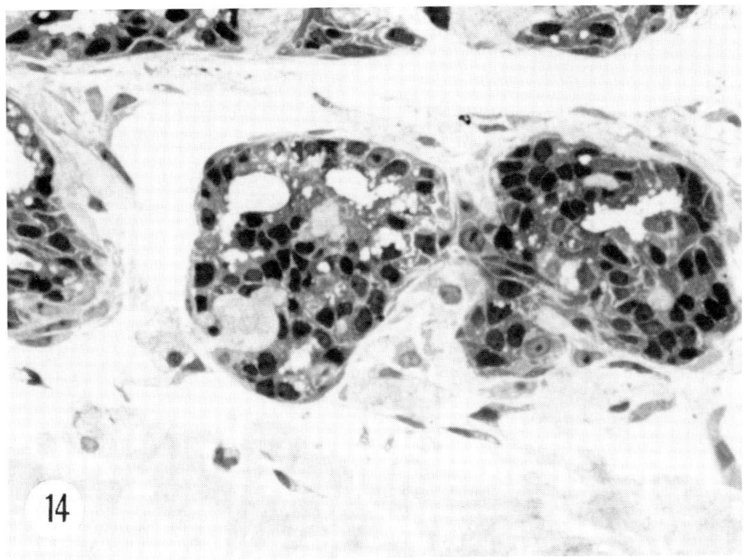

Figure 14.

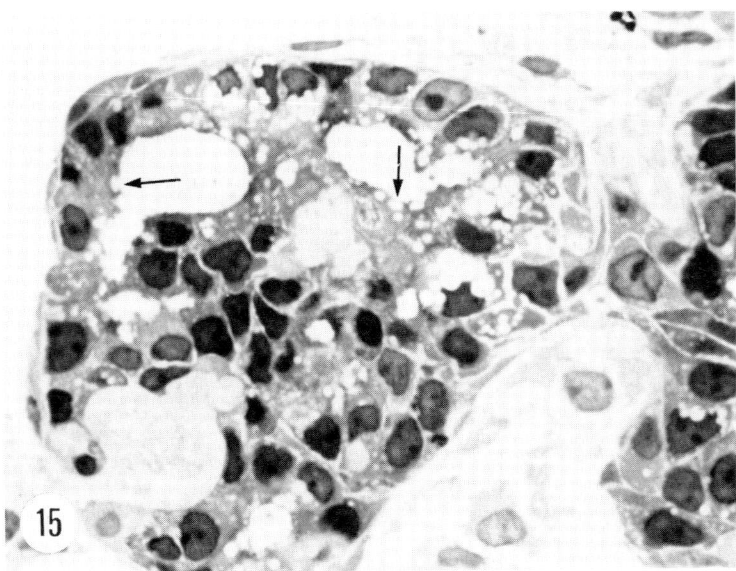

Figure 15.

cell lines that have been produced were poorly differentiated with regard to acinar characteristics (Brown et al., 1989; He et al., 1990). A much more refined characterization of the cells and their responses to a variety of agents, e.g., secretagogues and hormones, now is warranted. These should include immuno-histochemistry for mucins and other secretory proteins, especially the glutamine-glutamic acid-rich proteins that are specific to the acinar cells of the rat submandibular gland (Moreira et al., 1989) and for specific patterns of microfilaments (Marshak and Leitner, 1987). Detection and quantitation of mRNAs for some of these proteins, and for protooncogenes, also is now possible (Kousvelari et al., 1988; Mirels et al., 1990).

REFERENCES

1. Amsterdam, A. and Jamieson, J. D., Structural and functional characterization of isolated pancreatic exocrine cells, *Proc. Natl. Acad. Sci. U.S.A.*, 69, 3028, 1972.
2. Amsterdam, A. and Jamieson, J. D., Studies on dispersed pancreatic exocrine cells. I. Dissociation technique and morphologic characteristics of separated cells, *J. Cell Biol.*, 63, 1037, 1974.
3. Ball, W. D., Development of the rat salivary glands. III. Mesenchymal specificity in the morphogenesis of the embryonic submaxillary and sublingual glands of the rat, *J. Exp. Zool.*, 188, 277, 1974.
4. Baum, B. J. and Kuyatt, B. L., Protein production and release by dispersed rat submandibular cells *in vitro* after adrenergic stimulation, *Life Sci.*, 29, 1143, 1981.
5. Bdolah, A., Ben-Zvi, R., and Schramm, M., The mechanism of enzyme secretion by the cell. II. Secretion of amylase and other proteins by slices of rat parotid gland, *Arch. Biophys.*, 104, 58, 1964.
6. Beck, K., Hunter, I., and Engel, J., Structure and function of laminin anatomy of a multidomain glycoprotein, *FASEB J.*, 4, 148, 1990.
7. Bernfield, M., Banerjee, S. D., Koda, J. E., and Rapraeger, A. C., Remodeling of the basement membrane: morphogenesis and maturation, *M. Ciba Found. Symp.*, 108, 179, 1985.
8. Borghese, E., The development *in vitro* of the submandibular and sublingual glands of mus musculus, *J. Anat. (London)*, 84, 287, 1950.
9. Brannon, P. M., Hirschi, K., and Korc, M., Effects of growth factor, insulin and insulin-like growth factor I on rat pancreatic acinar cells cultured in serum-free medium, *Pancreas*, 3, 41, 1988.
10. Brannon, P. M., Orrison, B. M., and Kretchmer, N., Primary cultures of rat pancreatic acinar cells in serum-free medium, *In Vitro*, 21, 6, 1985.
11. Brown, A. M., A method for the initiation and maintenance of permanent rat submandibular gland epithelial cell cultures, *Arch. Oral Biol.*, 19, 343, 1974.
12. Brown, A. M., Rusnock, E. J., Sciubba, J. J., and Baum, B. J., Establishment and characterization of an epithelial cell line from the rat submandibular gland, *J. Oral Pathol. Med.*, 18, 206, 1989.
13. Castle, J. D., Jamieson, J. D., and Palade, G. E., Secretion granules of the rabbit parotid gland. Isolation, subfractionation, and characterization of the membrane and content subfractions, *J. Cell Biol.*, 64, 182, 1975.

14. Cunha, G. R., Support of normal salivary gland morphogenesis by mesenchyme derived from accessory sexual glands of embryonic mice, *Anat. Rec.*, 173, 205, 1972.

15. Cutler, L. S. and Chaudhry, A. P., Differentiation of the myoepithelial cells of the rat submandibular gland *in vivo* and *in vitro*: an ultrastructural study, *J. Morphol.*, 140, 343, 1973.

16. Cutler, L. S. and Christian, C. P., Inhibition of rat salivary gland adenylate cyclase by glycosaminoglycans and high molecular weight polyanions, *Arch. Oral Biol.*, 29, 629, 1984.

17. Cutler, L. S., Christian, C. P., and Rendell, J. K., Glycosaminoglycan synthesis by adult rat submandibular salivary-gland secretory units, *Arch. Oral Biol.*, 32, 413, 1987.

18. Dembinski, A., Gregory, H., Konturek, S. J., and Polanski, M., Trophic action of epidermal growth factor on the pancreas and gastroduodenal mucosa, *J. Physiol. (London)*, 325, 35, 1982.

19. Denny, P. C. and Denny, P. A., Non-developmental modifications in submandibular glands of young rats induced by chronic isoproterenol administration, *Arch. Oral Biol.*, 26, 297, 1981.

20. Deschodt-Lanckman, M., Robberecht, P., Camus, J., Baya, C., and Christophe, J., Hormonal and dietary adaptation of rat pancreatic hydrolases before and after weaning, *Am. J. Physiol.*, 226, 39, 1974.

21. Durban, E. M., Extracellular matrix effects on mammary cell behavior, in *Cellular and Molecular Biology of Mammary Cancer*, Medina, D., Kidwell, W., and Heppner, G., Eds., Plenum Press, New York, 1987, pp. 81–104.

22. Durban, E. M., Mouse submandibular epithelial cell growth and differentiation in long-term culture: influence of the extracellular matrix, *In Vitro Cell Dev. Biol.*, 26, 33, 1990.

23. Garrett, J. R. and Parsons, P. A., Alkaline phosphatase and myoepithelial cells in the parotid gland of the rat, *Histochem. J.*, 5, 463, 1973.

24. Grobstein, C., Epithelio-mesenchymal specificity in the morphogenesis of mouse submandibular rudiments *in vitro*, *J. Exp. Zool.*, 124, 383, 1953.

25. Ham, R. G., Survival and growth requirements of nontransformed cells, *Handbook Exp. Pharmacol.*, 57, 13, 1981.

26. Hammer, M. O. and Sheridan, J. D., Electrical coupling and dye transfer between acinar cells in rat salivary glands, *J. Physiol.*, 275, 495, 1978.

27. He, X., Frank, D. P., and Tabak, L. A., Establishment and characterization of 12s adenoviral E1A immortalized rat submandibular gland epithelial cells, *Biochem. Biophys. Res. Commun.*, 170, 336, 1990.

28. Hoshino, K. and Lin, C. D., Hyperplasia-inducing factor in mouse salivary gland isografts, *Cancer Res.*, 28, 2556, 1968.

29. Johnson, D. A., Changes in rat parotid salivary proteins associated with liquid diet-induced gland atrophy and isoproterenol-induced gland enlargement, *Arch. Oral Biol.*, 29, 215, 1984.

30. Jones, R. O., The *in vitro* effect of epithelial growth factor on rat organ tissues, *Exp. Cell Res.*, 43, 645, 1966.

31. Justice, J. D. and Brannon, P. M., Synthesis of amylase by cultured rat pancreatic acinar cells: effects of antecedent diet, *J. Nutr.*, 119, 805, 1989.

32. Kallman, F. and Grobstein, C., Fine structure of differentiating mouse pancreatic exocrine cells in transfilter culture, *J. Cell Biol.*, 20, 399, 1964.

33. Kanamura, S. and Barka, T., Short term culture of dissociated rat submandibular cells, *Lab. Invest.*, 32, 366, 1975.

34. Kiser, C. S., Rahemtulla, F., and Mansson-Rahemtulla, B., Monolayer culture of rat parotid acinar cells without basement membrane substrates, *In Vitro Cell. Dev. Biol.*, 26, 878, 1990.

35. Kousvelari, E. E., Louis, J. M., Huang, L.-H., and Curran, T., Regulation of proto-oncogenes in rat parotid acinar cells *in vitro* after stimulation of adrenergic receptors, *Exp. Cell Res.*, 179, 194, 1988.

36. Kumegawa, M., Yajima, T., Maeda, N., Takuma, T., Minamide, C., and Machino, M., Synergistic effects of diet, thyroxine, and glucocorticoid hormones on amylase activity in parotid glands of growing rats, *Arch. Oral Biol.*, 26, 631, 1981.

37. Lamey, P. J., Marshall, W., and Ferguson, M. N., A quantitative study on growth and cell population identification in murine salivary gland culture, *Arch. Oral Biol.*, 27, 367, 1982.

38. Lauren, P. A. and Sorvari, T. E., The histochemical specificity of mucicarmine staining in the identification of epithelial mucosubstances, *Acta Histochem.*, 34, 263, 1969.

39. Lawson, K. A., Morphogenesis and functional differentiation of the rat parotid gland *in vivo* and *in vitro*, *J. Embryol. Exp. Morphol.*, 24, 411, 1970.

40. LeBlond, C. P. and Inoue, S., Structure, composition, and assembly of basement membrane, *Am. J. Anat.*, 185, 367, 1989.

41. Logsdon, C. D. and Williams, J. A., Pancreatic acini in short-term culture: regulation by EGF, carbachol, insulin and corticosterone, *Am. J. Physiol.*, 244, G675, 1983.

42. Lucas, D. R., The effect of hydrocortisone, oxygen tension and other factors on the survival of the submandibular, sublingual, parotid and exorbital lacrimal glands in organ culture, *Exp. Cell Res.*, 55, 229, 1969.

43. Lundberg, A., Anionic dependence of secretion and secretory potentials in the perfused sublingual gland, *Acta Physiol. Scand.*, 40, 101, 1957.

44. Mangos, J. A., McSherry, N. R., Butcher, F., Irwin, K., and Barber, T., Dispersed rat parotid acinar cells. I. Morphological and functional characterization, *Am. J. Physiol.*, 229, 553, 1975.

45. Marshak, G. and Leitner, O., Cytokeratin polypeptides in normal and metaplastic human salivary gland epithelia, *J. Oral Pathol.*, 16, 442, 1987.

46. Martinez, J. R. and Cassity, N., Effects of ouabain and furosemide on saliva secretion induced by sympathomimetic agents in isolated, perfused rat submandibular glands, *Experientia*, 40, 557, 1984.

47. Martinez, J. R. and Cassity, N., ^{36}Cl Fluxes in dispersed rat submandibular acini: effects of acetylcholine and transport inhibitors, *Pfluegers Arch.*, 403, 50, 1985.

48. Martinez, J. R., Holzgreve, H., and Frick, A., Micropuncture study of submaxillary glands of adult rats, *Pfluegers Arch.*, 290, 124, 1966.

49. Martinez, J. R., Martinez, A. M., and Cooper, C., cAMP stimulates active K+ uptake in rat submandibular slices, *Experientia*, 39, 362, 1983.

50. McKeehan, W. L., Barnes, D., Rewid, L., Stanbridge, E., Murakami, H., and Sato, G. H., Frontiers in mammalian cell culture, *In Vitro Cell. Dev. Biol.*, 26, 9, 1990.

51. McPherson, M. A. and Dormer, R. L., Mucin release and calcium fluxes in isolated rat submandibular acini, *Biochem. J.*, 224, 473, 1984.

52. Mercante, M. L., On the *in vitro* behavior of mouse submaxillary gland cells, *J. Cell Sci.*, 13, 441, 1973.

53. Mirels, L., Kopec, L., Yagil, C., Dickinson, D. P., Dziejman, M. and Tabak, L., Expression of glutamine/glutamic acid-rich proteins in rat submandibular glands, *Arch. Oral Biol.*, 35, 1, 1990.

54. Moreira, J. E., Tabak, L. A., Bedi, G. S., Culp, D. J., and Hand, A. R., Light and electron microscopic immunolocalization of rat submandibular gland mucin glycoprotein and glutamine/glutamic acid-rich proteins, *J. Histochem. Cytochem.*, 37, 515, 1989.

55. Mowry, R. W., The special value of methods that color both acidic and vicinyl hydroxyl groups in the histochemical study of mucins, with revised directions for the colloidal iron stain, the use of alcian blue 8Gx and their combinations with the periodic acid-schiff reaction, *Ann. N.Y. Acad. Sci.*, 106, 402, 1963.

56. Nielsen, S. P. and Peterson, O. H., Transport of calcium in the perfused submandibular gland of the cat, *J. Physiol. (London)*, 223, 685, 1972.

57. Oliver, C., Isolation and maintenance of differentiated exocrine gland cells *in vitro*, *In Vitro*, 16, 292, 1980.

58. Oliver, C., Waters, J. F., Tolbert, C. L., and Kleinman, H. K., Growth of exocrine acinar cells on a reconstituted basement membrane gel, *In Vitro Cell. Dev. Biol.*, 23, 465, 1987.

59. Pictet, R. L., Clark, W. R., Williams, R. H., and Rutter, W. J., An ultrastructural analysis of the developing embryonic pancreas, *Dev. Biol.*, 29, 436, 1972.

60. Quissell, D. O., Secretory response of dispersed rat submandibular cells. I. Potassium release, *Am. J. Physiol.*, 238, C90, 1980.

61. Quissell, D. O. and Barzen, K. A., Secretory response of dispersed rat submandibular cells. II. Mucin secretion, *Am. J. Physiol.*, 238, C99, 1980.

62. Quissell, D. O., Barzen, K. A., and Lafferty, J. L., Role of calcium and cAMP in the regulation of rat submandibular mucin secretion, *Am. J. Physiol.*, 241, C76, 1981.

63. Quissell, D. O., Deisher, L. M., and Barzen, K. A., Role of protein phosphorylation in regulating rat submandibular mucin secretion, *Am. J. Physiol.*, 245, G44, 1983.

64. Quissell, D. O., Mawhinney, T. P., Barzen, K. A., and Deisher, L. M., Comparison *in vitro* of the incorporation of D-[2-3H(N)]-mannose and D-[1-14C]-glucosamine into glycoproteins of dispersed rat submandibular gland cells, *Arch. Oral Biol.*, 28, 827, 1983.

65. Quissell, D. O. and Redman, R. S., Functional characteristics of dispersed rat submandibular cells, *Proc. Natl. Acad. Sci. U.S.A.*, 76, 2789, 1979.

66. Quissell, D. O., Redman, R. S., and Barzen, K. A., Serum-free primary culture of rat parotid acinar cells, *J. Dent. Res.*, 68, 956, 1989.

67. Quissell, D. O., Redman, R. S., and Mark, M. R., Short-term primary culture of acinar-intercalated duct complexes from rat submandibular glands, *In Vitro Cell. Dev. Biol.*, 22, 469, 1986.

68. Redman, R. S., Development of the Salivary glands, in *The Salivary System*, Sreebny, L. M., Ed., CRC Press, Boca Raton, FL, 1987, chap 1.

69. Redman, R. S., Quissell, D. O., and Barzen, K. A., Effects of desamethasone, epidermal growth factor and retinoic acid on rat submandibular acinar-intercalated duct complexes in primary culture, *In Vitro Cell. Dev. Biol.*, 24, 734, 1988.

70. Redman, R. S., Sweney, L. R., and McLaughlin, S. T., Differentiation of myoepithelial cells in the developing rat parotid gland, *Am. J. Anat.*, 158, 299, 1980.

71. Robinovitch, M. R., Smuckler, E. A., and Sreebny, L. M. Protein synthesis in a cell-free system derived from the rat parotid gland, *J. Biol. Chem.*, 244, 5361, 1969.

72. Rufo, M. B. and Barka, T., Cell differentiation in the terminal tubule of fetal rat submandibular gland in organ culture, *Anat. Rec.*, 184, 301, 1976.

73. Ruoff, N. J. M. and Hay, R. J., Metabolic and temporal studies on pancreatic exocrine cells in culture, *Cell Tissue Res.*, 204, 243, 1979.

74. Sasaki, R., Mura, M., Takeuchi, T., Furihata, C., Matushima, T., and Sugimura, T., Premature induction of amylase in pancreas and parotid gland of growing rats by dexamethasone, *Biochim. Biophys. Acta*, 428, 619, 1976.

75. Schneyer, C. A. and Hall, H. D., Autonomic regulation of changes in rat parotid amylase during postnatal development, *Am. J. Physiol.*, 223, 172, 1972.

76. Sharawy, M. and O'Dell, N. L., Regeneration of acini in submandibular gland autografts, *Anat. Rec.*, 195, 431, 1979.

77. Slavkin, H. C., Molecular determinants of tooth development: a review, *Crit. Rev. Oral Biol. Med.*, 1, 1, 1990.

78. Srinivasan, R. and Chang, W. W. L., Effect of neonatal sympathectomy on the postnatal differentiation of the submandibular gland of the rat, *Cell Tissue Res.*, 180, 99, 1977.

79. Srinivasan, R., Chang, W. W. L., and Van der Noen, H., The effect of isoproterenol on the postnatal differentiation and growth of the rat submandibular gland, *Anat. Rec.*, 177, 243, 1973.

80. Tapp, R. L., An attempt to maintain cultures from the submandibular gland of the adult rat *in vitro*, *Exp. Cell Res.*, 47, 536, 1967.

81. Trelstad, R. L., Glycosaminoglycans: mortar, matrix, mentor, *Lab. Invest.*, 53, 1, 1985.

82. Wicha, M. S., Lowrie, G., Kohn, E., Bagavandoss, P., and Mahn, T., Extracellular matrix promotes mammary epithelial growth and differentiation *in vitro*, *Proc. Natl. Acad. Sci. U.S.A.*, 79, 3213, 1982.

83. Wigley, C. B. and Franks, L. M., Salivary epithelial cells in primary culture: characterization of their growth and functional properties, *J. Cell Sci.*, 20, 149, 1976.

84. Williams, J. A., Korc, M., and Dormer, R. L., Action of secretagogues on a new preparation of functionally intact isolated pancreatic acini, *Am. J. Physiol.*, 235, F517, 1978.

85. Yagil, C., Naito, Y., and Barka, T., Immunocytochemical localization of a developmentally regulated, isoproterenol-inducible protein (LM Protein) in rat submandibular gland, *Am. J. Anat.*, 177, 55, 1986.

86. Yang, J., Lawson, L., and Nandi, S., Three dimensional growth and morphogenesis of mouse submandibular epithelial cells in serum-free primary culture, *Exp. Cell Res.*, 137, 481, 1982.

87. Yeh, C. K., Mertz, P. M., Oliver, C., Baum, B. J., and Kousvelari, E. E., Cellular characteristics of long-term cultured rat parotid acinar cells, *In Vitro Cell. Dev. Biol.*, 27A, 707, 1991.

88. Young, J. A. and Schogel, E., Micropuncture investigation of sodium and potassium excretion in rat submaxillry saliva, *Pfluegers Arch.*, 291, 85, 1966.

Chapter

14

Culture of Parotid Acinar Cells

Constance Oliver

INTRODUCTION

Exocrine acinar cells in general, and salivary gland acinar cells in particular, have proven to be very difficult to establish *in vitro*. Parotid acinar cells are a highly differentiated cell type, specialized for protein secretion. *In vivo*, the cells are polarized with numerous secretory granules filling the apical cytoplasm and parallel arrays of rough endoplasmic reticulum occupying the basal portion of the cells. A prominent Golgi apparatus is located in a juxtanuclear position. Additionally, the cells are joined at their apices by tight junctions and are electrically coupled (Peterson et al., 1981) Because of these properties, the acinar cells have not been amenable to establishment in culture.

Early attempts to culture acinar cells produced cells which were viable for only short periods of time *in vitro* (Barka and Van Der Noen, 1975; Kanamura and Barka, 1975; Mangos et al., 1975). These initial isolation procedures, as well as all subsequent ones, were based on the method developed by Amsterdam and Jamieson (1972) for producing isolated pancreatic acinar cells. With this technique, the tissue was treated with chelating agents and digested in collagenase and hyaluronidase solutions. Although the method was modified somewhat for use with the parotid gland, the results were not entirely satisfactory. The isolation procedure yielded a population of cells which was 90 to 95% acinar cells, with a viability of 80 to 90%. Immediately following isolation, the acinar cells retained their characteristic morphology. However, with some procedures (Barka and Van Der Noen, 1975; Kanamura and Barka, 1975), the isolated cells failed to release amylase following stimulation with secretagogues. The secretory response was not recovered with time in culture and in no case were the cells

307

viable for long periods of time. Refinements in the isolation and culture procedures (Oliver, 1980) produced acinar cells which could be kept in culture for up to 1 month and which would respond to secretagogues, although not as efficiently as acinar cells *in vivo*.

The ability of the parotid acinar cells to survive for extended periods in culture proved to be dependent, at least in part, on the substratum to which they were attached (Oliver, 1987). Modifying the culture medium also was critical to the survival of differentiated acinar cells.

When isolated parotid acinar cells were plated onto a reconstituted basement membrane gel and grown in an enriched medium, the cells not only divided *in vitro* but also remained differentiated. Other epithelial cell types (Bissell et al., 1982; Gospodarowicz and Tauber, 1980; Hay, 1984; Kleinman et al., 1981) are known to require attachment to various substrata for growth and differentiation. Specifically, in mammary epithelial cells (Emerman et al., 1979; Emerman and Pitelka, 1977) attachment to extracellular matrix promotes both maintenance of cell differentiation and cell division. Additionally, pancreatic acinar cells can be induced to divide *in vitro* if attached to collagen gels (Logsdon and Williams, 1986). Therefore, it is possible to culture differentiated parotid acinar cells for extended periods if the cells are plated onto a reconstituted basement membrane matrix and are grown in a highly supplemented medium.

CULTURE METHOD

Parotid glands from young, 50- to 150-g, male Sprague-Dawley (Harlan-Sprague Dawley, Indianapolis, IN) rats were used as a source of acinar cells. All rats were certified free of sialodacro adenitis virus and rat corona virus. These two viruses are the most common viruses infecting the parotid glands. Cultures initiated from infected animals gave rise to many giant cells, and other cells with morphologies more characteristic of transformed cells. It was difficult to identify typical acinar cells in such cultures.

The isolated acinar cells were prepared for culture as outlined in Table 1. Anesthetized animals were exsanguinated by cutting the left ventricle of the heart. The parotid glands and surrounding tissue were immediately placed in a sterile 35-mm tissue culture dish containing 3 ml of Ham's Nutrient Mixture F-12 (GIBCO, Grand Island, NY) supplemented with penicillin (100 U/ml; GIBCO)-streptomycin (100 μg/ml; GIBCO)(F-12) which had been oxygenated with 95% O_2 and 5% CO_2. The glands were transferred to a fresh dish containing F-12 and, under sterile conditions, using a dissecting microscope, the acini were freed from surrounding tissue, nerves, and blood vessels. The acini were then transferred to a clean dish without F-12 and minced finely with a pair of sterile curved scissors. The minced tissue was placed in a Spinner flask (BELLCO, Vineland, NJ) with F-12 and an atmosphere of 95% O_2 and 5% CO_2. Each pair of parotid glands was minced and transferred to the Spinner flask before the next

TABLE 1. Disassociatioin Procedure for Parotid Acinar Cell Treatment

Treatment	Medium	Time (min)	Temperature (°C)
Remove tissue surrounding gland	F-12		25
Mince tissue			25
Rinse two times	Ca^{++}, MG^{++}-free HBSS	5	25
Chelation of divalent cations	EDTA (7.6 mg/ml), Ca^{++}, Mg^{++}-free HBSS	15	35
Replacement of divalent cations	F-12	3	25
Enzymatic digestion	Collagenase (12.5 mg/ml)-hyaluronidase (150 U/ml), F-12	45–60	35
Rinse two times by centrifugation	F-12	15	25
Filter through 500 µm and 100 µm Nytex filters	F-12	5	25
Sedimentation of acinar cells	F-12 + 30% heat inactivated fetal bovine serum	10	25
Rinse by centrifugation	F-12	5	25
Sedimentation of acinar cells	F-12 + 30% heat inactivated fetal bovine serum	10	25
Rinse two times by centrifugation	F-12	5	25

pair was removed from the animal. Glands which were deprived of oxygen for any length of time rapidly degenerated.

Glands were normally collected from 6 to 12 animals at a time. Once all of the glands were minced and in Spinner flasks, the tissue was rinsed twice in Ca^{++}, Mg^{++} free Hank's Balanced Salt Solution (HBSS) and resuspended in 7.6 mg/ml tetrasodium EDTA in Ca^{++}, Mg^{++} free HBSS in order to chelate divalent cations. The Spinner flask was gassed with 95% O_2 and 5% CO_2, sealed, and placed on a magnetic stir plate in a 35°C incubator for 15 min. A 37°C incubator may be used, but cells of epithelial origin frequently prefer a slightly lower temperature.

After treatment with the EDTA, much of the surrounding connective tissue was removed and the cells of the acini were only loosely associated. If the chelation step was omitted, it became much more difficult to separate the acinar cells. The steps following the EDTA all had to be lengthened, resulting in a lowered viability in the final cell preparation.

The tissue was then rinsed twice in F-12 to replace the divalent cations and digested with collagenase-hyaluronidase. After the addition of the enzyme solution, the flask was returned to the incubator for 45 to 60 min. At the end of this period, most of the acinar cells were either single cells or in small clumps. The majority of the ducts were not dissociated. The amount and type of collagenase needed for dissociation was dependant upon the lot. In early studies with the parotid glands, 1.0 mg/ml of Type II collagenase (Worthington Biochemical Corporation, Freehold, NJ) was found to give excellent results. However, recently, using Type II collagenase alone has not been as satisfactory as using an equal mixture of Types II and IV collagenase. Although the acinar cells are sensitive to trypsin and trypsin-like enzymes, as long as the clostripain levels were moderate in the collagenase preparations, the clostripain did not seem to have any deleterious effects on the acinar cells. Furthermore, some clostripain activity was necessary to achieve adequate digestion of the tissue. The type of hyaluronidase used was also important. The hyaluronidase must be chromatographically purified. The impurities in the cruder preparations resulted in a great deal of cell damage and death. It was difficult to maintain cell viability unless purified hyaluronidase was used.

Following digestion, the cells were rinsed once by centrifugation in F-12, and the single cells and small clumps of cells were separated from the undigested material by passing the cell suspension through a 500 μm Nytex filter (Tetko, Elmsford, NY). The filter also served to break up the larger clusters of acinar cells. The cell suspension was pipetted several times through a 10-ml pipette to mechanically break up the acini and then filtered through a 100 μm Nytex filter.

In order to remove the majority of contaminating cell types, i.e., fibroblasts, myoepithelial cells, duct cells, and blood cells, the resulting cell suspension was layered on a solution of F-12 + 30% heat-inactivated fetal bovine serum. The cells were allowed to settle by gravity for 10 min. At this time, the upper 75% of the medium was removed, and the cells rinsed twice by centrifugation in F-12. After the second rinse they were again allowed to settle through the F-12 + 30% serum. The exact concentration of serum must be determined for each lot, since the density may vary. As an alternative to the gravity sedimentation, the acinar cells may be collected after a low speed centrifugation through the F-12 + serum. While the yield of cells was greater following centrifugation, the purity was often less than with the gravity sedimentation.

At this stage, the preparation was approximately 95% acinar cells with a viability of about 90%. The cells were present either as single cells or as small clusters of 5 to 10 cells. The acinar cells may be put into culture at this stage or, if it is critical to have single cells, the preparation may be rinsed in Ca^{++}, Mg^{++}-free HBSS and treated with EDTA for an additional 15 min. In order to assure that the preparation was composed of single cells, the cell suspension was filtered through a 25-μm Nytex filter. After filtration, the cells were rinsed in F-12 and were placed in culture.

The acinar cells may be cultured as suspensions for short periods of time or attached to a substratum. *In vivo*, the acinar cells lie on a basement membrane. Plating the isolated cells onto a reconstituted basement membrane aided their ability to survive *in vitro* (Oliver et al., 1987). The substratum itself had an effect on the ability of the acinar cells to remain differentiated and to divide in culture. In suspension culture, although the acinar cells remained differentiated, there was no evidence that they divided (Oliver, 1980) (Figure 1). However, if attached, the cells both remained differentiated and divided and could be maintained for longer periods of time. The most suitable substrate tested, to date, was a reconstituted basement membrane gel (Matrigel, Collaborative Research), although Primaria dishes (Falcon Labware, Becton Dickinson Labware, Oxnard, CA) proved to be satisfactory. Recently, Kiser et al. (personal communication) have had success plating the cells onto Luxware.

To coat the dishes, the stock Matrigel was thawed on ice and diluted 1:2 in cold F-12 and the bottom of a 35-mm dish coated with 250 μl of solution. The dishes were placed in an incubator for 15 to 30 min to allow the solution to gel. 1×10^5 cells in 1 ml of F-12 were then plated in each dish. The dishes were placed in a 35°C humidified incubator with 5% CO_2 in air for 2 h to allow the cells to attach. After 2 h, the F-12 and unattached cells were removed and the F-12 replaced with 2 ml of growth medium (Table 2). The medium was then changed every day or every 2 days. At least in the initial stages of culture, there was an advantage to feeding the cells every day (Oliver et al., 1987a). When the cells were fed every day, they maintained a more constant level of DNA synthesis. Differentiated parotid acinar cells can be maintained for extended periods under these culture conditions (Figure 2).

When the cells were first plated onto the gel, they were spherical in appearance and attached at their lateral-basal cell surface (Figure 3). Almost immediately, some of the cells became elongated as they attached to the substratum. By 16 to 18 h, many cells, especially single cells and those at the periphery of larger clusters, had flattened and begun to spread over the surface of the basement membrane extract. Significant outgrowth from the cluster appeared at 3 to 4 days, coinciding with the initiation of DNA synthesis. Although the cultured acinar cells were flattened they still retained many of the morphologic features of acinar cells *in vivo*. There were numerous peroxidase positive secretory granules in the cytoplasm, and arrays of rough endoplasmic reticulum were present (Figure 4). In addition, junctional complexes could be seen between the cells (Oliver et al., 1987a)

Difficulty arises with the cultures, however, when attempts are made to subculture the cells. To date, it has not been possible to keep subcultured differentiated acinar cells viable. Several different protocols for removing the cells from the culture dish have been tried. In addition to collagenase or collagenase-hyaluronidase, dispase, papain, and trypsin have all been investigated. None of them proved satisfactory either alone or in combination. Addi-

TABLE 2. Growth Medium

Component	Concentration	Source
Ham's nutrient mixture F-12		GIBCO, Grand Island, NY
Penicillin	100 U/ml	GIBCO
Streptomycin	100 µg/ml	GIBCO
Dexamethasone	10 µg/ml	Collaborative Research, Lexington, MA
EGF	10 ng/ml	Irvine Scientific, Irvine, CA
Reduced glutathione	10 µg/ml	Sigma Chemical Co., St. Louis, MO
Putrescine	1 mM	Collaborative Research
L-ascorbic acid	50 µg/ml	GIBCO
Insulin	5 µg/ml	ITS
Transferrin	5 µg/ml	Collaborative Research
Selenium	5 mg/ml	
Isoproterenol	10^{-6} M	Sigma

tionally, treatment of the primary cultures with EDTA prior to or concomitant with enzyme digestion did not improve the viability of the passaged cells. Altering the substratum or modifying the culture medium was also without success. For the present, while the parotid acinar cells will adapt to primary culture, it has not been possible to passage them.

Although, the methods described here were developed for rat parotid acinar cells, they should be applicable to other species and other glands. With the same procedure, rat pancreatic and exorbital lacrimal gland acinar cells have been cultured successfully (Oliver, 1987a). Additionally, human pancreatic acinar cells have also been grown *in vitro* using this technique (Oliver, 1987b). With the human pancreatic acinar cells, fetal bovine serum was substituted for the rat serum. This method should prove useful for a wide variety of acinar cell types.

DISCUSSION

Although the cultured parotid acinar cells remain relatively differentiated *in vitro*, they are not as fully differentiated as their *in vivo* counterparts. In culture, the cells lose the strict polarity seen *in vivo*, fail to develop tight junctions between all of the cells, and amylase levels drop sharply (Oliver et al., 1987b). By 1 week in culture, the amylase content of the cells is only 0.1% of the value seen on day 0. The parotid acinar cells may be unusual in this respect, since amylase levels in similar preparations of pancreatic acinar cells only fall to 50% of day 0 at this same time point (Oliver et al., 1987a). Based on incorporation of ^{35}S-methionine, the levels of the other secretory enzymes are not as affected as

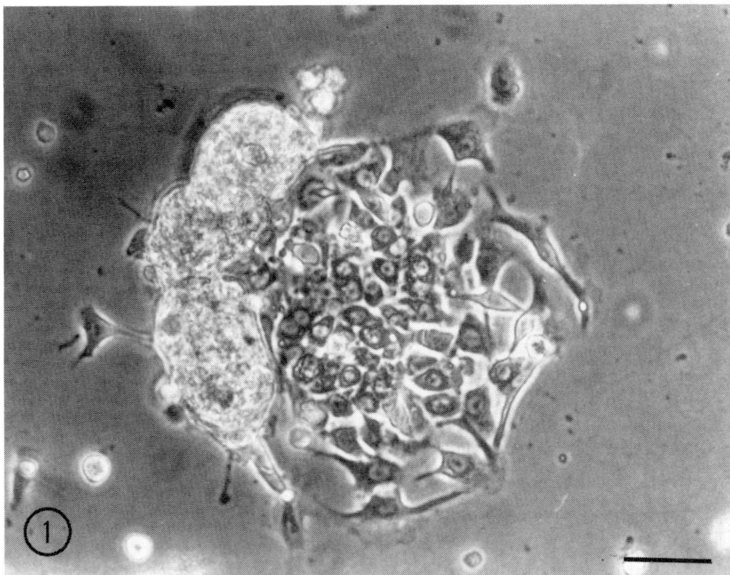

Figure 1. Rat parotid acinar cells. One day of culture. Cells at the edge of an aggregate have spread over the surface of the dish. × 150. Bar equals 100 µm.

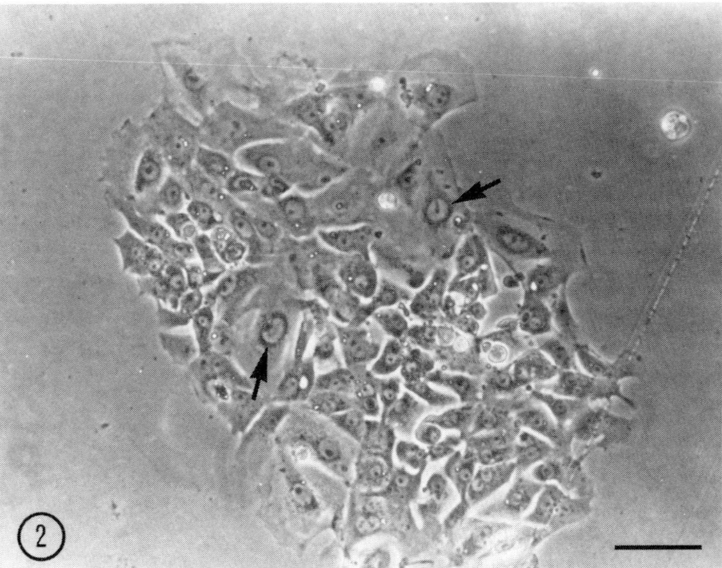

Figure 2. Rat parotid acinar cells. Four days of culture. All of the cells have flattened and spread over the surface of the culture dish. In some cells, secretory granules (arrows) are visible adjacent to the nucleus. × 150. Bar equals 100 µm.

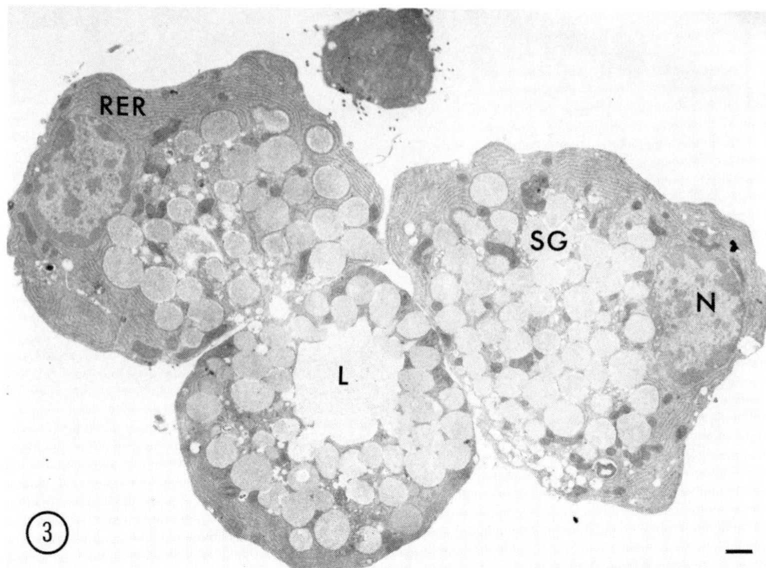

Figure 3. Rat parotid acinar cells. Freshly isolated. The small aggregate of acinar cells retains its polarity immediately following isolation. Secretory granules (SG) are present in the apical cytoplasm, and the basal cytoplasm is filled with parallel arrays of rough endoplasmic reticulum (RER). (L, lumen, N, nucleus) × 5600. Bar equals 1 μm.

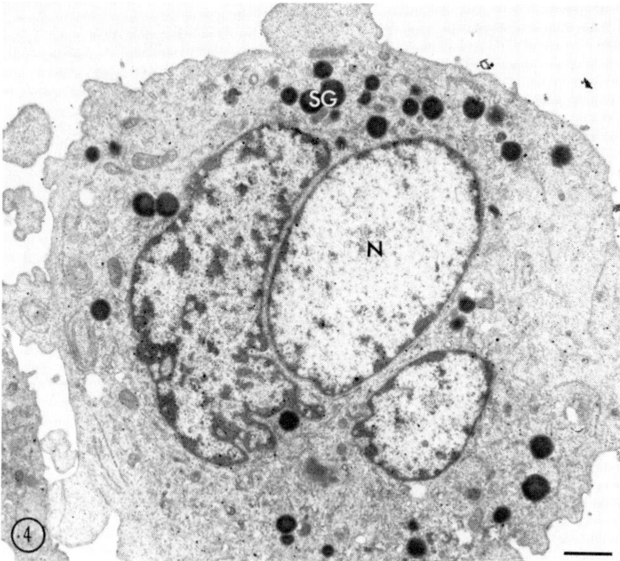

Figure 4. Rat parotid acinar cell. Four days of culture. Incubated in diaminobenzidine-H_2O_2. The cell has spread over the surface of the dish, and no longer retains its polarity. Peroxidase positive secretory granules (SG), and short segments of rough endoplasmic reticulum are present in the cytoplasm. (N, nucleus) × 11000. Bar equals 1 μm.

the amylase levels, and total protein synthesis does not diminish with time in culture (Oliver, 1980).

The medium composition directly affects the ability of the parotid acinar cells to remain differentiated and to divide *in vitro*. All of the medium components are known to influence cell division or differentiation in other cell types. Dexamethasone and EGF are both required for maintenance of pancreatic acinar cells in serum-free medium (Brannon et al., 1985), and insulin has been shown to increase amylase levels both in normal pancreatic acinar cells (Brannon et al., 1985) and in acinar cells from streptozotocin-treated rats (Korc, Iwamoto et al., 1981; Korc, Owerbach et al., 1981). In addition to their effects on exocrine acinar cells, both insulin (Barnes and Sato, 1980; Cherington, 1984; Pledger et al., 1984; Richman et al., 1976; Rillema and Linebaugh, 1977) and EGF (Barnes and Sato, 1980; Carpenter and Cohen, 1979; Cherington, 1984; Manjusri, 1982; Pledger et al, 1984; Rillema and Linebaugh, 1977) stimulate cell proliferation in other cell types. Putrescine may also be affecting DNA synthesis as well as helping to maintain the integrity of the cytoskeleton (Heby, 1981; Sunkara and Rao, 1981; Theoharides and Canellakis, 1975). Transferrin, required for transport of iron into cells, is essential for growth in a wide variety of cell types (Bottenstien et al., 1979). Reduced glutathione and selenium are both required for the action of glutathione peroxidase, a selenium-containing enzyme, in protecting the cells from damage from peroxides (Ham and McKeehan, 1978; Morris et al., 1984). Glucocorticoids, such as dexamethasone, play a role in potentiating the effects of other medium components such as insulin and EGF. They can affect not only DNA synthesis but the synthesis of cell-specific proteins (Germinario et al., 1983; Logsdon et al., 1985; Nicholas et al., 1984; Van Nest et al., 1983; Wakimoto and Oka, 1983). Lastly, the ascorbic acid is probably required for collagen synthesis (Kleinman et al., 1981) during basement membrane formation.

As mentioned previously, the presence of an extracellular matrix is one of the major factors in the ability to successfully culture differentiated parotid acinar cells. It is not clear what influence each of the matrix components may have in maintaining differentiation. The role of collagen in promoting differentiation and cell division is well established (Bissell et al., 1982; Hay, 1984; Kleinman et al., 1981; Mauch et al., 1988). Interaction with laminin has also been shown to be an important factor in maintaining differentiated functions in cultured cells (Runyan et al., 1988; Eckstein and Shur, 1989; Brunet-de Carvalho, 1989). The minor components, such as heparin sulfate, nidogen, and entactin, present in the basement membrane extract may also be required for maintaining differentiation in the parotid acinar cell cultures.

Although, differentiated parotid acinar cells can be successfully cultured for extended periods, it has not been possible to produce stable cell lines. The next phase of research with the acinar cell cultures should focus on the production of such cell lines. It should be possible to produce cell lines either by mutagenesis or transfection. The major obstacle so far has been keeping the cells differentiated. It is very likely that the conditions which favor growth and cell division do so at the expense of cell differentiation. However, at the present stage, the ability to culture parotid acinar cells, even for relatively short periods, provides a useful system for investigating the function of the parotid acinar cell.

REFERENCES

1. Amsterdam, A. and Jamieson, J. D., Structural and functional characterization of isolated pancreatic exocrine cells, *Proc. Natl. Acad. Sci. U.S.A.*, 69, 3028, 1972.
2. Barka, T. and Van Der Noen, H., Dissociation of rat parotid gland, *Lab. Invest.*, 32, 373, 1975.
3. Barnes, D. and Sato, G., Serum-free cell culture: a unifying approach, *Cell*, 22, 649, 1980.
4. Bissell, J. J., Hall, H. G., and Parry, G., How does the extracellular matrix direct gene expression?, *J. Theor. Biol.*, 99, 31, 1982.
5. Bottenstein, J., Hayashi, I., Hutchings, S., Masui, H., Mather, J., McClure, D. B., Okasa, S., Rizzino, A., Sato, G., Serrero, G., Wolfe, R., and Wu, R., The growth of cells in serum-free hormone-supplemented medium, in *Methods in Enzymology*, Jacoby, W. B. and Pastan, I. H., Eds., Academic Press, New York, 1979, 94.
6. Brannon, P. M., Orrison, B. M., and Kretchmer, N., Primary cultures of rat pancreatic acinar cells in serum-free medium, *In Vitro*, 21, 6, 1985.
7. Brunet-de Carvalho, N., Picart, R., Van de Moortele, S., Tougard, C., and Tixier-Vidal, A., Laminin induces formation of neurite-like processes and potentiates prolactin secretion by CH3 rat pituitary cells, *Differentiation*, 40, 106, 1989.
8. Carpenter, G. and Cohen, S., Epidermal growth factor, *Annu. Rev. Biochem.*, 48, 193, 1979.
9. Cherington, P. V., Regulation of fibroblast growth by multiple growth factors in serum-free medium, in *Mammalian Cell Culture*, Mather, J. P., Ed., Plenum Press, New York, 1984, 15.
10. Eckstein, D. J. and Schur, B. D., Laminin induces the stable expression of surface galactosyltransferase on lamellipodia of migrating cells, *J. Cell Biol.*, 108, 2507, 1989.
11. Emerman, J. T., Burwen, S. J., and Pitelka, D. R., Substrate properties influencing ultrastructural differentiation of mammary epithelial cells in culture, *Tissue Cell*, 11, 109, 1979.
12. Emerman, J. T. and Pitelka, D. R., Maintenance and induction of morphological differentiation in dissociated mammary epithelium on floating collagen membranes, *In Vitro*, 13, 316, 1977.
13. Germinario, R. J., McQuillan, A., Oliveria, M., and Manuel, S., Enhanced insulin stimulation of sugar transport and DNA synthesis by glucocorticoids in cultured human skin fibroblast, *Arch. Biochem. Biphys.*, 226, 498, 1983.
14. Gospodarowicz, D. and Tauber, J. P., Growth factors and the extracellular matrix, *Endocrinol. Rev.*, 1, 201, 1980.
15. Ham, R. G. and McKeehan, W. L., Development of improved media and culture conditions for clonal growth of normal diploid cells, *In Vitro*, 14, 11, 1978.
16. Hay, E. D., Cell-matrix interaction in the embryo: cell surface, cell skeletons, and their role in differentiation, in *The Role of Extracellular Matrix in Development*, Trelstad, R. L., Ed., Alan R. Liss, New York, 1984, 1.

17. Heby, O., Role of polyamines in the control of cell proliferation and differentiation, *Differentiation*, 19, 1, 1981.
18. Kanamura, S. and Barka, T., Short term culture of disassociated rat submandibular gland cells, *Lab. Invest.*, 32, 366, 1975.
19. Kleinman, H. K., Klebe, R. J., and Martin, G. R., Role of collagenous matrices in the adhesion and growth of cells, *J. Cell Biol.*, 88, 473, 1981.
20. Korc, M., Iwamoto, Y., Sankaran, H., Williams, J. A., and Goldfine, I. D., Insulin action in pancreatic acini from streptozotocin-treated rats. I. Stimulation of protein synthesis, *Am. J. Physiol.*, 240, G56, 1981.
21. Korc, M., Owerbach, D., Quinto, C., and Rutter W. R., Pancreatic islet-acinar cell interaction: amylase messenger RNA levels are determined by insulin, *Science*, 213, 351, 1981.
22. Logsdon, C. D. and Williams, J. A., Pancreatic acinar cells in monolayer culture: direct trophic effects of caerulein *in vitro.*, *Am. J. Physiol.*, 250, G440, 1986.
23. Logsdon, C. D., Moessner, A., and Williams, J. A., Glucocorticoids increase amylase mRNA levels, secretory organelles, and secretion in pancreatic AR42J cells, *J. Cell Biol.*, 100, 1200, 1985.
24. Mangos, J. A., McSherry, N. R., Butcher, F., Irwin, K., and Barber, T., Dispersed rat parotid acinar cells. I. Morphological and functional characterization, *Am. J. Physiol.*, 229, 553, 1975.
25. Manjusri, D., Epidermal growth factor: mechanisms of action, *Int. Rev. Cytol.*, 78, 233, 1982.
26. Mauch, C., Hatamochi, A., Scharffetter, K., and Krieg, G., Regulation of collagen synthesis in fibroblasts within a three-dimensional collagen gel, *Exp. Cell Res.*, 1978, 493, 1988.
27. Morris, J. G., Cripe, W. S., Chapman, Jr., H. L. Walker, D. F., Armstrong, J. B., Alexander, Jr., J. D., Miranda, R., Sanchez, Jr., A., Sanchez, B., Blair-West, J. R., and Denton, D. A., Selenium deficiency in cattle associated with heinz bodies and anemia, *Science*, 223, 491, 1984.
28. Nicholas, D. R., Sankaran, L., and Topper, Y. J., A unique and essential role for insulin in the phenotypic expression of rat mammary epithelial cells unrelated to its function in cell maintenance, *Biochim. Biophys. Acta*, 763, 309, 1983.
29. Oliver, C., Waters, J. F., Tolbert, C. L., and Kleinman, H. K., Growth of exocrine acinar cells on a reconstituted basement membrane gel, *In Vitro Cell Dev. Biol.*, 23, 465, 1987a.
30. Oliver, C., Waters, J. F., Tolbert, C. L., and Kleinman, H. K., Culture of parotid acinar cells on a reconstituted basement membrane substratum, *J. Dent. Res.*, 66, 594, 1987b.
31. Petersen, O. H., Iwatsuki, N., Philpott, H. G., Laugier, R., Pearson, G. T., Davison, J. S., and Gallacher, D. V., Membrane potential and conductance changes evoked by hormones and neurotransmitters in mammalian exocrine gland cells, in *Basic Mechanisms of Cellular Secretion. Methods in Cell Biology*, Vol. 23, Hand, A. R. and Oliver, C., Eds., Academic Press, New York, 1981, 513.
32. Pledger, W. J., Estes, J. E., Howe, R. H., and Leof, E. B., Serum factor requirement for the initiation of cellular proliferation, in *Mammalian Cell Culture*, Mather, J. P., Ed., Plenum Press, New York, 1984.
33. Richman, R. A., Claus, T. H., Pilkis, S. J., and Friedman, D. L., Hormonal stimulation of DNA synthesis in primary cultures of adult rat hepatocytes, *Proc. Natl. Acad. Sci. U.S.A.*, 73, 3589, 1976.

34. Sellema, J. A. and Linebaugh, B. E., Characteristics of the insulin stimulation of DNA, RNA and protein metabolism in cultured human mammary carcinoma cells, *Biochim. Biophys. Acta*, 475, 74, 1977.

35. Runyan, R. B., Versalovic, J., and Shur, B. D., Functionally distinct laminin receptors mediate cell adhesion and spreading: the requirement for surface galactosyltransferase in cell spreading, *J. Cell Biol.*, 107, 1863, 1988.

36. Sunkara, P. S. and Rao, P. N., Differential cell cycle response of normal and transformed cells to polyamine limitation, *Adv. Polyamine Res.*, 3, 347, 1981.

37. Theoharides, T. C. and Canellakis, Z. N., Spermidine inhibits induction of ornithine decarboxylase by cyclic AMP but not by dexamethasone in rat hepatoma cells, *Nature*, 255, 733, 1975.

38. Van Nest, G., Raman, R. K., and Rutter, W. J., Effects of dexamethasone and 5-bromodeoxyuridine on protein synthesis and secretion during *in vitro* pancreatic development, *Dev. Biol.*, 98, 295, 1983.

39. Wakimoto, H. and Oka, T., Involvement of collagen formation in the hormonally induced functional differentiation of mouse mammary gland in organ culture, *J. Biol. Chem.*, 258, 3375, 1983.

Established Salivary Cell Lines

Lauren L. Patton and Robert B. Wellner

INTRODUCTION

Saliva formation is a complex process involving the transport of water, electrolytes, and proteins by epithelial cells located in the acinar (secretory endpiece) and ductal regions of salivary glands (Young and Van Lennep, 1979; Van Lennep et al., 1986; Young et al., 1987). Because saliva performs numerous functions essential for the maintenance of oral health (Mandel, 1987, 1989), many investigators have focused their attention on the mechanisms by which saliva is formed in both normal and pathological states. There are several causes of abnormal saliva formation, including drug (Sreebny and Schwartz, 1986), radiation (Dreizen et al., 1976), and surgical therapies (May et al., 1973), and systemic diseases such as cystic fibrosis (Davis, 1987; Izutsu, 1987) and Sjögren's syndrome (Bloch et al., 1965; Fox et al., 1984; Talal, 1987). Some of these dysfunctions may involve aberrations in one or more of the different epithelial cell types located within the glands.

Salivary glands are composed of many distinct cell types which include the following: acinar (serous, special serous, seromucous, and mucous), myoepithelial, intercalated duct, granular duct, striated duct, and excretory duct (Pinkstaff, 1980; Young and Van Lennep, 1978, Van Lennep et al., 1986; see also Chapters 1 to 3). There exists marked structural diversity among glands (i.e., parotid, submandibular, sublingual, and minor salivary glands) within a given species, and among homologous glands of different species (Van Lennep et al., 1986). The contributions of specific cell types to normal saliva formation have been studied (with few exceptions) using acute acinar cell preparations from the secretory endpieces, or microperfusion or micropuncture systems of the main excretory ducts. Based on these studies, it has become clear that normal saliva formation occurs by a two-stage process (Thaysen et al., 1954; Young and Van Lennep, 1979; Van Lennep et al., 1986; Young et al., 1987) in which (1) a primary saliva is first formed by the action of acinar cells in the secretory endpieces of the glands and (2) the primary saliva is then modified by ductal epithelia as it is transported from secretory endpieces to the oral cavity. While these types of studies are essential and continue to provide important new information concerning the mechanism of saliva formation, other advances will require (or at least greatly benefit from) investigations using established salivary cell lines in culture. In this regard, the availability of cloned, well-characterized cell lines will (1) allow a better assessment of the contributions of specific cell types to saliva formation; (2) allow access to specific regions of the glands (e.g., intercalated and striated duct regions); and (3) broaden the applicability of somatic cell genetics, molecular biology and other approaches in studies concerning various aspects of salivary gland biology. During the past few years, several salivary cell lines of human, rat, and canine origin have been established in long term culture. Acinar cells, which are responsible for the initial formation of saliva, have not yet been imortalized. However, immortal cell lines derived from ducts or myoepithelia have provided new opportunities to investigate significant aspects of salivary gland biology. In this chapter, studies concerning the morphology and physiology of the currently available established salivary cell lines are reviewed, and directions for future research are discussed.

GENERAL CHARACTERISTICS OF ESTABLISHED SALIVARY CELL LINES

Thus far, at least seven epithelial cell lines of salivary gland origin have been established in long-term culture. Human cell lines have been derived from parotid, submandibular, and minor salivary glands. Cell lines of canine and rat origin have also been described. Five cell lines have been reported to be tumorigenic. In some cases, investigators have tentatively identified the region within the gland from which the cells were derived. Some general characteristics of the established salivary cell lines, discussed below, are summarized in Table 1.

HSY

This cell line was established from a human parotid adenocarcinoma composed primarily of intercalated duct cells (Yanagawa et al., 1986; Hayashi et al., 1987b). Transformed HSY cells were isolated by first transplanting the human tumor cells into nude mice, and then explanted cells obtained from the nude mouse tumors (adenocarcinoma) were cloned in semisolid agar. The most

TABLE 1. Established Salivary Cell Lines

Cell Line	Species	Gland	Tumorigenicity[a]	Cell Type
HSY	Human	PG	Adenocarcinoma	Epithelial
A253	Human	SMG	Epidermoid Carcinoma/ Undifferentiated Carcinoma[b]	Epithelial
HSG	Human	SMG	Adenocarcinoma	Epithelial
HPA	Human	MSG	Myoepithelioma	Myoepithelial
Nagoya-78	Human	MSG	Poorly differentiated malignant neoplasm	Epithelial
563	Canine	SMG	ND	Epithelial
RSMTx	Rat	SMG	ND	Epithelial

Note: Cell lines passaged 80 to 350 times in culture.

[a] Histological appearance of tumors formed by heterotransplantation of established cell lines.
[b] Epidermoid carcinoma observed in antithymocyte, serum-treated mice; undifferentiated carcinoma observed in nude mice. SMG: submandibular gland; MSG: minor salivary gland; PG: parotid gland; ND: not determined.

stable clone, HSY, was established in culture. Karyotype analysis revealed a near triploid chromosome number with at least two marker chromosomes. The ultrastructure was characterized by typical epithelial features, including desmosomes and tight junctions. Intercellular digitations formed by papillary infoldings were also observed. Interestingly, this cell line has been reported to contain secretory vesicles associated with moderately developed rough endoplasmic reticulum and Golgi (Hayashi et al., 1987b). Immunohistochemistry revealed the presence of amylase (acinar secretion product) in the HSY clone, and immunoelectron microscopy revealed positive reactivity to antiamylase serum by cytoplasmic secretory granules. Enzymatically active human amylase was also detected both within the cells and the culture medium. Immunoreactivity to antivasoactive intestinal polypeptide and cytokeratin was also observed in the cytoplasm. HSY cells were not reactive to antineurofilament antibody, however, suggesting that they are not of neural origin. Based on morphological and immunocytochemical considerations, it appears that HSY cells originate from either the intercalated duct or acinar region of the human parotid gland. Hayashi et al. (1987b) suggest that HSY cells are (1) derived from intercalated duct and (2) expression of some elements of the acinar phenotype might have occurred as a result of the transformation process. Although HSY cells do possess some acinar character, they clearly do not exhibit the full acinar phenotype as they lack α-adrenergic responsiveness (Table 2 and Patton et al., 1991a). In addition, the HSY clone appears to be able to differentiate into the myoepithelial phenotype since transplantation of these cells into nude mice (Yanagawa, 1986) results in the formation of tumors (adenocarcinoma) containing cell populations expressing either amylase (acinar cell marker) or myosin (myoepithelial cell marker).

TABLE 2. Demonstrated Responses of Established Salivary Cell Lines to Neuro-Agonist Stimulation

Cell Line	Stimulus	Receptor Subtype	Stimulatory Response
HSY	Muscarinic	ND	Ca^{2+} Mobilization $\uparrow$ K^+ Flux
	β-Adrenergic	ND	$\uparrow$ $[cAMP]_i$
A253	β-Adrenergic	β_2	$\uparrow$ $[cAMP]_i$ $\uparrow$ protein glycosylation
HSG	Muscarinic	ND	Ca^{2+} Mobilization $\uparrow$ $[IP_3]_i$ $\uparrow$ K^+ Flux
RSMTx	β-Adrenergic	β_2	$\uparrow$ $[cAMP]_i$ $\uparrow$ Cl^{1-} Flux $\downarrow$ Cell Volume
	α-Adrenergic	α	Ca^{2+} Mobilization $\uparrow$ $[IP_3]_i$

Note: Symbols used are as follows: $\uparrow$, increased; $\downarrow$, decreased; $[cAMP]_i$, intracellular cAMP content; ND, not determined; $[IP_3]_i$, intracellular inositol trisphosphate content. References are cited in text.

A253

This is the only established human salivary cell line currently available from the repository of the American Type Culture Collection (ATCC). Derived from an epidermoid carcinoma of the human submandibular gland, this cell line expresses the transformed phenotype, as judged by its ability to form colonies in soft agar, and to produce epidermoid carcinoma-like tumors when inoculated into ATS-treated mice (Giard et al., 1973), or undifferentiated carcinoma in nude mice (ATCC, 1988). In addition, A253 cells have been shown to lack contact inhibition of growth, i.e., they are able to grow over one another and form multilayers at least 8 to 10 cells thick when grown on plastic tissue culture dishes or polycarbonate filters (Marmary et al., 1989). The only information concerning the karyology of this cell line has been provided by the ATCC (ATCC, 1988), which reports that the cells have a near triploid chromosome number containing at least six marker chromosomes. The epithelial phenotype has been identified by electron microscopy, which revealed the presence of tight junctions, prominent desmosomes, occasional gap junctions, and a cytoplasm containing filament bundles, ribosomes, small mitochondria, occasional endoplasmic reticulum, and Golgi membranes. Large, abundant cytoplasmic deposits of electron-lucent material, presumably glycogen, have also been observed (Marmary et al., 1989). It was originally reported that similar deposits had not been observed in other established salivary cell lines. It has subsequently been noted, however, that periodic acid schiff-positive staining material was reported to occur in Nagoya-

78 cells (Kondo et al., 1971) and, more recently, in 563 cells (Strandstrom et al., 1989). Based on the available information, it is not possible to discern the region of the gland from which the A253 cell line was derived.

HSG

The HSG cell line (cloned in semisolid agar) was established from irradiated, histologically benign, submandibular gland tissue obtained from a patient treated for squamous cell carcinoma of the floor of the mouth (Shirasuna et al., 1981). HSG cells display the transformed phenotype as judged by their ability to form colonies in semisolid agar, exhibit fibrinolytic activity, and develop adenocarcinomas when inoculated into nude mice. In culture, these cells form occasional duct-like lumenal arrangements which are enhanced by co-culture with HPA myoepithelial cells (Sato et al., 1984). This result suggests that, *in vivo*, myoepithelial cells play a significant role in the formation of intercalated ducts. The karyology has not been reported. Ultrastructural examination reveals cuboidal cells joined by junctional complexes containing desmosomes and tight junctions. In addition, neighboring cells display intercellular digitations formed by papillary infoldings of the cytoplasmic processes. The mitochondria, endoplasmic reticulum, and Golgi are often located in one side of the cytoplasm. Tonofilaments, myofibrils, secretory granules (characteristic of acinar cells), and basal infoldings (characteristic of striated duct cells) are absent. Immunohistochemistry has shown that HSG cells express several specific antigens, including CEA (carcinoembryonic antigen), SC (secretory component), LF (lactoferrin), keratin, vimentin, desmin, and tropomyosin (Sato et al., 1984, 1985a). The secretion of human epidermal growth factor (hEGF) and transforming growth factor-β, as well as the expression of cell-surface hEGF receptors, has also been described (Sato et al., 1985b). Increased CEA expression, as well as S-100 protein expression (not normally detected in HSG cells), occurs in HSG cells co-cultured with HPA myoepithelial cells (Sato et al., 1984). Inasmuch as CEA has been reported to be expressed primarily in acinar and intercalated duct cells of the normal parotid salivary gland (Caselitz et al., 1981), it has been suggested that the presence of this antigen is further evidence that HSG cells are derived from intercalated duct. It has also been suggested (Sato et al., 1984) that expression of SC and LF (important components of salivary function; Mandel, 1989), and keratin (an epithelial antigenic marker; Franke et al., 1979; Sun et al., 1979) indicates that the HSG cells retain some differentiated properties normally expressed by intercalated duct cells *in vivo*. Based on the morphologic and immunohistochemical characterization, it appears that HSG cells are derived from the intercalated duct region of the human submandibular gland (Shirasuna et al., 1981; Sato et al., 1984, 1985a).

HPA

This cell line was isolated and cloned from a human pleomorphic adenoma arising in a minor salivary gland of the palate (Shirasuna et al., 1980). The primary tumor was composed of both intercalated duct and myoepithelial cells. Transformed cells from the primary tumor were selected for their ability to form colonies in semisolid agar. Five colonies were isolated, all of which ultrastructurally resembled myoepithelial cells. One clone, HPA, was selected for its stable growth properties. Transplantation of HPA cells into nude mice resulted in the growth of poorly differentiated neoplasms with an ultrastructure resembling myoepithelial cells. Viral particles, not found in the original tumor, appeared in the cytoplasm and intercellular spaces of the cultured cells and nude mouse tumors. The potential role of these viral particles in tumor formation is unknown. Karyology of this cell line has not been reported. Electron microscopy of the cultured cells revealed the presence of desmosomes, microvilli, myofibrils, membrane-bound vacuoles, dilated rough endoplasmic reticulum, well-developed Golgi, and large oval mitochondria. Immunohistochemistry has shown that HPA cells exhibit positive staining for S-100 protein and myosin (markers specific for myoepithelial cells), and a weak reaction with anticarcinoembryonic antigen, which is enhanced in co-culture with HSG cells (Sato et al., 1984). In addition, autoradiographic analysis has demonstrated the presence of oxytocin receptors in the cytoplasm (Sato et al., 1984). Overall, the morphologic and immunologic studies have been taken to suggest that HPA cells are of myoepithelial origin (Shirasuna et al., 1980; Sato et al., 1984).

Nagoya-78

Tissue obtained from a pleomorphic adenoma of a human labial minor salivary gland (Kondo et al., 1971) was used to establish this cell line. Although Nagoya-78 cells were derived from a benign tumor, the investigators believe that there are malignant histological features in the tumors formed by heterotransplantation of these cells into hamster cheek pouches. Chromosome analyses have revealed predominant sets of 62-65 chromosomes containing five markers and notable trisomy. Ultrastructurally, the Nagoya-78 cell line appears to be of epithelial origin, as it is reported to contain desmosomes. Periodic acid-Schiff staining has revealed the presence of abundant glycogen in the cytosol. The region of the gland from which these cells were derived is unknown. A stellate appearance of these cells under phase contrast microscopy (Kondo et al., 1971) has suggested the possibility of myoepithelial origin. Shirasuna et al. (1980) have questioned this notion, however, because (1) additional studies positively identifying the myoepithelial nature of these cells were not performed and (2) tumors formed in hamster cheek pouches were comprised of polygonal epithelial cells which did not resemble myoepithelioma.

563

This cell line was derived from the mandibular salivary gland of a male collie which had been euthanazied as a result of a malignant skin tumor of mesenchymal origin (Strandstrom et al., 1989). The investigators reported that the original salivary tissue lacked neoplastic character, and the established cell line was not virally infected. It was not directly determined, however, whether 563 cells might be transformed (e.g., ability to form colonies in soft agar or tumors in nude mice). Karyotype analysis of both early- and late-passage cells has revealed a large diploid and small tetraploid chromosome population. The modal chromosome number of 76 was reported to be of the normal canine karyotype; however, some chromosome structural abnormalities were noted. An ultrastructural examination revealed epithelial characteristics, including the presence of desmosomes, tonofilaments, tight junctions, and cellular processes within intercellular crypts. In addition, these cells showed the presence of intermediate filaments and lack of distinct secretory granules. Immunohisto-chemistry revealed additional epithelial character inasmuch as the 563 cell line was found to express cytokeratin and epithelial membrane antigen. Although these cells appear to be of glandular origin, the specific region from which they were derived is unknown.

RSMTx

Carcinogens have proven useful in establishing a number of different cell lines in culture (e.g., Babcock et al., 1983; Bennett et al., 1978; Stampfer and Bartley, 1985). RSMTx (Brown et al., 1989) is a cloned, salivary cell line that has been established in long-term culture as a result of carcinogenic treatment (3-methylcholanthrene) of normal tissue (weanling rat submandibular gland). Presumably, RSMTx is a transformed cell line; however, tumorigenicity and karyotyping studies have not been reported. The epithelial nature of this cell line has been confirmed by both morphological and immunocytochemical studies, although the expression of specific epithelial traits has been reported to vary with passage number. For example, early-passage cells (passages 4 to 12) displayed terminal bar-like structures containing tight junctions and desmosomes, while later-passage cells (passages 13 to 80) did not exhibit these features. Some epithelial character was retained by late-passage cells, however, as immunocy-tochemistry demonstrated the continued expression of both cytokeratin and epithelial membrane antigen. Immunocytochemistry, utilizing antiserum against RSMTx cell-surface antigens, was also used to demonstrate that RSMTx cells might be derived from ductal regions of the gland. Antiserum to early-passage cells preferentially reacted with intercalated, collecting, and excretory ducts; whereas, the reaction of antiserum to late-passage cells was limited to small (intercalated) ducts. Taken together, these results suggest that RSMTx is an immortal, epithelial cell line derived from a ductal region of the rat submandibular gland.

OTHER SALIVARY CELL CULTURES

Two groups (Knowles and Franks, 1977; Barka et al., 1980) have described the *in vitro* growth of transformed cells derived from mouse submandibular gland tissue treated with the carcinogen dimethylbenz(a)anthracene. This treatment allowed the maintenance of epithelial morphology (Knowles and Franks, 1977; Barka et al., 1980) and expression of epidermal growth factor, renin and peptidase (cell clones SCA-5, SCA-9, SCA-10, established by Barka et al., 1980; SCA-9 available from the ATCC) for at least 8 (Knowles and Franks, 1977) or 20 (Barka et al., 1980) cell passages. The phenotype of the SCA clones is interesting, and has led Barka et al. to consider the clones to be established epithelial cell lines which either (1) originate from granular convoluted tubule (GCT) or (2) are at least capable of expressing traits characteristic of the GCT cell phenotype.

More recently, the propagation of human submandibular gland (Kurth et al., 1988) and minor salivary gland (Kurth et al., 1989) ductal epithelia has been described. As opposed to carcinogen treatment, these investigators used a low concentration of calcium (0.1 mM) in the growth medium, which allowed the cells to be propagated in serum-free medium for at least 8 cell passages. Apparently, reduced extracellular calcium allowed epithelial duct cell propagation by preventing terminal keratinocyte formation.

DIFFERENTIATION STUDIES

In salivary histogenesis, progenitor stem-cell populations have the potential of giving rise to various cell types that comprise the normal salivary gland. Epithelial cells from excretory duct (Batsakis, 1980; Eversole, 1971), intercalated duct (Eversole, 1971; Alvares and Sesso, 1975; Chang, 1974; Batsakis, 1980; Zajicek, 1985), and acini (Redman and Sreebny, 1970; Kline, 1982) have been postulated to serve as stem-cell populations. Recently, it has been pointed out (Redman, 1987; Dardick et al., 1990) that all salivary gland cell types may be pluripotential. Investigators at the Tokushima University School of Dentistry have treated the putative intercalated duct cell clone, HSG, with several known differentiation-inducing agents in order to investigate mechanisms by which cytodifferentiation occurs in human salivary glands. Consistent with earlier observations of salivary tumor histogenesis (Eversole, 1970; Pierce, 1974; Batsakis, 1980), these studies have provided additional evidence that intercalated duct cells can act as progenitors for both the acinar and myoepithelial cell types (Figure 1).

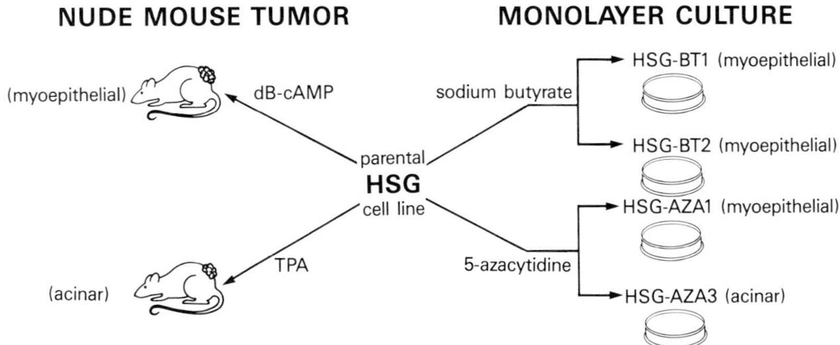

Figure 1. Parental HSG cells treated with specific agents express some traits characteristic of the acinar or myoepithelial phenotype in nude mouse tumors or monolayer culture. dB-cAMP = dibutyryl cyclic adenosine 3′,5′-monophosphate; TPA = 12-O-tetradecanoyl-phorbol-13-acetate.

Evidence for conversion of the HSG intercalated duct-like clone into myoepithelial-like cells was obtained by treating the cells with either dibutyryl cyclic AMP, sodium butyrate, or 5-azacytidine. In initial studies, 28 daily injections of 1 mM dibutyryl cyclic AMP into HSG-induced nude mouse adenocarcinomas resulted in the formation of isolated nests of myoepithelial-appearing cells surrounded by the intercalated duct-like cells of the original tumor (Hayashi et al., 1985). Expression of the myoepithelial phenotype included typical morphological features (spindle shape and presence of myofilaments and pinocytic vesicles) and antigenic markers (myosin and S-100 protein). In a subsequent study, 0.5 to 1.0 mM dibutyryl cyclic AMP treatment of HSG cells in monolayer culture for 10 days induced oxytocin receptor expression as well as the other myoepithelial traits previously observed in the tumor nest cells (Yoshida et al., 1986). In addition, dibutyryl cyclic AMP treatment resulted in tumor regression, decreased growth rate of cells in culture, and reduced colony formation by cultured cells in semisolid agar. As all of these changes required the continued presence of dibutyryl cyclic AMP (i.e., they are reversible), this agent was not useful for establishing permanent, HSG-derived, myoepithelial subclones.

In contrast to the results obtained with dibutyryl cyclic AMP treatment, exposure of HSG monolayers to 5 mM sodium butyrate for 28 days resulted in the generation of long-lived, continuous (at least 95 passages) populations of myoepithelial-like cells (Azuma et al., 1986). Only 2 of 40 sodium butyrate-derived HSG subclones, HSG-BT1 and HSG-BT2, could be maintained in monolayer culture. Both subclones were found to possess the same myoepithelial-like characteristics identified in the dibutyryl cyclic AMP-treated cells discussed above, and they also demonstrated reductions in growth rate, tumorigenicity, and the ability to form colonies in semisolid agar.

The proposed pluripotential nature (Eversole, 1971; Pierce, 1974; Batsakis, 1980) of salivary intercalated duct cells is supported by studies demonstrating that both myoepithelial and acinar-like phenotypes are induced in clones arising from an HSG cell population treated for 5 days with 5 μM 5-azacytidine (Sato et al., 1987). From 12 original isolates, 2 clonal cell lines exhibiting myoepithelial traits and 5 clonal cell lines exhibiting acinar traits could be maintained in monolayer culture. The most stable clone of each phenotype (AZA1, myoepithelial; AZA3, acinar) was characterized in detail. AZA1 exhibited several typical myoepithelial traits, including spindle shape, the presence of microfibrils and myosin, and the production of myoepithelioma upon transplantation into nude mice. Tests for the presence of oxytocin receptors and S-100 protein were not reported. AZA3 demonstrated such acinar traits as a polygonal shape, the presence of secretory granules containing amylase, and the production of acinic cell carcinoma upon heterotransplantation. Both clones demonstrated reductions in growth rate and the ability to form colonies in semisolid agar. The stability of either clone with respect to passage number was not reported.

As opposed to the effects of agents that induce either myoepithelial or mixed myoepithelial/acinar cell populations, phorbol ester (TPA, 12-O-tetradecanoyl-phorbol-13-acetate) treatment has been shown to induce solely the acinar phenotype in clones derived from the HSG cell line (Hayashi et al., 1987a). Injection of TPA (10^{-7} M, daily for 28 days) directly into nude mouse HSG-induced adenocarcinomas produced nests of neoplastic cells composed of pleomorphic, cuboidal, and large polygonal cells, the cytoplasm of which contained many secretory granules. Other notable ultrastructural features included intercellular digitations and junctional complexes containing desmosomes and tight junctions. In addition, neoplastic cells in the TPA-treated tumors showed immunohistochemically positive staining for amylase and lactoferrin.

Although the exact mechanisms by which the various agents induce salivary cell differentiation is unknown, it is interesting that the agents which activate protein kinase A (via an increase in cAMP) induce traits of the myoepithelial phenotype, while agents that activate protein kinase C induce expression of acinar traits. 5-Azacytidine treatment apparently acts in a less discriminatory manner, as hypomethylation of DNA (Razin and Riggs, 1980; Konieczny and Emerson, 1984) allows expression of genes encoding traits of either the acinar or myoepithelial phenotype.

NEUROTRANSMITTER SIGNAL TRANSDUCTION STUDIES

Saliva formation is a two-stage process, involving (1) an initial transport of water, electrolytes, and proteins by acinar cells into the secretory endpiece lumens of salivary glands, followed by (2) ductal epithelial cell modifications of the primary saliva during its passage to the oral cavity (Thaysen et al., 1954; Young and Van Lennep, 1979; Van Lennep et al., 1986; Young et al., 1987). Both stages of saliva formation are under the control of the autonomic nervous system.

Sympathetic and parasympathetic fibers innervate the acinar and ductal regions of salivary glands (Garrett, 1982), and drugs that either stimulate or block neuroreceptor action affect salivary secretion (Baum, 1987; Martinez, 1987; Young et al., 1987). Saliva formation is mediated by neuroreceptor activation of intracellular signaling systems, generating second messages (e.g., cyclic AMP or calcium) which may act alone or in concert to elicit specific cellular responses. Studies in the Clinical Investigations and Patient Care Branch of the National Institute of Dental Research have shown that at least four established salivary cell lines respond to either adrenergic or muscarinic-cholinergic stimulation. All of these transformed cell lines, three of which appear to be derived from ductal regions of salivary glands, possess different neuroreceptor compositions (Table 2 and discussion below). The availability of the HSG and HSY cell lines has allowed neurotransmitter response studies to be carried out on cells derived from otherwise inaccessible regions of the gland. If transformation has not altered the normal expression of neurotransmitter receptors in the established salivary cell lines, these variations in neuroreceptor composition may reflect the abilities of different epithelial cell types within salivary glands to respond to different types of neural stimulation.

Muscarinic Response

The primary response of most salivary glands to muscarinic stimulation involves changes in both water and electrolyte transport in secretory endpieces, and changes in ion transport in ductal regions of the glands (Baum, 1987; Martinez, 1987). Two epithelial cell lines, thought to be derived from the intercalated duct regions of human parotid and submandibular glands (HSY and HSG cell lines, respectively), have been shown to respond to muscarinic stimulation.

The muscarinic control of calcium and potassium flux in HSG cells has been investigated in some detail. The HSG cell line was initially reported to possess a single class of high affinity muscarinic receptor sites ($Kd = 0.17 \pm 0.07$ nM; Bmax $= 37 \pm 2$ fmol/mg protein, mean $\pm$ sem; He et al., 1989c) as judged by muscarinic antagonist binding studies. The receptors were shown to be functional as carbachol (muscarinic agonist) stimulation of HSG cells resulted in both the mobilization of calcium from intra- and extracellular stores and an increase in the incorporation of [^{3}H]-myoinositol into [^{3}H]-inositol trisphosphate. Initial inhibitor studies also suggested that the Ca^{2+} mobilization pathways are controlled by receptor-operated, voltage-insensitive, transport systems. Subsequent carbachol dose-response studies have revealed, however, that the extracellular uptake actually involves two entry pathways responding to either high- or low-affinity receptor stimulation (He et al., 1990b). In addition, it appears that Ca^{2+} entry controlled by the high-affinity site (but not the low-affinity site) might be voltage regulated. In other studies, it has been shown that agents such as phorbol myristate acetate (PMA, a protein kinase C activator; He et al., 1988) and W7 (N-

(6-aminohexyl)-5-chloro-1-naphthalenesulfonamide, a calmodulin antagonist; He et al., 1990b) can differentially inhibit Ca^{2+} release and entry pathways as a function of inhibitor concentration. W7 also reverses the increase in free intracellular calcium caused by maitotoxin-stimulated calcium entry (Wellner et al., 1990). These results have provided evidence that, in HSG cells, (1) both Ca^{2+} release and entry pathways are regulated by protein kinase C and perhaps calmodulin and (2) at least a portion of the extracellular Ca^{2+} might enter directly into the cytosol rather than initially entering an intracellular compartment (He et al., 1988).

In related studies, muscarinic stimulation of HSG intercalated duct-like cells has been shown to increase potassium (i.e., ^{86}Rb) influx and efflux (Ship and Wellner, 1989; Ship et al., 1990), resulting in only a small net decrease in cellular potassium content. These fluxes appear to be calcium activated and involve the participation of quinine- and charybdotoxin-sensitive K^+ channels. The effects of PMA, W7, and AlF_4^- have been studied in order to understand better how these K^+ channels are regulated.

Since PMA and W7 had been found to block muscarinic-stimulated increases in $[Ca^{2+}]_i$ (cited above), the ability of these agents to block muscarinic-stimulated K^+ fluxes was also tested. PMA was found to inhibit the muscarinic-stimulated K^+ fluxes, and this inhibition appeared to be largely due to an inhibition of the calcium signal (Ship et al., 1990). In contrast, the effect of W7 on muscarinic-stimulated K^+ fluxes was markedly different as W7 alone stimulated quinine- and charybdotoxin-sensitive K^+ (^{86}Rb) influx and efflux (Wellner et al., 1990; Patton et al., 1991b). W7-stimulated K^+ transport is different from muscarinic-stimulated K^+ transport, however, as it is not blocked by atropine and does not involve the generation of a typical calcium signal. Although W7 is known to act as a calmodulin (CaM) antagonist (Hidaka and Tanaka, 1985), the mechanism by which W7 stimulates HSG cell K^+ channel activity is unclear (Patton et al., 1991b).

GTP-binding proteins (G proteins) are known to be involved in many biochemical processes (Neer and Clapman, 1988; Bockaert et al., 1987), including signal transduction systems involved in calcium mobilization (Blackmore et al., 1988; Brown and Birnbaumer, 1988) and potassium transport (Logothetis et al., 1988; Brown and Birnbaumer, 1988). Studies utilizing AlF_4^- (Patton et al., 1990), a known G protein activator (Sternweis and Gilman, 1982), indicate that this agent stimulates both (1) HSG cell potassium (^{86}Rb) influx and efflux and (2) increases in $[Ca^{2+}]_i$. It is unclear whether or not these responses are due to G-protein activation, however, as a requirement for aluminum has not been firmly established. Increases in potassium influx and efflux are dose-dependent with respect to fluoride concentrationhowever, and inhibitor studies suggest that stimulated potassium transport is due at least in part to increases in potassium channel(s) (Patton et al., 1990) and Na/K/2Cl co-transport activity (Patton and Wellner, manuscript in preparation).

Studies of the HSY cell line (Patton et al., 1991a) suggest that these intercalated duct/acinar-like cells from the human parotid gland possess muscarinic responses similar to those of human submandibular intercalated duct (HSG) cells. In this regard, muscarinic stimulation of HSY cells results in calcium mobilization and calcium-triggered increases in potassium influx and efflux, with no associated change in net potassium content. Both fluxes are inhibited by PMA and the K^+ channel blockers quinine and charybdotoxin. PMA apparently inhibits the muscarinic response by blocking formation of the calcium signal. HSY cells do not respond, however, to α-adrenergic stimulation (as judged by a lack of Ca^{2+} mobilization), indicating that they lack this portion of the acinar phenotype.

α- and β-Adrenergic Responses

Stimulation of salivary glands by β-adrenergic agonists is known to increase protein synthesis and secretion by acinar cells in the secretory endpieces of the gland, and to affect ion transport properties of ductal epithelium. α-Adrenergic stimulation is generally associated with increased fluid and electrolyte secretion by acini, and altered ion tranport by ducts (reviewed in Young and van Lennep, 1979; Baum, 1987; Martinez, 1987; Young et al., 1987). In contrast to the seemingly ubiquitous nature of muscarinic-cholinegeric receptors in salivary gland secretory epithelia, the distribution of α- and β-adrenergic receptors is more variable with respect to species and salivary gland type (Young et al., 1986). Adrenergic receptors have been identified in three established salivary epithelial cell lines in culture. The human submandibular cell line A253 expresses β-adrenergic, but not α-adrenergic responsiveness; whereas the RSMT-A5 submandibular rat cell clone responds to both types of adrenergic stimulation. Neither A253 nor RSMT-A5 cell lines appear to possess muscarinic receptors. HSY cells possess β-adrenergic receptors, as isoproterenol increases intracellular cAMP content in a propranolol inhibitable manner (Patton et al., 1991a).

The β-adrenergic response in A253 cells is transmitted by a single class of high-affinity, β_2-subtype receptors (Kd = 0.34 ± 0.03 nM; Bmax = 167 ± 6.0 fmol/mg; Marmary et al., 1989), as judged by antagonist binding studies. These cells respond to β-adrenergic stimulation by exhibiting increases in both intracellular cAMP content and N-linked protein glycosylation. β-Adrenergic stimulation of protein glycosylation has been studied in acute rat and human salivary acinar cell preparations (Kousvelari et al., 1987). While the acinar cell response appears to be mediated by cAMP, the A253 cell response is different, as agents that increase intracellular cAMP content (cAMP analogues, forskolin, or cholera toxin; Marmary et al., 1989 and Wellner, unpublished observations) fail to increase the incorporation of sugars into proteins in these cells. Interestingly, Cook et al. (1988a, b) have also reported that agents that increase intracellular cAMP content fail to mimic the β-adrenergic stimulation of some secretory events occurring in acute rat submandibular epithelial cell preparations. These investigators have suggested that the β-adrenergic response might be mediated by another second messenger, calcium, rather than cAMP. Although it is unclear why agents that increase intracellular cAMP content fail to mimic the β-adrenergic response in A253 cells, it is possible that some human (i.e., A253) as well as rat salivary cell β-adrenergic responses might not be mediated by cAMP.

Sympathetic neurotransmitters are capable of stimulating both α- and β-adrenergic receptors *in vivo*. Thus far, RSMT-A5 is the only established salivary cell line known to possess both α- and β-adrenergic receptors. The binding characteristics and generation of second messengers have been documented for both receptor types in this cell line (He et al., 1989a,b). The α- and β-adrenergic receptor populations were identified as discrete, single classes of high-affinity binding sites; i.e., the α-receptor had a Kd of 0.052 ± 0.005 nM and a Bmax of 104 ± 12 fmol/mg, while the β-receptor had a Kd of 0.62 ± 0.03 nM and a Bmax of 101 ± 4 fmol/mg. Antagonist competition studies further indicated that RSMT-A5 cells contained primarily α_1- and β_2-subtype receptors. β-Adrenergic stimulation has been shown to increase maximally the intracellular cAMP content of these cells by ca. 130-fold. The α-adrenergic response was characterized by both an increased incorporation of $[^3H]$-myoinositol into $[^3H]$-inositol trisphosphate (He et al., 1989b), and Ca^{2+} mobilization when cells were stimulated while attached to a glass surface (Wellner, unpublished observations).

The β-adrenergic response has been studied in some detail (He et al., 1989a). Upon stimulation with isoproterenol (β-adrenergic agonist), light microscopy revealed that RSMT-A5 cells underwent a marked morphological change; i.e., they appeared more rounded and smaller than untreated cells. This response was accompanied by reductions in both intracellular water space and chloride content. In addition, both responses appeared to be cAMP mediated and to involve the participation of chloride channels, as judged by ion transport inhibitor studies. Overall, these results suggest that RSMT-A5 cells possess β-adrenergic regulated, cAMP-responsive, chloride channels which play an essential role in cell-volume regulation.

OTHER STUDIES

In addition to the differentiation and signal transduction studies cited previously, established salivary cell lines have proven useful for other *in vitro* studies. In this regard, anticancer drug efficacy has been evaluated in the A253 cell line (Lazo et al., 1985, 1986, 1989a,b; Darzynkiewicz et al., 1988; Darzynkiewicz and Carter, 1988; Pizzorno et al., 1989; Sebti et al., 1989), glucocorticoid receptors have been studied in the HSG cell line (Kurokawa et al., 1986, 1987; Kurokawa and Ota, 1988), and oncogene expression has been studied in both the RSMTx and HSG cell lines (Yeh et al., 1988). Investigations of cAMP-induced oncogene expression are important to the understanding of salivary epithelial cell proliferation. A discussion of these studies is not provided here, however, as this topic is reviewed in detail in Chapter 10.

In whole animal studies, glucocorticoids appear to be important in maintaining salivary gland integrity, as their supplementation reverses submandibular and parotid gland atrophy resulting from adrenalectomy (Johnson, 1987). However, the exact role of glucocorticoids in salivary biology is unclear as (1) glucocorticoid administration to adrenal-intact animals induces glandular atrophy and (2) there are conflicting reports concerning the effects of these hormones on amylase expression and saliva flow rates (Johnson, 1987). While the role of glucocorticoids seems unclear, high-affinity glucocorticoid receptors have been identified in the cytosol of human HSG cells in culture (Kurokawa et al., 1986), and translocation of the glucocorticoid receptor complex to the nucleus has been shown to result in an inhibition of DNA synthesis (Kurokawa et al., 1987; Kurokawa and Ota, 1988). These results suggest that the proliferation of tumor cells derived from the intercalated duct region of human submandibular glands may be able to be inhibited by glucocorticoid hormone action *in vivo* (Kurokawa et al., 1986).

Salivary gland malignancies that have spread to regional lymph nodes, or disseminated systemically at the time of diagnosis, have a poor prognosis. The efficacy of cytotoxic chemotherapeutic agents, particularly methotrexate, bleomycin, cis-platinum, and fluorouracil, has been actively explored in an effort to improve the treatment of head and neck cancers (Silverberg et al., 1985). As part of this effort, the human epidermoid carcinoma cell line A253 has been used to study chemotherapeutic agent toxicity *in vitro*. A253 cells are sensitive to bleomycin (a DNA antitumor antibiotic; Lazo et al., 1985, 1986), the mitochondrial antitumor dyes pyronin Y and toluidine blue (Darzynkiewicz and Carter, 1988), and P-30 protein (a novel anticancer agent; Darzynkiewicz et al., 1988); however, they are resistant to methotrexate, a folate antagonist (Pizzorno et al., 1989). The mechanism of bleomycin cytotoxicity in A253 cells has been studied in some detail (Lazo et al., 1989a, b; Sebti et al., 1989).

CONCLUDING REMARKS

Established cell lines derived from a variety of tissues have been sought for the purpose of investigating the roles of specific cell types in overall organ function. Cell lines can serve as useful models for studying biological processes in normal and pathological states, with cells of human origin being of particular interest. The recent establishment of immortal salivary cell lines, most of which are derived from human tissue, has provided new opportunities for studying mechanisms of salivary gland function and dysfunction. In this respect, putative intercalated duct-like cell lines isolated from human submandibular and parotid glands have proven useful for identifying neuroreceptor/hormone receptor composition and responsiveness in previously inaccessible regions of the glands. In addition, salivary cell lines have proven useful for studying differentiation, oncogene expression, and the efficacy of chemotherapeutic agents. A recent report also suggests that the human HSG submandibular cell line is useful for studying the salivary glandular dysfunction of Sjögren's syndrome (Atkinson et al., 1990).

Despite recent progess, the availability of specific salivary cell lines is limited to those derived from intercalated duct, myoepithelium, or undefined regions of the glands. The immortalization of fully differentiated acinar cell lines would significantly enhance our ability to study primary saliva formation, and fully differentiated cell lines from specific ductal elements (especially from striated and excretory ducts) would likewise improve our ability to study the ductal modification of primary secretion. The currently available, immortal salivary epithelial cell lines were derived from either neoplastic or carcinogen-treated tissue (the canine 563 cell line is a possible exception). Establishment of additional cell lines will probably involve (1) establishing growth conditions (e.g., medium components, matrices, permeable supports) which promote proliferation of specific cell types and/or (2) the use of specific immortalization techniques. With regard to the latter, primary or low passage salivary cell cultures (e.g., Kurth et al., 1989; Sens et al., 1985; Quissell et al., 1986; Barka et al., 1980; Wigley and Franks, 1976; see also Chapters 13 and 14) might be subjected to various *in vitro* transformation stagies used successfully in other cell immortalization efforts (e.g., Stamper and Bartley, 1989; Brown et al., 1989; Cone et al., 1988; Rhim et al., 1988; Scholte et al., 1989; Jetten et al., 1989). These strategies include the use of viruses, oncogenes, radiation, and carcinogens. In addition, primary or low passage salivary cell cultures might be fused to established cell lines, thereby creating hybrid cells suitable for long-term studies of specific salivary cell functions. This method has been used to maintain the expression of a differentiated acinar trait (inducible expression of the proline-rich protein multigene family; Wright and Carlson, 1988) for at least 20 cell passages.

Once established, defined salivary epithelial cell lines can be used not only to determine the contributions of specific cell types to saliva formation, but they can also be used to define the sites of action and mechanisms by which various pharmacological agents and other moieties affect salivary gland function. New techniques will be more readily applied to these investigations, including somatic cell genetic and molecular biological approaches, and the use of permeable supports to better define the polarized distribution of transport proteins, enzymes, cell surface receptors, and other moieties in the apical and basolateral membranes. These latter studies, which have been carried out successfully in other systems (e.g., Lang et al., 1990; Dharmsathaphorn et al., 1984), are particularly important as they will provide new information about the functioning of vectorial transport processes during both normal and abnormal saliva formation (Baum, 1987; Martinez, 1987; Young and Van Lennep, 1979; see also Chapters 5 and 7). Ultimately, information gained from studies of cultured salivary cell lines will contribute to a better understanding of the biology of salivary glands in health and disease, which is essential for the development of new therapeutic modalities.

REFERENCES

1. Alvares, E. P. and Sesso, A., Cell proliferation, differentiation and transformation in the rat submandibular gland during early postnatal growth. A quantitative and morphological study, *Arch. Histol. Jpn.*, 38, 177, 1975.
2. ATCC Catalogue of Cell Lines and Hybridomas, 6th ed., *American Type Culture Collection*, Rockville, MD, 1988, 208.
3. Atkinson, J. C., Royce, L. S., Bermudez, D. K., Wellner, R., Pillemer, S. R., and Fox, P. C., Anti-salivary duct cell autoantibodies in primary Sjogren's syndrome, *J. Dent. Res.*, 69, 150, 1990.
4. Azuma, M., Hayashi, Y., Yoshida, H., Yanagawa, T., Yura, Y., Ueno, A., and Sato, M., Emergence of differentiated subclones from a human salivary adenocarcinoma cell clone after treatment with sodium butyrate, *Cancer Res.*, 46, 770, 1986.
5. Babcock, M. S., Marino, M. R., Gunning, W. T., III, and Stoner, G. D., Clonal growth and serial propagation of rat esophageal epithelial cells, *In Vitro Cell. Dev. Biol.*, 19, 403, 1983.
6. Barka, T., Van Der Noen, H., Michelakis, A. M., and Schenkein, I., Epidermal growth factor renin, and peptidase in cultured tumor cells of submandibular gland origin, *Lab Invest.*, 42, 656, 1980.
7. Batsakis, J. G., Salivary gland neoplasia: an outcome of modified morphogenesis and cytodifferentiation, *Oral Surg. Oral Med. Oral Pathol.*, 49, 229, 1980.
8. Baum, B. J., Neurotransmitter control of secretion, *J. Dent. Res.*, 66, 628, 1987.
9. Bennett, D. C., Peachey, L., Durbin, H., and Rudland, P. S., A possible mammary stem cell line, *Cell*, 15, 283, 1978.
10. Blackmore, P. F., Lynch, C. J., Uhing, R. J., Fitzgerald, T., Bocckino, S. B., and Exton, J. H., Role of guanine nucleotide regulatory proteins and inositol phosphates in the hormone induced mobilization of hepatocyte calcium, *Adv. Exp. Med. Biol.*, 232, 169, 1988.
11. Bloch, K. J., Buchanan, W. W., Wohl, M. J., and Bunim, J. J., Sjogren's Syndrome. A clinical, pathological, and serological study of sixty-two cases, *Medicine*, 44, 187, 1965.
12. Bockaert, J. Homburger, V., and Rouot, B., GTP binding proteins: a key role in cellular communication, *Biochimie*, 69, 329, 1987.
13. Brown, A. M. and Birnbaumer, L., Direct G protein gating of ion channels, *Am. J. Physiol.*, 254, H401, 1988.
14. Brown, A. M., Rusnock, E. J., Sciubba, J. J., and Baum, B. J., Establishment and characterization of an epithelial cell line from the rat submandibular gland, *J. Oral Pathol. Med.*, 18, 206, 1989.

15. Caselitz, J., Seifert, G., and Jaup, T., Presence of carcinoembryonic antigen (CEA) in the normal and inflamed human parotid gland, *J. Cancer Res. Clin. Oncol.*, 100, 205, 1981.

16. Chang, W. W. L., Cell population changes during acinus formation in the postnatal rat submandibular gland, *Anat. Rec.*, 178, 187, 1974.

17. Cook, D. I., Day, M. L., Champion, M. P., and Young, J. A., Ca^{2+}, not cyclic AMP, mediates the fluid secretory response to isoproterenol in the rat mandibular salivary gland: whole-cell patch-clamp studies, *Pflugers Arch.*, 413, 67, 1988a.

18. Cook, D. I., Gard, G. B., Champion, M., and Young, J. A., Patch-clamp studies of the electrolyte secretory mechanism of rat mandibular gland cells stimulated with acetylcholine or isoproterenol, in *Molecular Mechanisms in Secretion*, Alfred Benzon Symposium 25, Thorn, N. A., Treiman, M., and Petersen, O. H., Eds., Munksgaard, Copenhagen, 1988b, 133.

19. Cone, R. D., Grodzicker, T., and Jaramillo, M., A retrovirus expressing the 12S adenoviral E1A gene product can immortalize epithelial cells from a broad range of rat tissues, *Mol. Cell Biol.*, 8, 1036, 1988.

20. Dardick, I., Byard, R. W., and Carnegie, J. A., A review of the proliferative capacity of major salivary glands and the relationship to current concepts of neoplasia in salivary glands, *Oral Surg. Oral Med. Oral Pathol.*, 69, 53, 1990.

21. Darzynkiewicz, Z. and Carter, S. P., Photosensitizing effects of the tricyclic heteroaromatic cationic dyes pyronin Y and toluidine blue O (tolonium chloride), *Cancer Res.*, 48, 1295, 1988.

22. Darzynkiewicz, Z., Carter, S. P., Mikulski, S. M., Ardelt, W. J., and Shogen K., Cytostatic and cytotoxic effects of Pannon (P-30 Protein), a novel cancer agent, *Cell Tissue Kinet.*, 21, 169, 1988.

23. Davis, P. B., Pathophysiology of cystic fibrosis with emphasis on salivary gland involvement, *J. Dent. Res.*, 66, 667, 1987.

24. Dharmsathaphorn, K., McRoberts, J. A., Mandel, K. G., Tisdale, L. D., and Masui, H., A human colonic tumor cell line that maintains vectorial transport, *Am. J. Physiol.*, 246, G204. 1984.

25. Dreizen, S., Brown, L. R., Handler, S., and Levy, B. M., Radiation-induced xerostomia in cancer patients, *Cancer*, 38, 273, 1976.

26. Eversole, L. R., Histogenic classification of salivary tumors, *Arch. Pathol.*, 92, 433, 1971.

27. Fox, R. I., Howell, F. V., Bone, R. C., and Michelson, P., Primary Sjogren syndrome: clinical and immunopathologic features, *Sem. Arthritis Rheum.*, 14, 77, 1984.

28. Franke, W., Appelhaus, B., Schmid, E., Freudenstein, C., Osborn, M., and Weber, K., Identification and characterization of epithelial cells in mammalian tissues by immunofluorescence microscopy using antibodies to prekeratin, *Differentiation*, 15, 7, 1979.

29. Garrett, J. R., Adventures with autonomic nerves. Perspectives in salivary glandular innervations, *Proc. R. Micr. Soc.*, 17, 242, 1982.

30. Giard, D. J., Aaronson, S. A., Todaro, G. J., Arnstein, P., Kersey, J. H., Dosik, H., and Parks, W. P., *In vitro* cultivation of human tumors: establishment of cell lines derived from a series of solid tumors, *J. Natl. Cancer Inst.*, 51, 1417, 1973.

31. Hayashi, Y., Yanagawa, T., Azuma, M., Yura, Y., Yoshida, H., and Sato, M., Induction of cells with a myoepithelial phenotype by treatment with dibutyryl cyclic AMP in human salivary adenocarcinoma cells grown in athymic nude mice, *Virchows Arch. (Cell Pathol.)*, 50, 1, 1985.

32. Hayashi, Y., Yoshida, H., Nagamine, S., Yanagawa, T., Yura, Y., Azuma, M., and Sato, M., Induction of cells with acinar cell phenotype including presence of intracellular amylase. Treatment with 12-O-tetradecanoyl-phorbol-13-acetate in a neoplastic human salivary intercalated duct cell line grown in athymic nude mice, *Cancer*, 60, 1000, 1987a.

33. Hayashi, Y., Yanagawa, T., Yoshida, H., Azuma, M., Nishida, T., Yura, Y., and Sato, M., Expression of vasoactive intestinal polypeptide and amylase in a human parotid gland adenocarcinoma cell line in culture, *J. Natl. Cancer Inst.*, 79, 1025, 1987b.

34. He, X., Ship, J., Wu, X., Brown, A. M., and Wellner, R. B., β-Adrenergic control of cell volume and chloride transport in an established rat submandibular cell line, *J. Cell Physiol.*, 138, 527, 1989a.

35. He, X., Wu, X., and Baum, B. J., Protein kinase C differentially inhibits muscarinic receptor operated Ca^{2+} release and entry in human salivary cells, *Biochem. Biophys. Res. Commun.*, 152, 1062, 1988.

36. He, X., Wu, X., and Baum, B. J. The effect of N-(6-aminohexyl)-5-chloro-1-naphthalenesulfonamide (W-7) on muscarinic receptor-induced Ca^{2+} mobilization in a human salivary epithelial cell line, *Pfluger's Arch.*, 1990, in press.

37. He, X., Wu, X., Brown, A. M., Wellner, R. B., and Baum, B. J., Characteristics of α_1-adrenergic receptors in a rat salivary cell line, RSMT-A5, *Gen. Pharmacol.*, 20, 175, 1989b.

38. He, X., Wu, X., Wellner, R. B., and Baum, B. J., Muscarinic receptor regulation of Ca^{2+} mobilization in a human salivary cell line, *Pfluegers Arch.*, 413, 505, 1989c.

39. Hidaka, H. and Tanaka, T., Modulation of Ca^{2+}-dependent regulatory systems by calmodulin antagonists and other agents, in *Calmodulin Antagonists and Cellular Physiology*, Hidaka, H. and Hartstone, D. J., Eds., Academic Press, Orlando, FL, 1985, 13.

40. Izutsu, K. T., Salivary electrolytes and fluid protection in health and disease, in *The Salivary System*, Sreebny, L. M., Ed., CRC Press, Raton, FL, 1987, 95.

41. Jetton, A. M., Yankaskas, J. R., Stutts, M. J., Willumsen, N. J., and Boucher, R. C., Persistence of abnormal chloride conductance regulation in transformed cystic fibrosis epithelia, *Science*, 244, 1472, 1989.

42. Johnson, D. A., Regulation of salivary glands and their secretions by masticatory, nutritional, and hormonal factors, in *The Salivary System*, Sreebny, L. M., Ed., CRC Press, Boca Raton, FL, 1987, 135.

43. Kline, R. M., Acinar cell proliferation in the parotid and submandibular glands of the neonatal rat, *Cell Tissue Kinet.*, 15, 187, 1982.

44. Knowles, M. A. and Franks, L. M., Stages in neoplastic transformation of adult epithelial cells by 7, 12-dimethylbenz (a)anthracene *in vitro*, *Cancer Res.*, 37, 3917, 1977.

45. Kondo, T., Muragishi, H., and Imaizumi M., A cell line from a human salivary gland mixed tumor, *Cancer*, 27, 403, 1971.

46. Konieczny, S. F. and Emerson, C. P., 5-Azacytidine induction of stable mesodermal stem cell lineages from 10T1/2 cells: evidence for regulatory genes controlling determination, *Cell*, 38, 791, 1984.

47. Kousvelari, E. E., Fox, P. C., and Baum, B. J., Regulatory aspects of N-linked glycoproteins, *J. Dent. Res.*, 66, 552, 1987.

48. Kurokawa, R., Hatakeyama, S., Kayakumoto, S., and Ota, M., Cytosol glucocorticoid receptor in the neoplastic epithelial duct cell line from human salivary gland (HSG), *Biochem. Int.*, 13, 671, 1986.

49. Kurokawa, R., Kayakumoto, S., Hatakeyama, S., and Ota, M., Subcellular distribution of glucocorticoid receptor in the neoplastic epithelial duct cell line from human salivary gland, *Biochem. Int.*, 14, 37, 1987.

50. Kurokawa, R. and Ota, M., Nonactivated and activated glucocorticoid receptor complexes from human salivary gland adenocarcinoma cell line, *Biochim. Biophys. Acta*, 970, 292, 1988.

51. Kurth, B. E., Hazen-Martin, D. J. Sens, M. A., and Sens, D. A., Ultrastructural and immunohistochemical characterization of submandibular duct cells in culture and modification of outgrowth differentation by manipulation of calcium ion concentration, *In Vitro Cell Dev. Biol.*, 24, 593, 1988.

52. Kurth, B. E., Hazen-Martin, D. J., Sens, M. A., DeChamplain, R. W., and Sens, D. A., Cell culture and characterization of human minor salivary gland duct cells, *J. Oral Pathol. Med.*, 18, 214, 1989.

53. Lang, F., Friedrich, F., Paulmichl, M., Schobersberger, W., Jungwirth, A., Ritter, M., Steidl, M., Weiss, H., Wöll, E., Tschernko, E., Paulmichl, R., and Hallbrucker, C., Ion channels in Madin-Darby canine kidney cells, *Renal Physiol. Biochem.*, 13, 82, 1990.

54. Lazo, J. S., Braun, I. D., Larabee, D. C., Schisselbauer, J. C., Meandzija, B., Newman, R. A., and Kennedy, K. A., Characteristics of bleomycin-resistant phenotypes of human cell sublines and circumvention of bleomycin resistance by liblomycin, *Cancer Res.*, 49, 185, 1989a.

55. Lazo, J. S., Braun, I. D., Meandzija, B., Kennedy, K. A., Pham, E. T., and Smaldone, L. F., Lidocaine potentiation of bleomycin A_2 cytotoxicity and DNA strand breakage in L 1210 and human A-253 cells, *Cancer Res.*, 45, 2103, 1985.

56. Lazo, J. S., Chen, D.-L., Gallicchio, V. S., and Hait, W. N., Increased lethality of calmodulin antagonists and bleomycin to human bone marrow and bleomycin-resistant malignant cells, *Cancer Res.*, 46, 2236, 1986.

57. Lazo, J. S., Schisselbauer, J. C., Meandzija, B., and Kennedy, K. A., Initial single-strand DNA damage and cellular pharmacokinetics of bleomycin A_2, *Biochem. Pharmacol.*, 38, 2207, 1989b.

58. Logothetis, D. E., Kim, D., Northup, J. K., Neer, E. J., and Clapman, D. E., Specificity of action of guanine nucleotide-binding regulatory protein subunits on the cardiac muscarinic K^+ channel, *Proc. Natl. Acad. Sci. U.S.A.*, 85, 5814, 1988.

59. Mandel, I. D., The functions of saliva, *J. Dent. Res.*, 66, 623, 1987.

60. Mandel, I. D., The role of saliva in maintaining oral homeostasis, *J. Am. Dent. Assoc.*, 119, 298, 1989.

61. Marmary, Y., He, X., Hand, A. D., Ship, J. A., and Wellner, R. B., β-Adrenergic responsiveness in a human submandibular tumor cell line, *In Vitro Cell Dev. Biol.*, 25, 951, 1989.

62. Martinez, J. R., Ion transport and water movement, *J. Dent. Res.*, 66, 638, 1987.

63. May, M., Lucente, F. E., Harvey, J. E., and Marovitz, W. F., Salivation testing in traumatic facial paralysis, *Ann. Oto. Rhino. Laryngol.*, 82, 17, 1983.

64. Neer, E. J. and Clapman, D. E. Roles of G protein subunits in transmembrane signalling, *Nature* 333, 129, 1988.

65. Patton, L., Horn, V., Taylor, S., and Wellner, R., AlF_4^- activation of K^+ (^{86}Rb) fluxes in a human submandibular epithelial cell line, *Fed. Proc.*, 4, A1965, 1990.

66. Patton, L. L., Pollack, S., and Wellner, R. B., Responsiveness of a human parotid epithelial cell line (HSY) to autonomic stimulation: muscarinic control of K^+ transport, *In Vitro Cell. Dev. Biol.*, 27A, 779, 1991a.

67. Patton, L., Ship, J., and Wellner, R., N-(6-Aminohexyl)-5-chloro-1-naphthalene-sulfon-amide (W7) stimulation of K^+ transport in a human salivary epithelial cell line, *Biochem. Pharmacol.*, 42, 1039, 1991b.

68. Pierce, G. B., Neoplasms, differentiations and mutations, *Am. J. Pathol.*, 77, 103, 1974.

69. Pinkstaff, C. A., The cytology of salivary glands, *Int. Rev. Cytol.*, 63, 141, 1980.

70. Pizzorno, G., Chang, Y. M., McGuire, J. J., and Bertino, J. R., Inherent resistance of human squamous carcinoma cell lines to methotrexate as a result of decreased polyglutamylation of this drug, *Cancer Res.*, 49, 5275, 1989.

71. Quissell, D. O., Redman, R. S., and Mark, M. R., Short-term primary culture of acinar-intercalated duct complexes from rat submandibular glands, *In Vitro Cell Dev. Biol.*, 22, 469, 1986.

72. Razin, A. and Riggs, A. D., DNA methylation and gene function, *Science*, 10, 604, 1980.

73. Redman, R. S., Development of the salivary glands, in *The Salivary System*, Sreebny, L. M., Ed., CRC Press, Boca Raton, Fl, 1987, 1.

74. Redman, R. S. and Sreebny, L. M., Proliferative behavior of differentiating cells in the developing rat parotid gland, *J. Cell Biol.*, 46, 81, 1970.

75. Rhim, J. S., Park, J. B., and Kawakami, T., Techniques for establishing human epithelial cell cultures: sensitivity of cell lines for propagation of herpes viruses, *J. Virol. Meth.*, 21, 209, 1988.

76. Sato, M., Azuma, M., Hayashi, Y., Yoshida, H., Yanagawa, T., and Yura, Y., 5-Azacytidine induction of stable myoepithelial and acinar cells from a human salivary intercalated duct cell clone, *Cancer Res.*, 47, 4453, 1987.

77. Sato, M., Hayashi, Y., Yoshida, H., Yanagawa, T., Yura, Y., and Nitta, T., Search for specific markers of neoplastic epithelial duct and myoepithelial cell lines established from human salivary gland and characterization of their growth *in vitro*, *Cancer*, 54, 2959, 1984.

78. Sato, M., Hayashi, Y., Yanagawa, T., Yoshida, H., Yura, Y., Azuma, M., and Ueno, A., Intermediate-sized filaments and specific markers in a human salivary gland adenocarcinoma cell line and its nude mouse tumors, *Cancer Res.*, 45, 3878, 1985a.

79. Sato, M., Yoshida, H., Hayashi, Y., Miyakami, K., Bando, T., Yanagawa, T., Yura, Y., Azuma, M., and Ueno, A., Expression of epidermal growth factor and transforming growth factor-β in a human salivary gland adenocarcinoma cell line, *Cancer Res.*, 45, 6160, 1985b.

80. Scholte, B. J., Kansen, M., Hoogeveen, A. T., Willemse, R., Rhim, J. S., Van Der Kamp, A. W. M., and Bijam, J., Immortalization of nasal polyp epithelial cells from cystic fibrosis patients, *Exp. Cell Res.*, 182, 559, 1989.

81. Sebti, S. M., DeLeon, J. C., Ma, L. T., Hecht, S. M., and Lazo, J. S., Substrate specificity of bleomycin hydrolase, *Biochem. Pharmacol.*, 38, 141, 1989.

82. Sens, D. A., Hintz, D. S., Rudisill, M. T., Sens, M. A., and Spicer, S. S., Explant culture of human submandibular gland epithelial cells: evidence of ductal origin, *Lab. Invest.*, 52, 559, 1985.

83. Ship, J. A. and Wellner, R. B., Muscarinic regulation of potassium transport in a human salivary epithelial cell line, *J. Cell Biol.*, 107, 784a, 1989.

84. Ship, J. A., Patton, L. L., and Wellner, R. B., Muscarinic regulation of potassium transport in a human submandibular epithelial cell line, *Am. J. Physiol.*, 259, C340, 1990.

85. Shirasuna, K., Sato, M., and Miyazaki, T., A myoepithelial cell line established from a human pleomorphic adenoma arising in minor salivary gland, *Cancer*, 45, 297, 1980.

86. Shirasuna, K., Sato, M., and Miyazaki, T., A neoplastic epithelial duct cell line established from an irradiated human salivary gland, *Cancer*, 48, 745, 1981.

87. Silverberg, I. J., Carter, S. K., and Cadman, E. C., Treatment-Chemotherapy, in *Oral Cancer*, 2nd ed., Silverman, S., Jr., Ed., American Cancer Society, New York, 63, 1985.

88. Sreebny, L. M. and Schwartz, S. S., A reference guide to drugs and dry mouth, *Gerodontology*, 5, 75, 1986.

89. Stampfer, M. R. and Bartley, J. C., Induction transformation and continuous cell lines from normal human mammary epithelial cells after exposure to benzo [α] pyrene, *Proc. Natl. Acad. Sci. U.S.A.*, 82, 2394, 1985.

90. Sternweis, P. C. and Gilman, A. G., Aluminum: a requirement for activation of the regulatory component of adenylate cyclase by fluoride, *Proc. Natl. Acad. Sci. U.S.A.*, 79, 4888, 1982.

91. Strandstrom, H. V., Madewell, B. R., Deitch, A. D., and Gumerlock, P. H., Canine 563 cells: an established canine salivary gland cell line, *J. Vet. Med.*, A36, 361, 1989.

92. Sun, T. T., Shih, C., and Green, H., Keratin cytoskeletons in epithelial cells of internal organs, *Proc. Natl. Acad. Sci. U.S.A.*, 76, 2813, 1979.

93. Talal, N., Overview of Sjogren's Syndrome, *J. Dent. Res.*, 66, 672, 1987.

94. Thaysen, J. H., Thorn, N. A., and Schwartz, I. L., Excretion of sodium, potassium, chloride and carbon dioxide in human parotid saliva, *Am. J. Physiol.*, 178, 155, 1954.

95. Van Lennep, E. W., Cook, D. I., and Young, J. A., Morphology and secretory mechanisms of salivary glands, in *Molecular and Cellular Basis of Digestion*, Desnuelle, P., Sjostrom, H., and Noren, O., Eds., Elsevier Science Publishers B. V. (Biomedical Division), Amsterdam, 435, 1986.

96. Wellner, R. B., Ship, J. A., and Patton, L. L., Effect of the calmodulin (CaM) antagonist W7 on K^+ (^{86}Rb) transport in a human submandibular epithelial cell line, *Fed. Proc.*, 4, A1959, 1990.

97. Wellner, R. B., Patton, L. L., and Pace, J. G., A comparison of the abilities of maitotoxin (MTX), carbachol (Cch), and 4Br-A23187 (A23) to increase free intracellular calcium ($[Ca^{2+}]_i$) in a human submandibular epithelial cell line (HSG-PA), *Am. Chem. Soc. Symposium*, in press.

98. Wigley, C. B. and Franks, L. M., Salivary epithelial cells in primary culture: characterization of their growth and functional properties, *J. Cell Sci.*, 20, 149, 1976.

99. Wright, P. S. and Carlson, D. M., Regulation of proline-rich protein and α-amylase genes in parotid-hepatoma hybrid cells, *FASEB J.*, 2, 3104, 1988.

100. Yanagawa, T., Hayashi, Y., Nagamine, S., Yoshida, H., Yura, Y., and Sato, M., Generation of cells with phenotypes of both intercalated duct type and myoepithelial cells in human parotid adenocarcinoma clonal cells grown in athymic nude mice, *Virchows Arch. [Cell Pathol.]*, 51, 187 1986.

101. Yeh, C.-K., Louis, J. M., and Kousvelari, E. E., β-Adrenergic regulation of c-fos expression in an epithelial cell line, *FEBS Lett.*, 240, 118, 1988.

102. Yoshida, H., Azuma, M., Yanagawa, T., Yura, Y., Hayashi, Y., and Sato, M., Effect of dibutyryl cyclic AMP on morphological features and biological markers of a human salivary gland adenocarcinoma cell line in culture, *Cancer*, 57, 1011, 1986.

103. Young, J. A., Cook, D. I., van Lennep, E.W., and Roberts, M., Secretion by the major salivary glands, in *Physiology of the Gastrointestinal Tract*, 2nd ed., Johnson, L. R., Ed., Raven Press, New York, 1987, 773.

104. Young, J. A. and Van Lennep, E. W., *The Morphology of Salivary Glands*, Academic Press, New York, 1978.

105. Young, J. A. and Van Lennep, E. W., Transport in salivary and salt glands, in *Membrane Transport in Biology*, Giebisch, G., Tosteson, D. C., and Ussing, H. H., Eds., Springer-Verlag, Berlin, 1979, 563.

106. Zajicek, G., Yagil, C., and Michaeli, Y., The streaming submandibular gland, *Anat. Rec.*, 213, 1985.

Regulation of Salivary Gland Development

Leslie S. Cutler

INTRODUCTION

The complete development of the major salivary glands can be divided into four separate but partially linked developmental phases: (1) morphogenesis — the development of the characteristic branched architecture of the glands, (2) cytodifferentiation — the development of specialized cells (secretory and ductal) within the branching structure, (3) development of the stimulus/secretion coupling system which transduces external stimuli into a secretory response by the secretory cells, and (4) anatomic coupling of sympathetic nerves with the secretory cells. Full, functional, physiologic development of salivary glands is attained when the secretory cells of the gland are anatomically connected to the nerves which activate secretion and the secretory cells have the surface receptors which enable them to undergo exocytosis in response to a neural stimulus.

These developmental phases are partially coupled in that they follow each other sequentially. For example, it appears that differentiation of secretory cells cannot begin until branching morphogenesis has been established (Cutler, 1980; Spooner et al., 1977) and appearance of the cell surface receptors that regulate secretion does not occur until well after secretory cell development is established (Cutler et al., 1981; Bottaro and Cutler, 1984; Cutler et al., 1989 for full review). On the other hand, these developmental phases are independent in that once initiated they proceed concurrently and are unaffected by the progression of the other processes. In this regard, for example, cytodifferentiation of secretory cells can continue even if morphogenesis is disrupted (Cutler, 1980) and the β-adrenergic receptors which regulate secretion appear at the cell surface whether in the presence or absence of neural connections (Cutler et al., 1985, 1988).

There is a growing body of data which suggest that each process is regulated by different molecules, many of which appear to reside in the extracellular matrix. The purpose of this discussion is to review the current information regarding the molecular regulators of the morphogenesis of and secretory cell differentiation in developing salivary glands. The reader is referred to a review by Cutler (1989) for more information on the development of the stimulus-secretion coupling system within the secretory cells and the development of functional neural connections between the sympathetic nervous system and the secretory cells.

REVIEW OF SALIVARY GLAND DEVELOPMENT

Salivary gland development begins in the developing fetus when specific cells of the oral epithelium are induced to proliferate and begin the production of secretory proteins which are specific to these glands. The cells in this salivary gland anlage continue to proliferate and initiate the development of the gland's characteristic branching pattern (morphogenesis). Once the branching pattern has been set, the process of secretory cell development (differentiation) is initiated. In this process there is an increased synthesis of secretory proteins which are packaged into secretory granules leading to the development of the typical exocrine cell structure (see Cutler 1989 for review) by those cells which are destined to become columnar in shape and develop a cytoplasmic architecture which both facilitates the synthesis and secretion of their exocrine products (cytodifferentiation). Once both of these processes have been set in motion the basic developmental pattern for the salivary glands has been established. Growth of the glandular rudiment continues although there appears to be a modulation of the type of secretory material produced by the cells as the gland matures.

Thus, glandular morphogenesis and differentiation of the secretory cell compartment is established during the prenatal period as is the intracellular arm of the stimulus-secretion coupling system (Cutler et al., 1981, Cutler 1989). Although the form of the gland and the production of secretory protein by the developing secretory cells is established prior to birth, functional development of the stimulus-secretion coupling systems of the gland is not complete until well after birth. The β-adrenergic receptors which regulate acinar cell protein secretion do not appear on the cell surface of the developing acinar cells of the rat submandibular gland until 5 to 6 days after birth. Functional, neural control of acinar cell secretion in the gland is not established until about this same time (Cutler et al., 1981; Bottaro and Cutler, 1984; Cutler et al., 1985). Interestingly, the muscarinic cholinergic and α-adrenergic receptors which regulate fluid and ion movement in the gland are functional at the time of birth (Cutler et al., 1981; Bylund et al., 1982; Bottaro and Cutler, 1984).

It should be noted that similar observations have been made with regard to the development of morphogenesis, cytodifferentiation, and stimulus-secretion coupling systems in the maturing parotid gland although the exact timing sequences appear to be different in the two glands (see Cutler 1989 for review).

MOLECULAR REGULATION OF SALIVARY GLAND MORPHOGENESIS

There is essentially no information available concerning the molecules involved in the induction and regulation of the initial events in salivary gland development. The earliest evidence of glandular development is the appearance of a cluster of oral epithelial cells capped by a dense mesenchymal cell network. As these epithelial cells proliferate and grow into the underlying connective tissue to form a stalk (the precursor of the main excretory duct system) which terminates in a lobular mass of cells referred to as the primary end-bud (the precursors of the intralobular ductal system and the acinar compartment). The primary end-bud is surrounded by a dense mesenchymal capsule which is distinct from the adjacent connective tissue. Controlled growth from the tip of the end-bud leads to the formation of three to five primary branches. In turn, each primary branch undergoes regulated growth from its most distal end giving rise to the characteristic arborized pattern which is specific to salivary glands. Continued differential growth at the ends of the newly forming branches leads to glandular enlargement and, eventually, the formation of lobules within the glandular parenchyma (Banerjee et al., 1977; Bernfield and Wessels, 1970; Bernfield et al., 1972; Borghese, 1950; Grobstein, 1953a; Spooner and Wessels, 1970, 1972). Interestingly, a recent study (Spooner et al., 1989) indicates that branching morphogenesis can occur even when rudiment growth is significantly inhibited suggesting that there are several separable regulatory activities involved in the complex process of glandular morphogenesis.

It has been clearly demonstrated that the mesenchymal capsule surrounding the epithelial portion of the salivary gland anlage directly controls the development of the branching pattern seen in the developing glands (Bernfield and Banerjee, 1972; Cohn et al., 1977; Grobstein, 1953a,b,c; Lawson, 1970, 1972, 1974). The mesenchyme appears to exercise this control through the selective production and destruction of specific extracellular matrix molecules (Bernfield et al., 1984; Grobstein, 1953 a,b,c; Spooner et al., 1985; Thompson and Spooner, 1982, 1983) These extracellular matrix molecules are deposited in and near the basal lamina which serves as the interface between the epithelial cells and their investing mesenchymal capsule (Figure 1). Similarly, the enzymes which selectively digest the extracellular matrix components appear to work at specific sites along the basal lamina.

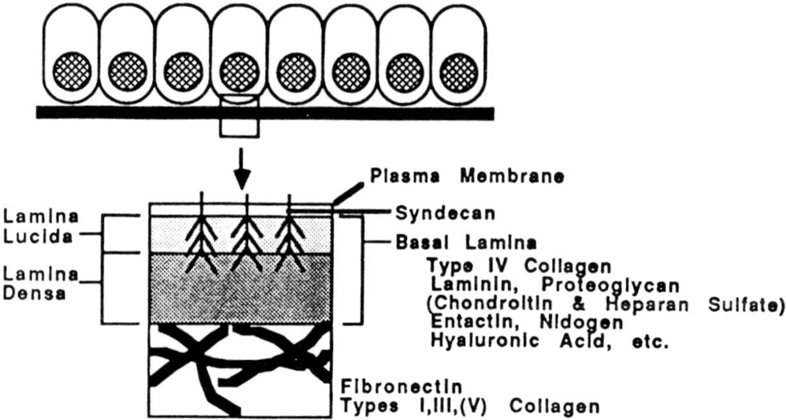

Figure 1. Schematic representation of the basal lamina.

The important role of the connective tissue capsule in the regulation of the development of the epithelial part of the submandibular gland was first noted by Borghese (1950) when he observed that branching morphogenesis and growth of the submandibular gland rudiment *in vitro* was inhibited if the surrounding mesenchymal capsule was removed. This observation was confirmed and expanded by Grobstein (1953a,b,c) who showed that the potential for branching morphogenesis resided predominantly in the end-buds while the stalk or duct had only limited capacity for branching. In addition, in now classical *in vitro*, transfilter experiments, Grobstein (1953c) demonstrated that the induction of branching morphogenesis appeared to be dependent upon a diffusible macromolecule which was produced by the capsular mesenchyme.

Subsequent experiments suggested (Grobstein and Cohen, 1965; Kallman and Grobstein, 1966) that collagen, derived from the capsular mesenchyme, was important in regulating and maintaining the branching pattern of the submandibular gland. Collagen appeared to serve as a stabilizing component which inhibited cell growth in specific areas of the growing rudiment while other areas were devoid of collagen and thus able to grow. Thus, the stabilized areas served as branch points from which morphogenetically active cells gave rise to new branches. This work (Grobstein and Cohen, 1965) suggested that collagen could reduce cell growth at the branch point and serve as an anchor point from which cells not inhibited by collagen could undergo mitosis and develop into new branches. It was also postulated that some type of localized collagenolytic activity might be involved at the points of growth to free some cells from the stabilizing effect of collagen.

Subsequent work from several laboratories has confirmed and expanded Grobstein's early work and hypotheses. Data has been developed which indicates that both types I and III collagen synthesis and deposition are required for salivary gland branching morphogenesis (Spooner and Faubion, 1980). There appears to be some question about the absolute requirement for type I collagen in salivary gland morphogenesis (Kratochwill et al., 1986). However, based on the work of Spooner and Faubion (1980) and more recent data from Fukuda and associates (1988) it appears that branching morphogenesis is affected by both types I and III collagen.

The potential importance of glycosaminoglycans in the regulation of salivary gland morphogenesis was first suggested by Kallman and Grobstein (1966). This observation also suggested that, in addition to collagen, the epithelial basal lamina might be important in the regulation of salivary gland morphogenesis.

Data which showed the significant role of the basal lamina and its components was developed by Bernfield and his colleagues (Banerjee et al., 1977; Bernfield et al., 1972; Cohn et al., 1977). Glycosaminoglycans (proteoglycans) and type I collagen were shown, by histochemical, electron microscopic, and biochemical studies, to be concentrated in the clefts or branch points while these molecules appeared to be absent or sparsely distributed at the growing ends of the branches (Banerjee et al., 1977; Bernfield et al., 1972; Bernfield and Banerjee, 1972; Banerjee and Bernfield, 1979; Cohn et al., 1977).

Radioautographic studies suggested that there were regional differences in glycosaminoglycan turnover and that there was an increased degradation of the basement membrane which occurred at the growing end of the branches. This degradative activity seemed to be mediated by the investing connective tissue since there was an activation of a neutral hyaluronidase with specificity for hyaluronic acid and chondroitin sulfate proteoglycan in the tissue at the time branching morphogenesis was initiated (Bernfield and Banerjee, 1982; Bernfield et al. 1984). Coincident with the activation of this hyaluronidase the pattern of glycosaminoglycan produced by the salivary epithelium changed. The glycosaminoglycans synthesized by the early rudiment were rich in hyaluronic acid and chondroitin sulfate. However, as morphogenesis progressed and became more complex the glycosaminoglycans produced showed an increased amount of heparin sulfate and a reduced content of hyaluronic acid and chondroitin sulfate (Banerjee and Bernfield, 1977; Cohn et al., 1977; Cutler et al., 1991). In the mature gland, heparin sulfate is the predominant glycosaminoglycan synthesized with only small amounts of hyaluronic acid and virtually no chondroitin sulfate is produced (Cutler et al., 1987).

The convergence of the observations suggesting the importance of both collagen and glycosaminoglycans in the regulation and stabilization of salivary gland morphogenesis came from data indicating that types I and III collagen appear to protect the basal lamina from breakdown by hyaluronidase and other enzymes by binding with heparan sulfate proteoglycan. This binding stabilizes the basal lamina and reduces the breakdown of the heparan sulfate-rich proteoglycan (Spooner and Faubion, 1980; Bernfield et al., 1984) which is the dominant proteoglycan of the maturing salivary epithelial basal lamina (Cutler et al., 1987, in press). Deposition of type III collagen at the cleft or branch point areas seems to play a key role in regulating both assembly of basal lamina components and stabilizing these points so that branching may proceed.

The direct proof that basement membrane proteoglycans were of critical importance in the control of salivary gland branching morphogenesis came from studies in which β,d-xyloside was used to block the assembly of glycosaminoglycan chains to the protein core of the proteoglycan (Spooner et al., 1985; Thompson and Spooner, 1982, 1983). Xyloside treatment of cultured submandibular gland rudiments blocked further morphogenesis in a reversible manner. That is, morphogenesis resumed after removal of the xyloside from the culture medium. At the concentration used, xyloside inhibited chondroitin sulfate proteoglycan synthesis but did not affect heparan sulfate proteoglycan production (Spooner et al., 1986; Cutler et al., submitted). This data strongly indicates that chondroitin sulfate proteoglycan plays the major role in regulating the morphogenetic branching of salivary glands.

REGULATION OF SECRETORY CELL DIFFERENTIATION

Information regarding the molecular factors involved in the regulation of secretory cell differentiation in the salivary glands is not as advanced as the that for salivary gland morphogenesis. The basic morphogenetic branching pattern is established prior to the initial events in secretory cell development (Cutler, 1980). As the branching pattern of the rudiment is being established, cells producing low levels of secretory proteins can be identified within the rudiment. However, these cells do not demonstrate any structural signs of differentiation, such as secretory granules at this time (Yamashina and Barka, 1973; Cutler, 1973; Cutler and Chaudhry, 1974). By the time secretory granules, the first fine structural sign of secretory cell differentiation, can be seen in developing secretory cells branching morphogenesis is very advanced (Cutler, 1973; Cutler and Chaudhry, 1973, 1974; Redman and Ball, 1978; Redman and Sreebny, 1970, 1971; Yamashina and Barka, 1972, 1973, 1974).

Direct epithelial-mesenchymal contact between the salivary epithelium and the surrounding capsular mesenchyme (Cutler, 1973, 1977; Cutler and Chaudhry, 1973) appears to have an important role in initiating the sequence of events (amplification of secretory protein synthesis and the subsequent structural changes at the cellular level) that lead to secretory cell cytodifferentiation within the salivary glands (Cutler, 1980). These contacts are seen only after the primary branching pattern has been established and only involve interaction between the epithelial cells in the end-buds of the initial 4 to 12 branches of the early rudiment and the surrounding mesenchyme (Cutler and Chaudhry, 1973; Cutler, 1977). Contacts of this type are not seen before or after this stage of development of the gland. If these direct epithelial-mesenchymal interactions are prevented by separating the developing epithelium from its investing mesenchyme differentiation of secretory cells does not occur. However, if the epithelial-mesenchymal contacts have taken place, many of the cells go on to develop into secretory cells (Cutler, 1980).

The continued presence of the mesenchyme does not seem to be required for secretory cell development to occur once the inductive signals for secretory cell differentiation have been passed. This observation points out a significant difference between morphogenesis of the rudiment and secretory cell development since the continued presence of the mesenchymal capsule is required in order for morphogenesis to occur. Of note is the observation that the attainment of apical-basal polarity of the secretory cells seems to require factors produced by the mesenchyme. Secretory cells differentiating in the absence of the capsular mesenchyme produce secretory material and package the material into secretory granules. However, the cells do not demonstrate the apical-basal polarity typical of salivary gland exocrine cells. Further, the clustering of cells into acinar units is not required for the differentiation of secretory cells. These observations suggest that the development of the typical glandular morphogenetic branching pattern and the differentiation of secretory cells within developing salivary glands are partially coupled but independently regulated processes (Cutler, 1980). The mesenchyme seems to regulate both the gross, glandular morphology and the fine, cytologic form (columnar shape, apical-basal polarity) of the individual cells within the gland.

It was noted that there was a fibrillar material between the opposing membranes of the direct epithelial-mesenchymal contacts involved in secretory cell differentiation (Cutler and Chaudhry, 1973; Cutler, 1977). The fibrillar material was structurally similar to the lamina densa of the basal lamina and suggested the possibility for a role for extracellular matrix molecules in the regulation of secretory cell differentiation. Recent work (Cutler et al., 1991), as part of a study investigating the role of glycosaminoglycans/proteoglycans in the differentiation of salivary gland secretory cells, has confirmed the earlier reports by (Spooner, 1982, 1983) which indicated that chondroitin sulfate proteoglycan plays a major role in the regulation of salivary gland morphogenesis. In these studies (Cutler, 1988; Cutler et al., 1991) 16-day rat embryonic submandibular gland rudiments were cultured in the presence or absence of β,d-xyloside under conditions known to permit secretory cell differentiation. Morphogenesis was inhibited in greater than 90% of the rudiments cultured in the presence of xyloside analogous to the findings of Spooner and associates. However, secretory cell differentiation was seen in more than 70% of the rudiments in which branching morphogenesis had been inhibited. Analysis of the effect of β,d-xyloside on glycosaminoglycan/proteoglycan synthesis by rat salivary gland rudiments indicated that xyloside specifically inhibited chondroitin sulfate synthesis but had little effect on heparan sulfate proteoglycan production. This observation is consistent with those of Spooner and colleagues (Spooner et al., 1986).

Proteoglycans play a very central role in the regulation of the morphogenesis of salivary glands. Chondroitin sulfate proteoglycan clearly appears to be the major proteoglycan in this regard. Interestingly, chondroitin sulfate proteoglycan does not appear to be involved in the regulation of secretory cell differentiation within the developing rudiment. This observation supports earlier work (Cutler, 1980) that indicated that morphogenesis and cytodifferentiation of salivary glands are separately regulated processes.

Recent studies (Cutler, 1991) have examined the effects of culturing developing (16-day embryonic) rat submandibular gland rudiments in the presence of polyclonal or monoclonal antibodies directed against extracellular matrix components. These studies have provided some interesting insights into the regulation of glandular morphogenesis and secretory cell differentiation. It appears that type IV collagen and laminin have roles in the regulation of salivary gland morphogenesis in addition to the roles already established for types I and III collagen and chondroitin sulfate proteoglycan. To date, with regard to the control of the differentiation of salivary gland secretory cells, type IV collagen is the only extracellular matrix molecule which was shown to play a role in this process. When embryonic submandibular gland rudiments were cultured in the presence of anti-type IV collagen antibody virtually all of the rudiments were prevented from undergoing further branching morphogenesis and secretory cell differentiation was not observed in any of these rudiments.

SUMMARY

The processes of morphogenesis and cytodifferentiation are partially linked, independently regulated processes. The full expression of both processes is modulated or controlled by epithelial-mesenchymal interactions involving both direct contacts and indirect molecular interactions. It appears that, at least in part, molecular constituents of the extracellular matrix may mediate the regulatory effects of both direct and indirect interactions. Data indicate that epithelial-mesenchymal interactions and various extracellular matrix molecules are intimately involved in the induction, control, and maintenance of salivary gland morphogenesis and cytodifferentiation.

With regard to the role of extracellular matrix molecules in the regulation of salivary gland morphogenesis and cytodifferentiation it appears that types I, III, and IV collagen, laminin, and chondroitin sulfate proteoglycan play roles in the control of glandular morphogenesis. With the exception of type IV collagen, and possibly laminin, these molecules do not appear to be involved in the regulation of cytodifferentiation of salivary gland secretory cells. On the other hand, of the extracellular matrix molecules tested so far, only type IV collagen appears to play a significant role in the regulation of salivary gland secretory cell differentiation.

REFERENCES

1. Banerjee, S. D., Cohn, R. H., and Bernfield, M. R., Basal lamina of embryonic salivary epithelia. Production by the epithelium and role in maintaining lobular morphology, *J. Cell Biol.*, 73, 445, 1977.
2. Bernfield, M. R. and Wessels, N. K., Intra- and extracellular control of epithelial morphogenesis, *Dev. Biol.*, (Suppl.) 4, 195, 1970.
3. Bernfield, M., Banerjee, S. D., and Cohn, R. H., Dependence of salivary epithelial morphology and branching morphogenesis upon acid mucopolysaccharide-protein (proteoglycan) at the epithelial surface, *J. Cell Biol.*, 52, 674, 1972.
4. Bernfield, M. R. and Banerjee, S. D., Acid mucopolysaccharide (glycosaminoglycan) at the epithelial-mesenchymal interface of mouse embryo salivary glands, *J. Cell Biol.*, 52, 664, 1972.
5. Bernfield, M. and Banerjee, S. D., The turnover of basal lamina glycosaminoglycan correlates with epithelial morphogenesis, *Dev. Biol.*, 90, 291, 1982.
6. Bernfield, M., Banerjee, S. D., Koda, J., and Raprager, A. C., Remodelling of the basement membrane: morphogenesis and maturation, in *Basement Membranes and Cell Movement*, Porter, R. and Whelan, J., Eds., Pittman Press, Bath, 1984, 179.
7. Borghese, E., Explanation experiments on the influence of the connective tissue capsule on the development of the epithelial part of the submandibular gland of *Mus musculus*, *J. Anat. (London)*, 84, 303, 1950.
8. Bottaro, B. and Cutler, L. S., An electrophysiological study of the postnatal development of the autonomic innervation of the rat submandibular, *Arch. Oral Biol.*, 29, 237, 1984.
9. Bylund, D. B., Martinez, J. R., Camden, J., and Jones, S. B., Autonomic receptors in the developing submandibular glands of neonatal rats, *Arch. Oral Biol.*, 27, 945, 1982.
10. Cohn, R. H., Banerjee, S. D., and Bernfield, M., Basal lamina of embryonic salivary epithelia. Nature of glycosaminoglycan and organization of extracellular materials, *J. Cell Biol.*, 73, 464, 1977.
11. Cutler, L. S., Morphogenesis and Functional Differentiation of the Rat Submandibular Gland *In Vivo* and *In Vitro*: Ultrastructural and Biochemical Studies, Ph.D. Thesis, State University of New York at Buffalo, Buffalo, NY, 1973.
12. Cutler, L. S., Intercellular contacts at the epithelial-mesenchymal interface of the developing rat submandibular gland *in vitro*, *J. Embryol. Exp. Morph.*, 39, 71, 1977.
13. Cutler, L. S., The dependent and independent relationships between cytodifferentiation and morphogenesis in developing salivary gland secretory cells, *Anat. Rec.*, 196, 341, 1980.
14. Cutler, L. S., Functional differentiation of salivary glands, in *Handbook of Physiology, Salivary, Pancreatic, Gastric, and Hepatobiliary Secretion*, Forte, J., Ed., American Physiologic Society Press, New York, 1989, 93.
15. Cutler, L. S., The role of extracellular matrix in the morphogenesis and differentiation of salivary glands, *Adv. Dent. Res.*, 4, 27, 1991.
16. Cutler, L. S. and Chaudhry, A. P., Intercellular contacts at the epithelial-mesenchymal interface during the prenatal development of the rat submandibular gland, *J. Morph.*, 33, 229, 1973.
17. Cutler, L. S. and Chaudhry, A. P., Cytodifferentiation of the acinar cells of the rat submandibular gland, *Dev. Biol.*, 41, 31, 1974.

18. Cutler, L. S. and Chaudhry, S. P., Cytodifferentiation of striated duct cells and secretory cells of the convoluted granular tubules of the rat submandibular gland, *Am. J. Anat.*, 143, 201, 1975.

19. Cutler, L. S. and Rodan, S. B., Biochemical and cytochemical studies on adenylate cyclase in the developing rat submandibular gland: differentiation of the acinar secretory compartment, *J. Embryol. Exp. Morphol.*, 36, 291, 1976.

20. Cutler, L. S., Christian, C. P., and Bottaro, B., Development of stimulus secretion coupling in salivary glands, in *Methods in Cell Biology*, Hand, A. R. and Oliver, C., Eds., Academic Press, New York, 23, 531, 1981.

21. Cutler, L. S., Schneyer, C., and Christian, C. P., The influence of the sympathetic nervous system on the development of β–adrenergic receptors in the rat submandibular gland, *Arch. Oral Biol.*, 30, 341, 1985.

22. Cutler, L. S., Christian, C. P., and Rendell, J. R., Glycosaminoglycan synthesis by adult rat submandibular gland secretory units, *Arch. Oral Biol.*, 32, 413, 1987.

23. Cutler, L. S., Christian, C. P., and Rendell, J. R., Sulfated glycosaminoglycan synthesis by developing rat submandibular gland secretory units, *Arch. Oral Biol.*, 36, 389, 1991.

24. Cutler, L. S., Christian, C. P., and Rozenski, D., Xylose-linked proteoglycan synthesis does not have a primary role in the control of secretory cell differentiation in salivary glands, *Dev. Biol.*, 147, 14, 1991.

25. Fukuda, Y., Masuda, Y., Kishi, J., Hashimoto, Y., Taro, H., Nogawa, H., and Nakanishi, Y., The role of interstitial collagens in cleft formation of mouse embryonic submandibular gland during initial branching, *Development*, 103, 259, 1988.

26. Grobstein, C., Analysis *in vitro* of the early organization of the rudiment in the mouse submandibular gland, *J. Morph.*, 93, 19, 1953a.

27. Grobstein, C., Epithelial-mesenchymal specificity in the morphogenesis of mouse submandibular rudiments *in vitro*, *J. Exp. Zool.*, 124, 383, 1953b.

28. Grobstein, C., Morphogenetic interaction between embryonic mouse tissues separated by a membrane filter, *Nature (London)*, 172, 869, 1953c.

29. Grobstein, C. and Cohen, J., Collagenase: effect on morphogenesis of embryonic salivary epithelium *in vitro*, *Science*, 150, 626, 1965.

30. Kallman, F. and Grobstein, C., Localization of glycocyamine-incorporating materials at epithelial surfaces during salivary epithelial-mesenchymal interactions *in vitro*, *Dev. Biol.*, 14, 52, 1966.

31. Lawson, K. A., Morphogenesis and functional differentiation of rat parotid gland *in vivo* and *in vitro*, *J. Embryol. Exp. Morph.*, 32, 411, 1970.

32. Lawson, K. A., The role of mesenchyme in the morphogenesis and functional differentiation of rat salivary epithelium, *J. Embryol. Exp. Morph.*, 27, 497, 1972.

33. Lawson, K. A., Mesenchyme specificity in rodent salivary gland development: the response of salivary epithelium to lung mesenchyme *in vitro*, *J. Embryol. Exp. Morph.*, 32, 469, 1974.

34. Redman, R. S. and Ball, W. D., Cytodifferentiation of secretory cells in the sublingual gland of the prenatal rat: a histochemical and ultrastructural study, *Am. J. Anat.*, 153, 367, 1978.

35. Redman, R. S. and Sreebny, L. M., The prenatal phase of morphosis of the rat parotid gland, *Anat. Rec.*, 168, 127, 1970.

36. Redman, R. S. and Sreebny, L. M., Morphologic and biochemical observations on the development of the rat parotid gland, *Dev. Biol.*, 25, 248, 1971.
37. Spooner, B. R. and Faubion, J. M., Collagen involvement in branching morphogenesis of embryonic lung and salivary gland, *Dev. Biol.*, 77, 84, 1980.
38. Spooner, B. S., Bassett, K., and Stokes, B., Sulfated glycosaminoglycan deposition and processing at the basal epithelial surface in branching and β-D-xyloside-inhibited embryonic salivary glands, *Dev. Biol.*, 109, 177, 1985.
39. Spooner, B. S., Bassett, K., and Spooner, B. S., Jr., Embryonic salivary gland epithelial branching activity is experimentally independent of epithelial expansion activity, *Dev. Biol.*, 133, 569, 1989.
40. Spooner, B. S., Cohen, H. I., and Faubion, J., Development of the embryonic mammalian pancreas: the relationship between morphogenesis and cytodifferentiation, *Dev. Biol.*, 61, 119, 1977.
41. Spooner, B. S. and Wessels, N. K. Effects of cytochalasin B upon microfilaments involved in morphogenesis of salivary epithelium, *Proc. Natl. Acad. Sci. U.S.A.*, 66, 360, 1970.
42. Spooner, B. S. and Wessels, N. K., An analysis of salivary gland morphogenesis: role of cytoplasmic microfilaments and microtubule, *Dev. Biol.*, 27, 38, 1972.
43. Spooner, B. S., Thompson-Pletscher, H. A., Stokes, B., and Bassett, K. E., Extracellular matrix involvement in epithelial branching morphogenesis, in *Developmental Biology*, Vol. 3, Steinberg, M. S., Ed., Plenum Publishing, New York, 1986, 225.
44. Thompson, H. A. and Spooner B. S., Inhibition of branching morphogenesis and alteration of glycosaminoglycan biosynthesis in salivary glands treated with β-D-xyloside, *Dev. Biol.*, 89, 417, 1982.
45. Thompson, H. A. and Spooner, B. S., Proteoglycan and glycosaminoglycan synthesis in embryonic mouse salivary glands: effects of β-D-xyloside, an inhibitor of branching morphogenesis, *J. Cell Biol.*, 96, 1443, 1983.
46. Yamashina, S. and Barka, T., Localization of peroxidase activity in the developing submandibular gland of normal and isoproterenol-treated rats, *J. Histochem. Cytochem.*, 20, 855, 1972.
47. Yamashina, S. and Barka, T., Development of endogenous peroxidase in fetal rat submandibular gland, *J. Histochem. Cytochem.*, 21, 42, 1973.
48. Yamashina, A. and Barka, T., Peroxidase activity in the developing rat submandibular gland, *Lab. Invest.*, 31, 82, 1974.

Cell-Restricted Secretory Proteins as Markers of Cellular Phenotype in Salivary Glands

William D. Ball

INTRODUCTION

Salivary Glands as a Model for Developmental Regulation

The salivary glands of mammals are a generally similar but specifically diverse group of exocrine organs that originate through the interaction of oral epithelium with investing mesenchyme and undergo extensive branching morphogenesis to form complex glandular structures (see Redman, 1987 for review). The major salivary glands, i.e., the parotid, submandibular, and sublingual glands, are large paired organs that lie outside the oral cavity proper, and which communicate with the oral mucosa through major ducts. Minor salivary glands are found in the submucosa of most regions of the oral cavity (See Hand, 1985; Klein, 1987, for general description of salivary glands). The content of mucous and serous cells varies among the salivary glands, and within each of these major classes, further diversity in the expression of different secretory proteins is evident (see below). In a given species, a particular secretory protein may be expressed in one specific gland, while other species may lack this protein or express it in a different gland (Sreebny et al., 1967; Junqueira et al., 1973; Dickinson et al., 1989b; Mirels et al., 1990). Similarly, the distribution of mucous and serous cells among the developmentally and anatomically defined major glands shows considerable taxonomic variability (Reviewed in Young and van Lennep, 1978). It appears, then, that evolutionary constraints mandating the presence of a particular secretory protein in a given gland are less than exacting. It may be that the diversity of exocrine cells in the different salivary glands provides a flexibility that ensures the synthesis of protein molecules essential for salivary functions among one or the other cell types. The requirement for the synthesis of any one salivary protein may be sufficiently nonstringent that no serious insult to the organism attends its absence. One may speculate that this permits a diversity of regulatory mechanisms capable of "fine tuning" the overall output of salivary proteins, providing a unique array of regulatory elements that may be accessible to experimental discovery and manipulation.

Cell-Restricted Proteins as Markers of Differentiated Cellular Phenotype

The concept of marker proteins for cellular phenotypes implies a restricted distribution of certain proteins such that their presence in a cell constitutes a description and definition of a specific type or of a class of cellular differentiation. The desirability of having protein molecules as signatures of cellular phenotype is apparent, given their direct correspondence to the levels of expression of their genes. It can be argued that the type and quantity of proteins synthesized by a cell constitute an accurate and complete description of that cell. Contemporary studies on the cellular biology of protein synthesis, processing, and transport have provided considerable data in support of this idea. The sequestration of proteins to the rough endoplasmic reticulum by a signal peptide, the dependence of glycosylation pattern on amino acid sequence, the existence of sequence-specific mechanisms of proteolysis, the role of sorting signals in determining specific membrane associations, all in the context of membrane flow from the rough endoplasmic reticulum, make evident the direct determination of cellular structures by amino acid sequence (see Alberts et al., 1989 for review).

Proteins vary, of course, in the extent to which they are ubiquitous or cell-type restricted. Rutter and Weber (1965) classified cellular proteins as primary, secondary, or tertiary proteins, depending on whether they are present in all cells, in a restricted subset of cells, or in only a single cell type. The expression of cell-restricted proteins was considered the best single criterion for defining a cellular phenotype. Implied in this formulation is that this gene expression is intimately linked to the determinative regulatory events in development (Rutter et al., 1967a). These, then, specify a pattern of gene expression in which the "correct" genes for a cellular phenotype are expressed and others maintained in a state of repression. Although the simplest statement of this idea implies all-or-nothing gene expression, it is clear, from consideration of the many ways in which protein synthesis is controlled (Kozak, 1988), that quantitative differences in protein expression also may be genetically regulated.

Scope, Emphasis, and Approach of the Review

This report will cover, primarily, the salivary glands of non-human mammals, principally rats and, to a lesser extent, mice, since it is in these species that most of the analytical and experimental work has been done. All descriptions refer to the laboratory rat (*Rattus norvegicus albinus*), except where noted. The substantial literature on the biochemistry, molecular biology, and genetics of human salivary proteins has been recently reviewed (Minaguchi and Bennick, 1989) and will not be covered here. Emphasis will be on the proteins of serous cells, since these have been more thoroughly characterized than have the highly specialized mucins, although some coverage of mucins is included. Recent reviews of salivary mucins are available (Nugent and O'Connor, 1984; Quissell and Tabak, 1990; Tabak, 1990)

The treatment of salivary marker proteins will emphasize cellular, biochemical and developmental studies; some consideration will be given to the molecular biology of selected salivary proteins.

MAJOR SALIVARY GLANDS

Parotid Gland

The parotid is the simplest of the major salivary glands in its histological organization. In the adult, all acinar cells are serous and account for the vast majority of secretory proteins. The parotid gland develops relatively slowly; little secretory differentiation is evident by birth, and well differentiated, functional secretory cells are not present until five days postpartum (Redman and Sreebny, 1970, 1971; Taga and Sesso, 1979; Ball et al., 1988a). Similarly, the acquisition of the mature set of secretory proteins is not complete until several weeks postpartum in the rat (Ball, 1971; Redman and Sreebny, 1971) and in the mouse (Shaw et al., 1986).

Digestive Enzymes

Amylase. The most thoroughly studied of the parotid secretory proteins is amylase. The rat parotid enzyme was purified by Loyter and Schramm (1962) by ethanolic precipitation of an amylase-glycogen complex and subsequently by chromatographic procedures (e.g., Ball, 1974a; Keller et al., 1975; Yamashita, 1981). On native gel electrophoresis under both anionic and cationic conditions, and on isoelectric focusing separation, four active forms of the enzyme have been reported; two of these appear to be derivatives of the major component (Robinovitch and Sreebny, 1970, 1972). Parotid gland amylase is different from pancreatic amylase in the rat (Sanders and Rutter, 1972), rabbit (Malacinski and Rutter, 1968) and mouse (Sick and Nielsen, 1964; Schibler et al., 1980). In the mouse, parotid and liver amylase are identical but are synthesized from two different mRNAs containing a common coding sequence and different 5′ nontranslated sequences (Young et al., 1981). A strong promoter, located 7.5 kb upstream from the common coding sequence, is utilized only in the parotid gland, while a weak promoter, 4.5 kb upstream is active in both parotid gland and liver (Schibler et al., 1983). During development of the parotid gland, the weak promoter is utilized first and is active shortly after birth; the strong promoter becomes active after about 2 weeks and subsequently directs the bulk of mRNA transcription (Shaw et al., 1985). This gene is one of an interrelated family of amylase genes; several of the others express pancreatic amylase, while still others have significant nonhomology to both the pancreatic and parotid amylase genes (Crerar et al., 1983). Among the major salivary glands of the rat, amylase appears to be restricted to cells of the parotid gland. Measurable levels of amylolytic activity can be found in other organs and, probably, in all organs at some level. For example in the submandibular gland, the sublingual gland, and in the lung, amylase activity is measurable at specific activities approximately 10,000-fold less than in the parotid gland, and this probably reflects the activity of nonsecretory carbohydrases. Only the pancreas, parotid gland and liver have mRNAs that hybridize extensively with the amylase genes (Crerar et al., 1983).

DNAase.   Salivary gland DNAase has been described primarily in the mouse and in the rat parotid gland (Schramm, 1964; Sreebny et al., 1965, 1967; Rutter et al., 1967b; Ball and Rutter, 1971; Ball, 1971). The organ distribution of secretory DNAase varies considerably among different mammalian taxa; it can be found exclusively in the pancreas (dog, hog), the parotid gland (horse, murine rodents), primarily in the submandibular gland (guinea pig), or in none of these organs (squirrel) (Sreebny et al., 1967). Some level of DNA endonuclease activity is measurable in all organs using sensitive assay procedures, but in cases where the properties of the enzyme have been investigated, DNAase activity in the pancreas, liver, and submandibular glands of rats and mice appears to reflect the presence of enzymes different from the parotid secretory enzyme (Ball and Rutter, 1971; Ball, 1971). These data suggest that DNAase is a cell-specific protein of the parotid gland.

The parotid DNAase is similar to the well-described DNAase I of the bovine pancreas (Keller et al., 1958; Laskowski, 1961) in its heat and acid stability, pH optimum, cation requirement, and substrate preference for native DNA (Ball and Rutter, 1971; Ball, 1971). It has been purified to apparent homogeneity, and has major and minor active isomorphs, which are resolvable by native gel electrophoresis at pH 8.3 (Ball, 1974a; Keller at al., 1975; Yamashita et al., 1981). The enzyme has not been further characterized; however, it can be obtained in substantial quantity from stimulated parotid saliva or from isolated secretion granules (Keller et al., 1975; Yamashita, 1981), as well as from tissue homogenates (Ball, 1974a). Polyclonal antibodies to purified DNAase have been used to localize the enzyme, immunocytochemically, to the secretion granules of parotid acinar cells (Yamashita, 1981). During development, it appears very suddenly, at about 15 days postpartum (Redman and Sreebny, 1971; Ball, 1971). In the mouse, this is about the time at which amylase mRNA begins to be transcribed from the strong promoter (above; Shaw et al., 1985).

RNAase.   An enzyme similar to the RNAase A of the pancreas (Keller et al., 1958) has been described in the rat parotid gland (Schramm, 1964; Robinovitch et al., 1969). This enzyme has been purified from gland homogenates by routine chromatographic procedures (Ball, 1974a) and by affinity chromatography (Ball and Nelson, 1978b), and can also be obtained from saliva and isolated secretion granules (Keller et al., 1975; Yamashita, 1981). Like the well-studied bovine pancreatic enzyme, the parotid RNAase is a heat-stable molecule, active at alkaline pH, and having no or little activity against polyadenylic acid. It has more than threefold greater activity against polycytidilic acid than against native ribosomal RNA (Ball and Nelson, 1978b). Also like pancreatic RNAase, it appears to be a basic protein, eluting in the void volume when separated on DEAE-Sephadex (Ball and Nelson, 1978a; Keller et al., 1975; Ball and Nelson, 1978b), and migrating to the cathode on native gel electrophoresis at pH 4.5, with Rf = 0.52 in 12.5% gels (Ball, 1974a; Ball and Nelson, 1978b). Although it is a secretory protein, it is found at levels 25-fold less than in the pancreas (Ball and Nelson, 1978b).

As is the case with amylase and DNAase, low levels of enzyme activity can be detected in all organs measured, including the submandibular gland. In submandibular homogenates, RNAase activity is partially inhibited (Menaker and Miller, 1973; Ball and Nelson, 1978b) by an inhibitor similar to that found in liver and several other organs (Roth, 1956; Shortman, 1961); this inhibitor also can inhibit parotid and bovine pancreas RNAase. Similarly, the submandibular inhibitor is inactivated by acid-treatment or by reaction with p-hydroxy-mercuribenzoate (Ball and Nelson, 1978b). Even after release from inhibition, however, RNAase activity in the submandibular gland remains lower than in kidney, liver, lung, and muscle tissue and much lower than in the parotid gland (1/540) and pancreas (1/13,750). This RNAase is similar in heat stability and pH optimum to the parotid RNAase, and also is inactive against polyadenylic acid. Limited data indicate that the submandibular RNAase activity differs in its substrate specificity, showing lower activity against polycytidilic acid than does the parotid enzyme (Ball and Nelson, 1978b). This suggests the possibility that, like DNAase, this protein is a cell-specific marker for the parotid gland.

Proline-Rich Proteins

A family of unusual proteins termed proline-rich proteins (PRP) is a prominent product of parotid secretions. These contain up to 44% proline and 26% glutamic acid residues (Fernandez-Sorensen and Carlson, 1974; Keller et al., 1975; Muenzer et al., 1979b; Iversen et al., 1982). The PRP are dramatically induced by chronic stimulation with isoproterenol for 1 week, and were initially obtained in bulk from the void peak of a Sephadex G-100 column, showing a single, broad, diffuse band on native gel electrophoresis at pH 8.3 (Fernandez-Sorensen and Carlson, 1974). Subsequently, these were resolved into a single acidic PRP (Ipr-1A2) and six basic PRP by CM-cellulose chromatography (Muenzer et al., 1979a). By sedimentation equilibrium, these proteins showed Mr between about 15 and 25 kDa, although by gel filtration they gave Mr of >70 kDa and by SDS PAGE, Mr between 22 and 36 kDa. The anomalous behavior on gel filtration and PAGE analysis apparently reflects the high axial ratios (>25) of the proteins, which exist as highly elongate rods (Muenzer et al., 1979b; Ziemer et al., 1982). Cell-free translation of PRP mRNA showed that all six of the basic PRP (bPRP) were synthesized and each incorporated methionine; since the mature proteins lack methionine, this indicates that each protein is separately synthesized from a unique mRNA template. Each of the bPRP was able to completely bind polyclonal antibodies raised to a mixture of the six bPRP, indicating antigenic identicality of the six bPRP protein species (Ziemer et al., 1982).

The studies cited above did not observe the constitutively expressed proline-rich proteins of the rat PRG. Keller et al. (1975) fractionated parotid saliva from untreated rats by DEAE-Sephadex chromatography and found a fraction eluting at about 0.15 *M* NaCl that contained 31% pro, 22% glu, and 18% gly, very similar to the composition reported for the acidic PRP obtained from isoproterenol-induced rats (Fernandez-Sorensen and Carlson, 1974; Muenzer et al., 1979b). Subsequent studies revealed a basic proline-rich fraction in the void fraction of the DEAE-Sephadex column (Robinovitch et al., 1977; Iversen et al., 1982). With isoproterenol induction, both the acidic PRP and the basic PRP fractions were substantially increased, with at least three novel components appearing in the basic fraction at 68, 32, and 28 kDa on SDS-PAGE (Robinovitch et al., 1977). In addition, a new, acidic, proline-rich fraction eluted at 0.2 *M* NaCl; this appears, from its amino acid composition and electrophoretic behavior, to be the acidic PRP (Ipr-1A2) described by Muenzer et al. (1979a,b). The basic PRP from untreated animals were characterized and found to include at least 6 species, with Mr between 25 and 48 kDa on SDS-PAGE; the amino acid composition of three of these was similar to that described by Muenzer et al. (1979b) for the isoproterenol-induced basic PRP (Iversen et al., 1982). It is not clear from the published data to what extent the isoproterenol-induced basic PRP might be amplified levels of the constitutive proteins. It would be of considerable interest to determine the relationship of the genes for constitutively expressed versus induced PRP.

Isoproterenol-inducible PRP also are found in the rat submandibular gland, and the PRP soluble in 10% TCA have been compared with those of the parotid. At least one (PRP-38) was shown to be identical in the two glands, and the overall pattern of induced TCA-soluble components suggests that others may also be common to the two glands (Mehansho and Carlson, 1983). A large (158 kDa) glycoprotein, also induced in the SMG, was not seen in the PRG; on the other hand, the PRG has an induced glycoprotein of 220 kDa which is not seen in the SMG. Both the PRP and the larger inducible glycoproteins showed components of variable electrophoretic mobility among different strains of rats (Humphreys-Beher, 1985), and the PRP were variable among different strains of mice (Ann et al., 1987). Humphreys-Beher (1985) has reported that differences between the large glycoproteins of the PRG and SMG are due primarily to differential glycosylation of what may be an identical protein core. Carlson has reported that the inducible glycoprotein (GP158) of the SMG has a peptide chain identical to a glycoprotein of 180 kDa from the PRG and that the difference in molecular weight is due to different carbohydrate compositions (Uhlig and Carlson, unpublished data; cited in Carlson, 1988).

Molecular analysis of the genes for the PRP indicated that they constitute a multigene family sharing considerable intraspecies and interspecies homology. cDNA clones were prepared from rat parotid mRNA; one of the clones was sequenced and the protein encoded appears to be the acidic PRP, 1A2 (24.5 kDa) (Ziemer et al., 1984). A number of other PRP cDNA clones were prepared from the parotid glands of rats and also of mice and the several clones fell into two groups, showing substantial but variable homology on low and moderate stringency hybridization (Clements et al., 1985). A complete sequence has been provided for one of the mouse PRP genes (Ann and Carlson, 1985).

Proteins very similar to the rat and mouse PRPs are synthesized constitutively in humans (Oppenheim et al., 1971), nonhuman primates (Oppenheim et al., 1985), and rabbits (Spielman and Bennick, 1989) and, after isoproterenol induction, in hamsters (Mehansho et al., 1987). Substantial homology is seen among PRPs from these different species. PRP from human parotid glands have been shown to bind to hydroxyapatite, suggesting a role in the protection of tooth enamel (Hay, 1973). A subsequent study has shown that proline-rich proteins constitute part of the acquired pellicle of the tooth (Kousvelari et al., 1980). In the mouse, PRP map to chromosome 8 and thus are not linked to amylase (chromosome 3) or parotid secretory protein (chromosome 2) (Azen et al., 1984).

An inductive effect similar to that found with isoproterenol was obtained in the parotid gland of rats maintained on a high level of dietary tannins. Cell-free translations showed that the proteins and mRNAs induced by tannins were identical to those induced by isoproterenol. Curiously, the 220-kDa glycoprotein that is isoproterenol inducible was not enhanced by the high-tannin diet. Unlike the parotid gland, the SMG did not show the induced response to tannins. The PRP were shown to bind to tannins with high affinity, prompting the suggestion that they might have a protective effect by inactivating harmful dietary tannins (Mehansho et al., 1983).

Proteins rich in proline have been found in or associated with the membrane of rat PRG secretion granules (Amsterdam et al., 1971), and a substantial fraction of these can be extracted by washing with isotonic saline (Robinovitch et al., 1975; Wallach et al., 1975a), prompting the suggestion that these proline-rich components might be the constitutive PRP identified in saliva (Keller et al., 1975), and that these had bound to the membranes during their isolation (Robinovitch et al., 1975). A subsequent study has raised some question concerning the purity of all of the earlier preparations of secretion granules, making unclear the extent to which the proline-rich components of the membranes reflect absorbed secretory proteins (Robinovitch et al., 1980); however, the possibility that secretory PRP might have high binding affinity for the secretion granule membranes raises the possibility of their having a function in the packaging of secretory proteins into forming secretion granules. It is noteworthy that all three functions thus far suggested for the PRP, i.e., protection of enamel, protection from dietary tannins and packaging secretory proteins, all reflect the capacity of PRPs for binding to different types of chemical structures.

Unlike the major digestive enzymes discussed above, some of the PRP, at least, appear to be expressed in both parotid and submandibular glands, but not in sublingual glands (Carlson, 1992). Still another isoproterenol-induced protein, cystatin, is restricted to the salivary glands but is present in all three glands (Bedi, 1989a). It would be extremely interesting to know whether other isoproterenol-induced proteins can be found and if they are also present in more than one salivary gland.

Other Major Secretory Products

Parotid Secretory Protein (PSP).   In the rat, two major previously undescribed proteins had been identified by native PAGE as M1 and M2 (Robinovitch and Sreebny; 1969). When saliva or secretion granule content was chromatographed on DEAE-Sephadex in 0.05 M phosphate, pH 8.1, Fraction I (starting conditions) contained Ml, in addition to ribonuclease, while M2 was eluted with a salt gradient in Fraction V at 0.2 M NaCl (Keller et al., 1975). We have found Ml to be highly immunologically cross-reactive with Protein SMG-A (23.5 kDa), one of the Bl-immunoreactive proteins (Bl-IP); these are characteristic of the neonatal submandibular gland, but have been found only in some intercalated duct cells in the adult SMG (Ball and Nelson, 1978a; Ball and Redman, 1984; Ball et al., 1988a). The Ml protein is highly leucine-rich (Keller et al., 1975; Wallach et al., 1975b; Kousvelari et al., 1984; Ball et al., 1992). It constitutes a substantial fraction of the total parotid secretion; Fraction I, in which Ml is the predominant component, comprises 26% of the total Lowry protein of saliva (Keller et al., 1975). We have purified Ml to apparent homogeneity, and compared it to Protein SMG-A; the two proteins are identical in molecular weight on SDS gels, isoelectric point (pH 4.5) and behavior on Affi-Gel Blue chromatography, and polyclonal antibodies to the two proteins are strongly cross-reactive (Ball et al., 1992). We have found one difference in the immune reactivities of the two antibody preparations; anti-SMG-A but not anti-Ml is cross-reactive with Protein SMG-C of neonatal Type I cells (Ball et al; 1992).

In attempting to characterize SMG-A and M1, our attention was drawn to a similar protein, parotid secretory protein (PSP), that had been extensively studied in the mouse. The most abundantly transcribed gene of the mouse parotid gland encodes a leucine-rich protein of about 20 kDa (Owerbach and Hjorth, 1980). This protein maps to chromosome 2 and segregates independently from amylase (chromosome 3; Owerbach and Hjorth, 1980) and the PRP (chromosome 8; Azen et al., 1984). A cDNA for mouse PSP has been cloned and sequenced, and defines a deduced molecular weight of 22.9 kDa, with 22.8% leucine (Madsen and Hjorth, 1985). Shaw and Schibler (1986) did not detect PSP in 11 other organs, including pancreas and submandibular glands by dot blot analysis, although Poulsen et al. (1986) reported faint bands by Northern blot analysis in adult mouse SMG and SLG. During postnatal development of the parotid gland, the onset of PSP transcription occurs earlier than that of amylase (Poulsen et al., 1986; Shaw and Schibler, 1986).

We found that antibodies raised against neonatal rat Protein SMG-A reacted on Western blots with Ml from rat PRG sonicates, and with a major band from mouse PRG sonicates at the same Mr (23.5 kDa) (Ball et al., 1992). We also found that, as in the rat, the neonatal mouse submandibular gland contains an immunoreactive protein at apparently identical mobility (Matsuura et al., 1992). A mouse PSP cDNA probe (Clone PS5, generously provided by Dr. Phil Shaw, Shaw and Schibler, 1986) on Northern blot hybridization under moderately stringent conditions, recognized a 1-kb transcript in adult rat PRG and in neonatal rat SMG. Hybridization selection of neonatal RSMG mRNAs using this probe, followed by *in vitro* translation, produced a translation product immunologically reactive with antibodies to SMG-Bl and to PRG-M1. Sequence analysis of cDNA clones encoding SMG-A and M1 indicated that the latter has extensive homology with mouse PSP throughout its 5′ untranslated protein coding and 3′ untranslated regions. Clearly, M1 is the rat homologue to mouse PSP and should henceforth be referred to as rat PSP (RPSP; Mirels and Ball, 1992).

SMG-A is homologous to RPSP, having greatest identity in its signal peptide and 3′ untranslated region, suggesting that the two proteins are differently regulated members of a multigene family. Transcripts homologous to RPSP and SMG-A, but more closely related to the latter, were identified in rat sublingual gland by northern blot analysis, consistent with our previous report of B1-immunoreactive proteins in both neonatal and adult SLG (Ball et al., 1988a), and suggesting that still other members of this gene family may be discovered (Mirels and Ball, 1992; Mirels et al., 1992).

Acidic Epididymal Glycoprotein. The major PRG secreted protein, M2, was eluted from a DEAE-Sephadex column with 0.2 *M* NaCl, along with acidic proline-rich proteins (Keller et al., 1975). On SDS-PAGE, M2 has an apparent Mr of 31 kDa, and in Western blots shows no immune cross-reactivity with the

Bl-IP of the neonatal SMG (Ball et al., 1992). Recently, this protein has been found to be identical to acidic epididymal glycoprotein (AEG; Charest et al., 1992). AEG is encoded by two allelic genes, transcribing two distinct mRNA species of greater length in the epididymis than in the parotid gland (Charest et al., 1988, 1992).

The expression of AEG is androgen-dependent in the epididymis (Charest et al., 1988). In the parotid gland it is not androgen-dependent, but is down regulated in response to chronic isoproterenol treatment, along with DNAase, amylase, and RPSP (M1) (Robinovitch et al., 1977; Naiman et al., 1992). During parotid gland development, AEG mRNA is first evident at 20 days postpartum; this is close to the time of appearance of amylase mRNA transcribed from the strong (parotid) promoter in the mouse (Shaw et al., 1985).

The function of AEG is not known. It has sequence homology with three metal-binding proteins and it has been speculated that, in the epididymis, the metal-binding function may play a role in sperm maturation (Charest et al., 1988).

There is no information on the possible function of parotid gland AEG. It is, however, a prominent secretory component; DEAE-Sephadex Fraction V, consisting almost entirely of AEG, constitutes about 10% of total parotid salivary protein (Keller et al., 1975).

Sublingual Gland

The sublingual gland of the rat is a mixed gland, having mucous tubuloacini with associated serous demilunes (Young and van Lennep, 1978). Developmentally, the pattern of cellular differentiation is established early in fetal life, and by 18 days postconception both serous and mucous cells are evident. The basic structural arrangement is well established by birth, with acini and demilune cells in an apparently functional configuration that is essentially the same as that seen in the adult (Redman and Ball, 1978). At 5 days postpartum, the serous demilune cells respond to β-adrenergic agonists by secreting proteins that are the same as those found in the adult (Ball et al., 1988a). With the exception of the major mucin, the secretory products of the gland have not been isolated or well-characterized.

Mucins

Typically, sublingual and submandibular glands of mammals have mucous glycoproteins. These have a high content of serine and threonine residues that are O-glycosylated, having many carbohydrate chains terminating in sialic acid (Herp et al., 1979). The major sublingual mucin was purified by Moschera and Pigman (1975) and found to have a molecular weight of 2.2×10^6 Da with 81% carbohydrate content, 40% ser and thr residues, and 49 mol sialic acid per 100 mol amino acids; no sulfate was detected. A minor mucin of similar composition also was reported. Subsequent analysis of carbohydrate structure characterized 5 oligosaccharides containing between 9 and 15 sugar residues, all containing sialic acid (Slomiany and Slomiany, 1978). Slomiany and co-workers have since reported two oligosaccharides of five and seven residues that can be sulfated by an endogenous sulfotransferase. This report did not, however, document the presence of sulfate in the native protein (Slomiany et al., 1988a). The sublingual mucin has covalently bound lipid as well as lipid that is bound hydrophobically to nonglycosylated regions (Slomiany et al., 1985). Further studies suggest that this lipid may determine mucin hydrophobicity (Slomiany et al., 1988b). Interestingly, both initial O-glycosylation and palmitic acid acylation are early events in mucin biosynthesis, occurring cotranslationally in the endoplasmic reticulum (Slomiany et al., 1988c). In the mouse sublingual gland, purification of mucin yields a product very similar to that of the rat in amino acid composition, carbohydrate content, and percentage of sialic acid. As in the rat, the mouse SLG molecule is larger than that of the SMG. While molecular weight estimates were not presented, sedimentation values were given as s20, w = 10.9 at 1.5 mg/ml mucin and 6.7 at 3.5 mg/ml; the latter figure was essentially the same in the presence of SDS (Roukema et al., 1976); little sulfate was reported for this mucin.

The rat SLG acinar cells secrete mucin primarily in response to cholinergic stimulation (Culp et al., 1991). This is in contrast to acinar cells of the PRG (Abe and Dawes, 1978; Ikeno et al., 1983) and SMG (Quissell and Barzen, 1980) which respond most actively to β-adrenergic stimuli. SMG granular duct cells are stimulated most strongly by α-adrenergic agonists (Murphy et al., 1980; Maitra et al., 1986).

Acinar Proteins Related to Neonatal Submandibular Components

Protein SMG-D.   The mucous acinar cells contain material reactive with antibodies to Protein D, which was originally described in the neonatal submandibular gland (Ball and Redman, 1984; Ball et al., 1988a). Protein D was subsequently purified from 5-day old SMG and its distribution was characterized by immunocytochemistry (Ball et al., 1991). Protein D localizes in the content of the secretion granules of the mucous cells and also is seen in serous demilune cells, where it is associated with the membranes of secretion granules (see below). The presence and localization of Protein D is essentially identical in 5-day neonatal and adult glands.

Protein SMG-C.   Still another protein from the neonatal SMG (Protein SMG-C, 89 kDa) is cross-reactive with protein components present in the content of secretion granules in SLG mucous cells. On electrophoretic immunoblots of sublingual gland extracts, reactive material is seen around 89 kDa in a somewhat diffuse pattern similar but not identical to that seen in the SMG (Ball et al., 1988b).

Proteins of the Serous Demilunes

Relatively little is known about the secretory products of the demilunes. They do contain proteins detectable by their immunoreactivity with antibodies to Protein Bl of the neonatal rat SMG (Ball et al., 1988a). The major species, at 27 and 18.5 kDa, are not identical to the major Bl-immunoreactive proteins (Bl-IP) of the neonatal SMG. Minor components, however, may be identical to one or more of the Bl-IP. These proteins are not present in the mucous acinar cells; they too, are present in both neonatal and adult cells. Smith and Toms (1986) have reported a glucagon-like imunoreactivity in the serous demilune cells of the adult SLG.

Submandibular Gland

The submandibular gland of rodents is complex, both in its adult cellular organization and in its developmental history. The acini of the adult gland are composed of mucous cells, but this condition is preceded by a transient but functional neonatal stage, in which the acini are serous and consist of two cellular phenotypes, one of which differentiates into the mature mucous cells (Strum, 1971; Yamashina and Barka, 1972, 1973; Chang, 1974; Cutler and Chaudhry, 1974; Ball and Redman, 1984; Ball et al., 1988a,b; Moreira et al., 1990, 1991).

The adult SMG of rats and mice also contains a unique secretory cell type in the granular ducts (also called granular convoluted tubules); these are modified striated ducts which have differentiated postnatally into serous exocrine cells that synthesize and secrete a number of growth factors, proteases, and hormones (Sreebny et al., 1955; Jacoby and Leeson, 1959; Cutler and Chaudhry, 1975; Gresik and MacRae, 1975; Srinivasan and Chang, 1975; Gresik, 1980; Murphy et al., 1980). In mice, these ducts are known to be markedly sexually dimorphic, showing more abundant histological profiles and having a higher level of differentiation in males than in females (Gresik and MacRae, 1975). In rats, this histological dimorphism is not evident (Sreebny et al., 1955; Cutler and Chaudhry, 1975). The duct system of the submandibular gland is further complicated by the presence of intercalated ducts (ID) that have secretion granules (Qwarnström and Hand, 1983), and which contain proteins not found in acinar or granular duct (GD) cells (Smith and Toms, 1986; Ball et al., 1988a,b; Moreira et al., 1990a).

Proteins of the Acinar Cells

Mucin.   Characteristic of the acinar cells of the rat is a mucous glycoprotein containing numerous oligosaccharides that are O-linked to serine and threonine residues. Tabak and co-workers (1985) have reported the isolation of a major mucin, the most complete side chains of which manifest blood group A reactivity (Tabak et al., 1985; Moreira et al., 1989). This mucin has a fairly typical chemical composition, with high levels of serine, threonine, and proline residues; 64% of the total molecular mass is carbohydrate, predominantly GalNAc, gal, fucose, and sialic acid; the molecular weight was estimated to be 103 kDa (Tabak et al., 1985). Previous reports have described purified mucins considerably different from this in composition (Malinkowsky and Herp, 1981; Flemming et al., 1982) and in size (Flemming et al., 1982). It may be the case that this reflects the existence of more than one mucous glycoprotein, since the report of Tabak et al. (1985) described the resolution of several mucin-like fractions. The blood group A mucin is representative of a relatively simple class of mucins, in contrast to larger, more complex mucins like that of the rat sublingual gland. It has been suggested that the former class functions primarily to control the oral microflora and the latter by binding to the hydroxyapatite of the tooth surface (Tabak, 1990). A monoclonal antibody with specificity for blood group A reacted with only acinar cells in the SMG, and failed to label the SLG (Moreira et al., 1991). This mucin appears to be quite different from the SLG mucin, the latter having much more GlcNAc, more sialic acid, more total carbohydrate, less fucose, and a molecular weight of 2.2×10^6 Da (Moschera and Pigman, 1975). It is anticipated that the monoclonal antibody to the SMG mucin would label other mucins having blood group A reactivity.

The mucin purified from the SMG of the mouse differs from the rat SMG mucin, having no (Denny and Denny, 1980) or little (Roukema et al., 1976) fucose, more sialic acid, a significant amount of GlcNAc, and a significant amount of mannose (Roukema et al., 1976; Denny et al., 1980). A molecular weight of 140 kDa was reported based on SDS-PAGE (Denny and Denny, 1980), indicating that the mouse SMG mucin, like that of the rat, is one of the smaller mucins. Denny and Denny (1982a) have subsequently provided evidence that the mannose present in this mucin is present in an N-linked oligosaccharide chain, an unusual feature of this mucin, which also has the more typical O-linked oligosaccharides.

A polyclonal antibody raised to mouse SMG mucin detected reactivity by immunofluorescence only in the mucous acinar cells of the SMG and not in the duct system. No reaction was seen in SLG, trachea, stomach, or intestine cells, or in the SMG or SLG of the rat (Denny and Denny, 1982b). A later report using a quantitative RIA provided similar data, except that incomplete reactivity was found in duodenal tissue (Denny and Denny, 1984). Like the mucin of the rat SMG, the mouse SMG mucin appears to be a cell-specific protein.

Glx-Rich Proteins (GRP).　　An abundant population of mRNA molecules in the rat SMG encodes a family of GRP (Heinrich and Habener, 1987; Mirels et al., 1987). The native proteins were found to contain 35% Glx, 10% Asx, and 14% Pro, with no carbohydrate. Two cDNAs encoding constitutively expressed GRPs have been cloned and sequenced. The most striking feature of the sequence is that about half of the protein coding sequence consists of a 23-residue, 50% Glx, tandem repeat which is repeated 5 times. (Mirels et al., 1987). On SDS-PAGE, several bands were visualized, a prominent doublet being present at about 40 kDa Mirels et al., 1987; Moreira et al., 1991). On the basis of mRNA size and primary composition, however, the estimates of Mr derived from SDS gels were anomalously high. Northern blot analysis showed two mRNA species at 950 and 1050 nucleotides, and *in vitro* translation yielded two primary translation products. The GRP represent a multigene family consisting of at least two separate genes (Heinrich and Habener, 1987; Cooper et al., 1991; Cooper and Tabak, 1991). Interestingly, the two known GRP protein species are differently regulated by isoproternol: GRP-Ca transcripts decrease and GRP-Cb levels increase on chronic isoproterenol treatment (Cooper et al., 1991).

The secreted GRP proteins have no homology with PRP, but have similar structural features; approximately half of the coding sequences consist of a tandem repeat, and the proteins consist primarily of a few amino acids. On the other hand, PRP genes show extensive homology with GRP genes in the signal sequence, with 91% of the nucleotides and 92% of the encoded amino acids being identical. GRP were found to bind hydroxyapatite with high affinity, suggesting a possible function in contributing to the enamel pellicle (Mirels et al., 1987). Hay (1973) has reported that the PRP from human saliva bind to hydroxyapatite; thus, it appears possible that the GRP and PRP have at least one similar function.

As is the case with mucin, GRP are found in the acini but not in the ducts or interstitial tissue of the SMG, nor are they found in the sublingual gland (Heinrich and Habener, 1987; Moreira et al., 1991). The report of Heinrich and Habener (1987), while clearly documenting the specific localization of GRP protein and mRNA to the acinar cells, misidentified these as serous cells and, similarly, misidentified the granular ducts as mucous cells. Northern blot analysis with a cDNA probe showed no GRP mRNA in sublingual gland, parotid gland, or liver (Mirels et al., 1987); further studies have shown the absence of GRP transcripts in lacrimal gland, stomach, small intestine, large intestine, kidney, trachea, liver, or Harderian gland (Mirels et al., 1990). GRP transcripts were not detected in the SMG of the DBA mouse or other mammalian species, presenting another striking example of taxonomic variability among salivary proteins (Mirels et al., 1990). It appears, then, that the GRP may be cell-specific proteins of rat SMG acinar cells, presenting an interesting comparison with the more widely expressed PRP.

Cystatin.   In 1974, Menaker and co-workers described, in electrophoretic gels, a protein from isoproterenol-treated rat SMG and called this LM (large mobile) protein. This protein was orginally purified and characterized by Naito and co-workers (Naito, 1981, 1982; Naito and Suzuki, 1981; Naito and Saito, 1982). It was found to be very different in its amino acid composition from the PRP, having very little proline or glycine (Naito and Suzuki, 1981). In rats 6 days and older, antibodies to LM revealed reactivity in cells of the developing and mature acini after chronic isoproterenol stimulation; some reactivity was seen in the developing SMG acini of untreated rats between 6 and 21 days old, and some reactivity was reported in the granular ducts of uninduced adult rats (Yagil et al.,

1986). A cDNA complementary to LM mRNA has been cloned and sequenced, a high degree of homology to Family 2 of the cystatin superfamily was found. The protein was shown to inhibit the cysteine proteinases papain and ficin, but not kallikrein, a serine protease (Shaw et al., 1988). This cystatin was purified by Bedi (1989a) and the amino acid sequence reported (Bedi, 1989b). In agreement with the findings of Shaw et al. (1988), the protein inhibited the cysteine proteases papain and ficin, but not the serine proteases kallikrein and trypsin, and amino acid sequence homologies also showed this cystatin to be a member of Family 2. The amino acid composition is relatively high in Glx (18%) and Asx (13%); no carbohydrate was detected. It was reported that on Western blots, reactive bands were seen only in the SMG, SLG, and slightly in the PRG, but not in liver, spleen, pancreas, lacrimal glands or plasma (Bedi, 1989a). This cystatin is similar to the PRP in its regulation by isoproterenol and in its distribution in more than one salivary gland, and different in these respects from the mucin and GRP.

"Spot" Proteins.　One of the most abundant mRNA species of the rat encodes a protein of 6370 Da called the "spot" protein (Dickinson et al., 1989b). This protein is homologous to similar proteins in various species and strains of mice that show divergence in both size and composition, despite their evident homology. A number of cDNA clones for these proteins have been sequenced and the amino acid compositions deduced. All have a characteristic leader sequence of 18 amino acids with an apparent N-terminal cysteine residue, and thus they appear to be secretory proteins. The proteins are acidic, with positively charged residues in the central to C-terminal region and hydrophobic residues clustered near the C-terminus. On SDS-PAGE, the spot proteins migrate at anomalously high Mr. For example, the primary translation product of the mouse spt-l gene shows an Mr of 34 kDa on SDS-PAGE, and a deduced Mr of 11,603 Da (Dickinson et al., 1989b). In the mouse, one of these proteins, SPT-l, is expressed at significant levels in the lacrimal gland, while a second, SPT-2, is not; both may be expressed at a low level in the SLG; neither is expressed in the PRG or in lung, small intestine, kidney, or liver. All spt genes appear to be clustered at a locus on chromosome 15 (Dickinson et al., 1989a).

Proteins of the Granular Ducts

A large number of bioactive factors have been reported present in the submandibular glands of rodents (see Barka, 1980); only a few for which there is substantial information on cellular organization and distribution, biochemical properties and/or genomic organization, will be covered here.

Amylase.　While amylase catalytic activity can be detected in the rat submandibular gland (Schneyer and Schneyer, 1960; Junqueira et al., 1964; Ball, 1974b) its specific activity is 39,000-fold less than in the parotid gland, and more than 2-fold less than in the lung, which does not secrete digestive enzymes (Ball,

1974b). The adult submandibular level is about the same as that of the 16-day embryonic submandibular and parotid glands, which have just begun their branching morphogenesis (Ball, 1974b). There is no evidence that SMG amylase activity is released in response to secretory agonists, nor is there enzyme-cytochemical or immunocytochemical documentation of its localization. There is no reason to think that the amylolytic activity detectable in rat submandibular glands represents a secretory amylase encoded by the family of genes that includes parotid and pancreatic amylase (Crerar et al., 1983).

In the mouse, amylase specific activity in the SMG has been reported to be present at levels 200,000-fold less than in the parotid gland (Schneyer and Schneyer, 1960). Junqueira et al., (1964) have found amylase specific activity to be ten-fold higher in mouse than in rat SMG. The increase in amylase with postnatal development of the male mouse SMG has been reported to be about 20-fold (Gresik and MacRae, 1975). Amylase levels have been found to be sexually dimorphic, being higher in males than in females, and the onset of this dimorphism is approximately coordinated with the differentiation of the granular ducts (Smith et al., 1971; Gresik and MacRae, 1975), consistent with the finding, by starch substrate histochemistry, that amylase activity is localized in granular ducts. No activity was seen in sublingual glands, exorbital lacrimal glands, skeletal muscle, tracheal glands or brown fat (Smith et al., 1971). Amylase has been purified from the SMG of C3H mice, showing a peak at an apparent Mr of 60 kDa on Bio-Gel P-lOO chromatography (Frati and Caputo, 1970). No further characterization of the mouse SMG amylase activity is available.

Nerve Growth Factor.   Originally purified from snake venom (Cohen, 1959), nerve growth factor (NGF) was subsequently obtained from the submandibular glands of male mice (Cohen, 1960). Its localization in the granular ducts has been documented (Schwab et al., 1976). Like other secretory components of the granular ducts it is androgen regulated, appearing around the time of puberty and reaching much higher levels in males than in females (Gresik, 1980). The native mouse NGF is a 7S protein complex containing three subunits. Beta subunits have NGF activity while α and γ subunits do not (Varon et al., 1967, 1968). The γ subunit was shown to have potent esterase activity against BAEE (Greene et al., 1968, 1969). It was found to be a serine protease highly homologous to glandular kallikrein (Bothwell et al., 1979), and is encoded by a member of the kallikrein multigene family located on chromosome 7 (Mason et al., 1983; Shine et al., 1983). The βNGF precursor can be correctly cleaved to the active form of NGF by the γ subunit and also by trypsin (Edwards et al., 1988). Although NGF is made in many tissues, the submandibular differs in having a longer βNGF precursor encoded by a differently spliced mRNA template; this mRNA predominates only in the submandibular gland (Edwards et al., 1986; Selby et al., 1987).

Epidermal Growth Factor.   EGF was originally isolated from the SMG of male mice (Cohen, 1962) and has since been identified in rats; in both species it is localized in the granular ducts (Turkington, 1971; Gresik et al., 1979, 1985). Like NGF, EGF appears around the time of puberty, showing a pronounced sexual dimorphism in the mouse, and a lesser degree of sexual dimorphism in the rat (Mori et al.,1983). In the mouse, EGF is a polypeptide of 53 amino acid residues and Mr of 6,045, and has been sequenced (Savage et al., 1972). Native EGF exists as a complex (74 kDa) of two molecules of EGF and two molecules of binding proteins (29 kDa each; Taylor et al., 1970, 1974). The mRNA for the precursor EGF has been identified and its sequence determined. It is unexpectedly large and predicts a preproEGF of about 1200 amino acids and Mr of about 130 kDa (Gray et al., 1983; Scott et al., 1983). Examination of the sequence suggested the possibility that there might be seven cryptic peptides, similar but not identical to EGF, encoded by the region upstream from the EGF gene, as well as possibly other peptides formed from the remaining 66% of the mRNA nucleotide sequence (Scott et al., 1983). Both preproEGF and mRNA have been detected in many organs, with the highest level, by orders of magnitude, being found in the SMG. Significant quantities of preproEGF mRNA are also seen in the kidney, and lesser amounts in the pancreas, from which EGF is probably secreted (Rall et al., 1985). In the kidney, only the 130 kDa form is found, leaving unclear the origin of the 6- and 30-kDa species that are found in urine (Carpenter and Cohen, 1979). Beerstecher et al. (1988) were unable to detect EGF in the mouse kidney immunocytochemically, using a monoclonal antibody to mouse EGF, suggesting that the relevant epitope is blocked in the 130-kDa precursor.

Like other granular duct proteins, EGF is stimulated by testosterone and also by thyroid hormones in the SMG. Similarly, levels of preproEGF mRNA are dramatically increased in the SMG of female mice after administration of testosterone or T3, while these have no stimulatory effect on mRNA levels in the kidney (Gubits et al., 1986). Despite the difference in regulation, no indication of differential processing of the primary transcript was seen, both SMG and kidney showing a discrete mRNA band of the same size, indicating that regulation by these hormones may be entirely at the level of transcription.

Renin.   The best known source of secreted renin is the kidney, from which it performs the important function of catalyzing the hydrolysis of angiotensinogen to angiotensin I, prior to its hydrolysis to the vasoactive peptide angiotensin II. Renin is also found in the submandibular gland of rats and mice (Barka, 1980) where it is localized in the granular ducts, showing the usual sexual dimorphism of granular duct secretory proteins (Gresik et al., 1978).

The cellular pathway of renin synthesis and processing has been well documented in mice. Studies from three laboratories have shown that preprorenin (45 kDa) is converted to prorenin (43 kDa) by microsomes, and then rapidly processed to single-chain renin (38 kDa), which is biologically active; following this, a slow conversion in the secretion granules results in the excision of an arg-arg dipeptide and the resultant conversion to a two-chain renin, with 33- and 5-kDa polypeptides joined by a disulfide bond (Panthier et al., 1982; Catanzaro et al., 1983; Pratt et al., 1983). Both one-chain and two-chain forms are secreted and, surprisingly, the one-chain renin is six-fold more active than the two-chain form; moreover, one-chain renin is secreted rapidly and constitutively, while two-chain renin is stored in secretion granules which are released on phenylephrine stimulation (Pratt et al., 1983, 1988).

Although the kidney and SMG renins are functionally and structurally similar, they are not the products of the same gene, nor are they identical. The Ren 1 gene, present in all subspecies of mice, encodes the glycosylated, thermostable kidney renin, while the unglycosylated, thermolabile renin of high producer strains is encoded by a second gene, Ren 2. Ren 1 is expressed in kidney and at a low level in submandibular gland. Ren 2 is expressed in SMG and not in kidney; both are testosterone inducible. The genomic DNAs containing each gene have been cloned and sequenced; they are highly homologous in the protein coding region (96%) and up to 179 bases upstream from the start codon, and nonhomologous in the remaining 5′ nontranslated region (Holm et al., 1983; Panthier et al., 1984).

Proteases.   The rat submandibular gland has long been known to have high levels of protease activity, which increases postnatally from very low neonatal levels to high levels in the adult; the major increases accompany puberty and the concurrent differentiation of the granular ducts (Sreebny et al., 1955; Cutler and Chaudhry, 1975; Denny and Cope, 1977). As is the case for some other granular duct components, the protease levels were sexually dimorphic, being higher in males than in females (Sreebny et al., 1955). Riekkinen and co-workers have reported the presence of a trypsin-like protease (glandulain) which is androgen-induced (Riekkinen and Niemi, 1968) and another protease (salivain) which is not (Riekkinen, 1966; Riekkinen et al., 1966).

In contrast to the SMG, low levels of protease are found in both the sublingual and parotid glands, measuring 1/200 and 1/2000 of the SMG level, using benzoyl arginyl β-naphthylamide as substrate (Barrett and Ball, 1974). An array of proteases are found in the rat SMG; most are trypsin-like in their substrate specificity (Riekinnen and Hopsu, 1965); a chymotrypsin-like protease has been reported (Riekinnen and Hopsu, 1966), and several proteases are part of the kallikrein family of esteroproteases (Ekfors et al., 1967; see Orstavik, 1980; Shine et al., 1983).

The kallikreins are a family of serine proteases able to cleave kininogen, a plasma protein, releasing kinin, a vasoactive peptide (see Schachter, 1980). Kallikrein immunoreactivity has been localized to the granular ducts and to a lesser extent, the striated ducts of the SMG; in the SLG and PRG, much lower levels of activity are found in the striated ducts. In all three glands, kallikrein is secreted primarily by α-adrenergic stimuli. In the kidney, reactivity is localized to the distal tubule, and in the pancreas to acinar cells (Orstavik, 1978).

During development of the rat SMG, kallikrein was measured enzymatically against BAEE and with antiserum to urinary kallikrein, and increased as expected for a granular duct component, rising sharply at about the time of puberty (Orstavik et al., 1977). Unlike protease activity measured against protein substrates, however (see Sreebny et al., 1955), no difference between sexes was observed.

Three trypsin-like proteases from mouse SMG, NGF γ subunit, βNGF endopeptidase and β binding protein were found to have similar catalytic properties and immunological relatedness to kallikrein; indeed, βNGF endopeptidase appeared identical to submandibular gland kallikrein (Bothwell et al., 1979). This raised the possibility that the three enzymes are members of a larger family of enzymes, others of which may also be involved in the processing of hormones and growth factors. A cDNA clone encoding rat submandibular kallikrein was used to detect 34 clones, with varying sequence homology, from a rat genomic library (Chao et al., 1986). In the mouse, a cDNA probe with extensive homology to kallikrein identified clones that suggested 25 to 30 kallikrein genes. All genes detected are located on chromosome 7. The sequences of three of these predict amino acid sequences highly homologous to NGF, EGF binding protein, and renin, again suggesting the likelihood of an extensive family of polypeptide-processing enzymes (Mason et al., 1983; Shine et al., 1983). Consistent with this, Ashley and MacDonald (1985) determined the nucleotide sequence of four rat kallikrein-related mRNAs. One (PS) appeared to encode kallikrein and a second (Sl) shared 86% amino acid sequence homology with PS; another (S2) encodes tonin, a serine protease which directly cleaves angiotensinogen to produce angiotensin II (Lazure et al., 1984); the fourth predicted enzyme (S3) has 84% homology to tonin (Ashley and MacDonald, 1985). Recently, Shai et al. (1989) have reported the complete nucleotide sequence for a tonin gene, and found it identical to the previously described S2 (tonin) gene (Ashley and MacDonald, 1985), except for a one-base mismatch. They describe a second gene (RSKG-50) more closely related to kallikrein. There thus appear to be two groups of genes, one group more closely related to tonin and the other group more like kallikrein. Within each group, the sequence identity among different genes is as high as 85%. Two genomic clones for kallikrein-like genes have been identified in the kidney, one falling into each of the above groups (Chen et al., 1988).

A number of other possible members of the kallikrein protease family have been identified. In a study measuring catalytic activity against α-N-benzoyl-DL-arginine p-nitroanilide seven charge isomorphs were detected in the developing rat SMG, 2 appearing at 4 weeks and the rest appearing by 8 weeks; no sex difference was seen (Ikeno and Ishiguru, 1984). It was suggested, based on the early appearance of two peaks, that these protein species might be localized in acinar cells. One of the later peaks was subsequently purified and found to be similar to tonin but specifically different from both tonin and kallikrein (Ikeno et al., 1986). Several other trypsin-like proteases similar to kallikrein and/or tonin have been isolated and partially characterized (Khullar et al., 1986; Barlas et al., 1987; Kato et al., 1987; Xiong et al., 1990). The possible relationship of these to the kallikrein gene family remains to be established.

Most of the well-characterized products of the granular ducts appear to be bioactive molecules or proteins involved in processing or binding bioactive molecules. This is a very different secretory system than that of the acinar cells.

There appear to be no well-characterized secretory proteins known to be present in both types of secretory cells. This is curious, in view of the prevailing opinion that in the adult rat, the intercalated ducts have progenitor cells for cellular replacement in both secretory compartments. It also is of interest that none of the unique secretory proteins of the neonatal SMG appear to be present in the GD or in their progenitor ductal elements during neonatal development (Ball et al., 1988a,b).

Secretory Proteins of the Neonatal Acinar Cells

Embryonic and neonatal stages of the SMG are characterized by the presence, in the immature secretory end pieces, of two distinct types of exocrine cells. The cells with uniformly electron-dense granules have been called terminal tubule cells by Yamashina and Barka (1973), and Type I cells by Ball and Redman (1984), based on the classification of their granule type by Cutler and Chaudhry (1974). A second cell type was called proacinar cells (Yamashina and Barka, 1973) and Type III cells (Ball and Redman, 1984; Cutler and Chaudhry, 1974). The Type III secretion granules have electron-lucent substructures in a moderately electron-dense matrix and are reactive for peroxidase, in contrast to the unreactive Type I cells (Strum, 1971; Yamashina and Barka, 1973; Cutler et al., 1981).

Peroxidase.   Endogenous peroxidase activity has been described in the salivary glands of a variety of mammals including humans, rats, and mice (Herzog and Miller, 1970; Strum and Karnovsky, 1970; Novikoff et al., 1971; Carlsoo et al., 1971; Osugi, 1977). A peroxidase found in bovine SMG, SLG, and PRG was purified and found to be a lactoperoxidase (Morrison et al., 1965). In the developing rat, SMG peroxidase activity was used by Yamashina and Barka (1972, 1973, 1974) to monitor secretory differentiation during fetal and postnatal development. By 19 days postconception, peroxidase reactivity was seen in secretion granules of cells which were to become Type III (proacinar) cells. The specific activity was highest at 3 days postpartum and then decreased markedly to 14 days before increasing again to levels at 42 days that were somewhat lower than the neonatal peak (Yamashina and Barka, 1974). Curiously, with maturation of the acinar cells, secretion granules no longer exhibited peroxidase activity and at older stages only the rough ER cisternae were reactive. It is not clear whether this indicates the presence of different species of peroxidase, differential susceptibility to fixation denaturation or some other complication. Quantitative assays indicate that peroxidase activity is secreted in response to isoproterenol-stimulation both in the neonate and at 42 days of age (Yamashina and Barka, 1974; Cutler et al., 1981), although release of the enzyme was less at 42 days than in the neonate (Yamashina and Barka, 1974). A quantitative study of peroxidase activity in the mouse showed reactivity in the SMG, SLG, and PRG, using o-dianisidine as substrate, but measurement of peroxidase oxidation of thiocyanate revealed reactivity only in the SMG and PRG, and not in the SLG (Osugi, 1977). This, too, is consistent with the possibility of there being more than one type of peroxidase in the salivary glands. Resolution of these questions will require the isolation and characterization of the murine peroxidases and the preparation of specific antibodies.

Amylase and Ribonuclease. These enzymatic activities are extremely low in adult rat SMG, and probably represent some irreducible general tissue level of amylolytic and nucleolytic intracellular enzymes unrelated to the secretory enzymes found in the parotid gland. In 4-day old rats, however, RNAase (Menaker and Miller, 1973; Ball, 1974b) and amylase (Ball, 1974b) activities accumulate at significant levels; for RNAase the specific activity is 54-fold, and for amylase, 16-fold higher than in the adult (Ball, 1974b). Subsequent studies showed that these enzymes were secreted on β-adrenergic stimulation (Ball and Nelson, 1978a), indicating that they are content proteins of the secretion granules of Type III (proacinar) cells (Yamashina and Barka, 1972; Cutler et al., 1981; Ball and Redman, 1984).

The presence of these "parotid-specific" enzymatic activities in the neonatal submandibular gland raises the issue of the identity of these protein molecules. A subsequent study, based on an analysis of the catalytic properties of the several salivary RNAases, provided the suggestion that the adult SMG RNAase is different from both the neonatal RNAase and the adult PRG enzyme; however, the limitations of the methods used prevented a definitive conclusion (Ball and Nelson, 1978b). The relationship of the neonatal SMG enzymes to the well-known amylase and RNAase of the adult PRG presents an interesting question. Approaching this question probably will be most convenient for amylase, since the biochemistry of the parotid enzyme is well understood, and considerable information is available on the arrangement, number, and type of genes in the amylase multigene family (Crerar et al., 1983). The existence of still another source of secreted amylase in the serous lingual glands (Field and Hand, 1987; Field et al., 1989; see below), adds another dimension to this question. Since reasonable quantities of the lingual amylase can be obtained (Field et al., 1989), an immunochemical approach is feasible; with molecular analysis of the amylase genes it may be possible to assign the lingual amylase and the neonatal SMG enzyme to one or more of these genes, or to document their being encoded elsewhere.

Bl-Immunoreactive Proteins (Bl-IP). Bl-IP are proteins reactive with polyclonal antibodies to Protein SMG-Bl, and are the predominant secretory products obtained on isoproterenol-stimulation of neonatal glands; they consist of three major species, SMG-A (23.5 kDa), Bl (26 kDa), and B2 (27.5 kDa) (Ball and Nelson, 1978a; Ball and Redman, 1984; Ball et al., 1988a). In the fetal and neonatal SMG, these proteins are localized exclusively in Type III cells, and they disappear as these differentiate into mature mucous cells (Moreira et al., 1990,

1991). The function of the Bl-IP is yet unknown, and they have been characterized only in terms of electrophoretic, chromatographic and immunochemical
parameters (Ball and Redman, 1984; Ball et al., 1988a, 1992). The presence of
similar and/or identical proteins in the adult sublingual and parotid glands is one
of the most interesting aspects of their existence, since it indicates their
differential regulation among the salivary glands, with each gland showing an
unique pattern of expression (described above). Perhaps most importantly, the
occurrence of Bl-IP as transient gene products of the immature SMG acini, and
as major adult components in the SLG and PRG, may help to identify essential
regulatory elements involved in cell specific gene expression.

In considering the complex pattern of Bl-IP expression and localization,
three further complications must be stated here; these are described more
completely below.

1. Recently, we have found that occasional groups of acini containing Bl-
   IP reactivity are seen in adult glands (Man et al., 1992; see below, this
   section).
2. Following Bl-IP disappearance from the maturing acinar cells some
   highly reactive cells remain in the intercalated ducts (see below; Ball et
   al., 1988a,b; Moreira et al., 1990).
3. While a survey of adult organs showed no Bl-IP in stomach, intestine,
   pancreas, thymus, muscle, brain, testis, adrenal, pituitary or thyroid (Ball
   et al., 1988a), low levels of activity were seen in the exorbital lacrimal
   glands, tracheal glands and in serous lingual (von Ebner's) glands,
   primarily in the ducts (Hand and Ball, unpublished data; see below).

Protein SMG-C.   Cholinergic stimulation of the neonatal SMG elicits the
appearance of Protein SMG-C (89 kDa), which in contrast to the Bl-IP, is
exclusively localized in the Type I (terminal tubule) cells of fetal and early
neonatal glands (Ball and Redman, 1984; Ball et al., 1988b). The cell-specificity
of the Protein C and the Bl-IP immune reactivities is evident at their first
appearance, at 18 d and 19 days postconception, respectively (Moreira et al.,
1990). Following birth, Protein C reactivity (C-IR) begins to appear also in the
Type III cells, as these begin their differentiation into mucous cells, and with
maturation, low levels of C-IR remain in the mucous acinar cells. The onset of
C-IR in the Type III cells parallels both the disappearance of the Bl-IP, and the
accumulation of mucin and GRP during the course of Type III-mucous cell
differentiation (Moreira et al., 1991). The C-IR of Type I cells remains much
more intense as long as these cells are identifiable in the developing acini;
following this, strongly labeled cells are present in cells of the intercalated ducts,
particularly in those cells at the ID-acinar junctions (Ball et al., 1988b; Moreira
et al., 1990).

In the SLG and in the minor sublingual gland, low levels of C-IR are seen
in the mucous acinar cells (Ball et al., 1988b; see above). The adult PRG is a

purely serous gland and shows no C-IR; however, the neonatal PRG is known to have scattered mucous cells (Taga and Sesso, 1979) and these also show low levels of C-IR (Ball et al., 1988b). A survey of other adult organs showed no C-IR in mucous palatal glands, lacrimal gland, thyroid, trachea, esophagus, stomach, duodenum, colon, liver, kidney, pancreas, adrenal, prostate, seminal vesicle, skeletal muscle, or mucous lingual glands. Serous von Ebner's glands of the tongue showed slight reactivity in some of the ducts (Ball et al., 1988b).

Protein SMG-D.   In the neonatal SMG, a large protein (175 kDa) was secreted with both adrenergic and cholinergic stimulation, and thus appeared to be a product of *both* the Type I and Type III cells (Ball and Redman, 1984; Ball et al., 1988a). Subsequently, ultrastructural immunocytochemistry confirmed this assumption, and provided the additional finding that Protein D was membrane associated in the neonatal cells (Ball et al., 1991). This membrane association is more complete in Type I than in Type III cells, with the latter also having substantial Protein D signal in the content of the granules. During development, the membrane localization is seen at the first appearance of Type I secretion granules at 18 days postconception and of Type III granules at 19 days. With the postnatal maturation of Type III cells into mucous cells, the cells show much less Protein D labeling, and this label is in the content rather than associated with membranes. After the Type I cells disappear from the acini, Protein D label is seen in some cells of the ID, also localized to the granule content (Ball et al., 1991).

In the sublingual gland, Protein D label is strong in demilunes, where it is membrane associated, and in mucous acinar cells where it is content localized. Electrophoretic immunoblots show a reactive band at identical mobility to that of purified Protein D (Ball et al., 1991). In the parotid gland, reactivity is seen only in the neonate where it is localized to the scattered mucous cells; the adult is unreactive except for weak reactivity in some intercalated ducts (Ball et al., 1991; see below).

Reappearance of Bl-Reactive Acini in Adult SMG.   Earlier studies had found, using immunofluorescent and ultrastructural immunogold procedures, that with maturation of the acinar cells, Bl-IR was no longer seen, and that C-IR and D-IR were present at much lower levels in the adult mucous cells (Ball et al., 1988a,b; Moreira et al., 1990; 1991). Subsequently, a more extensive survey, using the light-microscopic ABC procedure, has revealed that clusters of acini reactive for the Bl-IP are occasionally seen in rats between 45 days and 9 months postpartum (Man et al., 1992). These differ from the neonatal acini, however, in that they do not have two cell types distinguishable by light microscopic immunocytochemistry; all the cells in these acini show Bl-IR, and none show the strong reactivity for Protein C seen in neonatal Type I cells. They also show the high levels of Protein D reactivity typical of neonatal Type III cells, and they lack reactivity for the GRP, which are found in abundance in the surrounding, more typical, adult acinar cells (Man et al., 1992). This is precisely the phenotypic profile of neonatal Type III cells, and is consistent with the hypothesis that these are acini that are newly differentiated from progenitor cells, perhaps the B1-reactive cells of intercalated ducts, rather than being developmentally retarded neonatal acini. The fact that the B1-positive adult acinar cells were seen as clumps of acini rather than as individual cells is particularly interesting in view of the proposal by Denny and co-workers (Denny et al., 1990b) that intercalated ducts may undergo an *en bloc* differentiation into secretory acini. The B1-reactive acini have not been seen and described electron microscopically due to their infrequency and the limited size of the fields available for electron microscopic visualization (Ball et al.; 1988a,b; Moreira et al., 1990).

Proteins of the Intercalated Ducts (ID) of the Major Salivary Glands

The acini of major salivary glands are drained by small ducts (ID) which, after joining with ID from other acini, connect the acini with the next ductal compartment. In the case of the parotid and sublingual glands, intercalated ducts empty into striated ducts; in the submandibular gland, the specialized granular ducts are interposed between the intercalated and striated ducts. The ID are longest in the parotid gland, intermediate in the SMG, and shortest in the SLG (Young and van Lennep, 1978). Evidence has been presented that the ID contain the stem cells for both acinar and granular duct cells during normal cellular replacement (Zajicek, 1985; Denny et al., 1990a,b) and after partial (Hanks and Chaudhry, 1971) or unilateral (Yagil et al., 1985) extirpation (see Denny and Denny, this volume). In the rat SMG, the intercalated duct cells often contain a few electron-dense secretory granules, different in morphology than the mucous granules of the acinar cells. At the junction of the ID and acini, another cell type is seen; it is larger than ID cells but smaller than acinar cells and has numerous secretion granules morphologically different than those of both acinar and ID cells. It was suggested that this cell might represent a stage in the differentiation of a progenitor cell for acinar and or intercalated duct cells (Qwarnström and Hand, 1983). Cells of similar location and appearance have been described in the SMG of the female mouse (Gresik and MacRae, 1975).

Hormones

Past reports have suggested that salivary glands might have an endocrine function (Godlowski, 1960; Barka, 1980); indeed, several laboratories have reported the presence of insulin-, glucagon-, and somatostatin-like molecules in salivary glands of humans and rats (Murakami et al., 1982; Bhatena et al., 1977). Evidence has been presented that the salivary insulin-like material is comparable to pancreatic insulin and that it is synthesized by parotid gland cells (Murakami et al., 1982; Smith and Patel, 1984). In an immunocytochemical study on insulin and glucagon-like peptides in rat salivary glands, Smith and Toms (1986) reported that insulin reactivity was seen in ID cells of the parotid gland; in the submandibular gland, reactivity was seen in cells at the acinar-ID junction, the apparent location of the granulated cells described ultrastructurally by Qwarnström and Hand (1983). The SLG showed no reactivity. With antiserum to the N-terminus of pancreatic glucagon, occasional labeled cells were seen in the acini of the PRG, and very scattered cells were seen in SMG acini. From the published pictures, both of these, as well, might be acinar-ID junctional cells. In the SLG, strong reactivity was seen in the serous demilunes. No cells were reactive with antibodies to the C-terminus of pancreatic glucagon, suggesting that the reactive component is glicentin or another glucagon precursor in which the C terminus is blocked, rather than the 3500 Da glucagon found in the pancreas.

No reactivity was seen, in any salivary gland, with antibodies to gastric inhibitory peptide, somatostatin, or pancreatic polypeptide (Smith and Toms, 1986).

Proteins Related to Neonatal Submandibular Components

Bl-IP and Protein C reactivities are seen in cells of the intercalated ducts in the adult SMG (see above). Simultaneous localization of both reactivities using Protein A-gold detection with differently sized gold particles showed, at the ultrastructural level, that the anti-C and anti-Bl antibodies labeled entirely different cells, never being colocalized in the same cell (Moreira et al., 1990). The cells that labeled with anti-C often appeared to be cells at the acinar-ID junction, and might well be the insulin-reactive cells of Smith and Toms (1986) and/or the granulated cells of Qwarnström and Hand (1983). It will be of interest to determine immunocytochemically, whether insulin and Protein C colocalize in the same cells.

Subsequently, it was found, using a light microscopic ABC technique on serial sections, that Protein D reactivity could be localized in either cell type (i.e., colocalization with Bl-IP or with Protein C; Man et al., 1992). The existence of at least two biochemically distinct cell types in the ID may be an important finding in view of the prevailing opinion that ID cells can differentiate into both acinar and granular duct cells. This is a particularly interesting hypothesis, since the acinar cells and the granular duct cells each have a number of cell restricted secretory proteins, but seemingly none in common. The Bl-IP, Protein C, and Protein D, for example, are not detectable in granular ducts (Ball et al., 1988a,b; 1991); moreover, the SMG mucin and Glx-rich proteins (GRP), which are major secretory proteins of adult acini, are found in no ductal compartment, including the ID (Moreira et al., 1991). All of this raises the question whether, if the ID are progenitor cells for both acinar and granular duct secretory compartments, they contain common progenitor cells for both acini and granular ducts, or different cells already committed to one or the other lineage. With respect to the Protein C and Bl-labeled ID cells, which is the important progenitor cell, or is still another cell type involved in cellular replacement?

MINOR SALIVARY GLANDS

Minor salivary glands are found in the submucosa throughout the oral cavity, (for general information and review, see Young and van Lennep, 1978; Hand, 1985; Klein, 1987). Most are mucous glands although the most extensive descriptions of secretory proteins are available for the serous von Ebner's glands of the tongue (see below).

Mucous Glands

Palatal and Buccal Glands

Sulfated glycoproteins have been reported in the palatal and buccal secretions of the rat and chromatographic separation produced a number of glycoproteins of varying molecular size, and extent and type of glycosylation. All major fractions showed blood-group A reactivity. The individual species have not been well characterized and it is not clear to what extent the mucous proteins of the buccal and palatal regions are similar to each other or to the mucins of the submandibular and sublingual glands (Green and Embery, 1987). An earlier study had shown, autoradiographically, intense uptake of radiolabeled sulfate over the posterior buccal glands of the rat (Davies and Young, 1954).

Anterior Buccal Gland

In the rat, a minor, purely mucous gland develops as a branch of the parotid duct (Stensen's Duct); the main duct of this anterior buccal gland (ABG) joins Stensen's Duct about 3 mm from the latter's oral terminus (Redman, 1972). Histochemically (Redman, 1972) and ultrastructurally (Nicholas and Redman, 1985), the ABG appears to be a fairly typical mucous gland. Its secretory products have not been characterized and the mode of regulation of its secretions is not known. It is mentioned here because its position relative to the parotid gland could mean that parotid saliva collected at the orifice of Stensen's Duct may have secretory components derived from the ABG. The common method of saliva collection, which involves cannulation of the parotid duct more than 1 cm "upstream" from the duct of the ABG, avoids this complication but also provides parotid saliva different from that which normally enters the oral cavity (Redman, 1972).

Serous Cells of the Tongue

Serous cells are prominent in the von Ebner's glands of the tongue as well as being present in serous demilunes associated with the acini of lingual mucous glands. The ultrastructure of these cells has been described in the rat (Hand, 1970).

Digestive Enzymes

Lingual Lipase. The best described component of the secretion of the rat lingual serous cells is a lipase which hydrolyzes triglycerides to mostly diglycerides and free fatty acids at pH 4.5 to 5.4 (Hamosh and Scow, 1973). Measurements of gastric lipolysis suggest that this lipase is the major digestive enzyme for gastric hydrolysis of the long-chain triglycerides of milk in the neonate (Liao et al., 1983). The lipase first appears in the late fetus; during early postnatal life, when digestion of milk is of paramount importance, it rapidly increases to functionally significant levels (Hamosh and Hand, 1978). The lipase has been purified; it shows an Mr of 51 kDa and a broad optimal pH between 4 and 6.5 (Field and Scow, 1983). It preferentially hydrolyzes long-chain triacylglycerol to diaglycerol and free fatty acids, and diacylglycerol and monoacylglycerol are cleaved at decreasing rates. The lingual lipase has 15% Asx and 12% gly residues and no cys; it has GlcNAc and GalNAc at 10 and 4 mol/100 mol, respectively (Field et al., 1989). Stimulation of secretion of lingual lipase is primarily cholinergic (Field and Hand, 1987).

Lipolytic activity was not found in the parotid, submandibular or sublingual glands or in the stomach wall; it is present at low levels, however, in the soft palate and oral pharynx (Hamosh and Scow, 1973).

Lingual Amylase. Along with lipase, an amylase is secreted by the serous von Ebner's glands in response to cholingeric stimulation (Field and Hand, 1987); the amylase has been purified and partly characterized (Field et al., 1989). On SDS-PAGE it has an Mr of 59 kDa and an isoelectric point of 5.0. The amino acid composition is similar to that of the parotid and pancreatic amylases, except that the lingual enzyme has more proline (molar ratio = 11%) and less valine (molar ratio = 4%); no hexosamine was detected.

Neonatal SMG Proteins: Possible Occurrence in Lingual, Lacrimal, and Tracheal Glands

Although not salivary glands, the serous lacrimal and tracheal glands are also considered here, because they are the only nonsalivary organs that have been found to have immune cross-reactivity with the major secretory proteins of the neonatal SMG.

Antibodies to the Bl-IP (Hand and Ball, unpublished), to Protein C (Ball et al., 1988b) and to Protein D (Ball et al., 1991) show some reactivity with cells of the serous von Ebner's glands of the adult rat, as detected by light microscopic immunofluorescence and electron microscopic immunogold labeling; the reactive components are associated primarily with ducts, the acini having no or little reactivity. Like the von Ebner's glands, the lacrimal glands have cross-reactivity with antibodies to the Bl-IP and Protein D; however, they are unreactive with antibodies to Protein C. Here, too, the reactive cells are found in the ducts. The serous tracheal glands also are unreactive for Protein C, but have reactivity for both B1-IP and Protein D in the ductal cells and also in the acini (Ball et al., 1991).

The immunoreactive components of the von Ebners, lacrimal and tracheal glands have not been identified and characterized; thus, it is not clear whether these reflect the presence of proteins identical to those of the neonatal SMG, or the existence of different proteins that, genetically or accidentally, are antigenically related. The fact that all three of these quite different protein components are present in the von Ebners glands, and that two (Bl-IR and D-IR) are present in the lacrimal and tracheal glands, suggests a developmental relatedness of these serous exocrine cells with those of the major salivary glands. The comparison with the SMG is particularly intriguing since, in the adult SMG as well, these immunoreactive components are present at high levels only in cells of the intercalated ducts. The intercalated ducts are considered to be progenitor cells for the replacement of adult acinar cells (see above) and so it is tempting to speculate that this relatedness may reflect a general type of serous differentiation, later to be overridden (in the SMG) or amplified (in the SLG or PRG) by the regulatory mechanisms that establish the cellular phenotype of the mature, specialized secretory cells. Any further development of this notion will depend on the identification of the protein molecules that are responsible for the immunocytochemical reactivities of the ductal cells. It is clear in the SMG, that the adult gland does have immunoreactive SMG-A (Moreira et al., 1991; Ball et al., 1992) and probably D (Ball et al., 1991), migrating at the appropriate mobility on Western blots. It is not clear, however, to what extent these components are contributed by the reactive ductal cells and/or by the reactive acini described above. No information is yet available for Protein C, or for any protein component in the minor glands.

ACKNOWLEDGMENTS

The author would like to thank Ms. Alyce R. Smith and Mrs. Charity Lancaster-Taylor for typing the manuscript, Ms. Sheila A. Mathias for compiling the bibliography, and Dr. Gregory B. Stanton for instruction in word processing. Thanks are due to Drs. Nancy A. Charest, David J. Culp, Paul C. Denny, Douglas P. Dickinson, Arthur R. Hand, Lily Mirels, Robert S. Redman, Murray R. Robinovitch, and Lawrence A. Tabak for helpful discussions on the content of the review. We are indebted to Dr. Philip Shaw for his generous gift of mouse PSP clone, PS5. Work from the author's laboratory was supported by NIH Grant DE-06635. I thank Ms. Wilda C. M. Gafney for technical assistance.

REFERENCES

1. Abe, K. and Dawes, C., The effects of electrical and pharmacological stimulation on the types of proteins secreted by rat parotid and submandibular glands, *Arch. Oral Biol.*, 23, 367, 1978.
2. Alberts, B., Bray, D., Lewis, J., Raff, M., Roberts, K., and Watson, J. D., *Molecular Biology of the Cell*, Garland Publishing, New York, 1989.
3. Amsterdam, A., Schramm, M., Ohad, I., Salomon, Y., and Selinger, Z., Concomitant synthesis of membrane protein and exportable protein of the secretory granule in rat parotid gland, *J. Cell Biol.*, 50, 187, 1971.
4. Ann, D. K. and Carlson, D. M., The structure and organization of a proline-rich protein gene of a mouse multigene family, *J. Biol. Chem.*, 260, 15863, 1985.
5. Ashley, P. L. and MacDonald, R. J., Kallikrein-related mRNAs of the rat submaxillary gland: nucleotide sequences of four distinct types including tonin, *Biochemistry*, 24, 4512, 1985.
6. Azen, E. A., Carlson, D. M., Clements, S., Lalley, P. A., and Vanin, E., Salivary proline-rich protein genes on chromosome 8 of mouse, *Science*, 226, 967, 1984.
7. Ball, W. D., Development of the rat salivary glands. I. Accumulation of parotid gland DNase activity, *J. Exp. Zool.*, 178, 331, 1971.
8. Ball, W. D., Isolation of salivary ribonuclease, deoxyribonuclease and amylase from the parotid gland of the rat, *Rattus norvegicus albinus*, *Biochim. Biophys. Acta*, 341, 305, 1974a.
9. Ball, W. D., Development of the rat salivary glands. IV. Amylase and ribonuclease activity during embryonic and neonatal development of the parotid and submaxillary glands, *Dev. Biol.*, 41, 267, 1974b.
10. Ball, W. D. and Nelson, N. J., A stage-restricted secretory system in the submandibular gland of the neonatal rat, *Differentiation*, 10, 147, 1978a.
11. Ball, W. D. and Nelson, N. J., Some properties of the ribonuclease activities from the parotid and submandibular glands of the neonatal and adult rat, *Arch. Oral Biol.*, 23, 243, 1978b.
12. Ball, W. D. and Redman, R. S., Two independently regulated secretory systems within the acini of the submandibular gland of the perinatal rat, *Eur. J. Cell Biol.*, 33, 112, 1984.
13. Ball, W. D. and Rutter, W. J., The DNase activities of the mouse, *J. Exp. Zool.*, 176, 1, 1971.
14. Ball, W. D., Hand, A. R., and Moreira, J. E., A neonatal secretory protein associated with secretion granule membranes in developing rat salivary glands, *J. Histochem. Cytochem.*, 39, 1693, 1991.
15. Ball, W. D., Hand, A. R., Iversen, J. M., and Robinovitch, M. R., The B1-immunoreactive proteins of the perinatal submandibular gland: similarity to the major parotid gland protein, RPSP, *Crit. Rev. Oral Biol. Med.*, in press, 1992.
16. Ball, W. D., Hand, A. R., and Johnson, A. O., Secretory proteins as markers for cellular phenotypes in rat salivary glands, *Dev. Biol.*, 125, 265, 1988a.

17. Ball, W. D., Hand, A. R., Moreira, J. E., and Johnson, A. O., A secretory protein restricted to Type I cells in neonatal rat submandibular glands, *Dev. Biol.*, 129, 265, 1988b.

18. Barka, T., Biologically active polypeptides in submandibular glands, *J. Histochem. Cytochem.*, 28, 836, 1980.

19. Barlas, A. B., Gao, X., and Greenbaum, L. M., Isolation of a thiol-activated T-kininogenase from the rat submandibular gland, *FEBS*, 218, 266, 1987.

20. Barrett, M. L. and Ball, W. D., Development of the rat salivary glands. II. The identification of a trypsin-like protease in the submaxillary gland, *Dev. Biol.*, 36, 195, 1974.

21. Bedi, G. S., Purification and characterization of an inducible cysteine proteinase inhibitor from submandibular glands of isoproterenol-treated rats, *Arch. Biochem. Biophys.*, 270, 335, 1989a.

22. Bedi, G. S., Amino acid sequence of an inducible cysteine proteinase inhibitor (cystatin) from submandibular glands of isoproterenol-treated rats, *Arch. Biochem. Biophys.*, 273, 245, 1989b.

23. Beerstecher, H. J., Huiskens-Van Der Meig, C., and Warnaar, S. O., An immuno-histochemical study performed with monoclonal and polyclonal antibodies to mouse epidermal growth factor, *J. Histochem. Cytochem.*, 36, 1153, 1988.

24. Bhathena, S. J., Smith, S. S., Voyles, N. R., Pehos, J. C., and Recant, L., Studies on submaxillary gland immunoreactive glucagon, *Biochem. Biophys. Res. Comm.*, 74, 1574, 1977.

25. Bothwell, M. A., Wilson, W. H., and Shooter, E. M., The relationship between glandular kallikrein and growth factor-processing proteases of mouse submaxillary gland, *J. Biol. Chem.*, 254, 7287, 1979.

26. Carlsöö, A., Kumilien, A., and Bloom, G. D., A comparative histochemical study of peroxidase activity in the submandibular glands of five mammalian species including man, *Histochemie*, 26, 80, 1971.

27. Carlson, D. M., Proline-rich proteins and glycoproteins: expression of salivary gland multigene families, *Biochimie*, 70, 1689, 1988.

28. Carlson, D. M., Salivary proline-rich proteins: biochemistry and regulation of expression, *Crit. Rev. Oral Biol. Med.*, in press, 1982.

29. Carpenter, G. and Cohen, S., Epidermal growth factor, *Annu. Rev. Biochem.*, 48, 193, 1979.

30. Catanzaro, D. F., Mullins, J. J., and Morris, B. J., The biosynthetic pathway of renin in mouse submandibular gland, *J. Biol. Chem.*, 258, 7364, 1983.

31. Chang, W. W. L., Cell population changes during acinus formation in the postnatal rat submandibular gland, *Anat. Rec.*, 178, 187, 1974.

32. Chao, L., Gerald, W., and Chao, J., *KININS IV*, Greenbaum, L. M. and Margolius, H. S., Eds., Plenum Press, New York, 1986.

33. Charest, N. J., Joseph, D. R., Wilson, E. M., and French, F. S., Molecular cloning of complementary deoxyribonucleic acid for an androgen-regulated epididymal protein: sequence homology with metalloproteins, *Mol. Endocrinol.*, 2, 999, 1988.

34. Charest, N. J., Petrusz, P., Ordronneau, P., Joseph, D. R., Wilson, E. M., and French, F. S., Developmental expression of an androgen regulated epididymal protein, *Endocrinology*, 125, 942, 1989.

35. Charest, N. J., Beck, C. M., Naiman, B., and Hand, A. R., Acidic epididymal glycoprotein is a major secretory protein of rat parotid gland, in preparation, 1992.

36. Chen, Y., Chao, J., and Chao, L., Molecular cloning and characterization of two rat renal kallikrein genes, *Biochemistry*, 27, 7189, 1988.

37. Clements, S., Mehansho, H., and Carlson, D. M., Novel multigene families encoding highly repetitive peptide sequences, *J. Biol. Chem.*, 260, 13471, 1985.

38. Cohen, S., Purification and metabolic effects of a nerve growth-promoting protein from snake venom, *J. Biol. Chem.*, 234, 1129, 1959.

39. Cohen, S., Purification of a nerve growth promoting protein from the mouse salivary gland and its neuro-cytoxic antiserum, *Proc. Natl. Acad. Sci. U.S.A.*, 46, 302, 1960.

40. Cohen, S., Isolation of a mouse submaxillary gland protein acceleration incisor eruption and eyelid opening in the new-born animal, *J. Biol. Chem.*, 237, 1555, 1962.

41. Cooper, L. F. and Tabak, L. A., Characterization of the gene encoding the salivary Glu/Glu-rich C-terminal variant A protein, *Gene*, 106, 261, 1991.

42. Cooper, L. F., Elia, D. M., and Tabak, L. A., Secretagogue-coupled changes in the expression of glutamine/glutamic acid-rich proteins (GRPs), *J. Biol. Chem.*, 266, 3532, 1991.

43. Crerar, M. M., Swain, W. F., Pictet, R. L., Nikovits, W., and Rutter, W. J., Isolation and characterization of a rat amylase gene family, *J. Biol. Chem.*, 258, 1311, 1983.

44. Culp, D. J., Graham, L. A., Latchney, L. R., and Hand, A. R., Rat sublingual gland as a model to study glandular mucous cell secretion, *Am. J. Physiol.*, 260, C1244, 1991.

45. Cutler, L. S. and Chaudhry, A. P., Cytodifferentation of the acinar cells of the rat submandibular gland, *Dev. Biol.*, 41, 31, 1974.

46. Cutler, L. S. and Chaudhry, A. P., Cytodifferentation of striated duct cells and secretory cells of the convoluted granular tubules of the rat submandibular gland, *Am. J. Anat.*, 143, 201, 1975.

47. Cutler, L. S., Christian C. P., and Bottaro, B., *Methods in Cell Biology*, Vol. 23, Academic Press, New York, 1981, 531.

48. Davies, D. V. and Young, L., Radioautographic studies of the digestive tracts of rats injected with inorganic sulphate labelled with sulphur-35, *Nature*, 173, 448.

49. Denny, P. A. and Denny, P. C., Purification and biochemical characterization of a mouse submandibular sialomucin, *Carbohydr. Res.*, 87, 265, 1980.

50. Denny, P. A. and Denny, P. C., A mouse submandibular sialomucin containing both N- and O-glycosidic linkages, *Carbohydr. Res.*, 110, 305, 1982a.

51. Denny, P. A. and Denny, P. C., Localization of a mouse submandibular sialomucin by indirect immunofluorescence, *Histochem. J.*, 14, 403, 1982b.

52. Denny, P. A., Pimprapaiporn, W., Kim, M. S., and Denny, P. C., Quantitation and localization of acinar cell-specific mucin in mouse submandibular glands during postnatal development, *Cell Tissue Res.*, 251, 381, 1988.

53. Denny, P. C. and Cope, P., Age-specific differentiations during the development of the rat submandibular gland, *Differentiation*, 8, 105, 1977.

54. Denny, P. C. and Denny, P. A., Diurnal variation of sialomucin concentration in female mouse submandibular glands measured by radioimmunoassay, *Arch. Oral Biol.*, 29, 1033, 1984.

55. Denny, P. C. and Denny, P. A., *Biology of the Salivary Glands*, Dobrosielski-Vergona, K., Ed., CRC Press, Boca Raton, FL, 1992, chap. 12.

56. Denny, P. C., Chai, Y., Pimprapaiporn, W., and Denny, P. A., Three-dimensional reconstruction of adult female mouse submandibular gland secretory structures, *Anat. Rec.*, 226, 489, 1990a.

57. Denny, P. C., Chai, Y., Klauser, D. K., and Denny, P. A., Three dimensional localization of DNA synthesis in secretory elements of adult female mouse submandibular gland, *Adv. Dent. Res.*, 4, 34, 1990b.

58. Dickinson, D. P., Abel, K., Near, J., Taylor, B. A., and Gross, K. W., Genetic and tissue-specific variation in the expression of a closely linked murine multigene family on chromosome 15 that encodes salivary and lacrimal proteins, *Biochem. Genet.*, 27, 613, 1989a.

59. Dickinson, D. P., Mirels, L., Tabak, L. A., and Gross, K. W., Rapid evolution of variants in a rodent multigene family encoding salivary proteins, *Mol. Biol. Evol.*, 6, 80, 1989b

60. Edwards, R. H., Selby, M. J., and Rutter, W. J., Differential RNA splicing predicts two distinct nerve growth factor precursors, *Nature*, 319 784, 1986.

61. Edwards, R. H., Selby, M. J., Garcia, P. D., and Rutter, W. J., Processing of the native nerve growth factor precursor to form biologically active nerve growth factor, *J. Biol. Chem.*, 263, 6810, 1988.

62. Ekfors, T. O., Riekkinen, J., Malmiharju, T., and Hopsu-Havu, V. K., Four isozymic forms of a peptidase resembling kallikrein purified from the rat submandibular gland, *Hoppe Seyler's Z. Physiol. Chem.*, 348, 111, 1967.

63. Fernandez-Sorensen, A. and Carlson, D. M., Isolation of a "proline-rich" protein from rat parotid glands following isoproterenol treatment, *Biochem. Biophys. Res. Comm.*, 60, 249, 1974.

64. Field, R. B. and Hand, A. R. Secretion of lingual lipase and amylase from rat lingual serous glands. *Am. J. Physiol.*, 253, G217, 1987.

65. Field, R. B. and Scow, R. O., Purification and characterization of rat lingual lipase, *J. Biol. Chem.*, 258, 14563, 1983.

66. Field, R. B., Spielman, A. I., and Hand, A. R., Purification of lingual amylase from serous glands of rat tongue and characterization of rat lingual amylase and lingual lipase, *J. Dent. Res.*, 68, 139, 1989.

67. Frati, L. and Caputo, A., Purification of α-amylase from C3H mouse submaxillary gland by gel filtration, *J. Chromatog.*, 48, 547, 1970

68. Gerald, W. L., Chao, J., and Chao, L., Immunological identification of rat tissue kallikrein cDNA and characterization of the kallikrein family gene, *Biochim. Biophys. Acta*, 886, 1, 1986a.

69. Gerald, W. L., Chao, J., and Chao, L., Sex dimorphism and hormonal regulation of rat tissue kallikrein mRNA, *Biochim. Biophys. Acta*, 867, 16, 1986b.

70. Godlowski, Z. Z. and Calandra, J. C., Salivary glands as endocrine organs, *J. Appl. Physiol.*, 15, 101, 1960.

71. Gray, A., Dull, T. J., and Ullrich, A., Nucleotide sequence of epidermal growth factor cDNA predicts a 128,000 molecular weight protein precursor, *Nature*, 303, 722, 1983.

72. Green, D. R. J. and Embery, G., Isolation, chemical and biological characterization of sulphated glycoproteins synthesized by rat buccal and palatal minor salivary glands *in vivo* and *in vitro*, *Arch. Oral Biol.*, 32, 391, 1987.

73. Greene, L. A., Shooter, E. M., and Varon, S., Enzymatic activities of mouse nerve growth factor and its subunits, *Proc. Natl. Acad. Sci. U.S.A.*, 60, 1383, 1968.

74. Greene, L. A., Shooter, E. M., and Varon, S., Subunit interaction and enzymatic activity of mouse 7S nerve growth factor, *Biochemistry*, 8, 3735, 1969.

75. Gresik, E., Michelakis, A., and Barka, T., Immunocytochemical localization of renin in the submandibular gland of the mouse during postnatal development, *Am. J. Anat.*, 153, 450, 1978.

76. Gresik, E., van der Noen, H., and Barka, T., Epidermal growth factor-like material in rat submandibular gland, *Am. J. Anat.*, 156, 83, 1979.

77. Gresik, E. W., Postnatal developmental changes in submandibular glands of rats and mice, *J. Histochem. Cytochem.*, 28, 860, 1980.

78. Gresik, E. W. and MacRae, E. K., The postnatal development of the sexually dimorphic duct system and of amylase activity in the submandibular glands of mice, *Cell Tissue Res.*, 157, 411, 1975.

79. Gresik, E. W., Gubits, R. M., and Barka, T., *In situ* localization of mRNA for epidermal growth factor in the submandibular gland of the mouse, *J. Histochem. Cytochem.*, 33, 1235, 1985.

80. Gubits, R. M., Shaw, P. A., Gresik, E. W., Muda-Onetti, A., and Barka, T., Epidermal growth factor gene expression is regulated differently in mouse kidney and submandibular gland, *Endocrinology*, 119, 1382, 1986.

81. Hamosh, M. and Hand, A. R., Development of secretory activity in serous cells of the rat tongue. Cytological differentiation and accumulation of lingual lipase, *Dev. Biol.*, 65, 100, 1978.

82. Hamosh, M. and Scow, R. O., Lingual lipase amd its role in the digestion of dietary lipid, *J. Clin. Invest.*, 52, 88, 1973.

83. Hand, A. R., The fine structure of von Ebner's gland of the rat, *J. Cell Biol.*, 44, 340, 1970.

84. Hand, A. R., *Orban's Oral Histology and Embryology*, Bhaskar, S. M., Ed., C. V. Mosby, St. Louis, 1985.

85. Hanks, C. T. and Chaudry, A. P., Regeneration of rat submandibular gland following partial extirpation: a light and electron microscopic study, *Am. J. Anat.*, 130, 195, 1971.

86. Hay, D. I., The interaction of human parotid salivary proteins with hydroxyapatite, *Arch. Oral Biol.*, 18, 1517, 1973.

87. Heinrich, G. and Habener, J. F., Genes encoding proteins with homologous contiguous repeat sequences are highly expressed in the serous cells of the rat submandibular gland, *J. Biol. Chem.*, 262, 5262, 1987.

88. Herp, A., Wu, A. M. and Moschera, J., Current concepts of the structure and nature of mammalian salivary mucous glycoproteins, *Mol. Cell Biochem.*, 23, 27, 1979.

89. Herzog, V. and Miller, F., Die Lokalisation endogener peroxydase in der glandula parotis der Ratte, *Z. Zellforsch. Mikrosk. Anat.*, 107, 403, 1970.

90. Holm, I., Ollo, R., Panthier, J., and Rougeon, F., Evolution of aspartyl proteases by gene duplication: the mouse renin gene is organized in two homologous clusters of four exons, *EMBO J.*, 557, 1983.

91. Humphreys-Beher, M. J., Strain-specific differences in the proline-rich proteins and glycoproteins induced in rat salivary glands by chronic isoprenaline treatment, *Biochem. J.*, 230, 369, 1958.

92. Ikeno, K. and Ishiguro, I., Developmental changes of trypsin-like protease activity in the submandibular glands of male and female young growing rats, *Arch. Oral Biol.*, 29, 669, 1984.

93. Ikeno, K., Ikeno, T., and Kuzuya, H., Purification and characterization of a trypsin-like protease in the submandibular gland of rats, *J. Biochem.*, 99, 1219, 1986.

94. Ikeno, T., Nasu, J., and Kuzuya, H., Quantitative changes in amylase activity in the salivary glands, pancreas saliva and serum after administration of isoproterenol, pilocarpine, and acetylcholine, *J. Dent. Res.*, 62, 56, 1983.

95. Iversen, J. M., Johnson, D. A., Kauffman, D. L., Keller, P. J., and Robinovitch, M. R., Isolation and characterization of the basic proline-rich proteins from rat parotid saliva, *Arch. Oral Biol.*, 27, 925, 1982.

96. Jacoby, F. and Leeson, C. R., The postnatal development of the rat submaxillary gland, *Am. J. Anat.*, 93, 201, 1959.

97. Junqueira, L. C. U., Toledo, A. M. S., and Saad, A., *Salivary Glands and Their Secretions*, Sreebny, L. M. and Myer, J., Eds., Macmillan, New York, 1964.

98. Junqueira, L. C. U., Toledo, A. M. S., and Doine, A. I., Digestive enzymes in the parotid and submandibular glands of mammals, *An. Acad. Brasil Cienc.*, 45, 629, 1973.

99. Kato, H., Nakanishi, E., Enjyoji, K., Hayashi, I., Oh-Ishi, S., and Iwanaga, S., Characterization of serine proteinases isolated from rat submaxillary gland: with special reference to the degradation of rat kininogens by these enzymes, *J. Biochem.*, 102, 1389, 1987.

100. Keller, P. A., Robinovitch, M., Iversen, J., and Kauffman, D. L., The protein composition of rat parotid saliva and secretory granules, *Biochim. Biophys. Acta*, 379, 562, 1975.

101. Keller, P. G., Cohen, E., and Neurath, H., The proteins of bovine pancreatic juice, *J. Biol. Chem.*, 233, 344, 1958.

102. Khullar, M., Scicli, G., Carretero, O. A., and Scicli, A. G., Purification and characterization of a serine protease(esterase B) from rat submandibular glands, *Biochemistry*, 25, 1851, 1986.

103. Klein, R. M., *Oral Development and Histology*, Williams and Wilkins, Baltimore, 1987.

104. Kousvelari, E. E., Baratz, R. S., Burke, B., and Oppenheim, F. G., Immunochemical identification and determination of proline-rich proteins in salivary secretions, enamel pellicle, and glandular tissue specimens, *J. Dent. Res.*, 59, 1430, 1980.

105. Kousvelari, E. E., Grant, S. R., Banerjee, D. K., Newby, M. J., and Baum, B. J., cAMP mediates β-adrenergic-induced increases in N-linked protein glycosylation in rat parotid acinar cells, *Biochemistry*, 222, 17, 1984.

106. Kozak, M., A profusion of controls, *J. Cell Biol.*, 107, 1, 1988.

107. Laskowski, M., Sr., *The Enzymes*, Vol. 5 (Part B), Boyer, P. D., Lardy, H., and Myrback, K., Eds., Academic Press, New York, 1961.

108. Lazure, C., Leduc, R., Seidah, N. G., Thibault, G., Genest, J., and Chrétian, M., Amino acid sequence of rat submaxillary tonin reveals similarities to serine proteases, *Nature*, 307, 555, 1984.

109. Liao, T. H., Hamosh, P., and Hamosh, M., Gastric lipolysis in the developing rat ontogeny of the lipases active in the stomach, *Biochim. Biophys. Acta*, 754, 1, 1983.

110. Loyter, A. and Schramm, M., The glycogen amylase complex as a means of obtaining highly purified α-amylases, *Biochim. Biophys. Acta*, 65, 200, 1962.

111. Madsen, H. O. and Hjorth, J. P., Molecular cloning of mouse PSP mRNA, *Nucleic Acids Res.*, 13, 1, 1985.

112. Maitra, S. R., Rabito, S. E., and Carretero, O. A., In *KININS IV*, Greenbaum, L. M. and Margolius, H. S., Eds., Plenum Press, New York, 1986.

113. Malacinski, G. M. and Rutter, W. J., Multiple molecular forms of α-amylase from the rabbit, *Biochemistry*, 8, 4382, 1969.

114. Man, Y., Moreira, J. E., Hand, A. R., and Ball, W. D., The differentiated state of intercalated duct cells in submandibular glands(SMG) of adult rats, in preparation, 1992.

115. Mason, A. J., Evans, B. A., Cox, D. R., Shine, J., and Richards, R. I., Structure of mouse kallikrein gene family suggests a role in specific processing of biologically active peptides, *Nature*, 303, 300, 1983.

116. Matsuura, S., Suzuki, K., Ball, W. D., and Hand, A. R., Immunocytochemistry of developing mouse submandibular glands(SMG), in preparation, 1992.

117. Mehansho, H., Hagerman, A., Clements, S., Butler, L., Rogler, J., and Carlson, D. M., Modulation of proline-rich protein biosynthesis in rat parotid glands by sorghums with high tannin levels, *Proc. Natl. Acad. Sci. U.S.A.*, 80, 3948, 1983.

118. Menaker, L. and Miller, S. A., Regulation of protein synthesis in the submandibular salivary gland of the developing rat. Ribonuclease and ribonuclease inhibitor activity, *Arch. Oral Biol.*, 18, 1303, 1973.

119. Menaker, L., Sheetz, J. H., Cobb, C. M., and Navia, J. M., Gel electrophoresis of whole saliva and associated histologic changes in submandibular glands of isoproterenol-treated rats, *Lab. Invest.*, 30, 341, 1974.

120. Minaguchi, K. and Bennick, A., Genetics of human salivary proteins, *J. Dent. Res.*, 68, 2, 1989.

121. Mirels, L. and Ball, W. D., Neonatal rat submandibular gland protein SMG-A and parotid secretory protein are alternatively regulated members of a salivary protein multigene family, *J. Biol. Chem.*, 267, 2679, 1992.

122. Mirels, L., Bedi, G. S., Dickinson, D. P., Gross, K. W., and Tabak, L. A., Molecular characterization of glutamic acid/glutamine-rich secretory proteins from rat submandibular glands, *J. Biol. Chem.*, 262, 7289, 1987.

123. Mirels, L., Kopec, L., Dickinson, D. P., Dziejman, M., and Tabak, L. A., Expression of glutamine/glutamic acid-rich proteins in rat submandibular glands, *Arch. Oral Biol.*, 35, 1, 1990.

124. Mirels, L., Girard, L. R., and Ball, W. D., Molecular cloning of developmentally regulated neonatal rat submandibular gland proteins, *Crit. Rev. Oral Biol. Med.*, in press, 1992.

125. Moreira, J. E., Tabak, L. A., Bedi, G. S., Culp, D. J., and Hand, A. R., Light and electron microscopic immunolocalization of rat submandibular gland mucin-glycoprotein and glutamine/glutamic acid-rich proteins, *J. Histochem. Cytochem.*, 37, 515, 1989.

126. Moreira, J. E., Hand, A. R., and Ball, W. D., Localization of neonatal secretory proteins in different cell types of the developing rat submandibular gland, *Dev. Biol.*, 39, 370, 1990.

127. Moreira, J. E., Ball, W. D., Mirels, L., and Hand, A. R., Accumulation and localization of two adult acinar cell secretory proteins during development of the rat submandibular gland, *Am. J. Anat.*, 191, 167, 1991.

128. Mori, M., Hamada, K., Naito, R., Tsukitani, K., and Asano, K., Immunohistochemical localization of epidermal growth factor in rodent submandibular glands, *Acta Histochem. Cytochem.*, 16, 536, 1983.

129. Morrison, M., Allen, P. Z., Bright, J., and Jayasinghe, W., Lactoperoxidase V. Identification and isolation of lactoperoxidase from salivary gland, *Arch. Biochem. Biophys.*, 111, 126, 1965.

130. Moschera, J. and Pigman, W., The isolation and characterization of rat sublingual mucus-glycoprotein, *Carbohydr. Res.*, 40, 53, 1975.

131. Muenzer, J., Bildstein, C., Gleason, M., and Carlson, D. M., Purification of proline-rich proteins from parotid glands of isoproterenol-treated rats, *J. Biol. Chem.*, 254, 5623, 1979a.

132. Muenzer, J., Bildstein, C., Gleason, M., and Carlson, D. M., Properties of proline-rich proteins from parotid glands of isoproterenol-treated rats, *J. Biol. Chem.*, 254, 5629, 1979b.

133. Murakami, K., Taniguchi, H., and Baba, S., Presence of insulin-like immunoreactivity and its biosynthesis in rat and human parotid gland, *Diabetologia*, 22, 358, 1982.

134. Murphy, R. A., Watson, A. Y., Metz, J., and Forssman, W. G., The mouse submandibular gland: an exocrine organ for growth factors, *J. Histochem. Cytochem.* 28, 890, 1980.

135. Naiman, B. E., Hand, A. R., and Charest, N. J., Tissue-specific hormonal regulation of rat acidic epididymal glycoprotein (AEG), in preparation, 1992.

136. Naito, Y., Studies on LM protein appearing in submandibular glands of isoproterenol-treated rats. I. Purification and characteristics of its appearance, *Chem. Pharm. Bull.*, 29, 1365, 1981.

136. Naito, Y., Studies on LM protein appearing in submandibular glands of isoproterenol-treated rats. III. Some aspects of its formation, *Chem. Pharm. Bull.*, 30, 1059, 1982.

137. Naito, Y. and Saito, M., Studies on LM protein appearing in submandibular glands of isoproterenol-treated rats. IV. Size and shape determination, *Chem. Pharm. Bull.*, 30, 1063, 1982.

138. Naito, Y. and Suzuki, I., Studies on LM protein appearing in submandibular glands of isoproterenol-treated rats, *Chem. Pharm. Bull.*, 30, 1373, 1981.

139. Nicholas, S. and Redman, R. S., Ultrastructure of the anterior buccal gland of the rat, *Anat. Rec.*, 213, 140, 1985.

140. Novikoff, A. B., Beard, M. E., Albala, A., Sheid, B., Quintana, N., and Biempica, L., Localization of endogenous peroxidases in animal tissues, *J. Microscopie*, 12, 381, 1971.

141. Oppenheim, F. G., Hay, D. I., and Franzblau, C., Proline-rich proteins from human parotid saliva. Isolation and partial characterization, *Biochemistry*, 10, 4233, 1971.

142. Oppenheim, F. G., Offner, G. D., and Troxler, R. F., Amino acid sequence of a proline-rich phosphoglycoprotein from parotid secretion of the subhuman primate *Macaca fascicularis*, *J. Biol. Chem.*, 260, 10671, 1985.

143. Orstavik, T., Nustad, K., and Brandtzaeg, P., A biochemical and immunohistochemical study of kallikrein in normal and isoproterenol-stimulated rat salivary glands during postnatal development, *Arch. Oral Biol.*, 22, 495, 1977.

144. Orstavik, T. B., The distribution and secretion of kallikrein in some exocrine organs of the rat, *Acta Physiol. Scand.*, 104, 431, 1978.

145. Orstavik, T. B., The kallikrein-kinin system in exocrine organs, *J. Histochem. Cytochem.*, 28, 881, 1980.

146. Osugi, T., Development of peroxidase-mediated antibacterial activity in embryonic salivary glands of the mouse, *Arch. Oral Biol.*, 22, 237, 1977.

147. Owerbach, D. and Hjorth, J. P., Inheritance of a parotid secretory protein in mice and its use in determining salivary amylase quantitative variants, *Genes*, 95, 129, 1980.

148. Panthier, J., Foote, S., Chambraud, B., Strosberg, A. D., Corvol, P., and Rougeon, F., Complete amino acid sequence and maturation of the mouse submaxillary gland renin precursor, *Nature*, 298, 90, 1982.

149. Panthier, J., Dreyfus, M., Tronik-Le Roux, D., and Rougeon, F., Mouse kidney and submaxillary gland renin genes differ in their 5′ putative regulatory sequences, *Proc. Natl. Acad. Sci. U.S.A.*, 81, 5489, 1984.

150. Poulsen, K., Jakobsen, B. K., Mikkelsen, B. M., Harmark, K., Nielsen, J. T., and Hjorth, J. P., Coordination of murine parotid secretory protein and salivary amylase expression, *EMBO J.*, 5, 1891, 1986.

151. Pratt, R. E., Ouellete, A. J., and Dzau, V. J., Biosynthesis of renin: multiplicity of active and intermediate forms, *Proc. Natl. Acad. Sci. U.S.A.*, 80, 6809, 1983.

152. Pratt, R. E., Carleton, J. E., Roth, T. P., and Dzau, V. J., Evidence for two cellular pathways of renin secretion by the mouse submandibular gland, *Endocrinology*, 123, 1721, 1988.

153. Quissell, D. O. and Barzen, K. A., Secretory response of dispersed rat submandibular cells. II. Mucin secretion, *Am. J. Physiol.*, 238, C99, 1980.

154. Quissell, D. O. and Tabak, L. A., *Handbook of Physiology*, Section 6, Vol. 3, Forte, J. E., Ed., Waverly Press, Baltimore, 1989.

155. Qwarnström, E. E. and Hand, A. R., A granular cell at the acinar-intercalated duct junction of the rat submandibular gland, *Anat. Rec.*, 206, 181, 1983.

156. Rall, L. B., Scott, J., Bell, G. I., Crawford, R. J., Penschow, J. D., Niall, H. D., and Coghlan, J. P., Mouse prepro-epidermal growth factor synthesis by the kidney and other tissues, *Nature*, 313, 228, 1985.

157. Redman, R. S., The anterior buccal gland of the rat: a mucous salivary gland which develops as a branch of Stensen's duct, *Anat. Rec.*, 172, 167, 1972.

158. Redman, R. S., *The Salivary System*, Sreebny, L. M., Ed., CRC Press, Boca Raton, FL, 1987.

159. Redman, R. S. and Ball, W. D., Cytodifferentiation of secretory cells in the sublingual gland of the prenatal rat: a histological, histochemical and ultrastructural study, *Am. J. Anat.*, 153, 367, 1978.

160. Redman, R. S. and Sreebny, L. M., The prenatal phase of the morphosis of the rat parotid gland, *Anat. Rec.*, 168, 127, 1970.

161. Redman, R. S. and Sreebny, L. M., Morphologic and biochemical observations on the development of the rat parotid gland, *Dev. Biol.*, 25, 248, 1971.

162. Riekkinen, P. J., Rat submandibular gland proteases, as influenced by diurnal rhythm, feeding, age, sex and isoprenaline treatment, *Ann. Med. Exptl. Fenn.*, 44, 549, 1966.

163. Riekkinen, P. J. and Hopsu, V. K., Studies in the salivary glands. I. Trypsin-like enzymes, *Ann. Med. Exptl. Fenn.*, 43, 6, 1965.

164. Riekkinen, P. J. and Hopsu-Havu, V. K., A chymotrypsin-like protease from the rat submandibular gland, *Acta Chem. Scand.*, 20, 2169, 1966.

165. Riekkinen, P. J. and Niemi, M., Androgen-dependent salivary gland protease in the rat, *Endocrinology*, 83, 1224, 1968.
166. Riekkinen, P. J., Ekfors, T. O., and Hopsu, V. K., Purification and characteristics of an alkaline protease from rat submandibular gland, *Biochim. Biophys. Acta*, 118, 604, 1966.
167. Robinovitch, M. R. and Sreebny, L. M., Separation of rat parotid isoamylases by preparative polyacrylamide gel electrophoresis, *Arch. Oral Biol.*, 15, 1381, 1970.
168. Robinovitch, M. R. and Sreebny, L. M., On the nature of the molecular heterogeneity of rat parotid amylase, *Arch. Oral Biol.*, 17, 595, 1972.
169. Robinovitch, M. R., Keller, P. J., Iversen, J., and Kauffman, D. L., Demonstration of a class of proteins loosely associated with secretory granule membranes, *Biochim. Biophys. Acta*, 382, 260, 1975.
170. Robinovitch, M. R., Keller, P. J., Johnson, D. A., Iversen, J. M., and Kauffman, D. L., Changes in rat parotid salivary proteins induced by chronic isoproterenol administration, *J. Dent. Res.*, 56, 290, 1977.
171. Robinovitch, M. R., Iversen, J. M., and Oberg, S. G., Isolation and partial characterization of secretory granule membranes from the rat parotid gland, *Arch. Oral Biol.*, 25, 523, 1980.
172. Roth, J. S., Studies on the properties and distribution of ribonuclease inhibitor in the rat, *Biochim. Biophys. Acta*, 21, 34, 1956.
173. Roukema, P. A., Oderkerk, C. H., and Salkinoja-Salonen, M. S., The murine sublingual and submandibular mucins: their isolation and characterization, *Biochim. Biophys. Acta*, 428, 432, 1976.
173. Rutter, W. J. and Weber, C. S., *Developmental and Metabolic Control Mechanisms in Neoplasia*, Williams and Wilkins, Baltimore, 1965, 195.
174. Rutter, W. J., Ball, W. D., Bradshaw, W. S., Clark, W. R., and Sanders, T. G., *Experimental Biology and Medicine*, Vol. 1, S. Karger, Basel, 1967a, 110.
175. Rutter, W. J., Ball, W. D., Bradshaw, W. S., Clark, W. R., and Sanders, T. G., *Secretory Mechanisms of Salivary Glands*, Academic Press, New York, 1967b, 238.
176. Sanders, T. G. and Rutter, W. J., Molecular properties of rat pancreatic and parotid α-amylase, *Biochemistry*, 11, 130, 1972.
177. Savage, C. R., Inagami, T., and Cohen, S., The primary structure of epidermal growth factor, *J. Biol. Chem.*, 247, 7612, 1972.
178. Schacter, M., Kallikreins (kininogenases) — a group of serine proteases with bioregulatory actions, *Pharmacological Reviews*, 31, 1, 1980.
179. Schibler, U., Tosi, M., Pittet, A. C., Fabiana, L., and Wellauer, P. K., Tissue-specific expression of mouse α-amylase genes, *J. Mol. Biol.*, 142, 93, 1980.
180. Schibler, U., Hagenbüchle, O., Wellauer, P. K. and Pittet, A. C., Two promoters of different strengths control the transcription of the mouse alpha-amylase gene amy-1 in the parotid gland and the liver, *Cell*, 33, 501, 1983.
181. Schneyer, L. H. and Schneyer, C. A., Regulation of salivary gland amylase activity, *Ann. N.Y. Acad. Sci.*, 85, 189, 1960.
182. Schramm, M., *Salivary Glands and Their Secretions*, Sreebny, L. M. and Meyer, J., Eds., Pergamon Press, Oxford, 1964.
183. Schwab, M., Stockel, K., and Thoenen, H., Immunocytochemical localization of nerve growth factor (NGF) in the submandibular gland of adult mice by light and electron microscopy, *Cell Tissue Res.*, 169, 289, 1976.

184. Scott, J., Urdea, M., Quiroga, M., Pescador-Sanchez, R., Fong, N., Selby, M., Rutter, W. J., and Bell, G. I., Structure of a mouse submaxillary messenger RNA encoding epidermal growth factor and seven related proteins, *Science*, 221, 236, 1983.

185. Selby, M. J., Edwards, R. H., and Rutter, W. J., Mouse nerve growth factor gene: structure and expression, *Mol. Cell Biol.*, 7, 3057, 1987.

186. Shai, S., Woodley-Miller, C., Chao, J., and Chao, L., Characterizaton of genes encoding rat tonin and a kallikrein-like serine protease, *Biochemistry*, 28, 5334, 1989.

187. Shaw, P. and Schibler, U., Structure and expression of the parotid secretory protein gene of mouse, *J. Mol. Biol.*, 192, 567, 1986.

188. Shaw, P., Sordat, B., and Schibler, U., The two promoters of the mouse α-amylase gene amy-1 are differentially activated during parotid gland differentiation, *Cell*, 40, 907, 1985.

189. Shaw, P., Sordat, B., and Schibler, U., Developmental coordination of amylase and psp gene expression during mouse parotid gland differentiation is controlled posttranscriptionally, *Cell*, 47, 107, 1986.

190. Shaw, P., Cox, J. L., Barka, T., and Naito, Y., Cloning and sequencing of cDNA encoding a rat salivary cysteine proteinase inhibitor inducible by β-adrenergic agonists, *J. Biol. Chem.*, 263, 18133, 1988.

191. Shine, J., Mason, A. J., Evans, B. A., and Richards, R. I., *Cold Spring Harbor Symposium on Quantitative Biology*, Vol. 158, 1983.

192. Shortman, K., Studies on cellular inhibitors of ribonuclease. I. The assay of the ribonuclease-inhibitor system and the purification of the inhibitor from rat liver, *Biochim. Biophys. Acta*, 51, 37, 1961.

193. Sick, K. and Nielsen, J. R., Genetics of amylase isozymes in the mouse, *Hereditas*, 51, 291, 1964.

194. Slomiany, A. and Slomiany, B. L., Structures of the acidic oligosaccharides isolated from rat sublingual glycoprotein, *J. Biol. Chem.*, 253, 7301, 1978.

195. Slomiany, A., Mizuta, G., Zalensa, G., Tsukada, H., and Slomiany, B. L., Co-translational processing and intracelluar transport of rat salivary mucus glycoprotein, *Arch. Oral Biol.*, 33, 807, 1988a.

196. Slomiany, A., Murty, V. L. N., Liau, Y. H., Carter, S. R., and Slomiany, B. L., Enzymatic sulphation of mucus glycoprotein in rat sublingual salivary gland, *Arch. Oral Biol.*, 33, 669, 1988b.

197. Slomiany, B. L., Murty, V. L. N., Takagi, A., Tsukada, H., Kosmala, M., and Slomiany, A., Fatty acid acylation of salivary mucin in rat submandibular glands, *Arch. Biochem. Biophys.*, 242, 402, 1985.

198. Slomiany, B. L., Murty, V. L. N., Sarosiek, J., Piotrowski, J., and Slomiany, A., Role of associated and covalently bound lipids in salivary mucin hydrophobicity: effect of proteolysis and disulfide bridge reduction, *Biochem. Biophys. Res. Comm.*, 151, 1046, 1988.

199. Smith, P. and Toms, B. B., Immunocytochemical localization of insulin- and glucagonlike peptides in rat salivary glands, *J. Histochem. Cytochem.*, 34, 627, 1986.

200. Smith, P. H. and Patel, D. G., Imunochemical studies of the insulin-like material in the parotid gland of rats, *Diabetes*, 33, 661, 1984.

201. Smith, R. J., Frommer, F., and Schiff, R., Localization and onset of amylase activity in mouse salivary glands determined by a substrate film method, *J. Histochem. Cytochem.*, 19, 310, 1971.

202. Spielman, A. I. and Bennick, A., Isolation and characterization of six proteins from rabbit parotid saliva belonging to a unique family of proline-rich proteins, *Arch. Oral Biol.*, 34, 117, 1989.

203. Sreebny, L. M., Meyer, J., Bachem, E., and Weinmann, J. P., Postnatal changes in proteolytic activity and in the morphology of the submaxillary gland in male and female rats, *Growth*, 19, 57, 1955.

204. Sreebny, L. M., Wanamaker, B. B., and Robinovitch, M. R., The DNAase of the rat parotid gland, *J. Dent. Res.*, 44, 463, 1965.

205. Sreebny, L. M., Ruark, J. W., and Tamarin, A., The distribution of deoxyribonuclease in the salivary glands and pancreas of various mammals, *Arch. Oral Biol.*, 12, 777, 1967.

206. Srinivasan, R. and Chang, W. W. L., The development of the granular convoluted duct in the rat submandibular gland, *Anat. Rec.*, 182, 29, 1975.

207. Strum, J., Unusual peroxidase-positive granules in the developing rat submaxillary gland, *J. Cell Biol.*, 51, 575, 1971

208. Strum, J. M. and Karnovsky, M. J., Ultrastructural localization of peroxidase in submaxillary acinar cells, *J. Ultrastruct. Res.*, 31, 323, 1970.

209. Tabak, L. A., *Molecular Genetics of the Mouth: Frontiers of Oral Physiology*, Vol. 8, Ferguson, D. B., Ed., Karger Press, Basel, Switzerland, 1990, 77.

210. Tabak, L. A., Mirels, L., Monte, L. D., Ridall, A. L., Levine, M. J., Loomis, R. E., Lindauer, F., Reddy, M. S., and Baum, B. J., Isolation and characterization of a mucin-glycoprotein from rat submandibular glands, *Arch. Biochem. Biophys.*, 242, 383, 1985.

211. Taga, R. and Sesso, A., Ultrastructural studies on developing parotid gland of the rat at early postnatal periods, *Arch. Histol. Jpn.*, 42, 427, 1979.

212. Taylor, J., Cohen, S., and Mitchell, W., Epidermal growth factor: high and low molecular weight forms, *Proc. Natl. Acad. Sci. U.S.A.*, 67, 164, 1970.

213. Taylor, J., Mitchell, W., and Cohen, S., Characterization of the binding protein for epidermal growth factor, *J. Biol. Chem.*, 249, 2188, 1974.

214. Turkington, R. W., Males, J. L., and Cohen, S., Synthesis and storage of epithelial-epidermal growth factor in submaxillary gland, *Cancer Res.*, 31, 252, 1971.

215. Varon, S., Nomura, J., and Shooter, E. M., The isolation of the mouse nerve growth factor protein in a high molecular weight form, *Biochemistry*, 6, 2202, 1967.

216. Varon, S., Nomura, J., and Shooter, E. M., Reversible dissociation of the mouse nerve growth factor protein into different subunits, *Biochemistry*, 7, 1296, 1968.

217. Wallach, D., Kirshner, N., and Schramm, M., Non-parallel transport of membrane proteins and content proteins during assembly of the secretory granule in rat parotid gland, *Biochim. Biophys. Acta*, 375, 87, 1975a.

218. Wallach, D., Tessler, R., and Schramm, M., The proteins of the content of the secretory granules of the rat parotid gland, *Biochim. Biophys. Acta*, 382, 552, 1975b.

219. Xiong, W., Chen, L., and Chao, J., Purification and characterization of a kallikrein-like T-kininogenase, *J. Biol. Chem.*, 265, 2822, 1990.

220. Yagil, C., Michaeli, Y., and Zajicek, G., Compensatory proliferative response of the rat submandibular salivary gland to unilateral extirpation, *Virchow's Arch.*, B 49, 83, 1985.

221. Yagil, C., Naito, Y., and Barka, T., Immunocytochemical localization of a developmentally regulated isoproterenol-inducible protein (LM protein) in rat submandibular gland, *Am. J. Anat.*, 177, 55, 1986.

222. Yamashina, S. and Barka, T., Localization of peroxidase activity in the developing submandibular gland of normal and isoproterenol treated rats, *J. Histochem. Cytochem.*, 20, 855, 1972.
223. Yamashina, S. and Barka, T., Development of endogenous peroxidase in fetal rat submandibular gland, *J. Histochem. Cytochem.*, 21, 40, 1973.
224. Yamashina, S. and Barka, T., Peroxidase activity in the developing rat submandibular gland, *Lab. Invest.*, 31, 82, 1974.
225. Yamashita, S., Immunohistochemical study of amylase and deoxyribonuclease in rat parotid gland, *Acta Histochem. Cytochem.*, 14, 236, 1981.
226. Young, J. A. and van Lennep, E. W., *The Morphology of Salivary Glands*, Academic Press, New York, 1978.
227. Young, R. A., Hagenbüchle, O., and Schibler, U., A single mouse α-amylase gene specifies two different tissue-specific mRNAs, *Cell*, 23, 451, 1981.
228. Zajicek, G., Yagil, C., and Michaeli, Y., The streaming submandibular gland, *Anat. Rec.*, 213, 150, 1985.
229. Ziemer, M. A., Mason, A., and Carlson, D. M., Cell-free translations of proline-rich protein mRNAs, *J. Biol. Chem.*, 257, 11, 176, 1982.
230. Ziemer, M. A., Swain, W. F., Rutter, W. J., Clements, S., Ann, D. K., and Carlson, D. M., Nucleotide sequence analysis of a proline-rich protein cDNA and peptide homologies of rat and human proline-rich proteins, *J. Biol. Chem.*, 259, 10475, 1984.

Changes in the Structure of Salivary Glands With Age

Anne Field and John Scott

INTRODUCTION

In this chapter we address the problem of the progressive, sequential changes in structure that occur in the salivary glands in the postmaturation period.

Aging is best considered as an interrelated series of biological processes which are cumulative and have the effect of decreasing the viability of the organism and increasing the risk of mortality within a finite period of time. By their nature aging changes lead to an increased vulnerability to disease so that the process of aging overlaps and interreacts with age-related disease. Aging itself may progressively lead to a breakdown in tissue integrity and functional capacity and in this way its effects can be synergistic with age related pathology. However, it is important to understand that aging itself is not pathological.

All tissues and organs of the body undergo changes with age, some (e.g., skin) more obviously than others. Salivary glands are no exception and undergo complex structural changes with age. The nature and extent of these changes vary widely and are not consistent for all types of salivary glands and all species. Nevertheless, information concerning structural changes in salivary glands across the adult life span is essential if we are to understand the functional impairments which are detectable in some but not all salivary glands with age (see Chapter 19 in this volume). We shall demonstrate in this chapter that the salivary glands in the elderly exhibit marked histological differences when compared with those of the mature young adult.

TECHNIQUES FOR THE ASSESSMENT OF STRUCTURAL AGE CHANGES IN SALIVARY GLANDS

Much of the evidence for structural age changes has come from observations in the major salivary glands of man and the laboratory rat. Early descriptive accounts of age-related alterations of salivary structure in man and animals focused on atrophic changes in parenchymal components together with increasing fibrosis and adiposity (Yamaguchi, 1925; Hamperl, 1931; Andrew, 1949a; Andrew, 1949b; Andrew, 1952). However, recent advances in histopathological techniques have contributed to our knowledge of structural derangements in human and animal salivary glands. In particular the use of stereological analysis has enabled detailed quantitative assessments to be made of the extent and rate of the volumetric tissue changes which affect salivary glands as age advances. While such studies generally point to a progressive replacement of secretory epithelium by connective tissues, interpretation of the results has to be guarded where proportional volumetric changes are unsupported by firm evidence about possible changes in the total gland volume. For example, an apparent age-dependent decrease in a particular tissue component may simply reflect an increase in other gland components in an enlarging gland. However, it is worth noting that for human salivary glands there is no evidence for an increase in size with age (Waterhouse et al., 1973; Scott, 1980; Drummond and Chisholm, 1984). Indeed, where gland weight or volume has been measured, the evidence suggests a tendency for a reduced volume beyond the age of 75 years (Scott, 1975). It is likely, therefore, that where systematic measurement has recorded relative losses in salivary parenchyma with increasing age, these studies do reflect an actual reduction in the true volume of tissue available for participation in secretory function.

Most of the quantitative studies on age changes in salivary structure have been carried out on human salivary glands obtained post-mortem. Inevitably, therefore, the populations studied both within and between different series have been widely heterogeneous, particularly in respect of medical therapeutic histories, causes of death, nutritional background, smoking and drinking habits, and dental status including the presence or absence of dentures. Although in most series reported in this chapter attempts have been made to control for some of these variables, e.g., by using "sudden death" cases and excluding particular drug medications and diseases, the elimination of such variables has probably at best been only partially achieved. Nevertheless, the strict application of excluding criteria employed in many of the investigations has probably avoided the seriously biased variability between age groups which characterized some of the earlier aging studies of salivary function and led to erroneous conclusions (see Baum, 1981a).

A COMPARISON OF THE YOUNG AND OLD ADULT SALIVARY GLAND (GENERAL STRUCTURE)

Depending on the species and the type of gland, the salivary parenchyma consists of serous, mucous, or mixed acini together with various types of ducts arranged into morphological units of lobules. These are separated and supported by connective-tissue septa from which a finely dispersed fibrovascular network extends to surround individual acini and ducts. In human glands there is a progressive disparity of size and shape of the lobules as age increases. Furthermore, as originally described by Hamperl in 1931, there is evidence of age-related atrophy of acini, an increase in the intralobular ducts, and an apparent increase in the proportional volume of connective tissues throughout the gland.

The characteristic structural feature of the young adult gland is its even and compact lobular structure and uniform appearance of the parenchyma (i.e., acini plus ducts). Figure 1 contrasts the histological appearances of the young and old human submandibular glands. The young gland shows compact lobules with uniform acini and a low proportion of ducts. The old gland shows more loosely structured lobules with acini of disparate sizes and a high proportion of intralobular ducts. Figure 2 illustrates the differences between young and old human labial salivary glands. In the old glands the acini are widely spaced and variable in size, shape, and staining capacity and the ducts are dilated and far more numerous than in the young. The parotid gland in man is typified at all ages by a fatty infiltration which is highly variable in intensity both between subjects in the same age group and between different lobules in the same gland. This infiltrate tends to obscure the compact appearance of the young adult state in the parotid (Figure 3) compared with other types of salivary gland. However, broadly similar aging changes occur and the old parotid gland shows acinar atrophy and ductal irregularities when compared with the young parotid (Figure 3).

The salivary glands of laboratory animals, particularly the rat, also evidently undergo age-related histological changes but these appear to be less severe than those seen in man. Changes at an ultrastructural level, on the other hand, have been better categorized in animals than in human glands and these are discussed in detail at the end of this chapter. The following sections describe the detailed qualitative and quantitative changes that occur in the human salivary glands from maturity to old age.

THE HUMAN SUBMANDIBULAR GLAND

Qualitative Age Changes in Human Submandibular Glands

Structural aging studies on the human submandibular gland are more numerous than on other major glands probably because its compact encapsulated nature and its superficial location facilitate removal of the intact gland at necropsy.

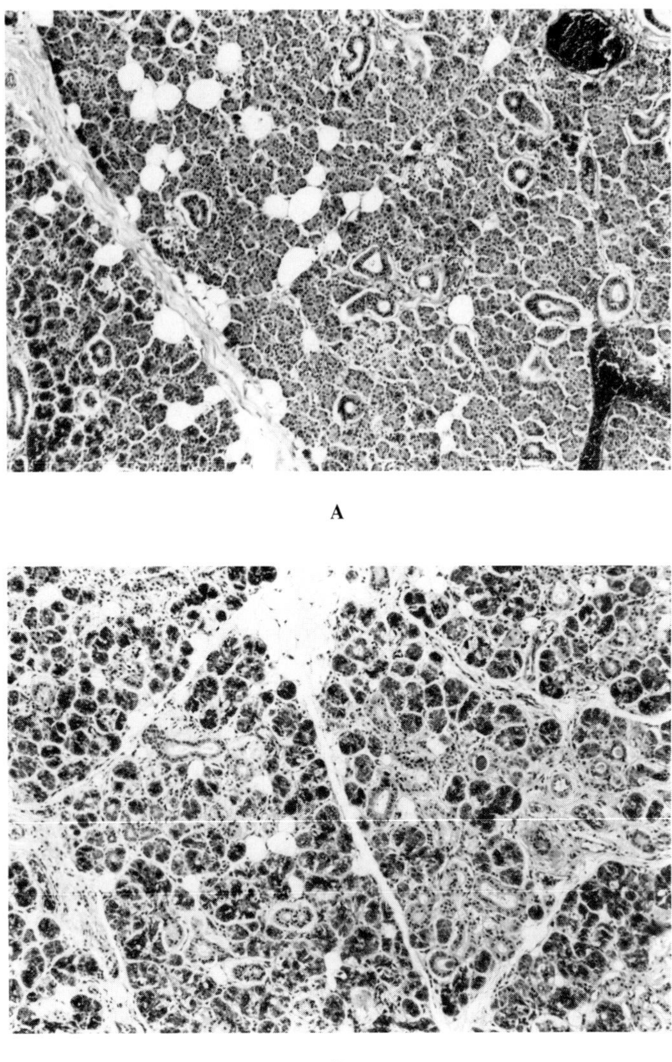

A

B

Figure 1. Submandibular glands from a male aged 22 years (A) and a male aged 80 years (B). The young gland has compact lobules with uniform acini and a low proportion of ducts. The old gland shows more loosely structured lobules with acini of disparate size and a high proportion of intralobular ducts. Both H & E; both × 87. (Reprinted from Scott, J., *Gerodontology*, 3, 149, 1986. With permission.)

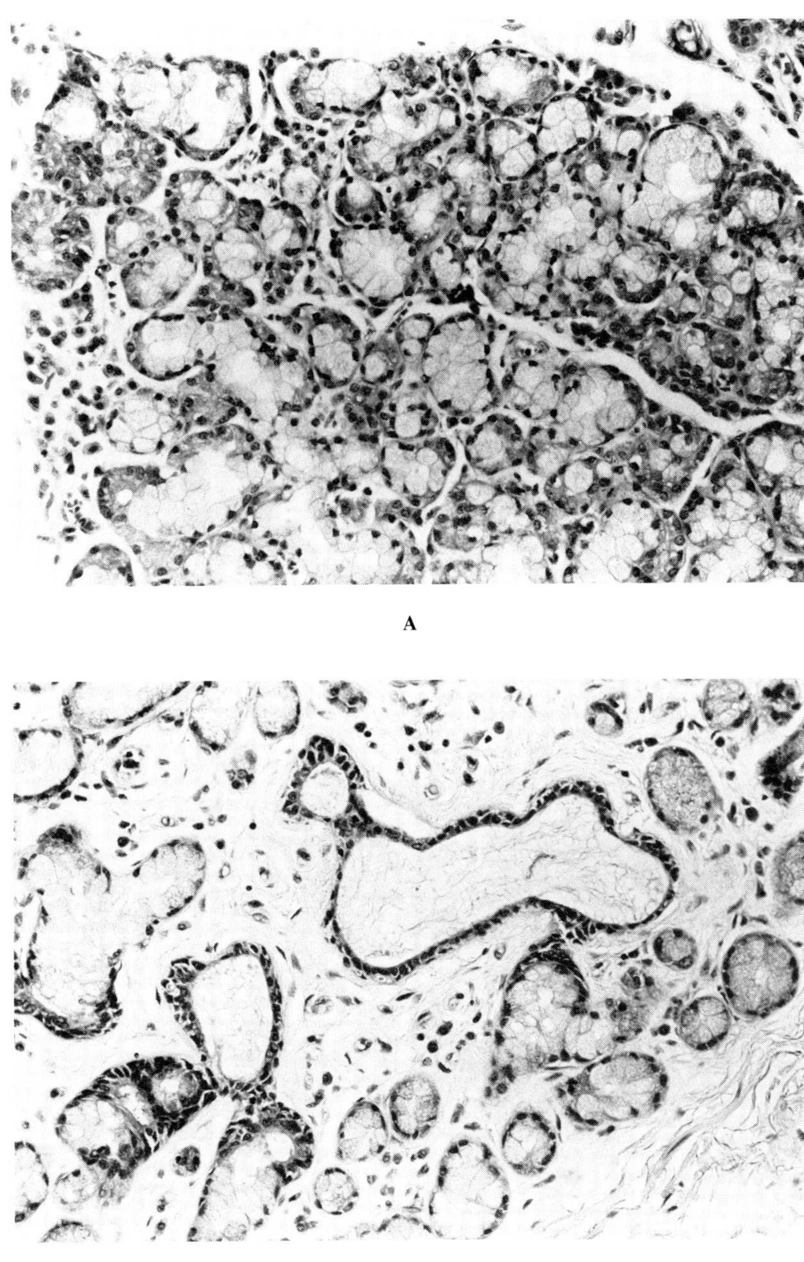

A

B

Figure 2. Labial glands from a male aged 29 years (A) and a male aged 60 years (B). Note the loss of acini, fibrosis, and ductal dilatation and hyperplasia in the older gland. Both H & E; × 240.

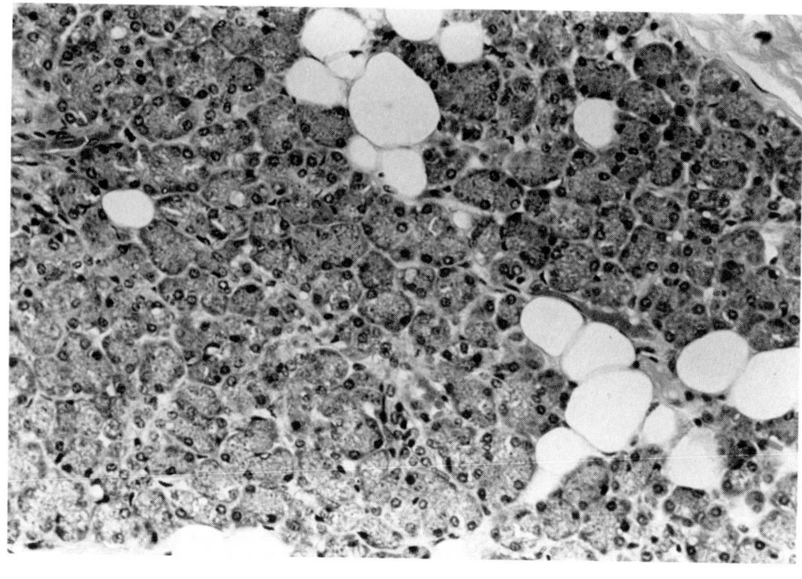

A

B

Figure 3. Parotid gland from a male aged 24 years (A) and a male aged 90 years (B). In the young gland the acini are even and compactly arranged without evidence of degenerative changes. In the old gland the acini are uneven in size, shape, and staining and there is more interstitial fibrous tissue. Both H & E; both × 220.

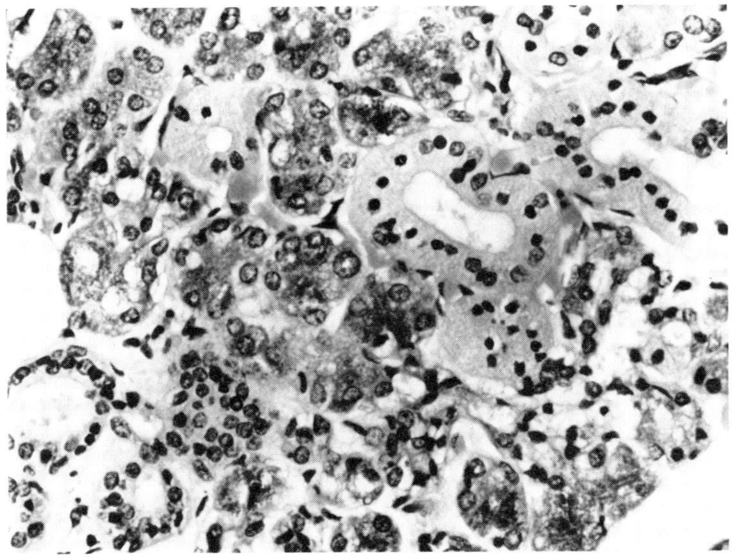

Figure 4. Oncocytes in a submandibular intralobular duct and adjacent acini in a female aged 89 years. Note their bulging, swollen outlines, finely granular cytoplasm, and predominantly pyknotic nuclei. H & E; × 375.

Parenchymal Changes

In the human submandibular gland, the acini may exhibit loss of granules and cell shrinkage with age so that their lumens become wider and they resemble primitive ducts, while the intralobular ducts themselves become hyperplastic and dilated (Hamperl, 1931; Scott, 1977a). Dilatation is also a feature of the extralobular ducts in old age. More rarely, these larger ducts may also exhibit basal metaplasia and hyperplasia with occasional obliteration of the lumens (Scott, 1977b).

Oncocytes

One of the characteristic features of aging salivary glands is the occurrence of morphologically distinctive, probably inactive, cells referred to by Hamperl (1931) as oncocytes. Oncocytosis affects the submandibular glands along with all salivary glands but is most frequent in the sublingual. These cells are of swollen appearance, often with bulging outlines (Figure 4). They have strongly eosinophilic, slightly granular, cytoplasm with a central, vesicular or pyknotic nucleus. Occasionally, binucleate forms occur. Their ultrastructure is characterized by dense cytoplasmic accumulations of mitochondria and the loss of specialized cytoplasmic structures. Although initially the abundant closely packed mitochondria are all of normal morphology, these organelles gradually assume a distorted appearance (Tandler, 1966; Bogart, 1970). Oncocytes occur singly or in clusters. Sometimes the latter consist of large hyperplastic foci and occasionally this change is sufficiently extensive to replace an entire lobule. Oncocytic foci are also sometimes seen at the centers of lobules otherwise replaced by adipose tissue. At a histological level oncocytes in human submandibular glands appear to arise frequently from intercalated or striated ducts (Scott, 1977a; Scott, 1980) but undoubtedly these cells also arise by transformations of acinar cells (Hamperl, 1931; Scott, 1977a). Ultrastructural studies of human submandibular glands, in fact, have shown oncocytes arising most often from the acini and have confirmed the occurrence of transitional forms (Tandler, 1966).

Although in man oncocytes are infrequent below age 30 years, they are almost invariably present after the age of 70 years (Hamperl, 1931; Scott, 1977a). Kurashima and Hirokawa (1986) found that the highest frequency of oncocytic change in the submandibular glands was in the 60 to 69 year age group. However, oncocytes are not solely related to aging. They are a prominent feature of salivary tumors, most notably adenolymphoma and oxyphilic adenoma, and they may also be associated with dilatation and cystic obstruction of the salivary ducts (Southam, 1974). Moreover, oncocytes also occur in many exocrine and endocrine glands besides the salivary glands where also they are predominantly associated with aging. The sporadic distribution of oncocytes alongside apparently functionally intact cells and their abnormal ultrastructural features attest to the nonfunctional character of oncocytes. They are regarded, therefore, as the end result of a specific type of degenerative change. Hamperl (1931) explained the increasing prevalence of oncocytes with increasing age by postulating that once oncocytes formed they underwent no further degenerative changes but retained a capacity for reproduction.

Age-Related Fibrosis

Throughout all salivary glands there is a fine network of supporting connective tissues extending from the capsule, where present, to interlobular septa and thence intralobularly to surround the individual parenchymal elements. Centrally, the main collecting ducts are also ensheathed in fibrous tissue which again is continuous with the fine intralobular fibrillar network. In many types of salivary gland there is an age-related increase in the amount and density of this fibrous skeletal component both around the ducts and in the septa, which thus appear widened (Figure 1), and intralobularly so that the acini become slightly more widely spaced. These changes are particularly well developed in man and in aging submandibular gland (Andrew, 1952; Garrett, 1962; Scott, 1977a; Kurashima and Hirokawa, 1986).

As well as increasing in amount with age the connective tissue component of the salivary glands also exhibits altered histological characteristics. The collagen fibers become denser, more fragmented, and less evenly orientated and

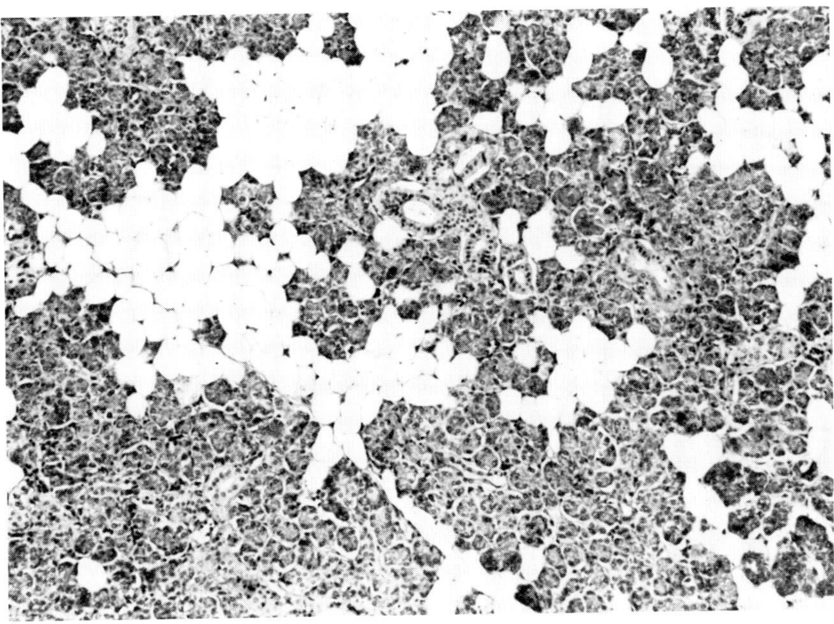

Figure 5. Fatty infiltration of the submandibular gland in old age. Note the apparent origin of this process in the septa and the mature adipose cells extending centrally to replace or displace intralobular parenchymal elements. H & E; × 79. (Reprinted from Scott, J., *Frontiers of Oral Physiology*, S. Karger, Basel, 1987, 46. With permission.)

elastic fibers are more numerous, thicker, and fragmented. Prominent elastosis of the connective tissue sheaths of the extralobular ducts is an occasional finding in human submandibular glands after the age of 75 years and intralobular elastic tissue, rarely seen in young glands, is frequently encountered especially around striated ducts (Scott, 1977a).

Lipid Accumulation and Adiposity with Increased Age

Undoubtedly there is an increase in the level of adiposity with increasing age in salivary glands. This fatty infiltration which consists of the accumulation of mature adipocytes is frequently seen in human submandibular glands in old age. The process appears to originate in the septa with the mature adipose cells extending centrally to replace or displace intralobular epithelial elements (Figure 5). The amount of adipose tissue, which is widely variable from gland to gland in any age group, has been shown to be independent of the general adiposity of the individual (Waterhouse et al., 1973).

A separate and distinctive feature of salivary parenchymal cells, namely the intracytoplasmic accumulation of lipid droplets, also increases with age (Yamaguchi, 1925) and although this led Andrew (1952) to argue that the accumulation and fusion of such droplets culminated in the large numbers of adipose cells seen in aging human salivary glands, this interpretation is now largely discounted (Garrett, 1962). Rather it is felt that fat may accumulate in aging salivary glands by two mechanisms: an increase of lipid droplets within the parenchymal cells and by an increase of mature adipocytes within the connective tissues of the glands. Intracellular lipid droplets have been observed at the light microscope level in the acini and ducts of all human major salivary glands (Garrett, 1962). They consist mainly of neutral fat and their presence has been confirmed by electron microscopy in man (Garrett, 1963). Some of the lipid droplets are associated with the pigment lipofuscin and this material gradually accumulates with age in many tissues and organs, generally without any impairment of function. In human salivary glands lipofuscin has been demonstrated in the striated ducts and gradually increases with age (Garrett, 1963).

Focal Chronic Inflammatory Changes in Human Submandibular Glands

Irrespective of age human salivary glands contain small numbers of lymphocytes, plasma cells, and histiocytes widely dispersed throughout the supporting tissues as a normal histological feature. In the submandibular glands two types of focal chronic inflammatory change have been recorded: focal aggregates consisting predominantly of lymphocytes; and limited intralobular areas of parenchymal change similar to those present in obstructive sialoadenitis. These have been termed focal lymphocytic adenitis (FLA) and focal obstructive adenitis (FOA) respectively (Scott, 1976a). Both these changes can be observed in the aging submandibular gland (Garrett, 1962; Scott, 1976a).

Focal Lymphocytic Adenitis in Human Salivary Glands.   This consists of collections of lymphocytes around small ducts or vessels (Hamperl, 1931; Andrew, 1952; Garrett, 1962; Waterhouse et al., 1973; Scott, 1977a; Kurashima and Hirokawa, 1986) and only infrequently leads to destruction and replacement of parenchyma (Figure 6). These foci are reported as being more prevalent in females, and in older age groups (Chisholm et al., 1970; Kurashima and Hirokawa, 1986). However, the present authors found no consistent pattern of age or sex differences either in the prevalence of glands exhibiting FLA or in the mean focal index score per decade (Scott, 1976a). Kurashima and Hirokawa noted a higher prevalence of affected cases after 70 years of age (80.3%) compared to the 40 to 70 year age group (53.8%) but they also found a fall in prevalence in the decade after 90 years, especially in males.

Focal Obstructive Adenitis.   In human salivary glands, focal obstructive adenitis consists of small, usually well-demarcated, areas of parenchymal degeneration with associated fibrosis and a variable degree of chronic inflammatory infiltration. These foci occur with increasing frequency in the major glands of man as age increases (Hamperl, 1931; Scott, 1976a). Because the

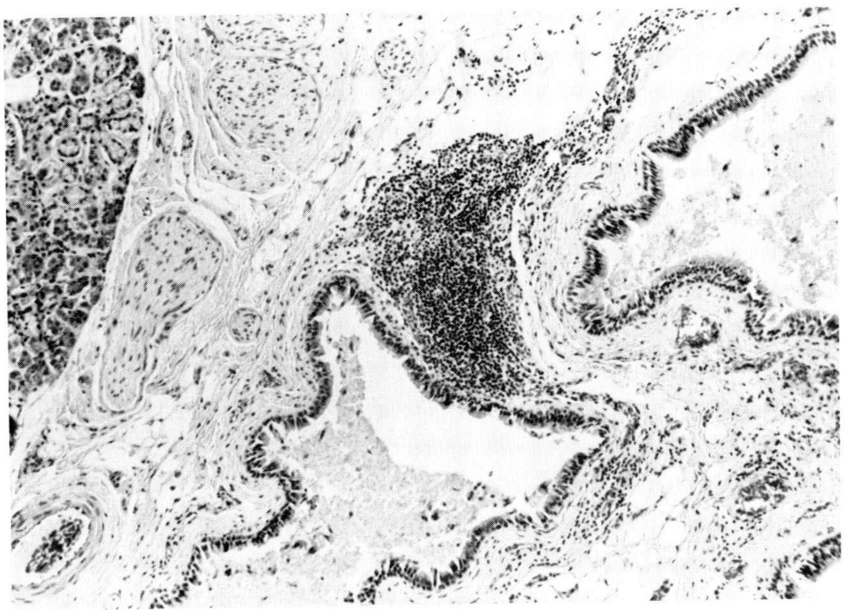

Figure 6. Focal lymphocytic adenitis in a submandibular gland. A large focus is present in the wall of a major excretory duct. Male aged 35 years. H & E; × 85.

histological appearances often resemble in microcosm those which occur throughout the gland after obstruction of a major duct, they have been referred to as focal obstructive adenitis in distinction from the focal lymphocytic adenitis of the previous section to which, however, they are unrelated (Scott, 1976a). The parenchymal changes in these lesions consist of acinar atrophy and/or dedifferentiation with loss of secretory granules, dilatation, and hyperplasia of the ducts and a chronic inflammatory infiltration of variable density (Figure 7). In long-standing foci, few remnants of the parenchyma can be found so that the affected area resembles a circumscribed fibrous scar often with few chronic inflammatory cells remaining (Scott, 1976a). These foci therefore reflect in miniature and in an intense form the generalized response of the submandibular glands to aging, namely, a loss of parenchyma and replacement fibrosis. The relationship of focal obstructive adenitis to age in human submandibular glands is shown in Figure 8. In this series the mean number of foci per unit sectional area increased almost 8-fold over 8 decades of human adult life.

Intraductal Deposits

In man, salivary calculi occur most frequently in the submandibular gland (Wakeley, 1948), where they consist of alternate organic and mineralized layers (Blatt et al., 1958) precipitated around a central organic nidus (Harrill et al.,

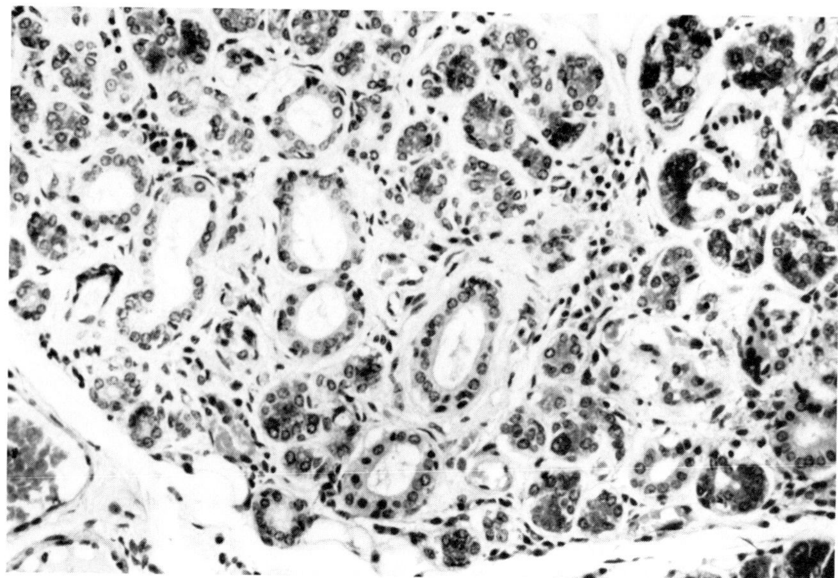

Figure 7. Focal obstructive adenitis in the submandibular gland of an 80-year-old man. Note the degeneration of the acini and the ductal dilatation and hyperplasia. Fibrosis and chronic inflammatory infiltration are also present. H & E; × 228.

1959) which is probably of cellular origin. Deposits of apparently solid material have been observed in the small salivary ducts of nondiseased submandibular glands obtained from necropsies. Although present in all age groups they are more common in old age (Scott, 1978), their mean prevalence increasing almost 5-fold between 16 and 95 years (Table 1). The deposits are of varied appearance. The majority are predominantly or entirely eosinophilic but some are haematoxyphilic and may be calcified. Most have a granular structureless appearance but some are concentrically laminated and appear to have formed about a central nidus (Figure 9). The deposits are mostly present in the lumens of the intralobular ducts but sometimes occur in the epithelial lining or even entirely in the connective tissues, as a result perhaps of atrophy of the original ductal epithelium. They also occur in the extralobular ducts and, rarely, in the acini where they may be intracellular (Epivatianos and Harrison, 1989). Sometimes these deposits are found in association with focal obstructive adenitis where they develop in the dilated ducts of the foci, possibly as a consequence of stagnant flow in these areas (Scott, 1976a; 1978).

The deposits probably form by localized precipitation of salivary glycoprotein which may itself result from altered salivary composition consequent on degenerative changes which occur in parenchyma with increasing age. It is of interest that similar structures have recently been described intracellularly in the acini of the feline submandibular gland both at the light microscope and ultrastructure levels (Epivatianos et al., 1986). Whether such deposits are

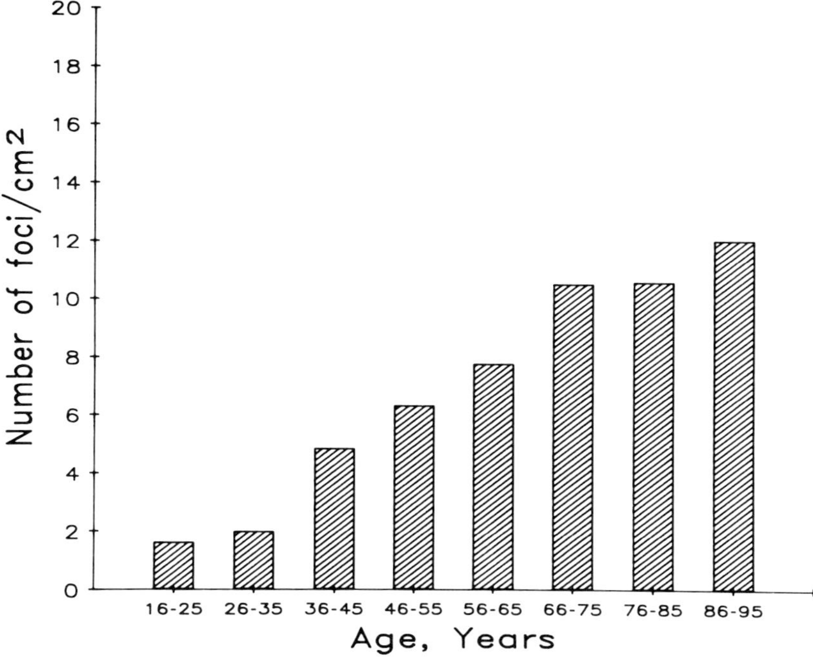

Figure 8. Increasing frequency of focal obstructive adenitis in human submandibular glands with age. Column heights show mean number of foci per cm^2 of histological section in age groups including at least five males and five females per group. (Data from Scott, J., *J. Oral Pathol.*, 5, 334, 1976a. With permission.)

TABLE 1. The Mean Prevalence of Intraductal Deposits in the Submandibular Glands From Subjects in Different Age Groups[44]

Age Group[a] (Years)	Number of Deposits per cm^2 (SEM)
16–35	1.12 (0.67)
36–55	2.39 (0.82)
55–75	3.49 (0.87)
76–95	5.47 (1.05)

[a] Each group contained 12 subjects.

forerunners of the microdeposits seen intraductally in human submandibular glands is speculative and they have not been related to age.

Although the histological structure of the laminated deposits in submandibular glands appears similar to that of salivary calculi, the large number of microdeposits in older age groups (Table 1) is not accompanied by an increased incidence of sialolithiasis in old age. Clearly, in all age groups, the great majority of microdeposits are easily voided in the saliva.

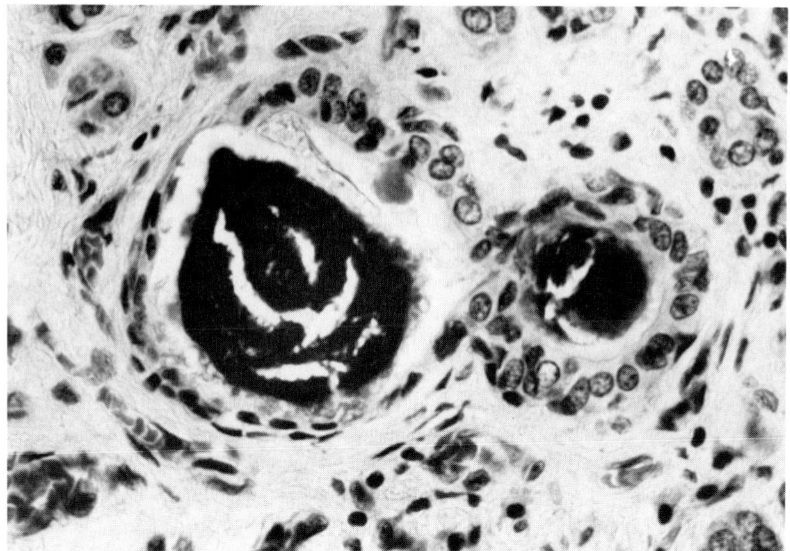

Figure 9. An intensely haematoxyphilic intraductal deposit in a submandibular gland showing laminated structure, fragmentation, and apparent mineralization. Note the atrophy of the epithelium in the larger transection. Male aged 56 years. H & E; ×580. (Reprinted with permission from Scott, J., *J. Oral Pathol.*, 7, 28, 1978.)

Degenerative Vascular Changes With Age

Degenerative vascular disease is an almost constant finding in middle-aged and older human subjects and has been described in the medium and small vessels of the submandibular gland at necropsy (Scott, 1977a). The extent to which this generalized vascular change might impair the capillary blood supply to individual parenchymal units has yet to be investigated, but it is reasonable to assume that it may give rise to some degree of ischemia within the submandibular glands. Whether ischemic hypoxia, resulting from degenerative arterial disease within the salivary glands, is an important factor in the parenchymal involution which accompanies old age is unknown. This type of hypoxia is far less amenable to compensating mechanisms (e.g., increased vascularity, hyperaemia) which develop extensively in response to hypoxic hypoxia in rats (Scott and Gradwell, 1989). It would be a mistake, however, to regard the extensive losses of salivary parenchyma in old age as being entirely the result of degenerative vascular disease. Parenchymal aging losses may also occur in occasional human glands where vascular degenerative changes are minimal.

Quantitative Age Changes in the Structure of Human Submandibular Glands

The submandibular is the only gland for which data are available for weight and volume over the human adult life span (Waterhouse et al., 1973; Scott, 1975).

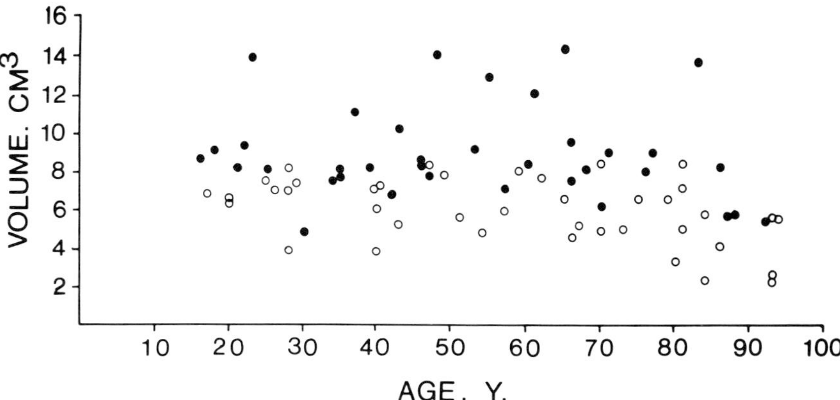

Figure 10. The volume of the left submandibular gland in 39 males (●) and 41 females (○) plotted against age. There is a wide range of values for each sex at all ages but note the tendency for shrinkage over the last two decades.

As can be seen in Figure 10 there is virtually no change in gland volume throughout most of adult life although the range at all ages is rather wide. However, beyond 75 years these data show a definite trend towards shrinkage of the glands as age increases further. Analysis of a series of 96 intact, nondiseased, formalin-fixed submandibular glands confirmed that for each of the sexes both the gland weight and volume were significantly reduced in the over-75-year age group but were unaffected by age differences in each of the three 20-year age groups below 75 years (Scott, 1976b).

Stereological analyses of healthy human submandibular glands have revealed that at around 20 years of age the parenchyma occupies approximately 70% of the gland volume. This then undergoes a steady decline with increasing age so that by about 90 years it occupies, on average, only some 50% of total gland volume (Waterhouse et al., 1973; Scott, 1977b). Since, as we have just seen, there is no enlargement of the gland throughout adult life these findings reflect a true loss of glandular epithelium with age. Moreover, it has also been shown that this loss occurs entirely within the acinar component (Figure 11) because concomitantly there is a progressive rise in the proportional volume of ducts with age (Table 2). These changes account for the prominence of ducts frequently observed in older glands. However, ducts account for only a small fraction of the whole gland and the severe loss of the main component, the acini, is compensated for largely by changes in the fibrovascular and adipose components (Table 2).

The age-related increase in the proportional volume of the ducts does not take place evenly in all categories of ducts. In the human submandibular glands a close analysis of volumetric changes in the ducts alone (Scott, 1977c) showed that most of the increase could be accounted for by changes in the volume proportion of non-striated intralobular ducts (Table 3), corresponding therefore with the hyperplasia, metaplasia, oncocytosis, and dilatation already noted in the

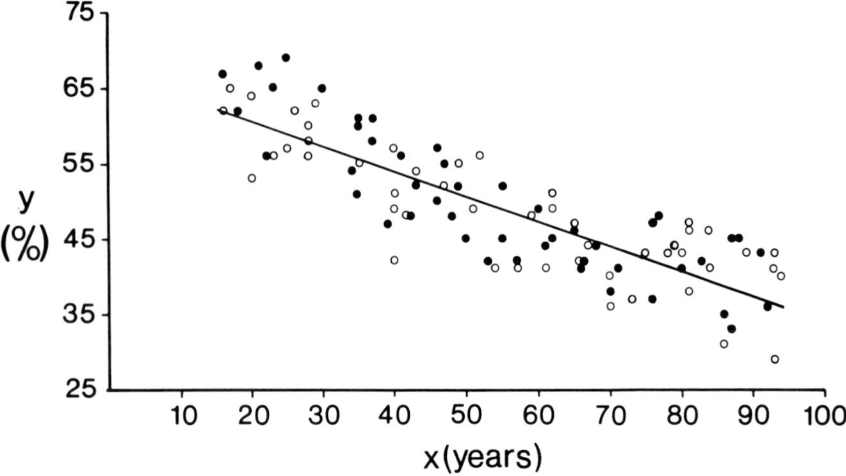

Figure 11. Age-related reduction in proportional volume of acinar tissue in human submandibular glands. The acinar volume (y), expressed as percentage of total gland volume, was obtained by point counting analysis and is plotted against age (x) in years for a total of 96 subjects, males (●) and females (○). The regression line, y = a + bx has been fitted by calculation of the constants a = 67.2 and b = –0.33, for which t = 15.6; *p* <0.001. (Reprinted from Scott, J., *Frontiers of Oral Physiology*, S. Karger, Basel, 1987, 54. With permission.)

TABLE 2. Proportional Volumes of Component Tissues (%) of Human Submandibular Glands in Decades of Age

Decade Years	No.	Acini Mean (SEM)	Ducts Mean (SEM)	Adipose Mean (SEM)	Other[a] Mean (SEM)
16–25	12	62.0 (1.50)	8.4 (0.37)	4.3 (0.99)	25.4 (1.03)
26–35	11	58.5 (1.26)	9.2 (0.43)	5.5 (1.74)	27.0 (1.35)
36–45	12	52.0 (1.57)	9.6 (0.34)	8.2 (1.65)	30.5 (1.02)
46–55	13	49.8 (1.44)	9.8 (0.42)	8.4 (1.99)	32.0 (2.04)
56–65	12	46.2 (1.10)	10.2 (0.49)	9.9 (1.51)	33.8 (0.98)
66–75	11	40.7 (0.80)	10.7 (0.60)	10.4 (2.08)	38.2 (2.08)
76–85	13	43.4 (0.97)	11.3 (0.53)	13.0 (2.45)	32.3 (1.81)
86–95	12	38.5 (1.59)	10.6 (0.55)	15.1 (1.47)	35.8 (2.38)

[a] Predominantly fibrous and vascular tissues.

descriptive accounts of age changes in these glands. An increase in proportional volume also occurred in the extralobular ducts although this was mostly restricted to the age groups above 65 years. An important finding emerging from this study was that the proportional volume of the striated ducts remained almost constant at about 5% of gland volume over all age groups (Figure 12). Thus, relative to the acinar volume, there is a higher volume of striated ducts which may be important in ensuring the maintenance of a hypotonic saliva in old age.

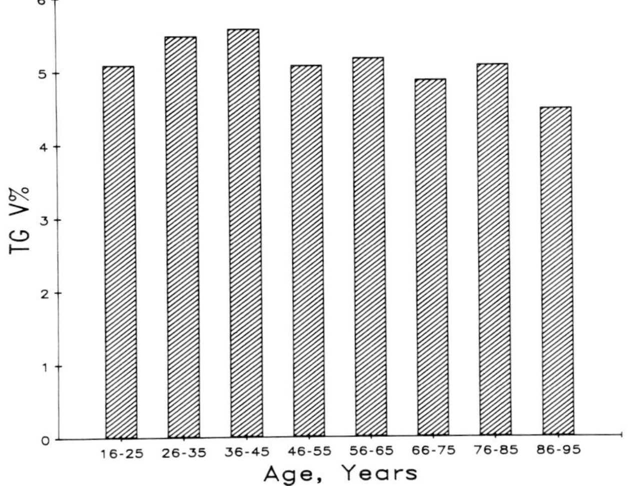

Figure 12. The mean proportion of total submandibular gland volume (TGV) occupied by striated ducts in eight successive decades of age. The volume proportions were obtained by point counting. There were at least nine subjects in each age group. Unlike other gland components, the striated ducts maintain a constant proportion of gland volume across the age groups.

TABLE 3. Proportion of Total Human Submandibular Duct Volume (%) Occupied by Different Categories of Ducts in Decades of Age

Decade Years	Number	Extralobular Mean (SEM)	Intralobular (Striated) Mean (SEM)	Intralobular (Nonstriated) Mean (SEM)
16–25	9	10.1 (0.94)	60.2 (2.33)	29.7 (1.57)
26–35	9	11.8 (0.92)	59.2 (1.82)	29.0 (2.08)
36–45	10	11.8 (1.03)	58.1 (2.09)	30.0 (2.50)
46–55	10	13.6 (1.30)	51.8 (1.74)	34.6 (1.84)
56–65	11	11.2 (0.66)	51.5 (2.06)	37.7 (2.03)
66–75	9	15.8 (2.56)	44.6 (2.37)	39.7 (1.80)
76–85	12	16.2 (1.98)	44.8 (2.69)	39.0 (2.61)
86–95	12	18.3 (1.39)	40.7 (2.18)	41.0 (1.55)

The stereological analyses of the human submandibular gland confirm that acinar loss, ductal proliferation, fibrosis and adiposity occur progressively throughout the whole of adult life. In a similar study of the submandibular glands in childhood and adolescence, however, no deteriorative features could be demonstrated in the parenchyma (Scott, 1979). Together these findings suggest that the onset of aging deterioration occurs in early adulthood soon after the development and growth of the gland is completed.

THE HUMAN PAROTID GLAND

Qualitative Age Changes in Human Parotid Glands

Structural changes in the human parotid gland are less evident than those of the submandibular gland. In a recent quantitative study using parotid samples from "sudden death" necropsies of both sexes, over four successive age groups (Table 4) Scott et al. (1987) showed that structural age changes within the parotid glands resemble those of the submandibular glands, although the overall loss of parenchyma may not be so severe. Thus, advancing age is accompanied by acinar atrophy and ductal irregularities and by increasing fibrosis and adiposity.

Parenchymal Changes

Although the parotid glands of younger individuals tend to show a more uniform appearance of parenchymal elements than in the elderly, this feature is frequently obscured by a high content of adipose tissue within the gland. This adiposity occurs in all age groups but is extremely variable between subjects. In many elderly, the parotid acini are uneven in size and shape, and in the density and staining capacity of their secretory granules. Atrophic acini are frequent especially beyond 75 years of age, both as a general feature throughout the lobes (Figure 3) and as discrete circumscribed areas in which acinar shrinkage and degranulation are accompanied by apparent hyperplasia of ducts. Such foci are rarely found below 35 years of age.

Nevertheless, glands from older age groups (>55 years) often retain the typical evenly compact appearance as illustrated in Figure 3 and may not be immediately recognizable as "old". Intralobular and collecting ducts in the younger age groups are variously numerous or scarce in different subjects, but have a generally uniform appearance, without the dilatation frequently encountered in the older age groups.

Oncocytes

As with other major human salivary glands there is increasing prevalence of oncocytes with increasing age (Bauer, 1950). Oncocytic change occurred frequently in the parotid glands of the over-75-year age group in the recent study by Scott et al. (1986).

Age-related Fibrosis and Adiposity

In the human parotid gland there is an age-related increase in fibrosis within the gland but this is generally overshadowed by increasing adiposity. On the other hand, the level of adiposity is such a variable feature of the human parotid gland that occasionally parotids from young adults are seen in which more than half the glands appear to be occupied by mature fatty tissue. Initially mature adipose cells are seen to accumulate in the septa but gradually they encroach on

TABLE 4. Mean Proportional Volumes of Constituents of Human Parotid Glands in Four Successive Adult Age Groups

Group (Year)	Acini Mean ± SEM (%)	Ducts Mean ± SEM (%)	Vascular Tissue Mean ± SEM (%)	Adipose Tissue Mean ± SEM (%)	Other Mean ± SEM (%)
16–35	48.1 ± 1.46	6.9 ± 0.56	4.7 ± 0.59	18.9 ± 1.61	21.4 ± 1.33
36–55	44.2 ± 2.15	6.7 ± 0.61	5.4 ± 0.38	22.5 ± 3.13	21.2 ± 1.18
56–75	38.7 ± 2.60	6.8 ± 0.53	5.4 ± 0.42	22.3 ± 2.81	27.7 ± 2.31
>75	32.5 ± 2.46	7.1 ± 0.51	6.1 ± 0.62	27.2 ± 3.56	27.1 ± 1.80

From Scott, J., Flower, E. A., and Burns, J., *J. Oral Pathol.*, 16, 505, 1987. With permission.

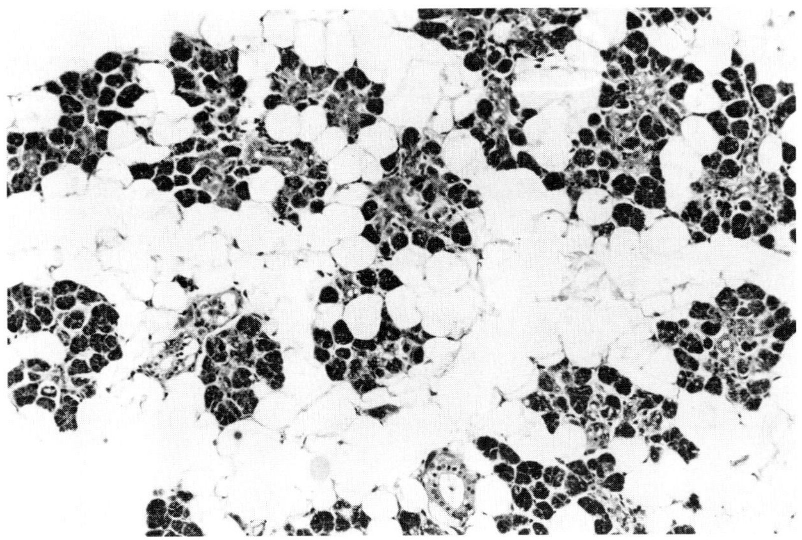

Figure 13. Parotid gland in a male aged 56 years showing intense infiltration by adipose tissue. H & E; × 120.

the parenchymal tissues which become replaced by fatty infiltration (Figure 13). The wide range of adipose infiltration between cases and across age groups suggests that any diagnosis of lipomatosis or adiposity in the parotid should be employed cautiously. The mere presence of extensive intralobular adiposity in the parotid cannot be considered diagnostic of any particular systemic disorder.

Other Changes

Lymphoid tissue occurring either as focal aggregates similar to those seen in the submandibular gland (Figure 14) or as organized follicles is normally present in the parotid glands of all age groups. Also small encapsulated nodes are

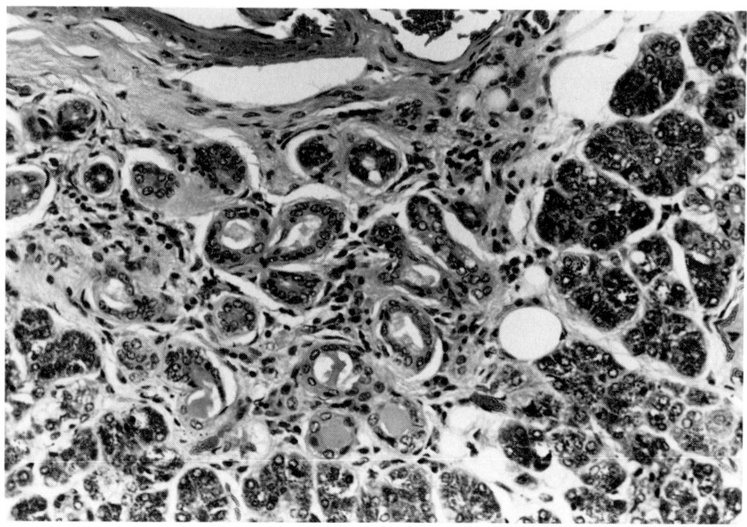

Figure 14. Focal obstructive adenitis in a parotid gland from a male aged 82 years. Acinar dedifferentiation to ductal forms is accompanied by fibrosis and a mild chronic inflammatory infiltrate. Note the typical well-demarcated wedge outline of this focus. H & E; × 228.

frequent within the parotid lobes. In general there is no clear relationship to aging although the total amount of lymphoid tissue appears to be reduced in the over-75-year age group (Scott et al., 1987). Degenerative arterial disease is present to some degree in all parotid glands from middle age onwards when it is usually accompanied by dilatation and tortuosity of blood and lymph vessels. Although small intraductal deposits occur in the parotid glands of elderly, they are much less numerous than in the submandibular glands and their relationship to aging remains unproved (Epivatianos and Harrison, 1989).

Quantitative Age Changes in Human Parotid Glands

Quantitative age changes in the human parotids have been examined by stereological analysis of 63 "sudden death" necropsies of both sexes (Table 4), aged 17 to 90 years (Scott et al., 1987). Table 5 shows that the mean proportional volume of acinar tissue exhibited a progressive and even reduction over the adult age span from 16 years upwards. This relationship was explored in greater detail by linear regression analysis and is depicted in Figure 15. While it is clear from these data that proportional acinar volume reduces with age the ductal component on the other hand remains stable over the adult life space (Table 5). As with the submandibular gland, increases in the adipose and fibrous content of the parotid provide the compensatory adjustment for the proportional losses suffered by the acinar tissue. It is of interest that the data in Table 5 show the increase in the fibrous component occurring mainly between the two middle-age groups

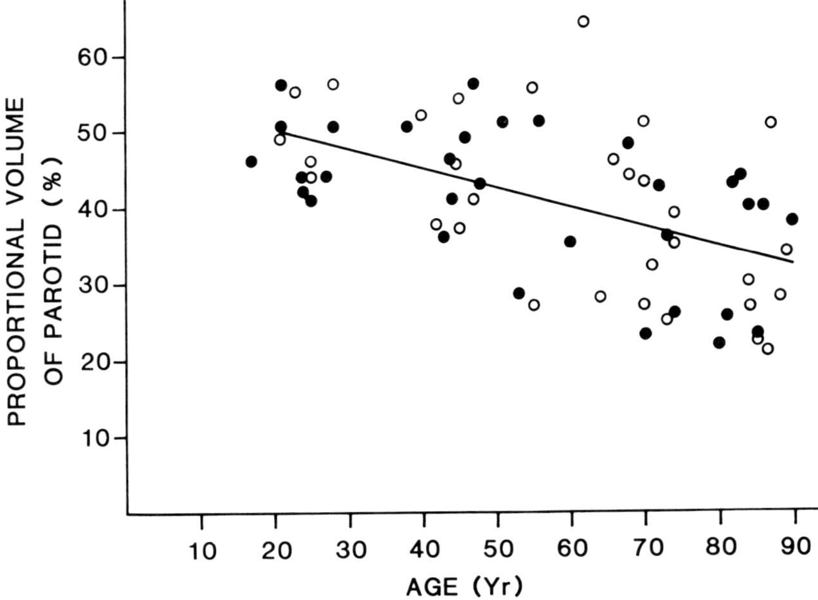

Figure 15. Scattergram for acini, as proportional volume of parotid tissue, and age for 63 necropsies, males (●) and females ( ) depicted separately. The regression line has been fitted by calculation of the constants, a = 52.75 and b = –0.21 for which p <0.01. (Reprinted from Scott, J., Flower, E. A., and Burns, J., *J. Oral Pathol.*, 16, 505, 1987. With permission.)

TABLE 5. Proportional Volumes of Component Tissues (%) of Human Labial Salivary Glands in Three Age Groups

Age Group Years	Number	Acini Mean (SEM)	Ducts Mean (SEM)	Other Tissue[a] Mean (SEM)
18–40	20	56.6 (1.93)	8.5 (0.92)	34.9 (1.59)
41–65	24	40.8 (2.81)	14.1 (1.58)	45.2 (2.15)
66–90	24	32.1 (2.40)	17.4 (1.10)	50.5 (1.71)

[a] Other tissues consist mainly of fibrous and vascular tissues and include adipose tissue. This was less than 5% in most glands.

From Scott, J., Flower, E. A., and Burns, J., *J. Oral Pathol.*, 16, 505, 1987. With permission.

(35 to 75 years) whereas the increase in the adipose tissue is predominately a feature of old age. These changes are shown in summary in Figure 16.

The quantitative differences revealed by stereological analysis of the parotid gland are relative only and must be interpreted with care, particularly in relation to age changes in overall gland size. There is indirect evidence from radiological assessment (Ericson, 1970) that there is a slight shrinkage of parotid size with age. Thus, the reduced proportional acinar volume with age represents

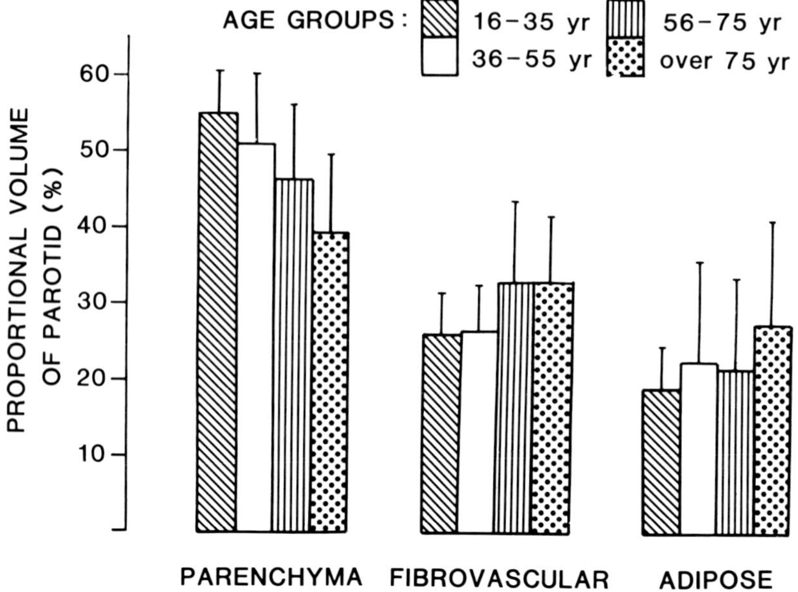

Figure 16. Summary of age changes in the major components of parotid glands in four age groups. Column heights show means with one standard deviation attached. (Reprinted from Scott, J., Flower, E. A., and Burns, J., *J. Oral Pathol.*, 16, 505, 1987. With permission.)

a true loss of secretory tissue which probably amounts to approximately one third of the acinar tissue present in adult maturity.

These structural age changes are of interest when compared with recent functional studies of parotid aging (Baum, 1981; Heft and Baum, 1984) which showed no significant loss of parotid outflow with age either under stimulation or at rest (see Chapter 19 this volume). The inconsistency of results between structural and functional studies of aging parotid requires some explanation. Either the reduced amount of parotid tissue in older people is working more efficiently than in the young or else in the latter group there may be a considerable reserve volume of secretory tissue which gradually depletes with age (Scott et al., 1987). If the second hypothesis is correct, the implication of a reserve of salivary tissue may hold considerable significance for the development of therapeutic strategies for patients with severe gland dysfunction.

THE HUMAN MINOR SALIVARY GLANDS

Qualitative Age Changes in Human Minor Salivary Glands

The minor salivary glands of man are widely distributed throughout all parts of the oral mucosa with the exception of the gingiva and hard palate. These glands are important in providing adequate lubrication and immunological and

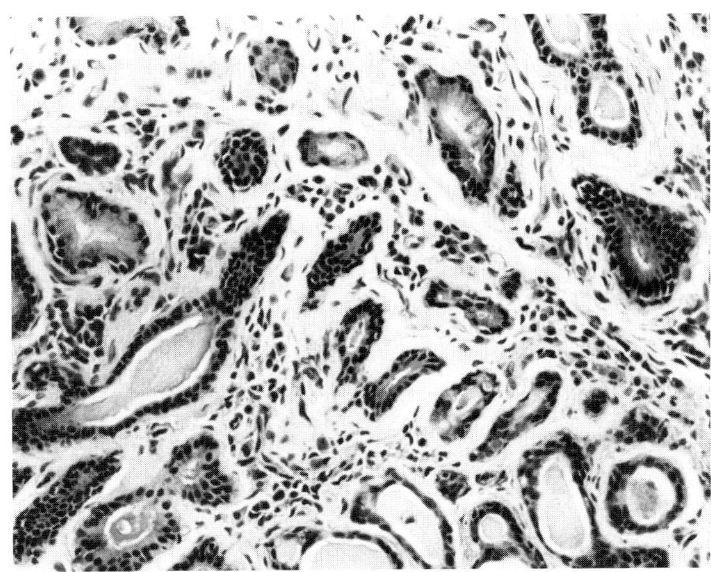

Figure 17. Severe ductal hyperplasia and acinar atrophy with fibrosis and chronic inflammation in a labial gland from a male aged 78 years. There is an increase in the number of ducts with multilayering of some duct walls. H & E; × 280.

antibacterial protection of the mucosal surfaces, particularly under resting conditions (Tabak et al., 1982). Any impairment in flow or composition of saliva from these glands in old age could adversely affect the oral mucosa which itself is atrophied by aging (Scott et al., 1983). The most extensive aging studies carried out on minor salivary glands have involved the lower labial glands (Scott, 1980; Drummond and Chisholm, 1984; Syrjanen, 1984; De Wilde et al., 1986). It is clearly important to establish the nature and extent of aging changes in these glands as they are frequently biopsied in the investigation of functional salivary disorders. The histology of human lingual salivary glands has also been studied through a range of ages (Soames, 1973; Drummond, 1986).

Parenchymal Changes

Acinar atrophy, ductal proliferation, and fibrosis are consistent age-dependent changes in human minor salivary glands (Figure 2). Ductal dilatation within the lobules (Figure 17) is also a feature of aging in these minor glands (Scott, 1980; Drummond and Chisholm, 1984) and is particularly marked in glands from individuals aged 50 years and older (Syrjanen, 1984). Severe levels of acinar atrophy which include degranulation and shrinkage of the cells with reversion to duct-like forms and reductions in both size and numbers of acini are frequently found in the labial glands of elderly subjects but rarely in the young (Scott, 1980; Syrjanen, 1984). Possibly, both the duct dilatation and the acinar atrophy in these superficially exposed glands may be as much a product of repeated minor trauma and partial obstruction over a prolonged period of time as an inherent feature of the aging processes. However, similar age changes develop in the palatal and lingual glands although not so severely as in the labial glands (Drummond, 1986).

Oncocytosis

Oncocytes increase in frequency with increasing age in human minor salivary glands (Hamperl, 1931; Soames, 1973; Shimono and Yamamura, 1975; Scott, 1980). In the labial glands dilatation of the ducts is often associated with oncocytosis of the lining cells, particularly in wide dilatations (Scott, 1980), and may thus be related to the oncocytic changes observed in mucous retention cysts and in the cystic forms of obstructive sialoadenitis (Southam, 1974).

Age-Related Fibrosis

An age-related increase in the amount and density of the fibrous component occurs in the labial glands in man. Age-dependent atrophy of the mucous glands in the tongue is also accompanied by fibrous replacement, but fibrosis is not a feature of aging lingual serous glands (Drummond, 1986).

Lipid Accumulation and Adiposity with Increasing Age

Accumulation of fat with aging in human labial glands tends to be less extensive than in the submandibular and parotid glands (Scott, 1980; Syrjanen, 1984; Drummond and Chisholm, 1984). This is also true for mucous glands in the dorsum of the tongue (Drummond, 1986). However, in lingual serous glands there is extensive replacement of acinar cells by adipose tissue. Whether this difference between the minor glands of the tongue is due to differing metabolism is not known, but Garrett (1962) has noted that serous cells contain larger droplets of intracellular fat than mucous glands.

Focal Inflammatory Changes in Human Minor Salivary Glands

A diffuse infiltrate of lymphoreticular cells of widely variable density occurs in the interparenchymal connective tissue of human minor salivary glands but is unrelated to age (De Wilde et al., 1986). Focal aggregates of lymphocytes also occur in the minor glands (Chisholm et al., 1970; Scott, 1980; Takeda and Komori, 1986). Although such foci were initially considered to occur exclusively in rheumatoid disease, several recent reports have suggested they may occasionally occur in otherwise normal glands in nonrheumatoid subjects (Scott, 1980; De Wilde et al., 1986) where they probably increase with age (Syrjanen, 1984; Takeda and Komori, 1986). Thus, labial focal lymphocytic adenitis can no longer be regarded as pathognomonic for autoimmune exocrinopathy (Scott, 1980; De Wilde et al., 1986).

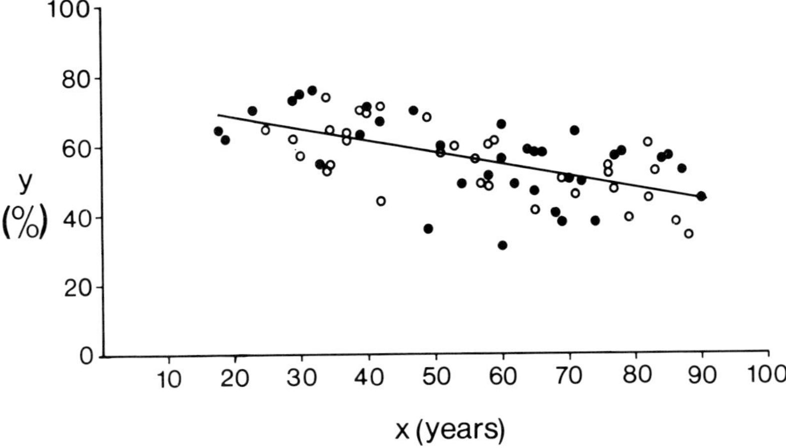

Figure 18. Age-related decrease in proportional volume of labial gland parenchyma (y) expressed as percentage of total gland volume plotted against adult age (x) in years. Males (●) and females (○) are depicted separately. The regression line, y = a + bx was calculated for 68 cases and gave a = 74.23 and b = –0.325 for which *p* <0.001. (Reprinted from Scott, J., *J. Frontiers of Oral Biology*, S. Karger, Basel, 1987, 58. With permission.)

Quantitative Age Changes in the Structure of Human Minor Salivary Glands

Quantitative age changes in the proportional volumes of component tissues of human labial glands have been assessed in several stereological investigations. In a series of 68 subjects aged 18 to 90 years examined at necropsy, after excluding drug medications or diseases likely to influence salivary structure, Scott (1980) reported a progressive decline in the proportional volume of labial salivary parenchyma with increasing age (Figure 18). This parenchymal change occurred entirely within the acinar component whereas the ductal proportional volume doubled between the young and old age groups (Table 5). The lost acinar tissue was replaced mainly by fibrovascular tissue while adipose infiltration was only rarely seen. These findings were generally confirmed in other post-mortem studies of the labial glands and of the lingual glands (Drummond and Chisholm, 1984; Drummond, 1986) and in a series of labial gland biopsies obtained from oral-surgery patients in whom systemic diseases had been excluded (De Wilde et al., 1986). In De Wilde's investigation the increasing ductal proportional volume in older age groups was shown to be due mainly to intralobular ductal hyperplasia rather than to dilatation which nevertheless also increases with age, occurring as a localized phenomenon (Figure 17).

The number of salivary lobules in standardized specimens of labial mucosa does not increase significantly with age (Scott, 1980). It is likely, therefore, that the reported age-dependent reduction in proportional acinar volume represents a true loss of secretory acinar tissue. This loss which amounts to more than 40%

of the acinar tissue present in the young age groups provides a structural explanation for the reports of an age-related decrease in labial salivary flow rate (Gandara et al., 1985).

AGE CHANGES IN THE SALIVARY GLANDS OF LABORATORY ANIMALS

A number of laboratory species have been used to study age-related morphological changes in salivary glands. However, most studies have been conducted on the submandibular glands of the rat. The laboratory rat has a life-span of approximately 30 months depending on the particular breed used and rats have been designated "old" at various ages conventionally corresponding to a mortality rate of 50%. The ductal system of the rat submandibular gland is complex and, in addition to intercalated, striated, and collecting ducts, includes a granular duct segment. Granular ducts differentiate from distal segments of the striated ducts at puberty, and exhibit sexual dichotomy, in their size and numbers being more prominent in males. Also, in rats, granular intercalated duct cells are present at the acinar-intercalated duct junction of the submandibular gland (Qwarnstrom and Hand, 1983).

Although submandibular function in the rat is apparently unimpaired by aging (Bodner and Baum, 1984), many earlier morphological studies of rat salivary glands indicated that aging in this animal is accompanied by degenerative changes in parenchymal structure. Recent studies have sought to demonstrate quantitative histological age-related changes in the submandibular and sublingual glands of the rat by stereological analysis and others have employed techniques for ultrastructural examination and cytochemistry to confirm a range of nuclear and cytoplasmic changes in acinar and ductal cells from old rats.

Qualitative Changes in Rodent Salivary Glands

Histological Changes in Parenchyma

Observations on rats and mice suggest that the histological changes with age in laboratory animals are less extensive than in humans. Acinar cells in parotid and submandibular glands of old rats are larger, more eosinophilic, have larger nuclei, and more prominent nucleoli and are more frequently binucleated than in young adult rats (Andrew, 1949a; Church, 1955; Bogart, 1967, 1970; Sashima, 1986; Scott et al., 1986). Intercalated ducts of rat submandibular glands are larger, more prominent, and sometimes dilated in old rats (Scott et al., 1986) and their granular cell content becomes more prominent with age (Bogart, 1970). The granular duct segment, on the other hand, exhibits regressive changes including nuclear pyknosis (Andrew, 1949b), cell shortening, and loss of granular density with age (Sashima, 1986). Similar age-dependent regression of the granular ducts of male mice were reported by Gresik and Azmitia (1980) who also showed a decrease in the number of granular duct cells immunocytochemically stained for epidermal growth factor and nerve growth in senescent mice.

Oncocytic changes were noted in the parotid and submandibular acinar cells of senile rats by Andrew (1949a, 1949b). Other authors have suggested that the oncocytes of rat salivary glands occur mainly by transformation of intercalated duct cells or, to a lesser extent, granular duct cells (Bogart, 1970; Sashima, 1986). Oncocytes also occur in the granular duct segment of aging mouse submandibular glands (Takeda et al., 1985) and in the striated duct segment of salivary glands of senile dogs (Goodpasture, 1918).

Intracellular lipid droplets have been observed at the light microscope level in the acini and ducts of parotid and submandibular glands of aged rats (Andrew, 1949a; Bogart, 1970). Lipofuscin granules occur in the parenchymal cells of submandibular glands of rats and increase in size and number with increasing age (Bogart, 1967; Sashima, 1986). Intraductal deposits of apparently solid material have been reported in the parotid glands of old rats (Andrew, 1949a) and similar (PAS positive) deposits have been found in the lumens of intercalated or small ducts of rats over 22 months of age (Sashima, 1986).

An increase in the amount of adipose tissue is a feature of aging in the rat parotid (Andrew 1949a; Kim et al., 1980) but the true extent of this change is difficult to assess because the gland itself is enclosed in adipose tissue so that the true margin of the original gland may be difficult to discern. Increased adiposity is not a feature of aging in rat submandibular or sublingual glands (Andrew, 1949b; Scott et al., 1986). A limited increase in the fibrous component of the submandibular glands of old rats was noted by Scott et al. (1986) who also reported occasional focal areas of acinar degeneration and fibrosis in these glands.

Ultrastructural Changes

Early studies investigated the extent of cytoplasmic changes in the serous acinar cells of senile rats. Abnormalities were demonstrated in the size, shape, and distribution of mitochondria and in the configuration and staining of the Golgi in the rat parotid gland (Kurtz, 1954) while reduced endoplasmic reticulum and increased lipofuscin pigment were reported in the submandibular glands of aged rats (Bogart, 1970). The increase in lipofuscin granules, often in association with lipid droplets, was later demonstrated in parotid acinar cells of aged rats (Kim, 1984). Both these structures and degenerating secretory granules were associated with acid phosphatase indicating their lysosomal nature.

More recent studies have focused on ductal changes both in the mouse and in the rat. They largely confirm and extend the findings reported at the light-microscopic level of shorter and fewer granular duct cells with depleted secretory granules (Gresik et al., 1982; Sashima, 1986). Similar regressive changes were also found in the intercalated granular duct cells of aged rats along with reduced RER and Golgi complexes (Sashima et al., 1988) which is contrary to the earlier finding (Bogart, 1970) of increased prominence of the juxta-acinar granular cells of the intercalated ducts. Mucinous and vacuolar secretory granules and intracytoplasmic crystalloids consisting of electron-dense, needle-shaped, or rectangular structures up to 800 nm long, were also reported in the granular ducts of aged rats by Sashima (1986). While interpretation of the detailed changes in the acinar and ductal cytoplasm of aged mice and rats is often incomplete, the inference to be taken from all these studies is that the synthesis of secretory products is likely to be impaired in senile animals (Kim et al., 1980).

Changes in Lymphocytic Tissues

Periductal infiltrations of lymphocytes and plasma cells and the occurrence of proliferating foci of lymphocytic cells plus increased prevalence of mast cells have been reported as age changes in the parotid and submandibular glands of the rat (Andrew, 1949a,b). Focal lymphocytic adenitis occurs in mice and increases with age both in control and autoimmune-disease-prone strains (Scott et al., 1990).

Aging in man and animals is accompanied by progressive degeneration of lymphoid tissue and systemic immune dysfunction (Makinoda and Kay, 1980). However, the cellular mechanism for immune deficiencies of T- and B-cell function that coincide with aging has not yet been delineated (Ebersole et al., 1988). A recent study has examined the distribution of lymphocytic phenotypes in various tissues, including submandibular and parotid salivary glands from rats at different ages from weanling to 20 months. Decreases in T-helper/T-suppressor ratios, as well as decreased numbers of plasma cell precursors in the secretory tissues of the older rats, suggest that alteration in normal secretory immune responses accompanies the aging process in rats (Ebersole et al., 1988).

Quantitative Age Changes in Rat Salivary Glands

There is no evidence of gross atrophy of rat salivary glands with age. As rats age their body weight continues to increase and the proportion of submandibular gland weight to total body weight remains constant throughout the life of the animal (Kuyatt and Baum, 1981).

Studies employing stereological analysis have been carried out recently on the submandibular and sublingual salivary gland of the rat (Scott et al., 1986). Their results largely confirm the findings of the qualitative changes with age recounted above. While there would appear to be a small but significant loss of proportional volume of acinar tissue in the submandibular gland between young (6 months) and old (24 months) rats (Figure 19), other studies have failed to confirm this age-related acinar loss (Sashima et al., 1986; Scott et al., 1989). In the sublingual gland there are no volumetric tissue changes between young and old rats (Scott et al., 1986). Where age-related volumetric tissue changes have been demonstrated, (e.g., Figure 19), the amounts of tissue losses sustained between young and old groups are far short of those described in human glands over the normal adult life span (Tables 2, 4, and 5). This discrepancy may reflect

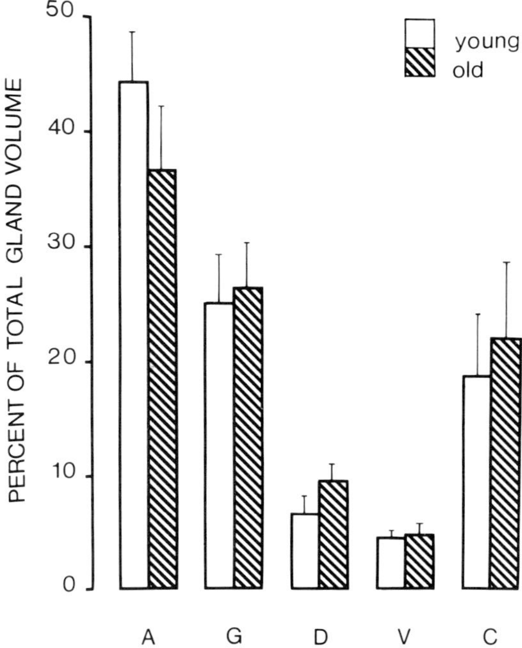

Figure 19. The proportional volumes of component tissues of the rat submandibular gland (mean ± SE) compared between two groups of young (n = 11) and old (n = 12) male rats. A = acini; G = granular ducts; D = ducts; V = vascular tissue; c = connective tissues. The age-related reduction in acinar volume and increase in ductal volume were significant at $p < 0.01$ and $p < 0.001$, respectively. (Reprinted from Scott, J., *Arch. Oral Biol.*, 311, 69, 1986. With permission.)

the comparatively young age of the oldest rats in these studies (24 months) which are more likely equivalent to the middle decades of the human life span rather than the 8th and 9th decades which are usually represented in series of human glands used in aging studies (Figures 11, 15, 18; Tables 2 and 4). Other stereological investigations have indicated that the submandibular granular duct component of rats undergoes a decrease in proportional volume between 6 and 22 months of age with a concomitant reduction in cells with PTAH positive granules. Using a similar quantitative technique, Gresik and Azmitia (1980) demonstrated that in mice also there is a pronounced fall in the ratio of granular tubular cells to acinar cells between 12 to 29 months.

SUMMARY

In this chapter we have presented a detailed account of the structural changes which occur in different salivary glands and different species during the course of adult life. Although in general these changes suggest a reduced volume of tissue available for secretory function, they should not automatically be regarded as being of a regressive or degenerative nature. Undoubtedly some tissue damage does occur with the passage of time and where, because of their superficial location, glands are vulnerable to minor irritation and mild trauma, it is not surprising that pathological changes — chronic inflammation, fibrous acinar degeneration, ductal metaplasia, and hyperplasia — are frequent. On the other hand, the bulk of tissue losses sustained by major salivary glands both in man and in the rat are not immediately evident as cytologically demonstrable defects. Rather, these losses are slow, insidious, and virtually undetectable without recourse to special techniques of histological measurement. Under these circumstances the structural changes may well reflect an equilibrium response to gradually changing functional requirements and metabolic demands in old age. The balance between structural integrity and functional ability is thus likely preserved in old age, but its operative level may be being continuously redefined. If so, then the reduced secretory tissue mass of old age is as much a feature of normality for the aged state as is the higher parenchymal volume encountered in the young adult.

REFERENCES

1. Andrew, W., Age changes in the parotid glands of Wistar Institute rats with special reference to the occurrence of oncocytes in senility, *Am. J. Anat.*, 85, 157, 1949a.
2. Andrew, W., Age changes in the salivary glands of Wistar Institute rats with particular reference to the submandibular glands, *J. Gerontol.*, 4, 95, 1949b.
3. Andrew, W., A comparison of age changes in salivary glands of man and of the rat, *J. Gerontol.*, 7, 178, 1952.
4. Bauer, W. H., Old age changes in human parotid glands with special reference to peculiar cells in uncommon salivary gland tumors (Abstract), *J. Dent. Res.*, 29, 686, 1950.
5. Baum, B. J., Research on aging and oral health: an assessment of current status and future needs, *Spec. Care Dent.*, 1, 156, 1981a.
6. Baum, B. J., Evaluation of stimulated parotid saliva flow rate in different age groups, *J. Dent. Res.*, 60, 1292, 1981b.
7. Blatt, I. M., Denning, R. M., Zumberge, J. H., and Maxwell, J. H., Studies in sialolithiasis. I. The structure and mineralogical composition of salivary gland calculi, *Ann. Otol. Rhinol. Laryngol. (St. Louis)*, 67, 595, 1958.
8. Bodner, L. and Baum, B. J., Submandibular gland secretory function in young adult and aged rats, *Comp. Biochem. Physiol.*, 77A, 235, 1984.
9. Bogart, B. I., The effect of aging on the histochemistry of the rat submandibular gland, *J. Gerontol.*, 22, 372, 1970.
10. Bogart, B. I., The effect of aging in the rat submandibular gland: an ultrastructural, cytochemical and biochemical study, *J. Morph.*, 130, 337, 1970.
11. Chisholm, D. M., Waterhouse, J. P., and Mason, D. K., Lymphocytic sialadenitis in the major and minor glands: a correlation in postmortem subjects, *J. Clin. Pathol.*, 23, 690, 1970.
12. Church, E. L., Age changes in the nucleus of salivary glands of Wistar Institute rats, *Oral Surg.*, 8, 301, 1955.

13. De Wilde, P. C. M., Baak, J. P. A., van Houwelingen, J. C., Kater, L., and Slootweg, P. J., Morphometric study of histological changes in sublabial salivary glands due to aging process, *J. Clin. Pathol.*, 39, 406,1986.

14. Drummond, J. R. and Chisholm, D. M., A qualitative and quantitative study of the ageing human labial salivary glands, *Arch. Oral Biol.*, 29, 151, 1984.

15. Drummond, J. R., The histology of the human lingual salivary glands, *I.R.C.S. Med. Sci.*, 14, 116, 1986.

16. Drummond, J. R. and Chisholm, D. K., A qualitative and quantitative study of the ageing human labial glands, *Arch. Oral Biol.*, 29, 151, 1984.

17. Ebersole, J. L., Steffen, M. J., and Pappo, J., Secretory immune responses in ageing rats II. Phenotype distribution of lymphocytes in secretory and lymphoid tissues, *Immunology*, 64, 28, 1988.

18. Epivatianos, A., Harrison, J. D., Garret, J. R., Davies, K. J., and Senkus, R., Ultrastructural and histochemical observations on intracellular and luminal microcalculi in the feline sublingual salivary gland, *J. Oral Pathol.*, 15, 513, 1986.

19. Epivatianos, A. and Harrison, J. D., The presence of microcalculi in normal human submandibular and parotid salivary glands, *Arch. Oral Biol.*, 34, 262, 1989.

20. Ericson, S., The normal variation of the parotid size, *Acta Otolaryngolica*, 70, 294, 1970.

21. Gandara, B. K., Izutsu, K. T., Truelove, E. L., Ensign, W. Y., and Sommers, E. E., Age-related salivary flow rate changes in controls and patients with oral lichen planus, *J. Dent. Res.*, 64, 1149, 1985.

22. Garrett, J. R., Some observations on human submandibular glands, *Proc. R. Soc. Med.*, 55, 488, 1962.

23. Garrett, J. R., The ultrastructure of intracellular fat in the parenchyma of human submandibular salivary glands, *Arch. Oral Biol.*, 8, 729, 1963.

24. Gresik, E. W. and Azmitia, E., Age related changes in NGF, EGF and protease in the granular convoluted tubules of the mouse submandibular gland. A morphological and immunocytochemical study, *J. Gerontol.*, 35, 520, 1980.

25. Gresik, E. W., Brennan, M., and Azmitia, E., Age related changes in EGF and protease in submandibular glands of C57BL/6J mice, *Exp. Aging Res.*, 8, 87, 1982.

26. Hamperl, H., Beitrage zur normalen and pathologischen Histologie menschlicher Speicheldrusen, *Z. Zellforsch.*, 27, 1, 1931.

27. Harrill, J. A., King, J. S., and Boyce, W. H., Structure and composition of salivary calculi, *Laryngoscope (St. Louis)*, 69, 481, 1959.

28. Heft, M. W., and Baum, B. J., Unstimulated and stimulated parotid salivary flow rate in individuals of different ages, *J. Dent. Res.*, 63, 1182, 1984.

29. Kim, S. K., Changes in the secretory acinar cells of the rat parotid gland during aging, *Anat. Rec.*, 209, 345, 1984.

30. Kim, S. K., Weinhold, P. A., Han, S. S., and Wagner, D. J., Age related decline in protein synthesis in the rat parotid gland, *Exp. Gerontol.*, 15, 77, 1980.

31. Kurashima, C. and Hirokawa, K., Age-related increase of focal lymphocytic infiltration in the human submandibular glands, *J. Oral Pathol.*, 15, 172, 1986.

32. Kurtz, S. M., Cytologic studies of the salivary glands of the rat in reference to the aging process, *J. Gerontol.*, 9, 421, 1954.

33. Kuyatt, B. L. and Baum, B. J., Characteristics of submandibular glands from young and aged rats, *J. Dent. Res.*, 60, 936, 1981.

34. Makinoda, T. and Kay, M. M. B., Age influence on the immune system, *Adv. Immunol.*, 29, 287, 1980.

35. Qwarnstrom, E. E. and Hand, A. R., A granular cell at the acinar intercalated duct junction of the rat submandibular gland, *Anat. Rec.*, 235, 295, 1983.

36. Sashima, M., Age-related changes of rat submandibular gland: a morphometric and ultrastructural study, *J. Oral Pathol.*, 15, 507, 1986.

37. Sashima, M., Hatakeyama, S., Satoh, H., and Suzuki, A., Age-related changes of the granular intercalated duct cells of male rat submandibular gland, *Arch. Oral Biol.*, 33, 71, 1988.

38. Scott, J., Age, sex and contralateral differences in the volumes of human submandibular salivary glands, *Arch. Oral Biol.*, 20, 885, 1975.

39. Scott, J., The incidence of focal chronic inflammatory changes in human submandibular salivary glands, *J. Oral Pathol.*, 5, 334, 1976a.

40. Scott, J., Aging in the Human Submandibular Gland: A Morphometric Study, Ph.D. Thesis, University of Liverpool, Liverpool, 1976b.

41. Scott, J., Degenerative changes in the histology of the human submandibular salivary gland occurring with age, *J. Biol. Buccale*, 5, 311, 1977a.

42. Scott, J., Quantitative age changes in the histological structure of human submandibular salivary glands, *Arch. Oral Biol.*, 22, 221, 1977b.

43. Scott, J., A morphometric study of age changes in the histology of the ducts of human submandibular salivary glands, *Arch. Oral Biol.*, 22, 243, 1977c.

44. Scott, J., The prevalence of consolidated salivary deposits in the small ducts of human submandibular glands, *J. Oral Pathol.*, 7, 28, 1978.

45. Scott, J., Qualitative and quantitative changes in the histology of the human submandibular salivary gland during post natal growth, *J. Biol. Buccale*, 7, 341, 1979.

46. Scott, J., Qualitative and quantitative observations on the histology of human labial salivary glands obtained post mortem, *J. Biol. Buccale*, 8, 187, 1980.

47. Scott, J., Valentine, J. A., St. Hill, C. A., and Balasooriya, B. A. W., A quantitative histological analysis of the effects of age and sex on human lingual epithelium, *J. Biol. Buccale*, 11, 303, 1983.

48. Scott, J., Structure and function in aging human salivary glands, *Gerodontology*, 3, 149, 1986.

49. Scott, J., Bodner, L., and Baum, B. J., Assessment of age-related changes in the submandibular and sublingual salivary glands of the rat using stereological analysis, *Arch. Oral Biol.*, 31, 69, 1986.

50. Scott, J., Flower, E. A., and Burns, J., A quantitative study of histological changes in the human parotid gland occurring with adult age, *J. Oral Pathol.*, 16, 505, 1987.

51. Scott, J. and Gradwell, E., A quantitative study of the effects of chronic hypoxia on the histological structure of the rat major salivary glands, *Arch. Oral Biol.*, 5, 315, 1989.

52. Scott, J., Baum, B. J., Woods, K., and Berry, M., Chloride handling by submandibular cells from young and old rats with and without duct ligation, *Mech. Ageing Dev.*, 48, 231, 1989.

53. Scott, J., Wolff, A., and Fox, P. C., Histological assessment of the submandibular glands in autoimmune-disease-prone mice, *J. Oral Pathol. Med.*, 19, 131, 1990.

54. Shimono, M. and Yamamura, T., Ultrastructure of the oncocyte in normal human palatine salivary glands, *J. Electron Miscrosc.*, 24, 119, 1975.

55. Soames, J. V., A review of the histology of the tongue in the region of the foramen caecum, *Oral Surg.*, 36, 220, 1973.

56. Southam, J. C., Retention mucoceles of the oral mucosa, *J. Oral Pathol.*, 3, 197, 1974.
57. Syrjanen, S., Age related changes in structure of labial minor salivary glands, *Ageing*, 13, 159, 1984.
58. Tabak, L. A., Levine, M. J., Mandel, I. D., and Ellison, S. A., Role of salivary mucins in the protection of the oral cavity, *J. Oral Pathol.*, 11, 1, 1982.
59. Takeda, Y., Suzuki, A., and Ishikawa, G., Nodular hyperplasia of oncocytes in mouse submandibular glands, *J. Oral Pathol.*, 14, 182, 1985.
60. Takeda, Y. and Komori, A., Focal lymphocytic infiltration in the human labial salivary glands: a post-mortem study, *J. Oral Pathol.*, 15, 83, 1986.
61. Tandler, B., Fine structure of oncocytes in human salivary glands, *Virchows Arch. Abt. A Path. Anat.*, 341, 317, 1966.
62. Waterhouse, J. P., Chisholm, D. M., Winter, R. B., Patel, M., and Yale, R. S., Replacement of functional parenchymal cells by fat and connective tissue in human submandibular salivary glands: an age-related change, *J. Oral Pathol.*, 2, 16, 1973.
63. Wakeley, C., The surgery of the salivary glands, *Ann. R. Coll. Surg. Engl.*, 3, 289, 1948.
64. Yamaguchi, S., Studien uber die Mundspeicheldrusen. I. Uber das Fett. *Beitr. Path. Anat. U. Allg. Path.*, 73, 113, 1925.

Age-Related Changes in the Function of Salivary Glands

Kathleen Dobrosielski-Vergona

INTRODUCTION

It is not clear how the aging process influences saliva production. Investigators report no change, as well as a decline in the flow of whole saliva in human subjects. The majority of publications appearing before 1980 generally support a decline in salivary flow from measurements of whole saliva. Following these studies came the realization that many medications taken by older adults have side effects that include a diminution of salivary flow. More recent studies, which employ subjects not taking medications fail to demonstrate the age-related decline in the flow rate of whole saliva. Authors of these reports comment on the wide variability in flow rate. This variation, in addition to the effects of medications, compound the problem of assessing the effect of age on the production and flow of whole saliva.

Several laboratories reported an effect of aging on both stimulated, as well as unstimulated, salivary flow. The major stimuli are chewing paraffin wax, drops of 2% citric acid on the tongue, and sucking a citrus flavored candy drop. The inconsistency of the stimulus further complicates the comparison of the data reported on salivary flow rate. Chewing, sucking, or tongue swabbing may each have different effects on the production and/or release of whole saliva with or without an age dependent factor. In addition, the psychological state of an individual affects salivary flow. Unless data from longitudinal studies on nonmedicated individuals who are stimulated by the same technique and free of psychological anomaly are used, there will remain a degree of uncertainty about the role of aging in salivary flow rate. Given the important role of saliva in the maintenance of oral health and general physiology, any alteration in the regulation of salivary flow and/or the composition of saliva with aging will have potential significance to the quality of life of older individuals.

FLOW RATES

Mixed or Whole Saliva

The literature includes evidence for a decrease in the salivary flow rate of whole saliva as well as data suggesting that there is no decline in whole saliva with aging. Those studies that consider salivary flow in humans, without excluding subjects on medications, generally conclude that ageing does impair salivary production (Gutman and Ben-Aryeh, 1974). However, current pharmacologic information on the drugs most frequently prescribed for older individuals (antidepressants, antihypertensives, and cardiovascular therapies) have side effects that alter saliva production (Sreebny and Schwartz, 1986). Several studies using subjects that were unmedicated do not show a statistically significant alteration in the flow rate of whole saliva (Parvinen and Larmas, 1982; Gandara et al., 1985; Baum, 1989).

Parotid Saliva

The Carlson-Crittendon cup simplified the collection of parotid saliva from Stenson's duct of human subjects (see Schneyer and Schneyer, 1967). Studies focusing on the function of the parotid gland do not show any changes in the flow rate during aging. No age-related differences in flow rate from the parotid with or without stimulation are reported in unmedicated persons (see Heft and Baum, 1984; Scott, 1986; Baum, 1981). A wide range in normal flow rates from the parotid suggest that these values may be more meaningful if the gland size and/ or volume are considered. Such incomplete data may mask an aging effect on the parotid glands. Age does alter the saliva production in rats. Several laboratories report this decline in function. For example a 50% decline in stimulated parotid salivary flow was reported in both male and female Wistar rats (see Bodner and Baum, 1985).

Submandibular/Sublingual Glands

Comparisons of the salivary flow rates under unstimulated and 2% citrate stimulated conditions from the submandibular/sublingual glands in human subjects reveal no decrease in function (Tylenda et al, 1988; Tylenda and Baum, 1988). An earlier report by Pedersen et al. (1985) reported a decline in the secretion rate from human submandibular/sublingual glands with age. This conflict may be due to differences in the collection technique. Two possible explanations for differences in flow rates with age are alteration of acinar elements and/or a reduced glandular volume and weight.

Bodner and Baum (1984) reported a generally constant flow rate from the submandibular glands regardless of age in male and female Wistar rats. Stimulation with pilocarpine had the same effect in young and old rats.

Minor Glands

An age-related decrease in the salivary flow rate from human labial minor glands was reported by Gandara et al. (1985). The authors collected saliva on 4-mm diameter discs of Whatman #41 filter paper. This technique may have underestimated the actual saliva that was released. Drummond and Chisholm (1984) described morphological changes during aging in human minor glands. The increased fibrocity of these glands observed by Drummond and Chisholm may contribute to the decrease in salivary flow reported above. Generally, the flow patterns and secretion rates of the minor glands are poorly understood. The techniques required for such evaluations are being refined.

RESPONSE TO STIMULI

α-Adrenergic

Salivary flow is regulated, in part, by α and β adrenergic stimuli. Aging does alter the response of rat acinar cells to α-adrenergic stimulation. Baum (1989) reported a deficiency in the α_1-adrenergenic receptor-stimulated K^+ efflux in old rats. However, following muscarinic-cholinergic stimulation, K^+ efflux and Ca^{2+} efflux were not deficient in acinar cells from old animals. A significant decline in α_1-adrenergic-stimulated calcium efflux was observed in rats after 12 months of age (Ishikawa, et al, 1988). These authors also reported a 40% reduction in intracellular free calcium levels after α_1-adrenergic stimulation with age.

Bodner et al. (1983) have reported an age-related marked change in the response of rat salivary acinar cells to α-adrenergic secretion. No apparent change occurs in the number of α-binding sites per cell (Ishikawa et al., 1989). The age-dependent event that declines with age may occur somewhere between the receptor occupancy phospholipid turnover and Ca^{2+} mobilization, as suggested by Ito et al. (1982).

Baum et al (1983) reviewed evidence for considerable impairment of the α-adrenergic-mediated secretion of rat parotid cells. Further evidence for an age-related alteration in calcium metabolism, relevant to α_1-adrenergic stimulation, can be found in Gee et al. (1986) and Ambudkar et al., (1988). Ishikawa et al. (1989) presented an hypothesis that related this decline in α_1-adrenergic responsiveness to a defect in the ability of inositol triphosphate to release calcium from intracellular stores.

Unlike the parotid cells used in studies reported above, acinar cells from the rat submandibular gland do not reveal an age-dependent decline in the response to α_1-adrenergic stimuli. Baum et al., (1985) observed similar responses in submandibular gland cells from young and old rats, as measured by the mobilization of Ca^{2+} and the inhibition of protein synthesis by epinephrine.

β-Adrenergic

Age does not alter the secretory response of salivary glands to β-adrenergic agents. For example, Kim et al., (1982) followed the release of α-amylase from parotid lobules in 2- and 24-month old Sprague Dawley rats treated with isoproterenol. No age-related difference in response was observed.

Although the isoproterenol stimulation of amylase is not altered by age, both the DNA, cell division and cellular uptake of isoproterenol are markedly age dependent (Roth et al., 1974). Additional changes in the β-adenoreceptor number and function in salivary glands with age have been reported by Mysliwski (1977) and Piantanelli et al. (1980).

SALIVA COMPONENTS

Anionic proline rich proteins (PRP) are found in the secretory granules of the parotid gland. Baum et al. (1982) used PRP as a marker for exocytosis in the human parotid. The concentrations of PRP in parotid saliva were not altered during aging. Chauncey et al. (1987) reported results from a 9 year longitudinal analysis of age-related compositional changes in human parotid saliva. An overall decrease in the electrolyte, protein, and osmolality values was observed. An increased urea level with age was attributed to altered kidney function. Amylase is the most abundant enzyme in saliva. Studies reported by Aquirre et al. (1987) found no significant differences in α-amylase levels in human parotid saliva in subjects 23 to 84 years old. It must be pointed out that authors report a wide range of variability when measuring saliva proteins. This emphasizes the need for longitudinal studies.

Salivary mucins are known to have a major role in lubricating and protecting surfaces in the oral cavity (Tabak et al., 1990). Using the carbocyanine dye, Stains-all, the concentrations of two human salivary mucins, MG1 and MG2, are significantly reduced between two age groups: 18 to 35 years of age and 65 to 83 years of age. This difference was observed in stimulated and unstimulated saliva. Age-associated alterations in protein processing may contribute to these quantitative changes in mucins (Denny et al., 1991).

Animal studies do suggest an age-dependent reduction in the synthesis of exportable proteins, e.g., amylase (see Baum et al., 1982a). Gresik (1989) observed that a 40% decrease in the content of amylase in mouse submandibular glands is accompanied by a decrease in the rate of synthesis of secreted proteins. Kousvelari et al. (1988) reported a 20% reduction in protein synthesis by dispersed parotid acinar cells from rats.

A pronounced difference in this age effect on salivary glands was observed by Kuyatt and Baum (1981) when they compared data from experiments using male and female Wistar rats. In female rats, age altered the glycosylation of proteins. Unlike female rats, several biochemical parameters displayed age-related changes in male rats, including a decline in sialic acid and carbohydrate content of gland proteins and the amount of total gland protein. Additional references describing age related modifications in protein synthesis can be found in Baum et al. (1982a).

Epidermal growth factor (EGF) and nerve growth factor (NGF) are products of the mouse submandibular gland. Significantly fewer cells contained nerve growth factor in old mice (Gresik, 1989). A 50% reduction in the production of EGF was reported in senescent mice by Gresik and Azmitia (1980).

REGULATION OF IONS

Aging is accompanied by a defect in calcium-dependent events in salivary gland cells. Gee et al., (1986) reported a twofold decrease in epinephrine-stimulated calcium efflux in male Wistar rats during the latter half of the life span. One possible explanation for this observation is a functional defect in the ATP-dependent Ca^{2+} extrusion pathway (Ambudkar et al., 1988).

Biochemical analysis of parotid saliva in human studies reveal an alteration with age in the levels of pH, Na^{2+}, and Cl^- (see Scott, 1986). This information is based on cross-sectional data, however. Chauncey et al. (1987) reported on data from a 9 year longitudinal study revealing an overall decrease in electrolyte and osmolality values in human parotid saliva. The secretion of sodium in human parotid saliva is also altered with age which is consistent with a reported decline in chlorine secretion (see Baum et al., 1984).

Animal studies have shown that the effect of age on ion concentrations in saliva is dependent on gender. For example, Bodner and Baum (1985) reported an age-dependent increase in rat parotid salivary K^+ concentration in females but not in males. In this same paper, a comparison is made between the effects of age on the electrolyte secretions of the submandibular glands vs. the parotid glands. While the secretion of electrolytes by the rat submandibular gland is relatively unaltered, human submandibular gland secretion of electrolytes is age dependent. Further, the rat parotid secretion of electrolytes is significantly changed with aging, while the electrolyte secretory events of human parotid glands do not appear to be altered with age.

IMMUNE FUNCTION

Human parotid saliva secretory IgA values were compared in subjects participating in the Baltimore Longitudinal Study on Aging (BLSA). No correlation between age and secretory IgA was found (Aquirre et al., 1987). These results confirmed similar findings from other laboratories (Finkelstein et al., 1984).

The antimicrobial proteins in saliva can prevent oral infection and systemic access to serious pathogens. Total protein content of parotid saliva from males shows little change with age. The levels of specific proteins have been analyzed. Fox et al. (1987) reported an increase in lysozyme in unstimulated parotid saliva with age in both men and women. An age-related change was not seen in the levels of secretory IgA. If the data were expressed per volume of saliva, both lactoferrin and lysozyme amounts increased with age in stimulated parotid saliva.

The phenotype distribution of lymphocytes is altered during aging in the rat. Ebersole et al. (1988) reported a decreased Th cell percentage in the salivary glands of senescent rats. A decline in Th/Ts ratios with age suggested an age-related effect on the secretory immune function of these animals. A hamster model provided evidence for deficiencies in the oral immune response with age (Smith et al., 1983). After either primary or secondary immunization, the mean salivary IgA antibody levels were depressed in old (24 month) animals.

SUMMARY

Accurate assessments of saliva flow rates and possible compositional changes due to the aging process require meticulous, standardized experimental design. Further, longitudinal studies provide the most reliable data, given the heterogeneity of the elderly and the wide range of values for many salivary parameters. Most investigators report no change in salivary flow rates in older humans. The flow rate in rodents, in contrast, is altered by the aging process. Significant differences are found in the effects of age on the specific glands.

The response of rodent salivary glands to adrenergic stimuli is altered during aging. Distinct differences are observed between the age-related effects of adrenergic agents on a variety of cellular activities. Compositional changes with age in saliva and salivary gland cells are also documented in both humans and rodents. Particular age-dependent alterations in calcium handling are under investigation. Finally, aging appears to compromise the immune function of the salivary glands by effects on the production of secretory IgA, antimicrobial proteins, and T cell populations.

Future research is necessary to identify the specific salivary gland cell targets of aging. Eventually, genetic engineering may allow the shielding of these attacked sites in the genetic army of the salivary gland cells that are stricken during senescence. Such defense mechanisms could maintain oral health throughout the lifespan and contribute to the quality of life for those suffering from various pathologies of the salivary glands.

REFERENCES

1. Ambudkar, I. S., Kuyatt, S., Roth, G. S., and Baum, B. J., Modification of ATP-dependent Ca^{2+} transport in rat parotid basolateral membranes during aging, *Mech. Ageing Dev.*, 43, 45, 1988.
2. Aquirre, A., Levine, M. J., Cohen, R. E., and Tabak, L. A., Immunochemical guantitation of α-amylase and secretory I_gA in parotid saliva from people of various ages, *Arch. Oral Biol.*, 32, 297, 1987.
3. Baum, B. J., Evaluation of stimulated parotid saliva flow rate in different age groups, *J. Dent. Res.*, 60, 1292, 1981.
4. Baum, B. G., Salivary gland fluid secretion during aging, *J.A.G.S.*, 37, 453, 1989.
5. Baum, B. J., Costa, P. T., and Izutsu, K. T., Sodium handling by aging human parotid glands is inconsistent with a two-stage secretion model, *Am. J. Physiol.*, 246, R35, 1984.
6. Baum, B. J., Kuyatt, B. L., Helman, Y., Gee, M. V., and Roth, G. S., Alpha$_1$ adrenergic responsiveness of young adult and aged rat submandibular cells *in vitro*, *Comp. Biochem. Physiol.*, 81C, 121, 1985.
7. Baum, B. J., Kuyatt, B. L., and Humphreys, S., Protein production and processing in young adult and aged rat submandibular gland cells *in vitro*, *Mech. Ageing Dev.*, 23, 123, 1983.
8. Baum, B. J., Kousvelari, E. E., and Oppenheim, F. G., Exocrine protein secretion from human parotid glands during aging: stable release of the acidic proline-rich proteins, *J. Gerontol.*, 37, 392, 1982.
9. Baum, B. J., Levine, R. L., Kuyatt, B. L., and Sogin, D. B., Rat parotid gland amylase:evidence for alterations in an exocrine protein with increased age, *Mech. Ageing Dev.*, 19, 27, 1982a.
10. Bodner, L. and Baum, B., Submandibular gland secretory function in young adult and aged rats, *Comp. Biochem. Physiol.*, 77, 235, 1984.
11. Bodner, L. and Baum, B. J., Characteristics of stimulated parotid gland secretion in the aging rat, *Mech. Ageing Dev.*, 31, 337, 1985.
12. Bodner, L., Hoopes, M. T., Gee, M., Sto, H., Roth, G. S., and Baum, B. J., Multiple transduction mechanisms are likely involved in calcium-mediated exocrine secretory events in rat parotid cells, *J. Biol. Chem.*, 258, 2774, 1983.
13. Chauncey, H. H., Feller, R. P., and Kapur, K. K., Longitudinal age-related changes in human parotid saliva composition, *J. Dent. Res.*, 66, 599, 1987.
14. Denny, P. C., Denny, P. A., Klauser, D. K., Hong, S. H., Navazesh, M., and Tabak, L. A., Age-related changes in mucins from human whole saliva, *J. Dent. Res.*, 70, 1320, 1991.
15. Drummond, J. R. and Chisholm, D. M., A qualitative and quantitative study of the ageing human labial salivary glands, *Arch. Oral Biol.*, 29, 151, 1984.
16. Ebersole, J. L., Steffen, M. J., and Pappo, J., Secretory immune responses in ageing rats. II. Phenotype distribution of lymphocytes in secretory and lymphoid tissues, *Immunology*, 64, 289, 1988.

17. Finkelstein, M. S., Tanner, M., and Freedman, M. L., Salivary and serum I$_g$A levels in a geriatric outpatient population, *J. Clin. Immunol.*, 4, 85, 1984.

18. Fox, P. C., Heft, M. W., Bowers, M. R., Mandel, I. D., and Baum, B. J., Secretion of antimicrobial proteins from the parotid glands of different aged healthy persons, *J. Gerontol.*, 42, 466, 1987.

19. Gandara, B. K., Izutsu, K. T., Truelove, E. L., Ensign, W. Y., and Sommers, E. E., Age-related salivary flow rate changes in controls and patients with oral lichen planus, *J. Dent. Res.*, 64, 1149, 1985.

20. Gee, M. V., Ishikawa, Y., Baum, B., and Roth, G. S., Impaired adrenergic stimulation of rat parotid cell glucose oxidation during aging: the role of calcium, *J. Gerontol.*, 41, 331, 1986.

21. Gresik, E. W., Changes with senescence in the fine structure of the granular convoluted tubule of the submandibular gland of the mouse, *Am. J. Anat.*, 184, 147, 1989.

22. Gresik, E. W. and Azmitia, E. C., Age related changes in NGF, EGF and protease in the granular convoluted tubules of the mouse submandibular gland. A morphological and immunocytochemical study, *J. Gerontol.*, 4, 420, 1980.

23. Gutman, D. and Ben-Aryeh, G., The influence of age on salivary content and rate of flow, *Int. J. Oral Surg.*, 3, 314, 1974

24. Heft, M. W. and Baum, B. J., Unstimulated and stimulated parotid salivary flow rates in individuals of different ages, *J. Dent. Res.*, 63, 1182, 1984.

25. Ishikawa, Y., Gee, M. V., Ambudkar, I. S., Bodner, L., Baum, B. J., and Roth, G. S., Age-related impairment in rat parotid cell α_1-adrenergic action at the level of inositol triphosphate responsiveness, *Biochem. Biophys. Acta*, 968, 203, 1988.

26. Ishikawa, Y., Gee, M. V., Baum, B. J., and Roth, G. S., Decreased signal transduction in rat parotid cell aggregates during aging is not due to loss of alpha-adrenergic receptors, *Exp. Gerontol.*, 24, 25, 1989.

27. Ito, H., Baum, B. J., Uchida, T., Hoops, M. T., Bodner, L., and Roth, G., Modulation of rat parotid cell α-adrenergic responsiveness at a step subsequent to receptor activation, *J. Biol. Chem.*, 257, 9532, 1982.

28. Kim, S. K., Calkins, D. W., and Weinhold, P. A., Secretion of α-amylase from parotid lobules of young and old rats, *Exp. Gerontol.*, 17, 387, 1982.

29. Kousvelari, E. E., Banerjee, D. K., Murty, L., and Baum, B. J., N-linked glycosylation in the rat parotid gland during aging, *Mech. Ageing Dev.*, 42, 173, 1988.

30. Kuyatt, B. L. and Baum, B. J., Characteristics of submandibular glands from young and aged rats, *J. Dent. Res.*, 60, 936, 1981.

31. Mysliwski, A., Age-related changes in course of restitution of secretory material in rat submandibular gland stimulated with isoproterenol, *Exp. Gerontol.*, 12, 81, 1977.

32. Parvinen, T. and Larmas, M., Age dependency of stimulated salivary flow rate, pH and lactobacillus and yeast concentrations, *J. Dent. Res.*, 61, 1052, 1982.

33. Pedersen, W., Schubert, M., Izutsu, K., Mersai, T., and Truelove, E., Age-dependent decreases in human submandibular gland flow rates as measured under resting and post-stimulation conditions, *J. Dent. Res.*, 64, 822, 1985.

34. Piantanelli, L., Fattoretti, P., and Viticchi, C., Beta-adreno-receptor changes in submandibular glands of old mice, *Mech. Ageing Dev.*, 14, 155, 1980.

35. Roth, G. S., Karoly, K., Britton, V. J., and Adelman, R. C., Age-dependent regulation of isoproterenol-stimulated DNA synthesis in rat salivary gland *in vitro*, *Exp. Gerontol.*, 9, 1, 1974.

36. Schneyer, L. H. and Schneyer, C. A., Inorganic composition of saliva, in *Handbook of Physiology*, Code, C. F., Ed., American Physiological Society, Washington, D.C., 1967, 497.
37. Scott, J., Structure and function in aging human salivary glands, *Gerodontology*, 5, 149, 1986.
38. Smith, D. J., Ebersole, J. L., and Taubman, M. A., Local and systemic immune responses in aged hamsters, *Immunology*, 50, 407, 1983.
39. Sreebny, L. M. and Schwartz, S. S., A reference guide to drugs and dry mouth, *Gerodontology*, 5, 75, 1986.
40. Tabak, L. A., Structure and function of human salivary mucins, in *Critical Reviews in Oral Biology*, Vol. 1, Alvares, O., Ed., CRC Press, Boca Raton, FL, 229, 1990.
41. Tylenda, C. A. and Baum, B. J., Oral physiology and the Baltimore longitudinal study of aging, *Gerodontology*, 7, 5, 1988.
42. Tylenda, C. A., Skip, J. A., Fox. P. A., and Baum, B. J., Evaluation of submandibular flow rate in different age groups, *J. Dent. Res.*, 67, 1225, 1988.

INDEX

A

AlF$_4^-$, 162, 170, 330

Abl, 230, 231–235

Accessory glands, 40

  parotid, 2

  submandibular glands, in bats, 43, 68, 70–72

    Miniopterus, 67

    plasminogen activator gene expression in bats, 65

Acetylcholine, 107, 153

  and bicarbonate secretion, 113

  calcium role in response, 154

  stimulus-exocytosis coupling mechanism, 182

N-Acetylglucosamine, 247

Acidic epididymal glycoprotein, 363–364

Acidic proline-rich proteins, 142

Acinar cells, 39, 366–369

  age-related changes, minor salivary gland, 419–420

  B1-positive, 378

  development, 344

  evolutionary divergence, 53–56, see also Evolutionary divergence of acinar cells

  intracellular transport organization, 81–82

  lectin binding, 17–19

  marker proteins, 366–369

  neonatal, 374–378

  neurotransmitter control, see Neurotransmitters, control of calcium mobilization

Acinar proteins, neonatal submandibular components, 365

Acinic cell carcinoma, 328

Acquired immune deficiency syndrome, 219–221

Acromegaly, 211

Actin, myoepithelial cells, 31

Actin-like filaments, exocytosis, 186

Actinomycosis, 220

Activator protein-1, 231

Acute sialadenitis, 206, 207

Additives, dietary, 143

Adenitis, see also Sialadenitis

  focal lymphocytic adenitis, 406, 420

  focal obstructive adenitis (FOA), 406–407

Adenoid cystic carcinoma, 254

Adenomatoid hyperplasia, 210–211

Adenosine triphosphatase

  calcium pump, 157–158

  sodium, potassium, 24, 26, 27, 53, 108

  myoepithelial cells, 31

Adenylate cyclase, 193

  and exocytosis, 195

  signal transduction, 159, 160

Adiposity, age-related

  in laboratory animals, 423

  minor salivary glands, 420

  parotid gland, 414

  submandibular gland, 405–406

Adiposity, oligomenorrhea, and parotid swelling (AOP) syndrome, 212, 214

Adrenergic agents, α_1

  calcium role in response, 154

Adrenergic agonists, α, 120

Adrenergic agonists, β, 106, see also Isoproterenol

Adrenergic antagonists, β, 143

  and amylase, 132

Adrenergic fibers, 12

Adrenergic regulation, 153

  age-related changes in, 433–434

  α-receptor-mediated, 155,

    development, 344

    stimulus-exocytosis coupling mechanism, 183, 195

  α_1-receptor mediated, 157, 160

  β-receptor-mediated, 61, 155

    calcium-dependent intracellular responses, 192, 193

    and cyclic AMP pathway, 187–191

    development, 343, 344

    and galactosyltranferases, 241

    signal transduction pathway of, 161, 235

    stimulus-exocytosis coupling mechanism, 182, 195

  β_2-receptor-mediated, 161

  in cell lines, 322, 331–33

  development, 343, 344

  liquid diets and, 144

  signal transduction, 159–161, 235

  stimulus-exocytosis coupling mechanism, 182, 183

African house bat, 49–51

Age, see also specific glands

  functional changes, 431–437

    flow rates of saliva, 432–433

Q

Quinine, 331

R

Rabies virus, 222
Radiation, 204, 206
Radiation sialadenitis, 209, 210
Radiation therapy, 204
Radiographs, 202
Raf, 230, 231
Ranula, 209
Ras, 230, 231
Rat, 17, 19, 29, see also Laboratory animals
  biologically active peptides, 23
  carbonic anhydrase in, 24
  DNA synthesis, 266, 267
  granular ducts, 22
  innervation, 12
  lectin labeling, 26
  myoepithelial cells, 31
  parotid acinar cells, neurotransmitter
     control, see Neurotransmitters,
     control of calcium mobilization
  parotid glands, fluid transport, 112
  secretory granules, 21
Receptor operated channel (ROC), 167
Receptor proteins, 230
Receptors
  adrenergic, see Adrenergic regulation
  1,25-dihydrocholecalciferol, 142
  and gene expression, 61
Recycling, membrane carriers, 93–94
Refill calcium, 169
Regeneration, 277–278
Regulation of gene expression, 62–66, 95
  evolutionary divergence
    lysozyme c, 62–64
    plasminogen activator, 64–66
Relaxed constraints, 58–59
Renal disease, 204
Renin, 23, 26, 326, 371–372
Repair, 210
Replacement of cells, 272–275
Reproductive behavior, see Sexual behavior
Retention cysts, 208–209
Ribonuclease, 60
  liquid diets and, 144
  parotid gland, 359
  in submandibular gland, neonatal, 375
RNA

  dietary restriction and, 138
  liquid diets and, 144
  zinc deficiency and, 142
Ro (antibody), 217
Rotundus, 55
Rough endoplasmic reticulum, see
    Endoplasmic reticulum
RSMT-A5 cells, 331, 332
RSMTX, cells 321, 322, 325–326, 332
Rubella, 222
Rubidium flux, 167–168

S

S-100 protein, 25, 31, 323, 324
Saliva, see also Disease
  age-related changes, 434–435
  biochemical changes in disease, 203
  virus secretion in, 221–222
Saliva collection, 203
Salivain, 372
Salivary flow rates, see Flow rate
Salivary gland depression, 220
Salivation, 215
Sarcoidosis, 204, 217
SCA clones, 326
Scent marking, 68, 72
Scintigraphy, 202–203
Scotophilus, 49–51, 60
  borbonicus, 50, 51
  dingani, 50, 51
  nux, 50, 51
Second messenger-operated channel
    (SMOC), 167
Second messengers, 156, see also Signal
    transduction
  calcium and, 193
  and exocytosis, 195
  stimulus-exocytosis coupling, 183–184
Secretatogues, submandibular mucin
    secretion, 184
Secretion
  ion transport and, see Ion transport
  protein, see Protein secretion, cell biology
    of
  stimulation of, see Stimulus-exocytosis
    coupling mechanisms
  unstimulated, 87–88
Secretomotor fibers
  postganglionic, 11
  sympathetic nerves, 12
Secretory component, 31, 323